Prime-Detecting Sieves

London Mathematical Society Monographs Series

The London Mathematical Society Monographs Series was established in 1968. Since that time it has published outstanding volumes that have been critically acclaimed by the mathematics community. The aim of this series is to publish authoritative accounts of current research in mathematics and high-quality expository works bringing the reader to the frontiers of research. Of particular interest are topics that have developed rapidly in the last ten years but that have reached a certain level of maturity. Clarity of exposition is important and each book should be accessible to those commencing work in its field.

The original series was founded in 1968 by the Society and Academic Press; the second series was launched by the Society and Oxford University Press in 1983. In January 2003, the Society and Princeton University Press united to expand the number of books published annually and to make the series more international in scope.

Senior Editor: Brian Davies (King's College, London)
Honorary Editors: Martin Bridson (Imperial College, London) and Peter Sarnak (Princeton University and Courant Institute, New York)
Editorial Advisers: J. H. Coates (University of Cambridge), W. S. Kendall (University of Warwick), and János Kollár (Princeton University)

Prime-Detecting Sieves

Glyn Harman

PRINCETON UNIVERSITY PRESS
PRINCETON AND OXFORD

Published by Princeton University Press
41 William Street, Princeton, New Jersey 08540

In the United Kingdom: Princeton University Press
3 Market Place, Woodstock, Oxfordshire OX20 1SY

Library of Congress Cataloging-in-Publication Data has been applied for.
ISBN-13: 978-0-691-12437-7 (alk. paper)

British Library Cataloging-in-Publication Data is available

This book has been composed in LATEX

The publisher would like to acknowledge the author of this volume for providing
the camera-ready copy from which this book was printed.

Printed on acid-free paper. ∞

press.princeton.edu

Printed in the United States of America

10 9 8 7 6 5 4 3 2 1

Dedication

To Ruth, Matthew, Jonathan and Christopher

Contents

Preface

Prime numbers, as the basic elements from which all integers are composed under multiplication, have held a fascination for mathematicians for several millennia. There are many simple to state conjectures concerning primes that are comprehensible to the "man in the street," yet which have defeated the efforts of the greatest minds in succeeding generations. Nevertheless, significant progress has been made in understanding the distribution of prime numbers in recent decades. Although we may appear to be no closer to resolving the hardest problems (like Goldbach's conjecture, for example), we have greater insight into difficulties that lie behind these questions, and many advances have been made on related matters. The purpose of this book is to describe how simple combinatorial identities lead to prime-detecting sieves when sufficient arithmetical information can be garnered for a problem. The sieves themselves are generally very simple. It is in the development of the arithmetical information to be fed into the sieve that much hard work needs to be done. In a nutshell this book can be summed up in a single phrase: If you have the right Type I and Type II information, you can find primes in a given set. When applied to some of the examples we consider, for example the Friedlander-Iwaniec work on primes of the form $m^2 + n^4$ or Heath-Brown's theorem on primes of the form $x^3 + 2y^3$, obtaining the right Type II information becomes a major achievement.

I remember the excitement I felt as a research student reading the paper [85] and realising there was more work to be done in using sieve methods to produce primes. This book charts my own mathematical journey and the adventures of others in using sieve methods to generate primes. I hope that research students, and possibly even research workers, might be inspired by these pages to continue the great project started by Eratosthenes in antiquity: the application of sieve methods to produce prime numbers. My own motivation for the pursuit of these results has been purely mathematical. However, there are several in the cryptographic community who also have an interest in these things, in particular the application of these ideas to certain problems involving primes in arithmetic progressions.

I must thank my former colleagues at Cardiff who patiently listened to me lecturing on some of the material in this book over several years, and my current colleagues at Royal Holloway who likewise have endured my ramblings. The discussions stimulated in these two contexts have helped me to improve the exposition of the ideas, and have clarified the connections between apparently different themes. Several colleagues have read these chapters in one form or another and pointed out many errors. In particular I am indebted to Kaisa Matomäki who worked through much of the text line by line, finding many small and large errors, along with some unclear passages. I must bear full responsibility for those errors that remain. A special thanks must go to my co-authors — Roger Baker, Angel Kumchev, Phil Lewis, János Pintz, Jöel Rivat — who worked with me on some of the material that formed the basis for several parts of this tome. There are many good ideas in this book that I hope I have fully attributed to their originators. I must apologize if, in the compression of detail necessary to keep this volume a reasonable size, I may have at times oversimplified some profound arguments. On the other hand, at times I may have stuck very closely to the line taken by my fellow intrepid explorers on our quest to determine the behaviour of primes. Along with the thoughts of others there are a couple of contributions of my own that I hope will be of interest.

Some important auxiliary results are included in the appendix, which is provided to help research students at the start of their careers. This should not be seen as an alternative to reading and benefitting from classical texts like [27, 131, 157], supplemented by a modern work like [106]. It would be impossible to prove all the results we need here, and the work [157] is recommended for some of the basic results on the Riemann zeta-function we shall need later. For some results we will need to refer to the original research papers since they require material going far beyond the remit of this tome. Readers may be disappointed that some recent developments in prime number theory have been omitted, for example, the work of Goldston and Yildirim [48] on small gaps between primes, and the work of Green and Tao [52] on arbitrarily long arithmetic progressions of primes. However, these results have their own distinct flavours and, although dependent on sieve ideas, only appear to require Type I arithmetical information for their successful execution. However, it should be noted that the "Möbius and nilsequences conjecture" used by Green and Tao [53] looks like the form of Type II information discussed at the end of Chapter 2. Readers aware of this latest work and armed also with the contents of this volume may be able to discover more connections that could turn out to be fruitful for future research.

I have included exercises at the end of some chapters. These are intended for beginning postgraduates who may want to do some mathematics themselves as soon as possible.

Royal Holloway,
October 2006

Notation

We use the standard number-theoretic notation:

$\mu(d)$ The Möbius function

$\tau(d)$ The divisor function. We write $\tau_r(d)$ for the number of ways of writing an integer as a product of r factors. We shall frequently need the fact that, on average over any interval $[x, 2x)$, any power of the divisor function is bounded by a power of $\log x$, while $\tau(n) \ll n^\epsilon$ holds for all n. To be precise, we have, for any sequence of positive integers $a_1, \ldots, a_k$, not necessarily distinct,

$$\sum_{n \le x} \prod_{r \le k} \tau_{a_r}(n) \ll x(\log x)^C,$$

where

$$C = \prod_{r=1}^{k} a_r - 1.$$

In particular, the common case $k = 2, a_1 = a_2 = 2$ (that is the mean of $\tau^2(d)$), leads to $C = 3$.

$\phi(q)$ Euler's totient function

γ in several places represents Euler's constant $\gamma = 0.5772\ldots$, which often arises in the expression $\exp(-\gamma) \approx 0.56146$.

$\Lambda(n)$ The von Mangoldt function

$\pi(x)$ The number of primes up to x.

$\pi(x; q, a)$ The number of primes p up to x with $p \equiv a \pmod{q}$
$\mathrm{Li}(x)$ The logarithmic integral

$$\mathrm{Li}(x) = \int_2^x \frac{1}{\log y} \, dy.$$

$\omega(u)$ Buchstab's function, defined in Section 1.4.

$e(x) = \exp(2\pi i x)$

$\zeta(s)$ The Riemann zeta-function, where s is a complex variable.

$\rho = \beta + i\gamma$ is sometimes used for a non-trivial zero of the Riemann zeta-function.

$L(s, \chi)$ is used for a Dirichlet L-function, and $L(s, \lambda^m)$ for a Hecke L-function.

$a \sim A$ usually signifies $A \le a < 2A$.

The letter p, with or without a subscript always denotes a prime.

Most often calligraphic capitals like $\mathcal{A}, \mathcal{B}$, and so on, are employed to denote sets of integers, and we then use modulus signs to denote their cardinality.

$S(\mathcal{A}, z) = |\{n \in \mathcal{A} : p|n \Rightarrow p \geq z\}|$

$\mathcal{E}_d = \{n : nd \in \mathcal{E}\}$

$[x]$ always represents the integer part of x, and $\{x\} = x - [x]$ is often used for the fractional part of x.

We often write

$$\|x\| = \min_{m \in \mathbb{Z}} |x - m|.$$

However, in some later chapters $\| \ \|$ will be used as a *norm*. This will be clearly explained at the time.

Prime-Detecting Sieves

Chapter One

Introduction

1.1 THE BEGINNING

Many problems in number theory have the form: *Prove that there exist infinitely many primes in a set $\mathcal{A}$* or *prove that there is a prime in each set $\mathcal{A}^{(n)}$ for all large n*. Examples of the first include:

The twin-prime conjecture. Here one takes $\mathcal{A} = \{p + 2 : p \text{ a prime}\}$.

Primes represented by polynomials. A typical problem here is whether the quadratic $n^2 + 1$ is infinitely often prime. So one takes $\mathcal{A} = \{n^2 + 1 : n \in \mathbb{Z}\}$.

Examples of the second problem include:

Goldbach's conjecture. In this case $\mathcal{A}^{(n)} = \{2n - p : p \text{ a prime}, p \leq n\}$.

Is there a prime between consecutive squares? For this problem we take $\mathcal{A}^{(n)} = \{m : n^2 \leq m \leq (n + 1)^2\}$.

I must stress that we will not be able to get as far as a proof of these results in this book! As is well known, the solution to these problems seems to be well beyond all our current methods. Nevertheless, the methods we present here have produced remarkable progress in our knowledge of the distribution of primes in "thin" sequences. It should be clear from reading this book just what information we lack to tackle the above problems.

If we write $\pi(x)$ for the number of primes up to x, the prime number theorem tells us that $\pi(x) \sim x/\log x$. Thus we might hope that if the integers in a set $\mathcal{A}$ are about x in size, and assuming that there are no obstacles preventing primes from belonging to $\mathcal{A}$ (like $\mathcal{A}$ consisting only of even numbers), then the number of primes in $\mathcal{A}$ is about $|\mathcal{A}|/\log x$ (perhaps times some factor depending on the likelihood of small primes dividing integers in $\mathcal{A}$). We shall discover that our hopes are realized so long as we have the two types of arithmetical information introduced in Section 1.6. We shall find that when the information available is strong, we can obtain an asymptotic formula for the number of primes in a given set. When the information is not quite so strong, we can often still obtain a non-trivial lower bound for this quantity. The common thread running through this book is the use of sieve methods: either identities or inequalities. Indeed our inequalities are simply identities where we bound below by zero sums that are in all probability positive. We shall see that the sieve method can be traced back to Eratosthenes in antiquity. Even modern formulations of sieve identities that apparently have no connection with the ancient Greek mathematician turn out to be intimately related through the ubiquitous identity (1.3.1), which underlies all our work. From another point of view, our work here could be regarded simply as the inclusion/exclusion principle pushed to the nth degree (with $n \to \infty$!).

There are four basic types of problem that we will consider and we introduce them now to whet the reader's appetite.

1. Diophantine approximation. This is the easiest problem to deal with since much progress can be made with relatively elementary arithmetical information. We know that if α is irrational, then there are infinitely many pairs of coprime integers m, n with

$$\left| \alpha - \frac{m}{n} \right| < \frac{1}{n^2}.$$

See [57, Theorem 171], for example. Indeed, the right-hand side above can be improved by a factor up to $5^{-1/2}$. Now what if we wanted to have fractions with prime denominator? This is the sort of question a number theorist naturally asks. We would hope to get infinitely many solutions to

$$\left| \alpha - \frac{m}{p} \right| < \frac{1}{p^{1+\theta}},$$

and, as θ increases from 0 (trivial: there is a solution in m for every p) to 1, (the result is false for one: see note below), the problem presumably increases in difficulty. Indeed, no one has any idea how to increase θ above $\frac{1}{3}$ unless one assumes very strong results on primes in arithmetic progressions (stronger than the Generalized Riemann Hypothesis, which gives the $\frac{1}{3}$ exponent). Taking a different perspective on this problem, αn is dense (mod 1) if α is irrational, and one can consider Kronecker's theorem [57, Theorem 440] in the form

$$|\alpha n - m + \beta| < 3n^{-1}.$$

One would then ask about obtaining infinitely many solutions to

$$|\alpha p - m + \beta| < p^{-\theta}.$$

Taking $\beta = 0$ we recover our original problem. It will be useful to write

$$\|x\| = \min_{m \in \mathbb{Z}} |x - m|.$$

Thus the above problems correspond to small values of

$$\|\alpha p\| \quad \text{or} \quad \|\alpha p + \beta\|.$$

Remark. In [61] it is shown that there are uncountably many α such that

$$\|\alpha p\| < \frac{\log p}{500 p \log \log p}$$

has only finitely many solutions in primes p.

2. Primes in short intervals. We would like to know that the interval $[n, n + n^{1/2})$ contains primes for all large n. There is no method known at present that could tackle this problem (unless we assume an extraordinary hypothesis on the existence of Siegel zeros [40]). If we knew the Riemann Hypothesis were true, then we would obtain the expected asymptotic formula for the number of primes in the interval

$[n, n + n^{1/2+\epsilon})$. Here is a case where ϵ makes all the difference! To be more precise, Dirichlet series/polynomial methods only work when the intervals have this greater length. If we are unable to prove that there are primes in the interval $[n, n + n^{1/2+\epsilon})$ without the Riemann Hypothesis, how much larger do we have to make the interval to get an unconditional result? To keep with the convention of the first problem that increasing θ means increasing difficulty, how big can we make θ and the interval $[n, n + n^{1-\theta})$ still contains primes for all large n ? Can we get the expected asymptotic formula

$$\pi(n + n^{1-\theta}) - \pi(n) \ (= \Pi(n, \theta), \text{say}) \ \sim \ \frac{n^{1-\theta}}{\log n}?$$

We can in fact obtain this formula for $\theta < \frac{5}{12}$ (by Huxley's work [97]) and obtain good upper and lower bounds for larger θ. For example, we can get

$$1.01 > \frac{\Pi(n, \theta) \log n}{n^{1-\theta}} > 0.99$$

for $\theta \leq \frac{9}{20}$, as will be demonstrated in Chapter 10.

Instead of making the intervals longer and asking for primes, we can keep the intervals short and look for "prime-like" numbers. There are two likely candidates for such numbers: almost-primes (with a limited number of prime factors) and numbers with a large prime factor. We shall consider the latter only since conventional sieve methods provide the best answers for almost-primes. We can ask for an integer $m \in [n, n + n^{1/2+\epsilon})$ to have a large prime factor, say $> n^\theta$. Again, increasing the value of θ increases the difficulty of the problem, but θ can be taken quite close to 1 ($> \frac{25}{26}$; see Chapter 5 here). For this problem one can reduce the size of the interval to $[n, n + n^\alpha)$ with $\alpha \leq \frac{1}{2}$, but the best exponent to date for θ, even with $\alpha = \frac{1}{2}$, is now substantially smaller (0.738 proved in [120], we give an improved result in Chapter 6 here).

Instead of considering *all* intervals, we can ask what happens almost always. That is, we consider intervals $[x, x + y(x))$ with $x \leq X$ but allow $o(X)$ exceptions if $x \in \mathbb{N}$. Equivalently, if $x \in [1, \infty)$, we allow a set of exceptional x of measure $o(X)$. Now the Riemann Hypothesis furnishes us with an almost perfect answer: One can take $y(x) = (\log x)^2$. Even without this hypothesis one can still take y as quite a small power of x. Later in this work we shall require both the "all" and the "almost-all" results with primes restricted to arithmetic progressions with small modulus in order to apply the circle method to consider the distribution of Goldbach numbers (numbers represented as the sum of two primes) in short intervals. We shall also generalize our method to discuss Gaussian primes in small regions.

3. Primes in arithmetic progressions. Let $a, q \in \mathbb{N}, (a, q) = 1$. By a famous result due to Dirichlet we know that there are infinitely many primes in the arithmetic progression $a \pmod q$. The next natural questions to ask are: How big is the smallest such prime? How many such primes are there up to N? The smallest known value of C such that $p < q^C, p \equiv a \pmod q$ for all large q has become known as Linnik's constant since Linnik was the first to show that such a C exists.

We would expect that

$$\sum_{\substack{p \leq N \\ p \equiv a \bmod q}} 1 \sim \frac{\pi(N)}{\phi(q)}, \tag{1.1.1}$$

where $\phi(q)$ is Euler's totient function. The Generalized Riemann Hypothesis implies this result for $N > q^{2+\epsilon}$. Mention should also be made of recent work by Friedlander and Iwaniec assuming the existence of Siegel zeros [39]. Unfortunately (1.1.1) is only known to be true unconditionally for N substantially larger than q. See [27] for a thorough discussion of this question. However, it is possible to show that (1.1.1) is true on average for $q \leq N^{1/2}(\log N)^{-A}$ for some A (the Bombieri-Vinogradov theorem, which we prove in Chapter 2 here). As is well known, it is our ignorance concerning possible zeros of Dirichlet L-functions near the line $\operatorname{Re} s = 1$ that causes a lot of trouble. We are able to replace (1.1.1) with upper and lower bounds over larger ranges of q with our methods. We can also show that for most q Linnik's constant is not much larger than 2. Indeed, for almost all q we can show that it is actually less than 2. These results, although having importance in themeselves, have significance for other problems. For example, we can use average results on primes in arithmetic progressions to give good lower bounds for the greatest prime factor of $p + a$ where p is a prime (see Chapter 8 here). Also, we can use results on primes in most arithmetic progressions to study Carmichael numbers [65]. The techniques we use in this monograph do not appear capable of improving the best known value of Linnik's constant (see [81]), however. The reader will see why later when we consider primes in individual arithmetic progressions subject to a certain condition.

4. Primes represented by additive forms. It is conjectured that if $f(x) \in \mathbb{Z}[x]$ is nonconstant with a positive leading term and irreducible, and if $f(n)$ has no fixed prime divisor for $n \in \mathbb{N}$, then $f(n)$ will take on infinitely many prime values. Dirichlet's theorem shows that this is true for linear polynomials, but there are no known results for higher degree. If one is allowed two variables, the case $p = m^2 + n^2$ is well understood, and the more general case of the sum of two quadratic polynomials has been studied [100]. The first progress toward analogues for higher-degree forms came with the Friedlander and Iwaniec result that $m^2 + n^4$ takes on prime values infinitely often [37]. Indeed they were even able to furnish an asymptotic formula for the number of prime values taken as the region allowed for (m, n) expands. Further work was performed by Heath-Brown [82], who showed a similar result for $x^3 + 2y^3$. We shall prove both of these results in this book, although we do not have the space to provide all the details.

1.2 THE SIEVE OF ERATOSTHENES

The ancient Greek mathematician and astronomer Eratosthenes is credited with being the first to observe that the primes up to a given number, say N, can be found simply as follows. We write down all the integers up to N and take 2 as the first prime; then we cross out all subsequent multiples of 2. In general, find the next uncrossed number as the next prime (so after 2 we find 3, of course) and cross out all of its multiples. This is easily demonstrated on a piece of paper or an overhead projector with the numbers up to 100, say, but I don't know how to make it look exciting in print! (See Tables 1.1 and 1.2 below) The reader with an internet search should soon find a site that will give an animated version of the sieve, or one could quickly write one's own program to do this task. By the 13th century A.D. it had been noticed that one needs only to cross out multiples of primes up to $\sqrt{N}$ since all composite numbers up to N must have at least one prime factor not exceeding $\sqrt{N}$. The reader will note the problem with the number 1 — we must not cross out all its multiples, yet it is not a prime! Sometimes, even with quite sophisticated arguments it is still necessary to deal with the number 1 separately. Historically, 1 was originally considered to be a prime, of course.

1	2	3	4	5	6	7	8	9	10
11	12	13	14	15	16	17	18	19	20
21	22	23	24	25	26	27	28	29	30
31	32	33	34	35	36	37	38	39	40
41	42	43	44	45	46	47	48	49	50
51	52	53	54	55	56	57	58	59	60
61	62	63	64	65	66	67	68	69	70
71	72	73	74	75	76	77	78	79	80
81	82	83	84	85	86	87	88	89	90
91	92	93	94	95	96	97	98	99	100

Table 1.1 The integers from 1 to 100 before sieving

1	2	3	x	5	x	7	x	x	x
11	x	13	x	x	x	17	x	19	x
x	x	23	x	x	x	x	x	29	x
31	x	x	x	x	x	37	x	x	x
41	x	43	x	x	x	47	x	x	x
x	x	53	x	x	x	x	x	59	x
61	x	x	x	x	x	67	x	x	x
71	x	73	x	x	x	x	x	79	x
x	x	83	x	x	x	x	x	89	x
x	x	x	x	x	x	97	x	x	x

Table 1.2 After the sieve of Eratosthenes has been applied

Clearly the simple principle inherent in the sieve of Eratosthenes is not restricted just to finding the primes up to N. One can similarly "sieve" any given set of integers $\mathcal{A}$ by crossing out multiples of primes less than the square root of each number concerned. Nor need one only sieve to obtain primes. One could strike out all multiples of primes that divide some given integer q and thereby obtain the integers in a set coprime to q. As another example, one could cross out multiples of primes congruent to $3 \pmod 4$ to obtain those members of a set that are properly represented as a sum of two squares.

1.3 THE SIEVE OF ERATOSTHENES-LEGENDRE

In 1808 Legendre showed how the Sieve of Eratosthenes could be used to count the number of primes up to x. The crucial point is that one needs to distinguish between numbers that are crossed off once, twice, three times, and so on. If one estimates $\pi(x)$ by

$$\sum_{n \le x} 1 - \sum_{p \le \sqrt{x}} \sum_{p \mid n \le x} 1,$$

one gets too small a number because many numbers are crossed off more than once. If one uses

$$\sum_{n \le x} 1 - \sum_{p \le \sqrt{x}} \sum_{p \mid n \le x} 1 + \sum_{p \le \sqrt{x}} \sum_{q < p} \sum_{pq \mid n \le x} 1,$$

one obtains too large a number because numbers crossed out three times, for example, are counted three times by the final sum. The formula Legendre produced needed a whole string of multiple sums that increase in number with x. Clearly this was in need of some notation to tidy up the expression. Some years later Möbius defined the function $\mu(n)$, which bears his name, by writing

$$\mu(n) = \begin{cases} (-1)^r & \text{if } n \text{ has } r \text{ distinct prime factors,} \\ 0 & \text{otherwise.} \end{cases}$$

We take $\mu(1) = 1$ since 1 has no prime factors and note that $\mu(n)$ is zero whenever n has a square factor exceeding 1. In other words, $\mu(n)$ is only nonzero for *square-free* n.

We then obtain

$$\pi(x) - \pi(x^{\frac{1}{2}}) + 1 = \sum_{\substack{d \le x \\ d \mid P(x^{1/2})}} \mu(d) \sum_{n \le x/d} 1 = \sum_{\substack{d \le x \\ d \mid P(x^{1/2})}} \mu(d) \left[\frac{x}{d}\right],$$

where

$$P(z) = \prod_{p < z} p.$$

Note that the left-hand side of the Eratosthenes-Legendre formula is $\pi(x) - \pi(x^{1/2}) + 1$ and not $\pi(x)$ since we "sieve out" the primes up to $x^{1/2}$ and the number 1 is not sieved out at all on the right-hand side.

The formula can be proved directly by noting that

$$\sum_{d \mid n} \mu(d) = \prod_{p \mid n} (1 - 1) = \begin{cases} 1 & \text{if } n = 1, \\ 0 & \text{if } n \ge 2. \end{cases} \tag{1.3.1}$$

We shall find that this simple formula lies at the heart of much that follows throughout this book.

Clearly we can modify this formula in many ways. Say we want to determine whether an integer n is free of prime factors from some finite set $\mathcal{P}$. Let Q be the product of the primes in $\mathcal{P}$. We then have

$$\sum_{d|(Q,n)} \mu(d) = \begin{cases} 1 & \text{if } p|n \Rightarrow p \notin \mathcal{P}, \\ 0 & \text{if there is some } p|n, p \in \mathcal{P}. \end{cases}$$

Hence

$$\sum_{\substack{n \in \mathcal{A} \\ (n,Q)=1}} 1 = \sum_{n \in \mathcal{A}} \sum_{d|(Q,n)} \mu(d)$$

$$= \sum_{d|Q} \mu(d) \sum_{\substack{n \in \mathcal{A} \\ d|n}} 1.$$

As a particular example consider the number of integers coprime to a positive integer q in a given interval. Since $\mu(d) = 0$ if d has a squared factor, it makes no difference whether we use q or

$$Q = \prod_{p|q} p$$

(the *square-free kernel* of q) in the above. We get

$$|\{n : x < n \le x + y, (n,q) = 1\}| = \sum_{d|q} \mu(d) \sum_{x < nd \le x+y} 1$$

$$= \sum_{d|q} \mu(d) \left(\left[\frac{x+y}{d} \right] - \left[\frac{x}{d} \right] \right)$$

$$= \sum_{d|q} y \frac{\mu(d)}{d} + O\left(\sum_{d|q} 1 \right)$$

$$= y \frac{\phi(q)}{q} + O(\tau(q)).$$

Here we have written $\tau(q)$ for the number of divisors of q and noted that $\phi(q)$ satisfies

$$\frac{\phi(q)}{q} = \prod_{p|q} \left(1 - \frac{1}{p} \right) = \sum_{d|q} \frac{\mu(d)}{d}.$$

Since $\tau(q) = O(q^\epsilon)$ for any $\epsilon > 0$ [57, Theorem 315], we thus obtain a good result for the numbers in an interval that are coprime to q once the interval length is larger than a small power of q (assuming q is sufficiently large). However, this approach is hopeless if one tries to employ it to find the number of primes up to x. We would get

$$\pi(x) - \pi(x^{\frac{1}{2}}) + 1 = \sum_{\substack{d \le x \\ d|P(x^{1/2})}} \mu(d) \frac{x}{d} + O\left(\sum_{\substack{d \le x \\ d|P(x^{1/2})}} 1 \right).$$

However, it is not difficult to show that the error term above is not even $o(x)$ — using Mertens' prime number theorem (Theorem 1.2) its size is seen to be asymptotic to $(6/\pi^2)(1 - \log 2)x$. Since $\pi(x) = o(x)$ is trivial (this can be obtained by the above argument by sieving only by the primes up to $\log \log x$, say), we obtain no information. Despite this disappointment we shall find that the formula

$$\pi(x) - \pi\left(\tfrac{1}{2}x\right) = \sum_{\substack{d \leq x \\ d \mid P(x^{1/2})}} \mu(d)\left(\left[\frac{x}{d}\right] - \left[\frac{x}{2d}\right]\right)$$

is fundamental to our work as we switch from primes in some given set to primes in the interval $[x/2, x)$ or $[x - y, x)$, where y is of a slightly smaller order than x.

1.4 THE PRIME NUMBER THEOREM AND ITS CONSEQUENCES

In 1860 Riemann [149] formulated a programme to establish the following result, which had been conjectured some 68 years earlier by Gauss.
    We have

$$\pi(x) = \mathrm{Li}(x)(1 + o(1))$$

as $x \to \infty$, where

$$\mathrm{Li}(x) = \int_2^x \frac{1}{\log y}\, dy.$$

We shall refer to this result, perhaps as stated below with an explicit error term, as the *Prime Number Theorem* and frequently abbreviate this simply to the PNT. In 1896 Hadamard [54] and de la Vallée Poussin [29] independently proved this result. The crux of Riemann's programme is the study of the behaviour of the Riemann zeta-function defined for $\mathrm{Re}\, s > 1$ by

$$\zeta(s) = \sum_{n=1}^{\infty} n^{-s}. \tag{1.4.1}$$

Riemann proved, by establishing a functional equation, that $\zeta(s)$ possesses an analytic continuation to the whole of the complex plane except for the simple pole at $s = 1$. It should be noted that Euler (1737) had previously used the fact that

$$\zeta(x) \to \infty \quad \text{as} \quad x \to 1^+$$

to provide a proof of the infinitude of the set of primes.
    From our perspective there are three important stages in the proof of the PNT:
    1) Replace $\pi(x)$ by

$$\psi(x) = \sum_{n \leq x} \Lambda(n),$$

where $\Lambda(n)$ is von Mangoldt's function given by

$$\Lambda(n) = \begin{cases} \log p & \text{if } n = p^k, \\ 0 & \text{otherwise.} \end{cases}$$

We can then use partial summation to deduce $\pi(x) \sim \mathrm{Li}(x)$ from $\psi(x) \sim x$, obtaining similar error terms.

2) Relate $\psi(x)$ to the logarithmic derivative of $\zeta(s)$ given by

$$\frac{\zeta'}{\zeta}(s) = -\sum_{n=2}^{\infty} \Lambda(n) n^{-s} \quad \text{for } \mathrm{Re}\, s > 1 \tag{1.4.2}$$

using contour integration. Since $\zeta(s)$ has a simple pole at $s = 1$, its logarithmic derivative has a simple pole with residue -1 at $s = 1$.

3) Move the integral inside the critical strip $0 < \mathrm{Re}\, s < 1$ and use bounds for $(\zeta'/\zeta)(s)$, together with a zero-free region for $\zeta(s)$ to establish that $\psi(x) \sim x$.

As soon as we start developing sieve methods we will need the PNT and one of the basic results from stage 2 above, namely Perron's formula. When we move to applications involving primes in short intervals, we shall need to use all the above stages, including a more powerful zero-free region than was available in 1896. Since we shall need all the details of the proof of the PNT at some point or other, we therefore depart from historical order and use the best results known today to prove the following.

Theorem 1.1. *For any $\epsilon > 0$ we have*

$$\pi(x) = \mathrm{Li}(x) + O\left(x \exp\left(-(\log x)^{\frac{3}{5} - \epsilon} \right) \right). \tag{1.4.3}$$

Before commencing to assemble the results we shall require to establish this result, we pause to consider the connection between $\zeta(s)$ and the prime numbers. After all, (1.4.1) apparently has no reference to primes. Well, for $\mathrm{Re}\, s > 1$, we have

$$\zeta(s) \sum_{n=1}^{\infty} \mu(n) n^{-s} = \sum_{n=1}^{\infty} n^{-s} \sum_{d|n} \mu(d)$$

by multiplying the terms one by one and gathering the terms in n^{-s} together. From (1.3.1) the coefficients of n^{-s} are all zero, except for $n = 1$. We thus obtain

$$\zeta(s) \sum_{n=1}^{\infty} \mu(n) n^{-s} = 1,$$

and so

$$\frac{1}{\zeta(s)} = \sum_{n=1}^{\infty} \mu(n) n^{-s} = \prod_{p} \left(1 - \frac{1}{p^s} \right).$$

It follows that, for $\mathrm{Re}\, s > 1$, we have

$$\zeta(s) = \prod_{p} \left(1 - \frac{1}{p^s} \right)^{-1}.$$

Then

$$\log \zeta(s) = -\sum_{p} \log\left(1 - \frac{1}{p^s} \right),$$

so that

$$\frac{\zeta'}{\zeta}(s) = \frac{d}{ds}\log\zeta(s) = -\sum_p \frac{p^{-s}\log p}{1 - p^{-s}}$$

$$= -\sum_p (\log p)\left(\sum_{k=1}^{\infty} p^{-ks}\right) \qquad (1.4.4)$$

$$= -\sum_{n=2}^{\infty} \Lambda(n)n^{-s}.$$

This establishes (1.4.2). On the other hand,

$$\zeta'(s)(\zeta(s))^{-1} = -\sum_{n=2}^{\infty}(\log n)n^{-s}\sum_{m=1}^{\infty}\mu(m)m^{-s}$$

$$= -\sum_{n=2}^{\infty} n^{-s}\sum_{de=n}\mu(e)\log d.$$

Now

$$\sum_{de=n}\mu(e)\log d = \log n\sum_{d|n}\mu(d) - \sum_{d|n}\mu(d)\log d$$

$$= -\sum_{d|n}\mu(d)\log d$$

$$= -\log\left(\prod_{d|n}d^{\mu(d)}\right)$$

$$= -\log\left(\prod_{p|n}p^{e(p,n)}\right),$$

where

$$e(p,n) = -\sum_{\substack{d|n/p \\ (d,p)=1}}\mu(d) = \begin{cases} 1 & \text{if } n = p^k, \\ 0 & \text{otherwise.} \end{cases}$$

We thus obtain a "different" proof that

$$\frac{\zeta'}{\zeta}(s) = -\sum_{n=2}^{\infty}\Lambda(n)n^{-s}.$$

The point we are making is that there are intimate connections between identities involving $\zeta(s)$ and elementary identities between finite sums involving arithmetical functions. Using the fact that the functions n^{-s} are linearly independent, we can immediately deduce from (1.4.4) that

$$-\sum_{d|n}\mu(d)\log d = \Lambda(n), \qquad (1.4.5)$$

an identity we proved directly above and which is often proved in textbooks as an example of Möbius inversion applied to the formula

$$\log n = \sum_{d|n} \Lambda(n) \tag{1.4.6}$$

(see [3, pp. 32–33]). We further note the central role played by the formula (1.3.1) in the Eratosthenes-Legendre sieve and in all of the above working. Finally we remark that (1.4.5) gives a simple decomposition of $\Lambda(n)$. We shall reconsider this in Chapter 2 when we consider Vaughan's identity and related expressions.

The first tool we need to prove the PNT will have many other applications. We give a complete proof in the appendix. In the rest of this section, and indeed in much of the rest of this book, we write $s = \sigma + it$, $\sigma, t \in \mathbb{R}$ for a complex variable. Unless otherwise stated, the variables σ, t, s will always be so related.

Lemma 1.1. *When $\sigma > 1$, let*

$$F(s) = \sum_{n=1}^{\infty} \frac{a_n}{n^s}.$$

Write

$$f(x) = \max_{x/2 < n < 2x} |a_n|.$$

Suppose that

$$\sum_{n=1}^{\infty} \frac{|a_n|}{n^\sigma} = O(|1 - \sigma|^{-\alpha})$$

as $\sigma \to 1^+$. Then, if $c > 0, \sigma + c > 1$, we have

$$\sum_{n \le x} \frac{a_n}{n^s} = \frac{1}{2\pi i} \int_{c-iT}^{c+iT} F(w + s) \frac{x^w}{w} dw + O\left(\frac{x^c}{T(\sigma + c - 1)^\alpha} \right)$$
$$+ O\left(\frac{f(x) x^{1-\sigma} \log x}{T} \right) + O\left(f(x) x^{-\sigma} \min\left(\frac{x}{T||x||}, 1 \right) \right). \tag{1.4.7}$$

Remark. If $F(s) = \zeta(s)$, then $f(x) \equiv 1$ and $\alpha = 1$. When

$$F(s) = \frac{\zeta'}{\zeta}(s)$$

then $f(x) \approx \log 2x$ but still $\alpha = 1$.

Lemma 1.2. *For $|t| < e, 0 < \delta < 1$, we have, for $\delta \le |\sigma - 1| \le 1$,*

$$\left| \frac{\zeta'}{\zeta}(s) \right| \ll \delta^{-1}.$$

Proof. This follows immediately from the fact that $\zeta(s)$ has a simple pole at $s = 1$ and no zeros in the region $|t| < 4, -1 < \sigma < 3$. □

Lemma 1.3. *Suppose that*

$$|t| \geq e, \quad -1 \leq \sigma \leq 2, \quad \min_{\rho} |s - \rho| > (\log |t|)^{-1},$$

where the minimum is over zeros ρ of $\zeta(s)$. Then

$$\left| \frac{\zeta'}{\zeta}(s) \right| \ll (\log |t|)^2.$$

Proof. See [27, p. 99]. The crucial formula, whose consequence we shall also need in Chapter 7, is

$$\frac{\zeta'}{\zeta}(s) = \sum_{\substack{\rho \\ |\gamma - t| < 1}} \frac{1}{s - \rho} + O(\log t) \tag{1.4.8}$$

for $s = \sigma + it$, with $t \geq e$ and not coinciding with the ordinate of a zero. □

Lemma 1.4. *For $|t| > e^e$ we have $\zeta(s) \neq 0$ for*

$$\sigma > 1 - \frac{A}{(\log |t|)^{\frac{2}{3}} (\log \log |t|)^{\frac{1}{3}}},$$

where A is an absolute constant.

Proof. See [157, p. 135]. This result is now known with $A > \frac{1}{100}$; see [32]. □

Proof. (PNT) Our immediate goal is to establish that

$$\psi(x) = x + O\left(x \exp\left(-(\log x)^{\frac{3}{5} - \epsilon} \right) \right). \tag{1.4.9}$$

From Lemma 1.1 we have

$$\sum_{n \leq x} \Lambda(n) = -\frac{1}{2\pi i} \int_{c-iT}^{c+iT} \frac{\zeta'}{\zeta}(s) \frac{x^s}{s} \, ds + O\left(\frac{x(\log x)^2}{T} \right), \tag{1.4.10}$$

with $c = 1 + (\log x)^{-1}$. We take the integral to the line $\sigma = \sigma'$, where

$$\sigma' = 1 - \frac{A}{2(\log T)^{\frac{2}{3}} (\log \log T)^{\frac{1}{3}}}.$$

The pole at $s = 1$ gives the main term x in (1.4.9). From Lemma 1.3 the horizontal line contours $|t| = T, \sigma' \leq \sigma \leq c$ contribute

$$\ll \frac{x}{T} (\log T)^2, \tag{1.4.11}$$

which is essentially the same error as given by Perron's formula itself if T and x are fixed powers of each other and leads to a slightly smaller order error for our choice of T below. From Lemmas 1.2 and 1.3 we obtain

$$\left| \int_{\sigma'-iT}^{\sigma'+iT} \frac{\zeta'}{\zeta}(s) \frac{x^s}{s} \, ds \right| \ll x^{\sigma'} \int_{-T}^{T} \frac{(\log(|t| + 2)^2}{1 + |t|} \, dt$$

$$\ll x(\log T)^3 \exp\left(-\frac{A \log x}{2(\log T)^{\frac{2}{3}} (\log \log T)^{\frac{1}{3}}} \right). \tag{1.4.12}$$

To balance the error terms, let $T = \exp\left((\log x)^\alpha\right)$ with $0 < \alpha < 1$. The error terms in (1.4.10) and (1.4.11) are then

$$O\left(x \exp\left(-(\log x)^\alpha\right)(\log x)^2\right),$$

while the error in (1.4.12) is

$$O\left(x \exp\left(-A(\log x)^{1-\frac{2\alpha}{3}}(\log\log x)^{-\frac{1}{3}}\right)(\log x)^{3\alpha}\right).$$

If we pick $\alpha = 1 - 2\alpha/3$, that is, $\alpha = \frac{3}{5}$, we get both error terms

$$\ll x \exp\left(-(\log x)^{\frac{3}{5}-\epsilon}\right)$$

for any $\epsilon > 0$, which completes the proof of (1.4.9).

Partial summation will play a big role in much that we do, so we shall now give the deduction of (1.4.3) from (1.4.9) explicitly. We recall the basic partial summation identity that, for arbitrary functions f, g and integers $A < B$, we have

$$\sum_{n=A}^{B} f(n)g(n) = \sum_{n=A}^{B} (f(n) - f(n+1)) \sum_{m=A}^{n} g(m) + f(B+1) \sum_{m=A}^{B} g(m).$$

We suppose $x \geq 2$ is an integer and begin by writing

$$\pi(x) = \sum_{2 \leq n \leq x} \frac{1}{\log n} + \sum_{2 \leq n \leq x} \left(\frac{\Lambda(n) - 1}{\log n}\right) - \sum_{\substack{2 \leq n \leq x \\ p^2 | n}} \frac{\Lambda(n)}{\log n}$$

$$= S_1 + S_2 - S_3 \quad \text{say}.$$

Let $L = [\log_2 x]$ ($\log_2$ indicating logarithm to base 2 here). Trivial bounds yield

$$S_3 \leq \sum_{k=2}^{L} x^{\frac{1}{k}} \leq x^{\frac{1}{2}} + L x^{\frac{1}{3}} \ll x^{\frac{1}{2}}.$$

For any positive function $f(x)$ that is monotonically decreasing we have

$$\sum_{U \leq n \leq V} f(n) = \int_{U}^{V} f(y)\,dy + O(f(U)),$$

so

$$S_1 = \text{Li}(x) + O(1).$$

Now write

$$\sum_{2 \leq m \leq n} (\Lambda(m) - 1) = E(n),$$

which gives $E(n) \ll x \exp\left(-(\log x)^{\frac{3}{5}-\epsilon}\right) = F(x)$, say, when $n \leq x$. Partial summation then gives

$$S_2 = \sum_{2 \leq n \leq x} \left(\frac{1}{\log n} - \frac{1}{\log(n+1)}\right) E(n) + \frac{E(x)}{\log(x+1)}.$$

Thus

$$|S_2| \ll F(x)\left(\sum_{2 \leq n \leq x}\left(\frac{1}{\log n} - \frac{1}{\log(n+1)}\right) + \frac{1}{\log(x+1)}\right) = \frac{F(x)}{\log 2}.$$

This completes the proof of (1.4.3). $\qquad\square$

Before developing the PNT to cover sums over primes and almost-primes, we pause to give a result that will be of crucial importance when we consider primes in short intervals and which is obtained by a slight modification of the proof given above. We shall often use a splitting-up argument, as is common in analytic number theory, to produce variables in ranges that are of the same order of magnitude. For this purpose we introduce the notation $a \sim A$ to mean $A \leq a < 2A$.

Lemma 1.5. *Let $V \geq e, P \geq 2$. Then, if $|t| \sim V$, we have*

$$\left| \sum_{p \sim P} p^{-s} \right| \ll P^{1-\sigma} \exp\left(-\frac{\log P}{(\log V)^{\frac{7}{10}}} \right) + \frac{P^{1-\sigma}}{V} (\log P)^3.$$

Remark. In applications V and P will be related in such a way that we will obtain

$$\left| \sum_{p \sim P} p^{-s} \right| \ll P \exp\left(-(\log P)^\beta \right)$$

for some $\beta > 0$.

Proof. We leave this as an exercise for the reader. The major differences are that we take $T = \frac{1}{2} V$ and there is no pole on crossing the line $\sigma = 1$.   □

We now need some more notation that will be used throughout the rest of this book. First we define Buchstab's function $\omega(u)$ by $\omega(u) = u^{-1}$ for $1 \leq u \leq 2$ and then use induction to define $\omega(u)$ for $k \leq u \leq k+1$ ($k \in \mathbb{N}, k \geq 2$) by

$$\omega(u) = \frac{1}{u} \left(k\omega(k) + \int_k^u \omega(v-1) \, dv \right).$$

It follows immediately from this that $\frac{1}{2} \leq \omega(u) \leq 1$ for all u and that ω is the continuous solution to the delay/differential equation

$$(u\omega(u))' = \omega(u-1)$$

with the boundary condition $\omega(u) = u^{-1}$ for $1 \leq u \leq 2$. We shall prove in the appendix that $\omega(u) \to \exp(-\gamma)$ as $u \to \infty$, where γ is Euler's constant.

Write $\mathcal{B} = \mathbb{Z} \cap [x/2, x)$, $\mathcal{A} \subseteq \mathcal{B}$, and put

$$S(\mathcal{A}, z) = |\{n \in \mathcal{A} : p|n \Rightarrow p \geq z\}|$$

$$= \sum_{\substack{d|P(z) \\ dn \in \mathcal{A}}} \mu(d).$$

Clearly the number of primes in $\mathcal{A}$ is $S(\mathcal{A}, x^{1/2})$ and

$$S(\mathcal{B}, x^{\frac{1}{2}}) = \pi(x) - \pi\left(\frac{x}{2}\right) + O(1),$$

where the $O(1)$ arises only if x or $x/2$ is a prime.

We write

$$\mathcal{E}_d = \{n : nd \in \mathcal{E}\}.$$

The inclusion/exclusion principle then quickly gives Buchstab's identity for any set $\mathcal{E}$ and positive reals $z > w > 1$:

$$S(\mathcal{E}, z) = S(\mathcal{E}, w) - \sum_{w \le p < z} S(\mathcal{E}_p, p). \tag{1.4.13}$$

By the PNT we have

$$S(\mathcal{B}, z) \sim \frac{x}{2 \log x} \quad \text{for } x^{\frac{1}{3}} < z < \frac{x}{2}.$$

Then, for $x^{1/3} \le w < x^{1/2}$, (1.4.13) gives

$$S(\mathcal{B}, w) = S(\mathcal{B}, x^{\frac{1}{2}}) + \sum_{w < p \le x^{1/2}} S(\mathcal{B}_p, p)$$

$$= S(\mathcal{B}, x^{\frac{1}{2}}) + \sum_{w < p \le x^{1/2}} S(\mathcal{B}_p, (x/p)^{\frac{1}{2}})$$

since

$$\frac{x}{p} > p > \left(\frac{x}{p} \right)^{\frac{1}{2}}$$

in this range. Now the PNT gives

$$S(\mathcal{B}_p, (x/p)^{\frac{1}{2}}) = \frac{x}{2p \log(x/p)} \left(1 + O((\log x)^{-1}) \right).$$

Hence

$$S(\mathcal{B}, w) = \frac{x}{2 \log x} \left(1 + \sum_{w < p \le z} \frac{\log x}{p \log(x/p)} \right) \left(1 + O((\log x)^{-1}) \right).$$

Partial summation as before then gives (the details are contained in the appendix where we consider the error terms in greater detail)

$$S(\mathcal{B}, w) = \frac{x}{2 \log x} \left(1 + \int_w^z \frac{\log x}{y(\log(x/y))(\log y)} dy \right) \left(1 + O((\log x)^{-1}) \right).$$

The change of variables $v = (\log x)/(\log y)$ gives

$$\frac{dy}{y \log y} = -\frac{dv}{v},$$

and so

$$S(\mathcal{B}, w) = \frac{x}{2 \log x} \left(1 + \int_2^{\frac{\log x}{\log w}} \frac{1}{v - 1} dv \right) \left(1 + O((\log x)^{-1}) \right)$$

$$= \frac{x}{2 \log x} \left(1 + \log \left(\frac{\log x}{\log w} - 1 \right) \right) \left(1 + O((\log x)^{-1}) \right).$$

We thus conclude that

$$S(\mathcal{B}, w) \sim \frac{x}{2 \log x} u\omega(u) = \frac{x}{2 \log w} \omega(u) \tag{1.4.14}$$

when $w = x^{1/u}, 2 \le u \le 3$. It does not take much imagination to see that an inductive argument should establish (1.4.14) for all $u > 1$, and this is indeed established in the appendix.

In general we will be faced with multiple sums over primes of the form

$$\sum_{p_1,\ldots,p_k} S(\mathcal{B}_{p_1\ldots p_k}, z(p_1,\ldots,p_k)),$$

where

$$p_j \in \mathcal{I}(p_1,\ldots,p_{j-1})$$

and $\mathcal{I}(p_1,\ldots,p_{j-1})$ is some interval like

$$\left[w, \min\left(p_{j-1}, \left(\frac{x}{p_1,\ldots,p_{j-1}} \right)^{\frac{1}{2}} \right) \right].$$

Working as above, such as sum will be asymptotically equal to

$$\frac{x}{2} \int \cdots \int \frac{1}{\alpha_1 \ldots \alpha_k \log z} \omega(u)\, d\alpha_k \ldots d\alpha_1,$$

where

$$u = (1 - \alpha_1 - \cdots - \alpha_k) \frac{\log x}{\log z},$$

and the integration range for α_j is

$$\left[\frac{\log A}{\log x}, \frac{\log B}{\log x} \right],$$

where $\mathcal{I} = [A, B]$. For example, the sum

$$\sum_{x^\nu \le p_2 < p_1 < x^{1/2}} S(\mathcal{B}_{p_1 p_2}, p_2)$$

becomes

$$\frac{x}{2 \log x} \int_\nu^{1/2} \int_\nu^{g(\alpha_1)} \omega\left(\frac{1-\alpha_1}{\alpha_2} - 1 \right) \frac{1}{\alpha_1 \alpha_2^2}\, d\alpha_1\, d\alpha_2. \qquad (1.4.15)$$

Here

$$g(\alpha) = \min\left(\alpha, \frac{1-\alpha}{2} \right),$$

where we have noted that $S(\mathcal{B}_{p_1 p_2}, p_2) = 0$ when $p_1 p_2^2 > x$. Thus the summation condition $p_2 < p_1$ becomes $p_2 < \min\left(p_1, (x/p_1)^{1/2} \right)$.

We will often need upper bounds for integrals of the above type, and the following approximation to $\omega(s)$ is then useful (compare Diagram 1.1):

$$\omega(u) \begin{cases} = 1/u & \text{if } 1 \le u \le 2, \\ = (1 + \log(u-1))/u & \text{if } 2 \le u \le 3, \\ \le \frac{1}{3}(1 + \log 2) & \text{if } u \ge 3. \end{cases} \qquad (1.4.16)$$

We note that $\frac{1}{3}(1 + \log 2) = 0.56438\ldots$, whereas $\lim_{u \to \infty} \omega(u) \approx 0.56146$. The behaviour of $\omega(u)$ for small u is illustrated below. With this scale, the graph is practically flat (local maxima and minima cannot be seen) for larger values of u.

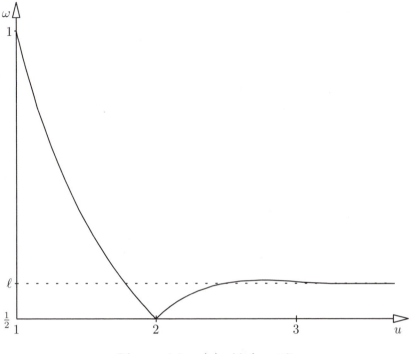

Diagram 1.1: $\omega(u)$ with $\ell = e^{-\gamma}$

Historically, Mertens' prime number theorem preceded the proof of the PNT proper by 22 years [127]. The two equivalent formulations of his result are still of great importance, and we shall need one or other version at various points in this work.

Theorem 1.2. (Mertens' Prime Number Theorem) *As $x \to \infty$, we have*

$$\sum_{p \leq x} \frac{1}{p} = \log \log x + C + o(1), \qquad (1.4.17)$$

where C is the constant

$$\gamma + \sum_{p} \left(\log \left(1 - \frac{1}{p} \right) + \frac{1}{p} \right)$$

and γ is Euler's constant. Rearranging and taking the exponential of the above formula then gives

$$\prod_{p \leq x} \left(1 - \frac{1}{p} \right) \sim \frac{e^{-\gamma}}{\log x}. \qquad (1.4.18)$$

Proof. See [57, Theorems 427–429]. Only real variable techniques are used in this proof with no reference to $\zeta(s)$.   □

1.5 BRUN, SELBERG, AND ROSSER-IWANIEC

Historically, the next advance on the sieve of Eratosthenes-Legendre came about from the work of Brun [20], developing ideas in a paper by Merlin [126]. If one takes the view that the problem with (1.3.1) is that d takes on values that are too large, then it is natural to consider upper and lower bounds for the basic sifting function

$$\sum_{d|n} \mu^-(d) \leq \sum_{d|n} \mu(d) \leq \sum_{d|n} \mu^+(d), \tag{1.5.1}$$

where $\mu^\pm(d) = 0$ for $d > D$. If this is possible, then, using $\mathcal{A}$ as in the previous section,

$$S^-(\mathcal{A}, z) \leq S(\mathcal{A}, z) \leq S^+(\mathcal{A}, z),$$

with

$$S^\pm(\mathcal{A}, z) = \sum_{\substack{d|P(z) \\ d \leq D}} \mu^\pm(d) \sum_{dn \in \mathcal{A}} 1$$

$$= \sum_{\substack{d|P(z) \\ d \leq D}} \mu^\pm(d)|\mathcal{A}_d|.$$

If, taking the simplest class of problems, for some X,

$$|\mathcal{A}_d| = \frac{X}{d} + R_d$$

and

$$\sum_{d \leq D} |R_d| = o\left(\frac{X}{\log x}\right), \tag{1.5.2}$$

then one might hope to get, as $x \to \infty$,

$$S^-(\mathcal{A}, z) \geq f\frac{X}{\log x}, \qquad S^+(\mathcal{A}, z) \leq F\frac{X}{\log x},$$

where f, F depend in some way on z and D.

This line of approach is taken in the standard text by Halberstam and Richert [55] and in its more recent successor by Greaves [51]. There are three basic methods to obtain (1.5.1). The original technique of Brun was based on combinatorial ideas (Chapter 2 in [55], Chapter 3 in [51]). Selberg's idea is, in the first instance, only applicable to obtaining the upper bound in (1.5.1) but is based on the simple idea of ensuring that

$$\sum_{d|n} \mu^+(d)$$

is a square. The reader will find expositions of this method in Chapter 2 of [51] or Chapter 3 in [55]. The third line of attack was provided by Rosser (unpublished) and developed by Iwaniec [102] (see also Chapter 4 of [51]). We shall give a new derivation of this sieve in Chapter 4 since its construction shares many features with the sieve method we present, indeed our method collapses to the Rosser-Iwaniec approach in the limit. We shall also need to apply this sieve in tandem with our method for two of the later problems (see Chapters 6 and 8).

The sieves of Brun, Selberg, and Rosser have enabled much progress to be made on some of the extremely difficult problems of number theory. Highlights include Chen's theorem [21] stating that all sufficiently large even integers are the sum of a prime and a number with at most two prime factors. Mention must also be made of the recent work on small differences between primes [48] and arithmetic progressions of primes [52]. However, by themselves, these sieves cannot generate primes. This will be discussed in more detail in Chapter 4. It is worthwhile to consider here what arithmetical information is being fed into these sieves. There is some "general" information (which can be used for many different problems) that is used to produce $\mu^{\pm}(d)$ and to estimate sums of the form

$$\sum_{\substack{d|P(z) \\ d \leq D}} \frac{\mu^{\pm}(d)}{d}$$

or, more generally,

$$\sum_{\substack{d|P(z) \\ d \leq D}} \frac{w(d)\mu^{\pm}(d)}{d},$$

where $w(d)$ is a multiplicative function, subject to certain natural constraints. Then there is more specific information (which varies according to the problems being considered) that is employed to give upper bounds of the form (1.5.2). For example, to consider the problem of integers in short integers without small prime factors, we take $\mathcal{A} = \mathbb{N} \cap [x - y, x)$. We put $X = y$ to give

$$R_d = \sum_{x-y \leq n < x} 1 - \frac{y}{d} = O(1).$$

It follows that it is possible to take D nearly as large as y. In many applications it is possible to replace (1.5.2) with a multiple sum

$$\sum_{\substack{m \leq M \\ n \leq N}} a_m b_n R_{mn}$$

and use more sophisticated ways of estimating this term. Indeed, Iwaniec and Jutila [105] were able to improve Huxley's prime number theorem by using flexible error terms in the Rosser-Iwaniec sieve along with extra information on the problem, namely, that one could directly estimate certain multiple sums of the form

$$\sum_{p_1 p_2 \in \mathcal{A}} 1, \qquad \sum_{p_1 p_2 p_3 \in \mathcal{A}} 1. \qquad (1.5.3)$$

This idea was further refined by Heath-Brown and Iwaniec [85]. In this last work, D was taken as large as $x^{0.92}$ when $y = x^{0.55+\epsilon}$. We shall need their results, which lead to this size of D, later in Chapter 10. Again it should be stressed that this is not sufficient in itself to detect primes — it is the additional information specific to the problem from (1.5.3) that is required to supplement the sieve methods.

1.6 ERATOSTHENES-LEGENDRE-VINOGRADOV

In the 1920s Hardy and Littlewood showed that every sufficiently large odd number is the sum of three primes assuming the Generalized Riemann Hypothesis for Dirichlet L-functions. This assumption was necessary to provide nontrivial bounds for exponential sums of the form

$$\sum_{p \leq N} e(p\alpha)$$

when α was not too well approximable by a rational with a small denominator. Here, as elsewhere in this book, we write $e(x) = \exp(2\pi i x)$. This sum arises naturally from an application of the circle method [161] to the problem. In this way we have

$$\sum_{p_1+p_2+p_3=N} 1 = \int_0^1 \left(\sum_{p \leq N} e(p\alpha) \right)^3 e(-\alpha N) \, d\alpha.$$

Vinogradov (see [163, Chapter 9]) found an ingenious argument that enabled the sieve of Eratosthenes-Legendre to be applied to this sum. We should note that Vinogradov was applying the sieve not to the immediate arithmetical situation (primes p_j such that $n = p_1 + p_2 + p_3$), as is the case for the sieves described in the previous section, but to the auxiliary functions that arise from the application of a standard number-theoretic method. As a consequence, Vinogradov's technique had to be "precise" — he could not throw away terms of the same size as the main term as happens in the other sieves. Of course, the Eratosthenes-Legendre sieve is just such a precise result.

Suppose we want to count primes with a weight $f(p)$. If $|f(n)| \leq 1$ for all n, then the sieve of Eratosthenes-Legendre becomes

$$\sum_{p \leq x} f(p) = \sum_{\substack{dn \leq x \\ d \mid P(x^{1/2})}} \mu(d) f(dn) + O(x^{\frac{1}{2}}).$$

At first sight it might seem that we have transformed one difficult problem (a sum over primes) into an equally difficult problem — the Möbius function seems no easier to handle than the characteristic function of the set of primes. Indeed we have turned a neat-looking sum into a rather messy-looking double sum. It is a general principle in number theory that if you cannot estimate a sum, you might be able to estimate it "on average," and the presence of two variables enables us to do just that, as we shall see.

Suppose that the numbers $f(p)$ were complex numbers on the unit circle, and we had some hope that these numbers were fairly evenly distributed, so that there would be some cancellation in the sum over p. We might then want to show that

$$\sum_{p \leq x} f(p) = O(F(x)),$$

where we would like $F(x)$ to grow as slowly as possible ($\pi(x)$ is the trivial bound, of course!). We will take the particular example $f(n) = e(\alpha n)$ which, as explained above, was Vinogradov's original application. This will help us to illustrate the principles at work in estimating double sums and will be applied later to investigate the distribution of αp modulo 1. Using the formula for the sum of a geometric progression, if $\alpha \notin \mathbb{Z}$, we have

$$\sum_{n=1}^{N} e(\alpha n) = \frac{e((N+1)\alpha) - e(\alpha)}{e(\alpha) - 1}.$$

Now,

$$|e(\alpha) - 1| = \left| e\left(\tfrac{1}{2}\alpha\right) - e\left(-\tfrac{1}{2}\alpha\right) \right| = 2|\sin(\pi\alpha)|.$$

Also, for $-\tfrac{1}{2} \le \alpha \le \tfrac{1}{2}$ we have $|\sin(\pi\alpha)| \ge 2|\alpha|$. Since $|\sin(\pi\alpha)|$ is periodic with period 1, and recalling the notation

$$\|x\| = \min_{n \in \mathbb{Z}} |x - n|,$$

we have $|\sin(\pi\alpha)| \ge 2\|\alpha\|$. Thus, using the trivial bound for small values of $\|\alpha\|$ we obtain

$$\left| \sum_{n=1}^{N} e(\alpha n) \right| \le \min\left(N, \frac{1}{2\|\alpha\|} \right),$$

with the obvious convention that N is taken as the minimum when $\|\alpha\| = 0$.

Now

$$\sum_{p \le x} e(\alpha p) = \sum_{\substack{dn \le x \\ d \mid P(x^{1/2})}} \mu(d) e(\alpha dn) + O(x^{\frac{1}{2}}).$$

Vinogradov's vital contribution to the analysis of this problem was to split the consideration of the sum on the right-hand side above according to the size of d. We shall not describe his method since the method we develop in Chapter 3 does essentially the same job with much greater clarity (the obscurity of Vinogradov's method was a great barrier to its application by other authors). The methods we describe in Chapter 2 also lead to equivalent results with far less pain than Vinogradov's procedure.

The first type of sum, which we will call a *Type I sum* makes use of the fact that n runs over consecutive integers. We therefore need to restrict the range of d, say, to $d \le D$. This D corresponds to the distribution level in the sieves described in the previous section. We thus obtain

$$\left| \sum_{\substack{dn \le x \\ d \mid P(x^{1/2}), d \le D}} \mu(d) e(\alpha dn) \right| \le \sum_{\substack{d \le D \\ d \mid P(x^{1/2})}} \left| \sum_{n \le x/d} e(\alpha dn) \right|$$

$$\le \sum_{d \le D} \min\left(\frac{x}{d}, \frac{1}{\|d\alpha\|} \right).$$

If α is irrational, or a rational with the right size denominator, we might hope that this final sum is smaller than x in size. Say $\alpha = a/q$ (and the following reasoning works with $\alpha = a/q + O(q^{-2})$ as well). Then, for $d \equiv 0 \pmod q$, we can only give the trivial bound x/d. On the other hand, as d runs through a set of nonzero residues $\pmod q$, $\|d\alpha\|$ runs through the numbers $\|h/q\|, h = 1, \ldots, q-1$ in some order. Hence each of the values $h/q, 1 \le h \le q/2$, is taken at most twice. Thus

$$\sum_{d=1}^{q-1} \frac{1}{\|d\alpha\|} \le 2 \sum_{d=1}^{q/2} \frac{q}{d} < 3q \log q.$$

Hence

$$\sum_{d \le D} \min\left(\frac{x}{d}, \frac{1}{\|\alpha d\|}\right) \le \sum_{\substack{d \equiv 0 \,(\mathrm{mod}\, q) \\ d \le D}} \frac{x}{d} + \sum_{\substack{d \not\equiv 0 \,(\mathrm{mod}\, q) \\ d \le D}} \frac{1}{\|\alpha d\|}$$

$$\le \frac{x}{q} \log\left(\frac{3D}{q}\right) + \left(\frac{D}{q} + 1\right) 3q \log q$$

$$\le 3\left(\frac{x}{q} + D + q\right) \log x.$$

We have thus established a nontrivial bound for this Type I sum if $x^\epsilon < q < x^{1-\epsilon}, D < x^{1-\epsilon}$.

More generally the above work shows that, for any given function such that $1 \le rf(r) \le N$, we have

$$\sum_{r \le R} a_r \sum_{n \le f(r)} e(\alpha nr) \ll \max_{r \le R} |a_r| \left(\frac{N}{q} + R + q\right) \log(NRq). \tag{1.6.1}$$

The much more difficult question is to deal with

$$\sum_{\substack{dn \le x \\ d|P(x^{1/2}), d \ge D}} \mu(d)e(\alpha dn).$$

Vinogradov's significant contribution was to convert such sums into sums of the following type which we shall call *Type II sums*:

$$\sum_{\substack{m \sim M, h \sim H \\ mh \le x}} a_m b_h e(mh\alpha) = S, \text{ say.}$$

Here a_m, b_n may be a bit messy, but the important feature is that they are bounded by something like the divisor function. That is, we can assume $|a_m| \le \tau(m), |b_n| \le \tau(n)$. We apply Cauchy's inequality, namely,

$$\left(\sum_{r=1}^{H} c_r d_r\right)^2 \le \left(\sum_{r=1}^{H} |c_r|^2\right) \left(\sum_{r=1}^{H} |d_r|^2\right),$$

to the Type II sum above to obtain

$$|S|^2 \le \sum_{m \sim M} |a_m|^2 \sum_{m \sim M} \left| \sum_{\substack{h \sim H \\ mh \le x}} b_h e(\alpha mh) \right|^2.$$

Now since a_m is bounded by the divisor function, we have

$$\sum_{m \sim M} |a_m|^2 \ll M \log^3 M.$$

We treat

$$\sum_{m \sim M} \left| \sum_{\substack{h \sim H \\ mh \leq x}} b_h e(\alpha mh) \right|^2$$

by squaring out and changing the order of summation (our idea here is to get a sum over consecutive integers from the m range that we can then treat like the Type I sum above) to get

$$\sum_{h_1, h_2 \sim H} b_{h_1} b_{h_2} \sum_{\substack{mh_j \leq x \\ m \sim M}} e(m(h_1 - h_2)\alpha)$$

$$\leq \sum_{h_1, h_2 \sim H} |b_{h_1} b_{h_2}| \left| \sum_{\substack{mh_j \leq x \\ m \sim M}} e(m(h_1 - h_2)\alpha) \right|$$

$$\ll \sum_{h_1, h_2 \sim H} |b_{h_1} b_{h_2}| \min\left(\frac{x}{H}, \frac{1}{\|\alpha(h_1 - h_2)\|} \right).$$

Now if we use the arithmetic mean/geometric mean inequality to obtain

$$|b_{h_1} b_{h_2}| \leq \tfrac{1}{2} \left(|b_{h_1}|^2 + |b_{h_2}|^2 \right),$$

we are left to estimate

$$\left(\sum_{h \sim H} |b_h|^2 \right) \max_{k \sim H} \sum_{\ell \sim H} \min\left(\frac{x}{H}, \frac{1}{\|\alpha(\ell - k)\|} \right).$$

The sum over h above leads to a term $\ll H \log^3 H$ since b_h is bounded by the divisor function. The sum over ℓ can be treated in a similar manner to the sum over d we had in Type I sums. There is the complication now, however, of the maximum over k. This only affects the terms corresponding to $d \equiv 0 \pmod{q}$ before. We thus obtain for the sum over ℓ the following expression:

$$\ll \left(\frac{H}{q} + 1 \right) \left(\frac{x}{H} + q \right) \log x.$$

Drawing all our information together we thereby conclude that

$$|S| \ll x^{\frac{1}{2}} \left(\frac{x}{q} + \frac{x}{H} + q + H \right)^{\frac{1}{2}} (\log x)^{\frac{7}{2}}.$$

This is nontrivial for $x^\epsilon < q < x^{1-\epsilon}$, $x^\epsilon < H < x^{1-\epsilon}$. Note that now we need a *lower* bound on one of the variables as well as an upper bound. Later we will want to apply this estimate with other bounds on the coefficients, so we note that we have actually proved that

$$|S| \ll \left(\sum_{m \sim M} |a_m|^2 \sum_{h \sim H} |b_h|^2 \right)^{\frac{1}{2}} \left(\frac{x}{q} + \frac{x}{H} + q + H \right)^{\frac{1}{2}} (\log x)^{\frac{1}{2}}. \qquad (1.6.2)$$

Exercises

1. Square-free numbers are easier to detect than primes, and only Type I information is required. Show that

$$|\mu(n)| = \sum_{d^2|n} \mu(d).$$

Hence deduce that the number of square-free integers between x and $x + y$ (where $y \leq x$) is

$$\frac{6y}{\pi^2} + O\left(\sqrt{x}\right).$$

2. Modify the proof of the PNT to show that

$$\left|\sum_{n \leq N} \mu(n)\right| \ll N \exp\left(-(\log N)^\alpha\right)$$

for some $\alpha > 0$.

3. The Liouville function is given by $\lambda(n) = (-1)^{t(n)}$, where $t(n)$ is the *total* number of prime factors of n (so $\lambda(n)$ is the totally multiplicative version of $\mu(n)$). Show that

$$\sum_{n=1}^{\infty} \frac{\lambda(n)}{n^s} = \frac{\zeta(2s)}{\zeta(s)}.$$

Hence, working similarly to Exercise 2, prove that

$$\sum_{n \leq X} \lambda(n) \ll X \exp\left(-(\log X)^{\frac{1}{2}}\right).$$

Chapter Two

The Vaughan Identity

2.1 INTRODUCTION

The crux of Vinogradov's adaptation of the Eratosthenes-Legendre sieve, as we explained in the last chapter, is to replace a sum over primes with double sums. There are Type I sums, where one variable has no weight attached, and Type II sums, where both variables may have a weight but neither variable is too small or too large and there is often a way of bounding such sums. In the 1970s R. C. Vaughan produced a major simplification to the study of sums over primes when he introduced the identity that now bears his name. The genesis of this formula lies in the idea of approximating infinite series by finite sums in multiplicative number theory. Such ideas had certainly been around since the 1930s, but the work of Montgomery and Gallagher in the late 1960s prepared the way for a major advance. Before Vaughan's work, it appears that only an approximation to $\zeta(s)^{-1}$ (and its L-function analogues) had been considered. While working on the Bombieri-Vinogradov theorem [158] Vaughan introduced a finite approximation to $-\zeta'(s)/\zeta(s)$ as well. This led him to the identity (actually stated first for L-functions)

$$\sum_{n=1}^{\infty} \Lambda(n)n^{-s} = -\frac{\zeta'}{\zeta}(s)$$

$$= F(s) - \zeta(s)F(s)G(s) - \zeta'(s)G(s) \qquad (2.1.1)$$

$$+ \left(-\frac{\zeta'}{\zeta}(s) - F(s)\right)(1 - \zeta(s)G(s)),$$

where

$$F(s) = \sum_{m \leq U} \Lambda(m)m^{-s}, \qquad G(s) = \sum_{d \leq V} \mu(d)d^{-s}.$$

Since the functions n^{-s} are linearly independent, we thus obtain Vaughan's identity:

$$\Lambda(n) = a_1(n) + a_2(n) + a_3(n) + a_4(n). \qquad (2.1.2)$$

Here

$$a_1(n) = \begin{cases} \Lambda(n) & \text{if } n \leq U, \\ 0 & \text{if } n > U, \end{cases} \qquad a_2(n) = -\sum_{\substack{mdr=n \\ m \leq U, d \leq V}} \Lambda(m)\mu(d),$$

$$a_3(n) = \sum_{\substack{hd=n \\ d \leq V}} \mu(d)\log h, \qquad a_4(n) = -\sum_{\substack{mk=n \\ m > U, k > 1}} \Lambda(m)\sum_{\substack{d|k \\ d \leq V}} \mu(d).$$

In view of (1.3.1), $a_4(n)$ can be rewritten as

$$\sum_{\substack{mk=n \\ m>U,k>1}} \Lambda(m) \sum_{\substack{d|k \\ d>V}} \mu(d). \tag{2.1.3}$$

It immediately follows that those values of m with $m > n/V$ are not counted since then we would have $n = mk > n$. Hence, if $|f(n)| \leq 1$, the choice $U = V$, together with partial summation to remove the $\log h$ factor arising from $a_3(n)$, gives

$$\sum_{n \leq x} \Lambda(n) f(n) = O(U) + \text{sums of Type I and II.}$$

Suppose we take $U = V = x^\beta$. Then the sums are as follows:

$$\text{Type I: } \sum_{\substack{m \leq M \\ mn \leq x}} a_m f(mn) \qquad \text{Type II: } \sum_{\substack{K \leq m < 2K \\ mn \leq x}} a_m b_n f(mn) \tag{2.1.4}$$

where

$$M \leq \max(x^{1-\beta}, x^{2\beta}), \qquad x^\beta \leq K \leq x^{1-\beta},$$

for some $\beta \in (0, \frac{1}{2})$. The coefficients a_m, b_n are clearly bounded by the divisor function. We thus have our desired decomposition of $\Lambda(n)$.

Now let us examine the coefficients in more detail. Suppose that $a_2(n)$, $a_3(n)$ are treated only as Type I sums. Then the coefficient a_m in (2.1.4) from $a_2(n)$ is in modulus

$$\left| \sum_{dh=m} \mu(d)\Lambda(h) \right| \leq \sum_{h|m} \Lambda(h) = \log m.$$

For the coefficient from $a_3(n)$ we get simply $\mu(m)$, which is bounded in absolute value by 1, although there is a log factor arising from the logarithmic weight attached to the Type I range. Moving to the Type II sums arising from $a_4(n)$ we note that one coefficient is $\Lambda(m)$ and the other is

$$\sum_{d|n, d \leq V} \mu(d),$$

and it is only this last factor that requires a divisor function to bound it. We may thus suppose in applications that we have

$$\sum_{m \sim M} |a_m|^2 \ll M \log M \quad \text{and} \quad \sum_{n \sim N} |b_n|^2 \ll N(\log N)^3. \tag{2.1.5}$$

It is worth noting that the implied constants in the above can all be evaluated easily and are relatively small. For example, when $|b_n| = \tau(n)$, we have

$$\sum_{n \sim N} |b_n|^2 \leq N(1 + \log N)^3.$$

Viewed in the form (2.1.2), we can see Vaughan's identity as a rearrangement of the formula

$$\Lambda(n) = \sum_{hd=n} \mu(d) \log h,$$

which we established implicitly in proving the equivalent formula (1.4.5) in Chapter 1. We now show how to derive Vaughan's identity directly from this formula. We suppose U, V are given such that we will be able to tackle the resulting Type I and Type II sums. Suppose that $n > U$. The first obvious move is to take out the terms with $d \leq V$, for this gives us a sum already of the right shape. We thus have

$$\Lambda(n) = \sum_{\substack{hd=n \\ d \leq V}} \mu(d) \log h + \sum_{\substack{hd=n \\ d > V}} \mu(d) \log h = S_1 + S_2, \quad \text{say.} \qquad (2.1.6)$$

We shall comment further on this initial step in Section 2.6. We immediately see that S_1 is $a_3(n)$ above. We apply (1.4.6) to S_2:

$$S_2 = \sum_{\substack{hd=n \\ d > V}} \mu(d) \sum_{m|h} \Lambda(m)$$

$$= \sum_{\substack{hd=n \\ m|h}} \mu(d)\Lambda(m) - \sum_{\substack{hd=n \\ d \leq V}} \mu(d) \sum_{\substack{m|h \\ m \leq U}} \Lambda(m) - \sum_{\substack{hd=n \\ d \leq V}} \mu(d) \sum_{\substack{m|h \\ m > U}} \Lambda(m)$$

$$= S_3 - S_4 - S_5,$$

say. Rewriting $h = mr$ we recognize that $-S_4$ is $a_2(n)$ above. Now the terms $m = n$ in S_3 and S_5 are identical and so cancel out. Writing the remainder of these sums as S_3', S_5', respectively, this gives

$$S_3' = \sum_{\substack{m|n \\ m < n}} \Lambda(n) \sum_{d|n/m} \mu(d) = 0,$$

and a simple rearrangement gives $-S_5' = a_4(n)$. In the case $n \leq U$ the sum S_5 disappears, so there is no term to cancel out with the term $m = n$ in S_3. This term is just $\Lambda(n)$, of course, which is $a_1(n)$. This establishes (2.1.2).

It seems paradoxical that (1.4.5) was known for so long yet this elementary rearrangement was not discovered until Vaughan wrote down (2.1.1). The optimist might take heart at this point: What further extremely useful identities are awaiting discovery, being currently hidden in a well-known formula? The pessimist might say that we have only illustrated the well-known saying: All identities are trivial — once written down by somebody else! The reader should note again the central role played in the above (explicitly and implicitly) by (1.3.1). Of course, this identity also underpins (2.1.1).

Before embarking on applications of Vaughan's identity we pause to consider when it may or may not be useful. We saw in Chapter 1 that the double sums that arise can be treated if $f(x) = e(\alpha x)$, and the next two sections treat this example in detail. In general Vaughan's identity works very well with $f(x) = e(g(x))$, where g is a polynomial or a polynomial-like function (for example, $g(x) = x^\lambda, \lambda > 0, \lambda \notin \mathbb{Z}$). However, if $f(x)$ is multiplicative, so that $f(xy) = f(x)f(y)$, then the conversion to double sums has no benefit since

$$\sum_{m,n} a_m b_n f(mn) = \left(\sum_m a_m f(m) \right) \left(\sum_n b_n f(n) \right),$$

and the new sums are no easier to deal with, in general, than the original sum weighted with the von Mangoldt function. Examples that arise in number theory are $f(n) = n^{-s}$ and $f(n) = \chi(n)$, where χ is a Dirichlet character. This is why Vaughan's identity cannot be used to bound

$$\sum_{n \leq N} \Lambda(n) n^{-s}, \tag{2.1.7}$$

which we saw in Chapter 1 is intimately connected with the zero-free region for $\zeta(s)$, nor for

$$\sum_{n \leq N} \Lambda(n) \chi(n), \tag{2.1.8}$$

which is related to the zero-free region of the L-function given as

$$L(s, \chi) = \sum_{n=1}^{\infty} \chi(n) n^{-s} \text{ for } \operatorname{Re} s > 1$$

and plays a crucial role in the study of primes in arithmetic progressions. Fortunately all is not lost (just as well since the sieve method we will describe later has exactly this same drawback!). In many applications one requires an average of (2.1.7) (integrated over t) or (2.1.8) (summed over χ), and the decomposition of a sum into a multiple sum has many advantages in this situation. We shall use this in Section 2.4 when we prove the Bombieri-Vinogradov theorem. The same principle occurs again when we later study primes in short intervals.

2.2 AN EXPONENTIAL SUM OVER PRIMES

We now give our first application of Vaughan's identity. As mentioned in the first chapter, exponential sums over primes arise naturally in additive problems involving primes via the Hardy-Littlewood circle method.

Theorem 2.1. *Suppose that $\alpha \in \mathbb{R}$ such that there are integers a, q with $(a, q) = 1$ and*

$$|q\alpha - a| < q^{-1}. \tag{2.2.1}$$

Then, for all $N \geq 2$ we have

$$\sum_{n \leq N} \Lambda(n) e(\alpha n) \ll (\log N)^{\frac{7}{2}} \left(\frac{N}{q^{\frac{1}{2}}} + N^{\frac{4}{5}} + (Nq)^{\frac{1}{2}} \right). \tag{2.2.2}$$

Remark. This bound is non-trivial for

$$(\log N)^{7+\epsilon} < q < N (\log N)^{-7-\epsilon}.$$

For smaller values of q a nontrivial bound can be obtained by a 'major arc' technique.

Proof. We take $U = V = N^{2/5}$ in (2.1.2). Our Type I sum is then of the form

$$\sum_{m \leq U^2} a_m \sum_{n \leq N/m} e(\alpha mn),$$

where $|a_m| \leq \log m$, or

$$\sum_{m \leq U} a_m \sum_{n \leq N/m} (\log n)e(\alpha mn),$$

where $|a_m| \leq 1$. For both of these sums (1.6.1) gives a bound

$$\ll N(\log N)^2 \left(\frac{U^2}{N} + \frac{1}{q} + \frac{q}{N} \right).$$

This is a suitable bound to establish (2.2.2).

Our Type II sums then may be split into $\ll \log N$ sums of the form

$$\sum_{m \sim M} \sum_{mn \leq N} a_m b_n e(\alpha mn),$$

where $U \ll M \ll N/U$ and a_m, b_n satisfy (2.1.5). From (1.6.2) we thus obtain a bound

$$\ll N^{\frac{1}{2}} (\log N)^{\frac{5}{2}} \left(\frac{N}{q} + M + \frac{N}{M} + q \right)^{\frac{1}{2}}.$$

Summing over the $\ll \log M$ values of M then completes the proof of (2.2.2). $\square$

2.3 THE DISTRIBUTION OF αp MODULO 1

The exponential sum considered above is obviously related to the distribution of αp modulo 1, that is, the way the fractional parts $\{\alpha p\}$ are distributed in sub-intervals of $[0, 1)$. Vinogradov was the first person to consider this problem (or at least he was the first to obtain unconditional results for this problem) [163], but he did not work in the most efficient way possible. With the advent of Vaughan's identity it became much easier to see what was going on, and this enabled Vaughan to prove the following result. We shall be especially interested in this question since the author first developed the sieve method given in Chapter 3 in this context and the relatively simple nature of the arithmetical information makes it a good "test case."

Theorem 2.2. *Let $\alpha \in \mathbb{R}$ and suppose that for integers a, q with $(a, q) = 1$ we have (2.2.1). Let $\delta \in (0, 1)$ be given. Write*

$$\chi(\theta) = \begin{cases} 1 & \text{if } ||\theta|| < \delta, \\ 0 & \text{otherwise.} \end{cases}$$

Then, for any real β and positive integers N, H, such that $N^{\epsilon} \ll q \ll N^{1-\epsilon}$, for some $\epsilon > 0$, we have

$$\sum_{n \leq N} \Lambda(n)\chi(n\alpha + \beta) = 2\delta \sum_{n \leq N} \Lambda(n) + O\left(\frac{N\delta}{H} \right) +$$

$$+ O\left(N(\log H)(\log N)^7 \left(\frac{q\delta}{N} + \frac{1}{q} + \frac{1}{N^{\frac{1}{2}}} + \frac{\delta}{N^{\frac{1}{3}}} \right)^{\frac{1}{2}} \right).$$

$$(2.3.1)$$

Remarks. We have not chosen all the parameters optimally to obtain (2.3.1) in general, but they are optimal in the case $N\delta^6 \leq 1$. In this situation, if $N^{1/2} < q < \delta^{-1}N^{1/2}$ we obtain

$$\sum_{n\leq N} \Lambda(n)\chi(n\alpha + \beta) = 2\delta \sum_{n\leq N} \Lambda(n) + O\left(\frac{N\delta}{H} + N^{\frac{3}{4}}(\log N)^7(\log H)\right).$$

If we let $\delta = CN^{-1/4}(\log N)^7$, this formula becomes

$$\sum_{n\leq N} \Lambda(n)\chi(n\alpha + \beta) = 2\delta N\left(1 + O(H^{-1}) + O((\log H)C^{-1}) + o(1)\right)$$

$$\geq N^{\frac{3}{4}}(\log N)^7(1 + o(1))$$

by taking H and C sufficiently large but fixed. Since there are infinitely many convergents to the continued fraction expansion of an irrational α, this gives infinitely many primes p with

$$||p\alpha + \beta|| \ll p^{-\frac{1}{4}}(\log p)^7. \tag{2.3.2}$$

Compared to Vaughan, we have saved a power of a logarithm in this case, but this comes from greater efficiency with the Fourier analysis. The reader should notice that this result is "modulo the PNT." That is, we have given a formula for the number of primes p in a certain set up to N (weighted with $\log p$) as a proportion of the total number of primes up to N. For various choices of parameters (for example, take $\delta = N^{1/5}, H = N^{1/20}$ in (2.3.1)), the error term currently known for the PNT is larger than the other errors arising. This *comparison principle* will be a main theme in this book: We compare the number of primes in some restricted set with either all the primes up to N or all the primes in a relatively long interval (for which we know an asymptotic formula from the PNT).

Before proving Theorem 2.2 we need auxiliary results that will have other applications. The natural way to tackle (2.3.1) is through the Fourier series expansion of the function

$$\chi(y) = \begin{cases} 1 & \text{if } ||y|| < \delta, \\ 0 & \text{otherwise.} \end{cases}$$

Since this series is not absolutely convergent, we could try to approximate χ by either a quickly convergent or finite Fourier series. Alternatively, one could simply switch to the Diophantine approximation a/q and work with congruences and the characters of the addition group (mod q), namely,

$$e\left(\frac{n\ell}{q}\right) \quad \ell = 0, 1, \ldots, q - 1.$$

However the transformation to exponential sums is achieved, the meat of the problem comes from a sum like

$$\sum_{\substack{\ell\neq 0 \\ |\ell|\leq L}} c_\ell e(\ell y), \tag{2.3.3}$$

where $|c_\ell| \ll \delta$ and L is about δ^{-1} in size. We will use the following finite Fourier series expansion here.

Lemma 2.1. *Let δ be as above and let $L \in \mathbb{N}$. Then there are coefficients $c_\ell^\pm$ such that*

$$2\delta - \frac{1}{L+1} + \sum_{0 < |\ell| \leq L} c_\ell^- e(\ell\theta) \leq \chi(\theta) \leq 2\delta + \frac{1}{L+1} + \sum_{0 < |\ell| \leq L} c_\ell^+ e(\ell\theta),$$

(2.3.4)

with

$$|c_\ell^\pm| \leq \min\left(2\delta + \frac{1}{L+1}, \frac{3}{2\ell}\right).$$

Remarks. In general we must choose L at least of size δ^{-1} to get a nontrivial result. Often we will take L larger than this, which brings in an extra $\log(L\delta)$ factor.

Proof. See [4, pp. 18–21]. □

Lemma 2.2. *Suppose $\rho, \gamma > 0, \gamma \neq \rho, T \geq 1$. Then*

$$\frac{1}{\pi} \int_{-T}^{T} e^{i\gamma t} \frac{\sin \rho t}{t} \, dt = \delta + O(T^{-1}|\rho - \gamma|^{-1}),$$

(2.3.5)

where

$$\delta = \begin{cases} 0 & \text{if } \gamma > \rho, \\ 1 & \text{if } \gamma < \rho. \end{cases}$$

Proof. This is an alternative version of Perron's formula, and its proof is given in the appendix. □

Proof. (Theorem 2.2) We apply Lemma 2.1 with $L = [H\delta^{-1}]$. It then suffices to prove that

$$\sum_{\ell=1}^{R} \left| \sum_{n \leq N} \Lambda(n) e(\ell n\alpha) \right| \ll RN(\log N)^7 \left(\frac{q}{RN} + \frac{1}{q} + \frac{1}{N^{\frac{1}{2}}} + \frac{1}{RN^{\frac{1}{3}}}\right)^{\frac{1}{2}}$$

(2.3.6)

for $\delta^{-1} \leq R \leq L$.

We apply (2.1.2) with $U = V = N^{1/3}$ and begin with an analysis of the Type I sums. We now have a triple sum, and we must "glue together" the variables ℓ and m. This introduces a divisor function. Since there will be a logarithmic factor arising either through a_m or from the sum over n, we are left to estimate a sum (compare (1.6.1))

$$(\log N) \sum_{r \leq RM} \tau(r) \min\left(\frac{N}{M}, \frac{1}{||\alpha r||}\right) \ll N^\eta \sum_{r \leq RM} \min\left(\frac{N}{M}, \frac{1}{||\alpha r||}\right)$$

$$\ll RN^{1+2\eta} \left(\frac{M}{N} + \frac{1}{q} + \frac{q}{RN}\right)$$

for any $\eta > 0$ (and the implied constant depends only on η). This is a satisfactory bound to establish (2.3.6) in view of the condition $N^\epsilon \ll q \ll N^{1-\epsilon}$ for some $\epsilon > 0$ (pick $\eta = \epsilon/4$, and we can assume that $\epsilon < \frac{1}{3}$).

When we consider the Type II sums that arise, we are faced with a problem that does not occur for Type I sums. We have sums like

$$\sum_{\ell \leq R} c_\ell \sum_{m \sim M} \Lambda(m) \sum_{n \leq N/m} b_n e(\ell m n \alpha) \tag{2.3.7}$$

with $|b_n| \leq \tau(n)$ as we have seen above, and $|c_\ell| \leq 1$. The difficulty is that the n range depends on the m range, so we cannot simply glue together the ranges ℓ and m or ℓ and n. For Type I sums we could sum over the inner range first before gluing the other two variables. We cannot do that if we are to apply Cauchy's inequality as we did to obtain (1.6.2). There are two standard ways around this problem: Either restrict variables to short intervals or apply a variant of Perron's formula (as we have given by Lemma 2.2). We will use the second method consistently in this book until we reach Chapters 12 and 13. Either way only introduces an extra logarithmic factor. We note that the value of the sum does not alter by changing N to $[N] + \frac{1}{2}$, and we henceforth assume that N takes this value. In (2.3.5) we put

$$\gamma = \log(mn), \qquad \rho = \log(N).$$

Our variation of N to half an odd integer implies that

$$|\rho - \gamma| \gg N^{-1}.$$

Hence (2.3.7) becomes $\mathfrak{I} + E$ with

$$\mathfrak{I} = \frac{1}{\pi} \int_{-T}^{T} \sum_{\ell \leq R} c_\ell \sum_{m \sim M} \Lambda(m) m^{it} \sum_{n \leq N/M} b_n e(\ell m n \alpha) n^{it} \frac{\sin(t \log N)}{t} \, dt,$$

$$E = O\left(\frac{N}{T} \sum_{\ell \leq R} c_\ell \sum_{m \sim M} \Lambda(m) \sum_{n \leq N/M} |b_n| \right).$$

We let $T = N^2$ in the above. The expression $\mathfrak{I}$ is just the integral of a Type II sum after combining the variables in some suitable way. The integral over t introduces an extra $\log N$ factor. The error E is trivially $\ll RN^\epsilon$. We have thus reduced (2.3.7) to a sum we can estimate plus suitably small errors.

Now we consider the coefficients that arise in the Type II sums. If $M < N^{1/2}$ we combine ℓ with m. Otherwise we combine ℓ with n. This ensures that the shortest summation range is

$$\geq \min\left(N^{\frac{1}{2}}, N^{\frac{1}{3}} V \right).$$

Combining ℓ and m gives a variable k with coefficient

$$\sum_{\substack{\ell m = k \\ m \sim M, \ell \leq V}} \Lambda(m) \leq \sum_{\ell m = k} \Lambda(m) = \log k.$$

Combining ℓ and n gives a new variable k with coefficient $d_k \ll \tau_3(k)$. It is this combination that leads to the highest powers of $\log N$ since

$$\sum_{k \leq K} \tau_3^2(k) \ll K(\log K)^8.$$

In this case, working as in Chapter 1, we get the log power

$$1 + 1 + \tfrac{1}{2}(1 + 8 + 1) = 7$$

with the 1's above coming from the summation over M, Perron's formula, the summation over m of $\Lambda(m)^2$, and the logarithm from the estimation of the simplest exponential sum. The estimate (2.3.6) then follows from (1.6.2).   □

We pause to consider what has set the limit of $\tfrac{1}{4}$ for the exponent in (2.3.2). The Type I sums were of a smaller order than the Type II sums. The Type II sums set the limit when $M \approx N^{1/2}$. These would also be a problem if M had to be as small as $N^{1/4}$, or as large as $N^{3/4}$. The reader might ask: First, is it possible to avoid getting such sums? Second, is it possible to have a different type of Fourier series expansion such that ℓ could be factorized in some way $\ell = h_1 h_2$ and glue h_1 to m and h_2 to n? In answer to the first question we note that $U < m < N/V$, so we need $U > N^{\frac{1}{2}}$ and $V > N^{1/4}$, or $V > N^{1/2}$ and $U > N^{1/4}$. In either case $UV > N^{3/4}$ and our treatment of the Type I sums then can save no more than $N^{-1/4}$. So the answer to the first question is "No, one cannot avoid the difficult sums by using Vaughan's identity." The avoidance of these sums (and analogues for other problems) will form the basis of much of the remainder of this monograph. The answer to the second question also seems to be no since the Fourier coefficients $c_\ell^{\pm}$ must approximate

$$\frac{\sin(2\pi\delta\ell)}{\pi\ell}.$$

2.4 THE BOMBIERI-VINOGRADOV THEOREM

The following result was first proved in 1965 by Bombieri [15] and Vinogradov [162] using zero-density estimates for L-functions. It is a very important result since it has the same strength in many applications as assuming that the Generalized Riemann Hypothesis is true on average. Later proofs removed the need to work via these density estimates, culminating in the work of Vaughan when he first derived (2.1.1). Later Vaughan showed that one could remove all references to L-functions [159] and work in a completely elementary manner except for a reference to Siegel's theorem (see Lemma 2.7). We shall follow Vaughan's proof here. It should be noted that the general principle of proof — Siegel-Walfisz for small q, large sieve for large q — resurfaces later in Chapters 12 and 13.

Theorem 2.3. *Let $A > 0$ be given. Suppose that $x^{1/2}(\log x)^{-A} \leq Q \leq x^{1/2}$. Then*

$$\sum_{q \leq Q} \max_{y \leq x} \max_{(a,q)=1} \left| \sum_{\substack{n \leq y \\ n \equiv a \,(\mathrm{mod}\, q)}} \Lambda(n) - \frac{y}{\phi(q)} \right| \ll_A x^{\frac{1}{2}} Q (\log x)^5. \tag{2.4.1}$$

Remarks. Since

$$\sum_{q \leq Q} \frac{x}{\phi(q)} \gg x \log Q,$$

(2.4.1) gives a nontrivial estimation for

$$\sum_{\substack{n \leq x \\ n \equiv a \,(\mathrm{mod}\, q)}} \Lambda(n)$$

on average over q up to just less than $x^{1/2}$. The maxima over y and $(a, q) = 1$ in (2.4.1) often arise in applications and do not present any serious difficulties in the proof.

By partial summation, as in Chapter 1, we can replace the sum over n weighted with $\Lambda(n)$ by a sum over primes, replacing $y/\phi(q)$ by $\mathrm{Li}(y)/\phi(q)$. We are thus showing that

$$\pi(x; q, a) \sim \frac{\mathrm{Li}(x)}{\phi(q)}$$

on average over q. The reader should note that the term

$$\frac{y}{\phi(q)}$$

arises as

$$\frac{1}{\phi(q)} \sum_{\substack{n \leq y \\ (n, q) = 1}} \Lambda(n),$$

but in the form stated this makes no difference to the error term. Nevertheless, the reader should note that the comparison principle introduced in the last section is at work here again: The number of primes in our "restricted" set $p \equiv a \,(\mathrm{mod}\, q)$ is being compared to the number of primes up to y.

Before embarking on the proof we need some more important auxiliary results. We shall need to use Dirichlet characters, and we write

$$\sideset{}{^*}\sum_{\chi}$$

to indicate summation over all primitive characters $(\mathrm{mod}\, q)$, where the modulus q will be clear from the context (usually an outer summation variable).

Lemma 2.3. (The Polya-Vinogradov Inequality) *For any $q \geq 2$ and $N \in \mathbb{N}$, we have, for any non-principal character χ mod q, that*

$$\sum_{n \leq N} \chi(n) \ll q^{\frac{1}{2}} \log q. \tag{2.4.2}$$

Proof. See [27, Chapter 23]. □

Lemma 2.4. *For any $M, N, Q \in \mathbb{N}$ and any sequence of complex numbers a_n, we have*

$$\sum_{q \leq Q} \frac{q}{\phi(q)} \sideset{}{^*}\sum_{\chi} \left| \sum_{n=M+1}^{M+N} a_n \chi(n) \right|^2 \ll (N + Q^2) \sum_{n=M+1}^{M+N} |a_n|^2. \tag{2.4.3}$$

Proof. This is an application of the large sieve and will be established in the appendix (A.3) with explicit constants. □

Lemma 2.5. *For any* $M, N, Q \in \mathbb{N}$ *and sequences of complex numbers* a_m, b_n *we have*

$$\sum_{q \leq Q} \frac{q}{\phi(q)} \sideset{}{^*}\sum_{\chi} \max_y \left| \sum_{m \leq M} \sum_{\substack{n \leq N \\ mn \leq y}} a_m b_n \chi(mn) \right| \tag{2.4.4}$$

$$\ll (M + Q^2)^{\frac{1}{2}} (N + Q^2)^{\frac{1}{2}} \left(\sum_{m \leq M} |a_m|^2 \right)^{\frac{1}{2}} \left(\sum_{n \leq N} |b_n|^2 \right)^{\frac{1}{2}} \log(MN).$$

Proof. We may treat the condition $mn \leq y$ with Lemma 2.2 as in the previous section, with $T = (MN)^2$. The proof is then completed by Cauchy's inequality and (2.4.3). □

For any character χ we write

$$\psi(u, \chi) = \sum_{n \leq u} \Lambda(n) \chi(n).$$

Lemma 2.6. *For all* $x \geq 1$, $q \geq 1$ *we have*

$$\sum_{q \leq Q} \frac{q}{\phi(q)} \sideset{}{^*}\sum_{\chi} \max_y |\psi(y, \chi)| \ll \left(x + Q x^{\frac{5}{6}} + Q^2 x^{\frac{1}{2}} \right) (\log(Qx))^4. \tag{2.4.5}$$

Proof. If $Q^2 \geq x$, then (2.4.5) follows immediately from (2.4.4) with $M = 1, a_1 = 1, b_n = \Lambda(N)$. We can thus suppose henceforth that $Q^2 < x$. We apply (2.1.2) with $U = V$, but we will not pick the optimum value for U until the choice is evident.

For a Type I sum like

$$\sum_{r \leq R} a_r \sum_{n \leq y/r} \chi(nr)$$

we apply (2.4.2) to obtain the bound

$$\ll R q^{\frac{1}{2}} \log q \max_r |a_r|.$$

If there is a logarithmic weight attached to n (that is, from $a_3(n)$ in (2.1.2)), the above bound is made worse only by a factor $\log y$ after partial summation. The contribution from all the Type I sums with $r \leq R$ is therefore (after summing over the primitive characters $(\bmod \, q)$ and over $q \leq Q$)

$$\ll R(\log Qx)^2 Q^{\frac{5}{2}} + x(\log x)^2, \tag{2.4.6}$$

where the second term comes from the trivial character when $q = 1$. It follows that we must have $R \leq \max((x/Q)^{1/2}, x^{5/6}Q^{-3/2})$ to establish (2.4.5). The reader can quickly verify that this can be quite small (around $x^{1/4}$) for Q approaching $x^{1/2}$. We therefore treat part of what we previously considered to be a Type I sum as a Type II sum, namely, the part with $R \geq U$. This comprises the whole of the contribution from $a_3(n)$ but only part of that from $a_2(n)$.

For the Type II sums coming, as in the previous sections, from $a_4(n)$ in (2.1.2), we may apply (2.4.4) with (2.1.5) to obtain the bound (summing over $m \sim M$)

$$\ll (Q^2 + M)^{\frac{1}{2}} \left(Q^2 + \frac{x}{M}\right)^{\frac{1}{2}} \left(\sum_{m \sim M} \Lambda(m)^2\right)^{\frac{1}{2}} \left(\sum_{n \leq x/M} \tau(n)^2\right)^{\frac{1}{2}} (\log x)$$

$$\ll \left(Q^2 x^{\frac{1}{2}} + QxM^{-\frac{1}{2}} + Q(xM)^{\frac{1}{2}} + x\right)(\log x)^3.$$

$$(2.4.7)$$

For those terms from $a_2(n)$ with $md > U$ we obtain in place of (2.4.7) a contribution

$$\ll \left(Q^2 x^{\frac{1}{2}} + QxU^{-\frac{1}{2}} + Qx^{\frac{1}{2}}U + x\right)(\log x)^2. \qquad (2.4.8)$$

Combining (2.4.6), (2.4.7) (summed over $\ll \log x$ values), and (2.4.8), we obtain an upper bound

$$\ll \left(Q^2 x^{\frac{1}{2}} + x + QxU^{-\frac{1}{2}} + Q^{\frac{5}{2}}U\right)(\log xQ)^4.$$

Noting that $Q^{3/2} x^{2/3} \ll Q^2 x^{1/2}$ if $q \geq x^{1/3}$, we therefore choose

$$U = \begin{cases} x^{\frac{1}{3}} & \text{if } Q \leq x^{\frac{1}{3}}, \\ x^{\frac{2}{3}} Q^{-1} & \text{otherwise.} \end{cases}$$

The reader can quickly verify that this choice also gives $R \geq U$, as needed for the Type I estimates. This establishes (2.4.5) and so completes the proof of this lemma.   □

Lemma 2.7. *Suppose that* $q \leq (\log x)^A$ *for some* $A > 0$. *Let* χ *be a primitive character (mod q). Then, for* $y \leq x$, *we have*

$$\psi(y, \chi) \ll x \exp\left(-C(A)(\log x)^{\frac{1}{2}}\right). \qquad (2.4.9)$$

In particular, we have

$$\psi(y, \chi) \ll_A x(\log x)^{-2A}. \qquad (2.4.10)$$

Remarks. The weakness of this result reflects our poor knowledge of primes in an arithmetic progression to a fixed modulus or, equivalently, our ignorance concerning possible real zeros of $L(s, \chi)$ near $s = 1$ (the Siegel zeros). It should also be noted that a bad distribution of primes $(\mathrm{mod}\, r)$ leads to a bad distribution $(\mathrm{mod}\, q)$ for all q divisible by r, and it is the weakness of the present result that means we can save no more than an arbitrary power of a logarithm on the trivial estimate in (2.4.1).

The constant in (2.4.9) is ineffective in the sense that, at present, we have no means of obtaining the implied constant for any given $A \geq 1$. This comes about because we cannot determine the value of $C'(\epsilon)$ in the proof below for any value of $\epsilon \leq 1$. See [27, Chapters 20 and 21]. This "ineffectiveness" carries over into the Bombieri-Vinogradov theorem itself and consequently all applications of that theorem.

Proof. This follows from Siegel's theorem that for any q there is at most one character (of necessity real) such that $L(\beta, \chi) = 0$ for $\beta > 1 - c(\log q)^{-1}$ for a certain constant c (see Chapter 21 in [27]) and that for any $\epsilon > 0$ there is $C'(\epsilon) > 0$ such that $\beta < 1 - C'(\epsilon)q^{-\epsilon}$. The reader will understand from the proof of the PNT, or Lemma 1.5, how such a deduction can be made using Perron's formula assuming certain bounds on $L'/L(s)$ in a narrow strip. The details are all in [27]. It should be noted that if x were nearer in size to q, then $\psi(y, \chi)$ would be very near to $y/\phi(q)$ in size should a Siegel zero exist. The reader should compare these remarks with the exercise at the end of Chapter 4. □

Proof. (Theorem 2.3) By the properties of characters,

$$\sum_{\substack{n \leq y \\ n \equiv a \bmod q}} \Lambda(n) - \frac{1}{\phi(q)} \sum_{\substack{n \leq y \\ (n,a)=1}} \Lambda(n) = \frac{1}{\phi(q)} \sum_{\chi \neq \chi_0 \bmod q} \bar{\chi}(a)\psi(y, \chi).$$

Hence

$$\max_{(a,q)=1} \left| \sum_{\substack{n \leq y \\ n \equiv a \bmod q}} \Lambda(n) - \frac{y}{\phi(q)} \right| \leq \frac{1}{\phi(q)} \sum_{\chi \neq \chi_0} |\psi(y, \chi)| + \frac{E(y) + O((\log y)^2)}{\phi(q)}.$$

Here

$$E(y) \ll y \exp\left(-(\log y)^{\frac{1}{2}}\right)$$

is the error term from the PNT (1.4.9), and the $O((\log y)^2)$ comes from prime power factors of q. Since

$$\sum_{q \leq Q} \frac{1}{\phi(q)} \ll \log Q,$$

it remains to prove that

$$\sum_{q \leq Q} \frac{1}{\phi(q)} \sum_{\chi \neq \chi_0} \max_{y \leq x} |\psi(y, \chi)| \ll x^{\frac{1}{2}} Q(\log x)^{-A}. \tag{2.4.11}$$

Our next task is to reduce the summation to primitive characters only. Suppose that χ is induced by the primitive character χ'. Then

$$\psi(y, \chi) - \psi(y, \chi') = \sum_{\substack{p|q \\ p^k \leq y}} \chi'(p^k) \log p \ll (\log yq)^2$$

after a simple calculation. Since each primitive character (mod q) induces characters to moduli that are multiples of q, the left-hand side of (2.4.11) is

$$\ll Q(\log x)^2 + \sum_{q \leq Q} \sideset{}{^*}\sum_{\chi} \max_{y \leq x} |\psi(y, \chi)| \left(\sum_{r \leq Q/q} \frac{1}{\phi(rq)} \right).$$

The first term is of a smaller order than the right-hand side of (2.4.11), so we may ignore it. Since $\phi(rq) \geq \phi(r)\phi(q)$ we obtain

$$\sum_{r \leq Q/q} \frac{1}{\phi(rq)} \ll \frac{1}{\phi(q)} \log\left(\frac{2Q}{q}\right).$$

We must therefore establish that

$$\sum_{q \leq Q} \frac{1}{\phi(q)} \sum_{\chi}^{*} \max_{y \leq x} |\psi(y, \chi)| \ll x^{\frac{1}{2}} Q (\log x)^4. \qquad (2.4.12)$$

By (2.4.5)

$$\sum_{q \sim W} \frac{1}{\phi(q)} \sum_{\chi}^{*} \max_{y \leq x} |\psi(y, \chi)| \ll \left(\frac{x}{W} + x^{\frac{5}{6}} + x^{\frac{1}{2}} W \right) (\log x)^4.$$

Summing over $W = 2^k$ with $(\log x)^A < W < Q$ gives a bound $\ll x^{1/2} Q (\log x)^4$ as required. For $W \leq 2(\log x)^A$ we use (2.4.10), which quickly gives an estimate $\ll x(\log x)^{-A}$, which establishes (2.4.12) and completes the proof. $\qquad \square$

An example of the application of the Bombieri-Vinogradov theorem is afforded by approximating the twin-prime problem. Suppose we wanted to find primes p such that $p + 2$ was almost-prime in some sense. Then, working as in Section 1.5, we would put $\mathcal{A} = \{p + 2 : x/2 < p < x\}$ and

$$R_d = \sum_{\substack{x/2 < p < x \\ p \equiv -2 \bmod d}} 1 - \frac{\mathrm{Li}(x) - \mathrm{Li}(x/2)}{\phi(d)}$$

(the change from d in our earlier simplified account to $\phi(d)$ in this application causes no problems). By Theorem 2.3 we have

$$\sum_{\substack{d < D \\ (d,2)=1}} |R_d| \ll x(\log x)^{-A},$$

provided that $D < x^{1/2}(\log x)^{-B}$ for some B. This is sufficient to prove that infinitely often $p + 2$ has all its prime factors $> p^{1/4 - \epsilon}$ (so it has at most four prime factors). By using a weighted sieve and a role reversal (for which one needs an analogue of the Bombieri-Vinogradov theorem), it is possible to show that infinitely often $p + 2$ has at most two prime factors (see [55, Chapter 11] for a similar application to the Goldbach problem). We shall give another application of Theorem 2.3 in Chapter 8.

2.5 LINNIK'S AND HEATH-BROWN'S IDENTITIES

After a few moments contemplation of Vaughan's identity, several questions naturally come to mind. We have triple sums that we treated as double sums: Can we make use of more variables? We have $\Lambda(m)$ as one of the coefficients: What if we apply Vaughan's identity again to this term? This would give multiple sums with more variables. Could we start with a different identity to get more flexible multiple sums? Heath-Brown produced what he called a generalized Vaughan identity [73] — although he admits in the paper where it was introduced that it is not strictly a generalization — by using the following formula, which is valid for all $k \in \mathbb{N}$ and any function $M(s)$:

$$\frac{\zeta'}{\zeta}(s) = \sum_{j=1}^{k}(-1)^{j-1}\,^{k}C_j\,\zeta(s)^{j-1}\zeta'(s)M(s)^j$$

$$+ \zeta(s)^{-1}(1-\zeta(s)M(s))^k\zeta'(s). \tag{2.5.1}$$

The formula follows immediately from the binomial theorem applied to the second term on the right-hand side of (2.5.1).

To apply (2.5.1) we follow the same philosophy as in (2.1.1) and let $M(s)$ be a partial sum of $\zeta(s)^{-1}$:

$$M(s) = \sum_{m\le M}\mu(m)m^{-s}.$$

For a given x let $M = [x^{1/k}]$. Then $(1-\zeta(s)M(s))^k$ is a series in n^{-s} with $n > x$. Using the linear independence of n^{-s} this implies an identity

$$\Lambda(n) = \sum_{j=1}^{k}(-1)^{j-1}\,^{k}C_j\,a_j(n) \tag{2.5.2}$$

where

$$a_j(n) = \sum_{\substack{n=r_1\dots r_{2j}\\ i>j\Rightarrow r_i\le x^{1/k}}}(\log r_1)\mu(r_{j+1})\dots\mu(r_{2j}).$$

We can thus transform a sum weighted with $\Lambda(n)$ to a multiple sum with as many as $2k$ variables. The advantage we have gained is that the new variables weighted with "difficult" coefficients, that is, $\mu(n)$, can be made as small as we like by taking k sufficiently large. So the only "long" sums have the Type I form.

If we take $k = 2$ in (2.5.2) we obtain

$$\Lambda(n) = 2\sum_{\substack{n_1 n_2=n\\ n_1\le x^{1/2}}}(\log n_1)\mu(n_2) - \sum_{\substack{n_1 n_2 n_3 n_4=n\\ n_3,n_4<x^{1/2}}}(\log n_1)\mu(n_3)\mu(n_4)$$

$$= 2b(n) - c(n), \quad\text{say.}$$

The reader can quickly verify that this is a result comparable to Vaughan's identity. For example, to use it in Section 2.2 note that $b(n)$ gives a Type I sum of the right form immediately. We split $c(n)$ into two parts $d(n) + e(n)$, where $\max(n_1, n_2) > x^{1/5}$ in $d(n)$, giving a Type I sum. For that part of $e(n)$ with $\max(n_3, n_4) > x^{2/5}$ we have a Type II sum of the required form. For the rest we note that $\max(n_1, n_2)\max(n_3, n_4) \in [n^{1/2}, x^{3/5})$, and this also leads to a suitable Type II sum (or these terms can be treated trivially for small n).

Heath-Brown first used (2.5.1) to give a new proof of Huxley's theorem [97].

Theorem 2.4. *For any fixed θ in the range $\frac{7}{12} < \theta \le 1$, and $y = x^\theta$, we have*

$$\sum_{x-y<n\le x}\Lambda(n) = y(1 + o(1)). \tag{2.5.3}$$

The first result of this type was given in 1930 by Hoheisel [92] (with a much weaker exponent), and Heath-Brown acknowledges this work as the inspiration for his identity. Work after Hoheisel tended to use zero-density estimates for the Riemann zeta-function rather than sieve identities. Although there were frequent improvements in this result up to 1972, there has been no substantive progress in the last 34 years for giving an asymptotic formula such as (2.5.3). All work in shorter intervals has only obtained lower bounds of the correct order of magnitude (see Chapter 7). We shall give our own proof of Huxley's result in Chapter 7, so we shall not reproduce Heath-Brown's proof here. Suffice it to say that he takes $k = 5$ and so has up to 10-fold sums to deal with.

One application of (2.5.2) that has been found useful (for example, see [5, 63]) is as follows. The reader should note that this result does *not* follow from Vaughan's identity.

Lemma 2.8. *Suppose that $u \leq x^{1/10}$. Then*

$$\sum_{N/2 \leq n \leq N} \Lambda(n) f(n)$$

can be written as $\ll (\log N)^5$ sums of Type I and Type II, where the Type I sums have the form

$$\sum_{m \leq M} a_m \sum_{N/2m \leq n \leq N/m} f(mn), \quad \text{with } M \leq 12Nu,$$

and the Type II sums have the shape

$$\sum_{m \sim M} \sum_{N/2m \leq n \leq N/m} a_m b_n f(mn), \quad \text{with } u^2 \leq M \leq 64N^{\frac{1}{3}}.$$

Proof. This follows from (2.5.2) with $k = 3$. Full proofs are given in [73, pp.1367–1368] and [5, Lemma 2] with slightly different parameters. The reader should easily be able to establish the result.   □

Heath-Brown's identity has the advantage that more flexible sums are produced, and there are situations, such as the theorem above, where use can be made of this (see Chapter 7). However, the disadvantage persists that if one makes a problem harder (for example, taking θ smaller in Theorem 2.4), the method collapses at a point: There is no "grey area" between an asymptotic formula and no result at all. Heath-Brown produced another identity that can be applied to remove this disadvantage. To see the motivation for this identity we recall that

$$\zeta(s) = \prod_p \left(1 - \frac{1}{p^s}\right)^{-1}.$$

From this representation it is clear that the effect of multiplying by a factor $(1-p^{-s})$ is to sieve out all multiples of p from the Dirichlet series for $\zeta(s)$. So, the factor

$$\Pi(s) = \prod_{p < z} \left(1 - \frac{1}{p^s}\right)$$

corresponds to sieving out all primes less than z. Thus

$$\zeta(s)\Pi(s) = \prod_{p \geq z}\left(1 - \frac{1}{p^s}\right)^{-1} = 1 + \sum_{\substack{n \geq z \\ p|n \Rightarrow p \geq z}} n^{-s}.$$

On the right-hand side above, the first nonzero terms after 1 are of the form p^{-s} for $z \leq p < z^2$. We now see that this is the zeta-function equivalent of the sieve of Eratosthenes.

In the zeta-function form the identity Heath-Brown derived takes the shape

$$\log\left(\zeta(s)\Pi(s)\right) = \sum_{j=1}^{\infty} \frac{(-1)^{j-1}}{j}\left(\zeta(s)\Pi(s) - 1\right)^j. \tag{2.5.4}$$

This is, of course, none other than the power series for $\log(1 + x)$, where $x = \zeta(s)\Pi(s) - 1$. The formula is therefore only valid for $|\zeta(s)\Pi(s) - 1| < 1$, but this holds in a half-plane, so we can still use the linear independence of the functions n^{-s} to produce an arithmetic identity.

Linnik [122] was the first to consider such identities. The factor $\Pi(s)$ plays an analogous role to the term $M(s)$ in (2.5.1), only instead of removing small n, it removes (as we have seen above) all powers of small primes. From the technical point of view it is neither easier nor harder to deal with than the $M(s)$ factor. To use (2.5.4) we note that

$$\log\left(\zeta(s)\Pi(s)\right) = -\sum_{p \geq z} \log\left(1 - p^{-s}\right) = \sum_{\ell=1}^{\infty} \frac{1}{\ell} \sum_{p \geq z} p^{-\ell s}.$$

Using the linear independence of the functions n^{-s} we thus arrive at the following identity, where we suppose that $z^k > x$:

$$\sum_{x/2 < p < x} f(p) + \sum_{\ell \geq 2} \frac{1}{\ell} \sum_{x/2 < p^\ell < x} f(p^\ell) = \sum_{j=1}^{k-1} \frac{(-1)^{j-1}}{j} \sum_{x/2 < n < x} a_j(n)f(n),$$
$$\tag{2.5.5}$$

with

$$a_j(n) = \sum_{\substack{n_1 \dots n_j = n \\ p|n_j \Rightarrow p \geq z}} 1.$$

We now have several differences from Vaughan's identity. First, we are summing over primes rather than integers weighted with the von Mangoldt function. More importantly, the sums that arise have a fixed sign. This is brought out more by taking $f(n)$ as the characteristic function of a set $\mathcal{A}$, and so (2.5.5) can then be rewritten as

$$S(\mathcal{A}, x^{\frac{1}{2}}) = \sum_{j=1}^{k} \frac{(-1)^{j-1}}{j} S(\mathcal{A}^k, z) + O\left(x^{\frac{1}{2}}\right), \tag{2.5.6}$$

where $\mathcal{A}^k = \{n_1 \dots n_k \in \mathcal{A}\}$. As one decreases θ in Theorem 2.4 using this identity with $k = 7$, it is found that all sums in (2.5.6) can be dealt with except some with $k = 6$. Since these have a negative weight attached, one only needs to obtain an upper bound for this term. Since the ranges of summation for the variables are quite restricted, it is not difficult to show that such sums give a smaller order contribution than the main term. In this way Heath-Brown [80] was able to establish that Theorem 2.4 holds with $\theta = \frac{7}{12}$. We shall outline a different proof of this result in Chapter 7.

The reader should note the similarity and contrasts between (2.5.6) and Buchstab's identity (1.4.13). Indeed, these two identities can play essentially the same roles. For $k = 2$, (2.5.6) is immediate. For $k = 3$ or 4, (2.5.6) can be deduced quickly from (1.4.13). For $k \geq 5$, the derivation of (2.5.6) from (1.4.13) is not so straightforward.

2.6 FURTHER THOUGHTS ON VAUGHAN'S IDENTITY

We now return to (2.1.6):

$$\Lambda(n) = \sum_{\substack{hd=n \\ d \leq V}} \mu(d) \log h + \sum_{\substack{hd=n \\ d > V}} \mu(d) \log h = S_1(n) + S_2(n). \qquad (2.6.1)$$

Since

$$\sum_{d|n} \mu(d) \log n = 0,$$

an alternative decomposition would be

$$\Lambda(n) = -\sum_{\substack{d|n \\ d \leq V}} \mu(d) \log d - \sum_{\substack{d|n \\ d > V}} \mu(d) \log d = T_1(n) + T_2(n).$$

Or, indeed,

$$\Lambda(n) = \sum_{\substack{d|n \\ d \leq V}} \mu(d) \log\left(\frac{V}{d}\right) + \sum_{\substack{d|n \\ d > V}} \mu(d) \log\left(\frac{V}{d}\right) = T_1^*(n) + T_2^*(n). \qquad (2.6.2)$$

The observation appears to have first been made by Heath-Brown (with a slightly different decomposition) [78] that the "main term" for $\Lambda(n)$ arises from T_1^*, while T_2^* contributes only to the error term. Goldston used (2.6.2) in [47], and we should also note that T_1^* is the case $k = 1$ of the expression

$$\frac{1}{k!} \sum_{\substack{d|n \\ d \leq V}} \mu(d) \left(\log\left(\frac{V}{d}\right)\right)^k, \qquad (2.6.3)$$

used by Goldston and Yildirim (and subsequent papers included Motohashi and Pintz) to great effect in their work on small differences between primes [48]. In that work (2.6.3) arises since

$$\mu(d) \left(\frac{\log(V/d)}{\log V}\right)^k$$

is an approximation to the coefficients $\lambda(d)$ in the Selberg sieve of dimension k where k is large but fixed. In the case $k = 1$ we note that (in the simplest case of one residue class sieved for each prime as we are considering here) we have (see [51, pp. 42–46] with $V = \sqrt{D}$)

$$\lambda(d) = \frac{d\mu(d)}{\phi(d)} \left(\sum_{d<V} \frac{\mu^2(d)}{\phi(d)} \right)^{-1} \sum_{\substack{k<V/d \\ (k,d)=1}} \frac{\mu^2(k)}{\phi(k)} \approx \mu(d) \frac{\log(V/d)}{\log V}.$$

There should be no surprise that there is this connection between the Selberg sieve and Vaughan's identity since the same arithmetical structure underpins all sieves and sieve identities.

Iwaniec in collaboration with others [36, 38] has also made much use of (2.6.1). Put another way, in certain situations cancellations caused by the presence of $\mu(n)$ in S_2 are crucial to the application of Vaughan's identity. This observation lies at the heart of the Friedlander-Iwaniec asymptotic sieve [38] discussed in Chapter 12.

To illustrate this we first consider

$$\sum_{n \leq N} T_1^*(n) = \sum_{d \leq V} \mu(d) \log\left(\frac{V}{d}\right) \sum_{\substack{n \leq N \\ n \equiv 0 \,(\mathrm{mod}\, d)}} 1$$

$$= \sum_{d \leq V} \mu(d) \log\left(\frac{V}{d}\right) \frac{N}{d} + O(V \log V)$$

$$= M(N, V) + E(N, V) \text{ say.}$$

By the variant of Perron's formula given as Lemma A.3 we have

$$M(N, V) = N \int_{c-i\infty}^{c+i\infty} \frac{1}{\zeta(s+1)} \frac{V^s}{s^2} \, ds.$$

We can take the integral back to the left of the line $\mathrm{Re}\, s = 0$ but remaining in the zero-free region for $\zeta(s)$. We encounter a pole at $s = 0$ with residue N. From standard calculations using the zero-free region it is then easy to show that

$$M(N, V) = N \left(1 + O\left(\exp\left(-(\log V)^{\frac{1}{4}} \right) \right) \right).$$

Hence, by the PNT, $T_1^*(n)$ does contribute the main term in this simple example. The reader should note that the main term has been obtained using only Type I information. Type II information is necessary to show that T_2^* is not too large on average.

Now suppose that a_n is a non-negative sequence: It might be the characteristic function of a set of interest. Then

$$\sum_{n \leq N} \Lambda(n) a_n = \sum_{d \leq V} \mu(d) \sum_{h \leq N/d} a_{hd} \log h + \sum_{h < N/V} \log h \sum_{\substack{d < N/h \\ d > V}} \mu(d) a_{hd}$$

$$= M + R, \text{ say.}$$

Suppose further that

$$\sum_{h \leq N/d} a_{hd} \log h = \lambda \sum_{h \leq N/d} \log h + O(J_d).$$

Then

$$\sum_{n\leq N} \Lambda(n)a_n = \lambda \sum_{n\leq N} \Lambda(n) + R - \lambda R' + O\left(\sum_{d\leq V} J_d\right),$$

where

$$R' = \sum_{\substack{h<N/V}} \log h \sum_{\substack{d<N/h \\ d>V}} \mu(d)$$

$$\leq (\log N) \sum_{h<N/V} \left|\sum_{V<d<N/h} \mu(d)\right|$$

$$\ll (\log N) \sum_{h<N/V} \frac{N}{h} \exp\left(-(\log N)^{\frac{1}{2}}\right)$$

by the methods used to establish the Prime Number Theorem, if $V > N^\epsilon$ for some $\epsilon > 0$ (Exercise 2 in Chapter 1). Hence, assuming that we have Type I information that leads to

$$\sum_{d\leq V} J_d < \lambda N \exp\left(-(\log N)^{\frac{1}{3}}\right),$$

we have

$$\sum_{n\leq N} \Lambda(n)a_n = \lambda\left(\sum_{n\leq N} \Lambda(n) + O(N\exp\left(-(\log N)^{\frac{1}{3}}\right))\right) + R.$$

It follows that if we are to obtain the correct asymptotic formula, then we must have $R = o(\lambda N)$. We shall illustrate this principle now with a couple of examples.

First, let $y = x^\theta$ and $y_1 = x(\log x)^{-A}$. In the following Y could be either y or y_1. Let $\mathfrak{I} = [x, x+y]$, $\mathfrak{J} = [x, x+y_1]$. We note that, for $d < Y$,

$$\sum_{hd\in[x,x+Y]} \log h = \frac{Y}{d}\log\left(\frac{x}{d}\right)\left(1 + O\left(\frac{Y}{x}\right)\right) + O(\log x).$$

Applying this with $Y = y$ and $Y = y_1$, and using a slight modification to the above working, then gives

$$\sum_{n\in\mathfrak{I}} \Lambda(n) = \frac{y}{y_1}\left(\sum_{n\in\mathfrak{J}} \Lambda(n) + O\left(x(\log x)^{-2A}\right)\right)$$

$$+ \sum_{h<x/V} \log h \sum_{\substack{dh\in\mathfrak{I} \\ d>V}} \mu(d) + O(V\log x)$$

$$= y\left(1 + O\left((\log x)^{-A}\right)\right) + \sum_{h<x/V} \log h \sum_{\substack{dh\in\mathfrak{I} \\ d>V}} \mu(d)$$

under the assumption that $V \leq y(\log x)^{-A-1}$. It follows that we must prove that the final sum is $o(y)$ in order to obtain the expected asymptotic formula. The reader should note that this holds for large ranges of the parameters θ and V.

Similarly, we may take the example $n \equiv a \pmod{q}$, where $(a, q) = 1$. If we assume that $q \leq (\log x)^C$, we obtain

$$\sum_{\substack{n \leq N \\ n \equiv a \pmod{q}}} \Lambda(n) = \frac{1}{\phi(q)}(1 + o(1)) \sum_{n \leq N} \Lambda(n) - \sum_{h \leq x/V} \sum_{\substack{d \equiv a\bar{h} \pmod{q} \\ V < d \leq x/h}} \mu(d).$$

We leave it to the reader to fill in the details as an exercise. The question arises: Are there examples where we can make use of the cancellation introduced by the $\mu(d)$ factor? From a different perspective, the problem facing the researcher is the *parity problem*. That is to say, we cannot detect primes in a given set using only Type I information. There must be some arithmetical input on the distribution of numbers with an odd or even number of prime factors. Of course, this is precisely what is happening in an estimation of the sum R. We are saying, in effect, that there are asymptotically the same number of integers with both an odd and an even number of prime factors in a given set. We shall return to these questions in later chapters. The reader should also note that we have used the comparison principle mentioned in Section 2.3 again. That is to say, we have related the number of primes in a given set (a short interval, for example) to the *known* number of primes in a long interval.

There is one final remark we must make on these final interpretations of Vaughan's identity. The crucial Type II sum to be estimated is

$$\sum_{h < x/V} \log h \sum_{\substack{dh \leq x \\ d > V}} \mu(d) U(dh),$$

where $U(n)$ is the characteristic function of a set. Usually this will be expanded out in terms of a Fourier or Dirichlet series. The saving that is required may then be thought of as the "orthogonality" of the function $\mu(d)$ to $U(dh)$ or to the Fourier or Dirichlet series associated with the problem. The reader might like to compare this with the terminology and work of Green and Tao [52, 53] as well as the work in Chapter 12.

Exercises

1. The Piatetski-Shapiro prime number theorem states that for $1 < c < \theta$ (where one would like to take θ as large as possible) we have

$$\sum_{n \leq N} \Lambda([n^c]) = N(1 + o(1)) \text{ as } N \to \infty.$$

Consider how Vaughan's identity (or the later identities) can be applied in tandem with Fourier analysis to convert this problem into the estimation of multiple exponential sums. It will be useful to write $\gamma = c^{-1}$ and use the approximation to $\{x\}$ (the fractional part of x) given by

$$\{x\} = \frac{1}{2} - \sum_{0 < |h| \leq H} \frac{e(xh)}{2\pi i h} + O\left(\min\left(1, \frac{1}{H\|x\|}\right)\right) \qquad (2.6.4)$$

(see [75] for a solution or turn to Section 3.7 for more details).

2. Using the identity

$$\frac{1}{\zeta} = 2F - F^2\zeta + \left(\frac{1}{\zeta} - F\right)(1 - \zeta F),$$

where F is a partial sum of $1/\zeta(s)$, or otherwise, produce a Vaughan-type identity for $\mu(n)$ (see [160] for a solution). Hence, give an analogue of the Bombieri-Vinogradov theorem for

$$\sum_{q \le Q} \max_{(a,q)=1} \max_{y \le x} \left| \sum_{\substack{n \le y \\ n \equiv a \pmod q}} \mu(n) \right|.$$

3. Use the Bombieri-Vinogradov theorem to show that all sufficiently large integers are the sum of a prime and a square-free number. (The identity proved in Exercise 1 in Chapter 1 is the starting point.)

Chapter Three

The Alternative Sieve

3.1 INTRODUCTION

As we have seen, Vaughan's method gives an *identity* most often applied to the auxiliary sums (trigonometric or character) that arise naturally from a number-theoretic question. Its limitation is that there is no grey area between it working (giving the correct asymptotic formula) and it failing (giving nothing). In this chapter we begin to consider a method introduced by the author in [59] that overcomes this disadvantage by two devices. First, the method is applied to the arithmetical information rather than the auxiliary results. Second, we can decompose our sums over primes, taking account of the signs of the sums. We can discard certain sums if they have the correct sign and, provided we have not thrown away too much, we still arrive at a positive lower bound for a sum over primes. We find we have a grey area between an identity and a trivial lower bound, which enables us to make more progress than was possible before.

We recall the notation used in Chapter 1:

$$\mathcal{B} = \mathbb{Z} \cap \left[\frac{x}{2}, x\right), \qquad \mathcal{A} \subseteq \mathcal{B},$$

and

$$S(\mathcal{A}, z) = |\{n \in \mathcal{A} : p|n \Rightarrow p \geq z\}|$$
$$= \sum_{\substack{d|P(z) \\ dn \in \mathcal{A}}} \mu(d)$$
$$= \sum_{d|P(z)} \mu(d)|\mathcal{A}_d|.$$

We introduce the function

$$\rho(n, z) = \begin{cases} 1 & \text{if } n \in \mathbb{N}, \; p|n \Rightarrow p \geq z, \\ 0 & \text{otherwise.} \end{cases}$$

It will be important that $\rho(n, z) = 0$ if $n \notin \mathbb{N}$. So

$$S(\mathcal{A}, z) = \sum_{n \in \mathcal{A}} \rho(n, z),$$

$$\rho(n, z) = \sum_{\substack{d|n \\ d|P(z)}} \mu(d),$$

and Buchstab's identity (1.4.13) gives, for $1 \leq w < z$,

$$\rho(n, z) = \rho(n, w) - \sum_{w \leq p < z} \rho(n/p, p). \qquad (3.1.1)$$

Also, $\rho(n, z(n))$ is the characteristic function of the set of primes if $n^{1/2} < z(n) \leq n$. It should be clear that whatever we write for $S(\mathcal{A}, z)$ has an analogue for $\rho(n, z)$, and vice versa. The former notation has an advantage when we are thinking of finding primes in a given set $\mathcal{A}$, the latter if we are dealing with a function summed over the primes or, as in later chapters, when we use a vector sieve.

Now we are going to consider how to break up the sum

$$\sum_{d | P(z)} \mu(d) g(d),$$

where, typically,

$$g(d) = |\mathcal{A}_d| \quad \text{or} \quad g(d) = \sum_{md \leq x} f(md).$$

The basic idea is very simple: Take out the largest prime factor of d, and then the next largest; keep on going until the conditions we require are met by some product of primes (or primes with other variables). Taking p as the largest prime factor of d we have (assuming $\mu(d) \neq 0$)

$$d = pd', \quad q | d' \Rightarrow q < p, \quad \mu(d') = -\mu(d).$$

The basic identity we use is thus

$$\sum_{d | P(z)} \mu(d) g(d) = g(1) - \sum_{2 \leq p < z} \sum_{d | P(p)} \mu(d) g(dp). \qquad (3.1.2)$$

The first term $g(1)$ on the right-side of (3.1.2) arises from the term $d = 1$ on the left-side. Of course, 1 has no prime factor! (Recall our comments on the sieve of Eratosthenes in Chapter 1.) From our work on Vaughan's identity we might now begin to perceive one central feature of our approach. If d is *small*, we know $g(d)$ fairly well as it involves an unrestricted sum over consecutive integers in a long range. If d is *large* and z is not too big, by repeated applications of (3.1.2) we can get a product of variables lying between R and Rz for some given R. Recall that our bounds for Type II sums in the previous chapter needed one variable neither too small nor too big. The pathway to our desired conclusion will be more convoluted because of the need to disentangle relationships between the summation variables.

We conclude this section with one particular case of (3.1.2). Given $1 < w < z$, write

$$g(d) = \begin{cases} 0 & \text{if } (d, P(w)) > 1, \\ S(\mathcal{E}_d, w) & \text{if } (d, P(w)) = 1. \end{cases}$$

Then

$$\sum_{d | P(z)} \mu(d) g(d) = \sum_{\substack{d | P(z) \\ (d, P(w)) = 1}} \mu(d) S(\mathcal{E}_d, w) = \sum_{\substack{n \in \mathcal{E} \\ p | n \Rightarrow p \geq w}} \sum_{d | n, d | P(z)} \mu(d) = S(\mathcal{E}, z).$$

Also,
$$g(1) = S(\mathcal{E}, w)$$
and
$$\sum_{2 \leq p < z} \sum_{d \mid P(p)} \mu(d) g(dp) = \sum_{w \leq p < z} \sum_{\substack{d \mid P(p) \\ q \mid d \Rightarrow q > w}} \mu(d) S(\mathcal{E}_{dp}, w)$$
$$= \sum_{w \leq p < z} S(\mathcal{E}_p, p).$$

We thus arrive, by a roundabout argument, at the Buchstab identity

$$S(\mathcal{E}, z) = S(\mathcal{E}, w) - \sum_{w \leq p < z} S(\mathcal{E}_p, p), \tag{3.1.3}$$

which we have already stated in Chapter 1 (1.4.13), claiming it to be a simple application of the inclusion/exclusion principle. However, it is important to note the relation between (3.1.2) and (3.1.3), that is, to see that the same process of taking out the largest prime factor is at work in both. We will use (3.1.2) on the *micro* scale to produce our fundamental building block in Section 3.3. We will use (3.1.3) on the *macro* scale to put the building blocks together. The same principle is at work in both though!

Our aim in the following sections will be to relate $S(\mathcal{A}, x^{1/2})$ to $S(\mathcal{B}, x^{1/2})$. We will be dealing with cases where the probability that a number in $\mathcal{A}$ belongs to $\mathcal{B}$ is (in some sense) λ, and we want to deduce that $S(\mathcal{A}, x^{1/2}) = \lambda S(\mathcal{B}, x^{1/2}) +$ smaller order.

3.2 COSMETIC SURGERY

We are going to get multiple sums with various cross-conditions on the variables. However, we will need to combine variables in various ways that make the cross conditions very awkward. The cross-conditions will have the form $p_i < p_j$ or $p_i p_j < t$ (or analogous expressions with more variables). The basic principle is this: Each cross-condition can be removed at the expense of a factor $\ll \log x$. We have already seen this for one condition in Section 2.3. These factors will only multiply error terms, not terms of the same magnitude as the main term. Hence, if the error terms are all $(\log x)^{-A}$ smaller than our main terms, we will succeed if we have fewer than $A - 1$ conditions to remove. It turns out that usually there are at most two conditions to be ironed out on the *micro* level and a bounded number of conditions on the *macro* level. The basic result we need is the alternative version of Perron's Formula given as Lemma 2.2. We now show how this can be applied to a typical sum in detail. Suppose we know that

$$\sum_{m \sim M} a_m \sum_{mn \in \mathcal{A}} b_n = \lambda \sum_{m \sim M} a_m \sum_{mn \in \mathcal{B}} b_n + O(Y) \tag{3.2.1}$$

for any coefficients a_m, a_n, with $|a_m| \leq \tau(m), |b_n| \leq \tau(n)$. Suppose we then have a sum

$$S = \sum_{m \sim M} \sum_{m = dp_1 \ldots p_h} \sum_{mn \in \mathcal{A}} \sum_{\substack{n = f q_1 \ldots q_r \\ p_j < q_k}} a_m b_n.$$

Here it is assumed that the variables p_u, q_v are restricted to certain ranges (see the next section where we use the current process). We then take $\gamma = \log p_j, \rho = \log(q_k - \frac{1}{2})$ in the lemma (recall that we used the same "half an odd integer" trick in Section 2.3), with $T = x^2 \lambda^{-1}$. It follows that

$$S = \frac{1}{\pi} \int_{-T}^{T} \sum_{m=dp_1 \dots p_h} a_m p_j{}^{it} \sum_{\substack{n=fq_1 \dots q_r \\ mn \in \mathcal{A}}} b_n \sin(t \log(q_k - \tfrac{1}{2})) \frac{dt}{t} + O(\lambda). \quad (3.2.2)$$

Now the integral between $-1/T$ and $1/T$ is trivially $O(x^2 T^{-1})$. We apply (3.2.1) to the integrand outside this small region of integration. Using the same argument for the new integral between $-1/T$ and $1/T$ therefore gives

$$S = \frac{\lambda}{\pi} \int_{-T}^{T} \sum_{m=dp_1 \dots p_h} a_m p_j{}^{it} \sum_{\substack{n=fq_1 \dots q_r \\ mn \in \mathcal{B}}} b_n \sin(t \log(q_k - \tfrac{1}{2})) \frac{dt}{t}$$

$$+ O(\lambda) + Y \int_{1/T}^{T} \frac{dt}{t}$$

$$= \lambda \sum_{m \sim M} \sum_{m=dp_1 \dots p_h} \sum_{mn \in \mathcal{B}} \sum_{\substack{n=fq_1 \dots q_r \\ p_j < q_k}} a_m b_n + O(\lambda) + O(Y \log x)$$

after another application of (2.3.5). Thus our remark that a cross-condition worsens the error term only by a factor $\log x$ has been justified. As mentioned in the previous chapter, it is possible to follow a more elementary approach by restricting p_j, q_k to short intervals initially, but the same conclusion is reached in the end.

3.3 THE FUNDAMENTAL THEOREM

We will work with $S(\mathcal{A}, z)$, but the results are easily modified to deal with

$$\sum_{\substack{n \in \mathcal{A} \\ p|n \Rightarrow p \geq z}} f(n) = \sum_{n \in \mathcal{A}} \rho(n, z) f(n).$$

Our final goal is to estimate $S(\mathcal{A}, x^{1/2})$; the immediate goal is to estimate $S(\mathcal{A}, z)$ (which will be our fundamental building block) for z some power of x. In the following sections we start assembling the blocks (the reader may notice the influence in my terminology of my three sons — when they were younger — playing with a certain well-known constructional toy of Danish origin!). The basic operations will involve establishing formulae like

$$S(\mathcal{A}, z) = \lambda S(\mathcal{B}, z) + E,$$

$$\sum_p c_p S(\mathcal{A}_p, z) = \lambda \sum_p c_p S(\mathcal{B}_p, z) + E,$$

where the errors E are not too large and λ represents the probability that a number belongs to $\mathcal{A}$. In this way we relate the "unknown" sums involving $\mathcal{A}$ to the "known" sums with $\mathcal{B}$ (we can calculate these as demonstrated in Section 1.4).

We are thus working "modulo the PNT" just as we did when applying Vaughan's identity. This is thus another instance of the comparison principle.

Theorem 3.1. (The Fundamental Theorem) *Suppose that for any sequences of complex numbers a_m, b_n that satisfy $|a_m| \leq 1, |b_n| \leq 1$ we have, for some $\lambda > 0$, $\alpha > 0$, $\beta \leq \frac{1}{2}$, $M \geq 1$, that*

$$\sum_{\substack{mn \in \mathcal{A} \\ m \leq M}} a_m = \lambda \sum_{\substack{mn \in \mathcal{B} \\ m \leq M}} a_m + O(Y) \tag{3.3.1}$$

and

$$\sum_{\substack{mn \in \mathcal{A} \\ x^\alpha \leq m \leq x^{\alpha+\beta}}} a_m b_n = \lambda \sum_{\substack{mn \in \mathcal{B} \\ x^\alpha \leq m \leq x^{\alpha+\beta}}} a_m b_n + O(Y). \tag{3.3.2}$$

Let c_r be a sequence of complex numbers such that

$$|c_r| \leq 1, \quad \text{and if } c_r \neq 0, \text{ then } p|r \Rightarrow p > x^\epsilon, \tag{3.3.3}$$

for some $\epsilon > 0$. Then, if $x^\alpha < M$, $2R < \min(x^{1-\alpha}, M)$, and $M > x^{1-\alpha}$ if $2R > x^{\alpha+\beta}$, we have

$$\sum_{r \sim R} c_r S(\mathcal{A}_r, x^\beta) = \lambda \sum_{r \sim R} c_r S(\mathcal{B}_r, x^\beta) + O(Y \log^3 x). \tag{3.3.4}$$

Remarks. In the above (3.3.1) is a Type I sum and (3.3.2) is a Type II sum. In [62] this result is stated with $|a_m|, |b_n|$ bounded by divisor functions but without the restriction that $p|r \Rightarrow p > x^\epsilon$. In all our applications we only need the result in its current form. If $\beta = \frac{1}{2}$ and we take $R = 1$, $c_1 = 1$, (3.3.4) becomes

$$S(\mathcal{A}, x^{\frac{1}{2}}) = \lambda S(\mathcal{B}, x^{\frac{1}{2}}) + O(Y \log^3 x),$$

a formula for the number of primes in $\mathcal{A}$. The reader should note the significance of the parameter β, which measures the "width" of the Type II sum. It is also interesting to compare the above with the fundamental lemma of familiar linear sieve theory [50, p. 92], which would require no Type II information but would supply (3.3.4) with $\beta = \epsilon$ and another error depending on ϵ. Here one could take $\epsilon \to 0$, and the extra error introduced would be of a smaller order than the main term. We shall discuss this further in the next chapter; in particular, see Theorem 4.3.

Proof. By the sieve of Eratosthenes-Legendre with $z = x^\beta$ we get

$$S(\mathcal{A}_r, z) = \sum_{\substack{d|P(z) \\ drn \in \mathcal{A}}} \mu(d).$$

Write

$$\psi(m) = \sum_{mn \in \mathcal{A}} 1 - \lambda \sum_{mn \in \mathcal{B}} 1;$$

that is, $\psi(m)$ is a remainder: the number of elements in $\mathcal{A}$ divisible by m less the expected amount. Thus

$$S(\mathcal{A}_r, z) = \lambda \sum_{\substack{d|P(z) \\ drn \in \mathcal{B}}} \mu(d) + \sum_{d|P(z)} \mu(d)\psi(dr)$$

$$= \lambda S(\mathcal{B}_r, x^\beta) + \sum_1(r) + \sum_2(r),$$

using Eratosthenes-Legendre again. Here we have written

$$\sum_1(r) = \sum_{\substack{d|P(z) \\ dr \leq M}} \mu(d)\psi(dr) \quad \text{and} \quad \sum_2(r) = \sum_{\substack{d|P(z) \\ dr > M}} \mu(d)\psi(dr).$$

Note that we have sidestepped the prime number theorem by the above approach of swapping back and forth with two applications of the sieve of Eratosthenes-Legendre.

Now

$$\sum_{r \sim R} c_r \sum_1(r) = \sum_{m \leq M} \sum_{\substack{d|P(z) \\ dr=m}} \mu(d)c_r\psi(m)$$

$$= \sum_{\substack{m \leq M \\ mn \in \mathcal{A}}} a_m - \lambda \sum_{\substack{m \leq M \\ mn \in \mathcal{B}}} a_m,$$

with

$$a_m = \sum_{\substack{dr=m \\ d|P(z) \\ r \sim R}} \mu(d)c_r.$$

Now, if $c_r \neq 0$, then the number of prime factors of r is $\leq \epsilon^{-1}$, so we have $|a_m| \leq 2^{1/\epsilon} = O(1)$. Thus, by (3.3.1)

$$\sum_{r \sim R} c_r \sum_1(r) = O(Y).$$

The meat of the proof is in establishing that

$$\sum_{r \sim R} c_r \sum_2(r) = O(Y \log^3 x). \tag{3.3.5}$$

First, suppose that $R < x^\alpha$. We apply (3.1.2) to $\sum_2(r)$ with

$$g(d) = \begin{cases} \psi(dr) & \text{if } dr > M, \\ 0 & \text{otherwise.} \end{cases}$$

Since $r \leq 2R < M$, we have $g(1) = 0$. Hence

$$\sum_2(r) = -\sum_{\substack{2 \leq p < z \\ pdr > M}} \sum_{d|P(p)} \mu(d)\psi(pdr). \tag{3.3.6}$$

Clearly we can apply (3.1.2) repeatedly to (3.3.6), but we need some algorithm for picking out the parts for which (3.3.2) is applicable and which parts need a further dose of (3.1.2). We therefore now describe the inductive step. Suppose we sum over a set of t primes: $p_t < \cdots < p_2 < p_1$, with $p_1 \ldots p_t r < M$. Then, writing $\pi_t = p_1 \ldots p_t$, we have

$$\sum_{\substack{p_1,\ldots,p_t \\ \pi_t dr > M}} \sum_{d | P(p_t)} \mu(d)\psi(dr\pi_t) = - \sum_{\substack{p_1,\ldots,p_{t+1} \\ \pi_{t+1} dr > M}} \sum_{d | P(p_{t+1})} \mu(d)\psi(dr\pi_{t+1})$$

$$= - \left(\sum_3 (r, t+1) + \sum_4 (r, t+1) \right), \quad \text{say,}$$

where $r\pi_{t+1} < x^\alpha$ in $\sum_3$ and $r\pi_{t+1} \geq x^\alpha$ in $\sum_4$. Note that the number $\pi_t d$ on the left-hand side above corresponds to $\pi_{t+1} d$ on the right-hand side since we have taken out p_{t+1} from the d on the left to get the d on the right. We only apply (3.1.2) to $\sum_3$ (in particular, we only apply it to that part of the right-hand side of (3.3.6) with $pr < x^\alpha$), so we will always have $p_1 \ldots p_u r < M$ when applying (3.1.2). Now the primes $p_1, \ldots, p_t$ are distinct, so we will apply (3.1.2) $\ll (\log x)/(\log \log x)$ times, using the well-known bound for the maximum number of distinct prime divisors of an integer. Since we began with $dr > M$, we eventually end up with only sums of the $\sum_4$ type to estimate. For these we have

$$x^\alpha \leq r\pi_{t+1} = (r\pi_t)(p_{t+1}) < x^\alpha x^\beta.$$

Thus one of the variables lies within the range $[x^\alpha, x^{\alpha+\beta}]$. We need a little more work before we can use (3.3.2) and the cosmetic surgery of the previous section. The problem that arises is that one group of variables involves p_{t+1}; the other involves d, which depends on p_{t+1} via the condition $d | P(p_{t+1})$.

Write $q_t = p_{t+2}$ (and we tacitly assume the condition $q_t < p_{t+1}$ in the following) and then note that by (3.1.2) we have

$$\sum_{r \sim R} c_r \sum_4 (r, t+1)$$

$$= \sum_{\substack{r \sim R \\ p_1,\ldots,p_{t+1}}} c_r \left(\sigma(r\pi_{t+1})\psi(r\pi_{t+1}) - \sum_{\substack{d | P(q_t) \\ dr\pi_{t+1}q_t > M}} \mu(d)\psi(rd\pi_{t+1}q_t) \right), \quad (3.3.7)$$

with

$$\sigma(u) = \begin{cases} 1 & \text{if } u > M, \\ 0 & \text{if } u \leq M. \end{cases}$$

From (3.3.2) we immediately obtain

$$\sum_{\substack{r \sim R \\ p_1,\ldots,p_{t+1}}} c_r \sigma(r\pi_{t+1})\psi(r\pi_{t+1}) = O(Y)$$

since

$$\sum_{r\pi_{t+1}=m} |c_r| \leq 2^{\frac{1}{\epsilon}} \quad (3.3.8)$$

from (3.3.3). For the rest of the right-hand side of (3.3.7) we can use Lemma 2.2 and (3.3.2) with (3.3.8) since there are just two joint conditions of summation:

$$q_t < p_{t+1}, \quad dr\pi_{t+1}q_t > M.$$

Thus we obtain

$$\sum_{r \sim R} c_r \sum_4 (r, t+1) \ll Y \log^2 x.$$

Since there are $\ll \log x$ values of t, we obtain (3.3.5) as required to complete the proof in this case.

Now suppose that $R > x^\alpha$. The proof follows immediately from (3.3.2) (with no need for Eratosthenes-Legendre at all!) unless $2R > x^{\alpha+\beta}$. For this case we work as before but combine the variable r with d and q_t rather than with π_{t+1}. The analogue of (3.3.8) holds for this case also. By the hypotheses of the theorem we have $x/dr < x^\alpha$, but $x/r > x^\alpha$. Hence, recalling the implicit variable n in the $\psi(\cdot)$ notation, we have $nd \geq x^\alpha$. Thus we obtain

$$x^\alpha \leq np_1 \ldots p_t < x^{\alpha+\beta}$$

as required to complete the proof. □

The reader at this stage may be asking the question: "At the end of the last chapter we saw that the main term seemed to come from Type I information only; the Type II information was only used to show that certain terms were of smaller order than the main term. Is there an analogous phenomenon here?" A careful look at the proof shows that the first term on the right-hand side of (3.3.2) is not a main term after all, owing to the cancellation introduced by a

$$\sum_d \mu(d)$$

factor implicit in the a_m coefficient (there are other conditions on d, but these do not stop cancellation from occurring). However, when proving the Type II information, the expression (3.3.2) is often the most natural to obtain. The reader should compare and contrast this with the work in Chapter 12.

3.4 APPLICATION TO THE DISTRIBUTION OF $\{\alpha p\}$

We now go back to the question discussed in Theorem 2.2 and apply our new technique. Since we are using the Greek letters α, β in our Fundamental Theorem, we shall consider $\|\xi p + \kappa\|$ where previously we had $\|\alpha p + \beta\|$. We will prove the following result, which is only marginally weaker (by a small power of a logarithm) than (2.3.2).

Theorem 3.2. *Given* $\xi, \kappa \in \mathbb{R}$ *with* ξ *irrational, there are infinitely many integers* x *such that*

$$\sum_{\substack{p \leq x \\ \|\xi p + \kappa\| < \delta}} 1 = 2\pi(x)\delta(1 + o(1)), \tag{3.4.1}$$

where $\delta = \delta(x) = x^{-1/4} \log^9 x$. *Indeed, if* ξ *is a quadratic irrational, then* (3.4.1) *holds for all* x.

Proof. Since ξ is irrational there are infinitely many convergents a/q to its continued fraction expansion. Pick x so that $q^2 = x^{4/3}$. We will apply the Fundamental Theorem with

$$\mathcal{A} = \{n \leq x : ||\xi n + \kappa|| < \delta\}$$

and with $\alpha = \frac{1}{4}, \beta = \frac{1}{2}, \lambda = 2\delta$, which suffices to establish an asymptotic formula for the number of primes in $\mathcal{A}$.

We first consider (3.3.1). By Lemma 2.1 with $L = x$ we have

$$\sum_{\substack{mn \in \mathcal{A} \\ m \leq M}} a_m = \lambda \sum_{\substack{mn \in \mathcal{B} \\ m \leq M}} a_m + O(1) + O\left(\lambda \sum_{\ell \leq x} \min\left(1, \frac{1}{\lambda \ell}\right) \left| \sum_{\substack{mn \in \mathcal{B} \\ m \leq M}} a_m e(\xi \ell mn) \right|\right).$$

The main error term above is essentially the same Type I sum we estimated in Section 2.3 (in fact we do not have the logarithmic factor that occurs there), and so we obtain a bound

$$\ll x^{1+\epsilon} \left(\frac{1}{q} + \frac{q\delta}{x} + \frac{M}{x}\right) (\log x)^2 \ll \lambda x^{1-\epsilon}$$

(where one power of the logarithm comes from the summation over ℓ between δ^{-1} and x), provided that $M \leq x^{3/4 - 3\epsilon}$. We thus have a satisfactory estimate for the Type I sum.

Now we turn to (3.3.2). As above,

$$\sum_{\substack{mn \in \mathcal{A} \\ m \sim M}} a_m b_n$$

$$= \lambda \sum_{\substack{mn \in \mathcal{B} \\ m \sim M}} a_m b_n + O(1) + O\left(\lambda \sum_{\ell \leq x} \min\left(1, \frac{1}{\lambda \ell}\right) \left| \sum_{\substack{mn \in \mathcal{B} \\ m \sim M}} a_m b_n e(\xi \ell mn) \right|\right).$$

Again we can work as in Section 2.3. We glue together ℓ, m if $M < x^{1/2}$ and glue ℓ, n otherwise. The coefficients for our Type II sum, say d_m, g_n, thus satisfy $d_m \ll \tau(m), g_n \ll 1$. The main error for a Type II sum (after summing over M and applying Lemma 2.2 as in Section 2.3) is thus

$$\ll x(\log x)^4 \left(\frac{1}{q} + \frac{1}{x^{\frac{1}{2}}} + \frac{q\delta}{x}\right)^{\frac{1}{2}}.$$

We can thus apply the Fundamental Theorem with $Y = x^{3/4}(\log x)^4$. Since the main term is of order $x\lambda/\log x$, this completes the proof. $\square$

The reader will note that we could reduce the power of a logarithm to 8 by changing (3.4.1) into a lower bound and altering δ to $Cx^{1/4}(\log x)^8$ for some sufficiently large C.

We now show that we did not need all the information that appeared above. We have, by (3.1.3),

$$S(\mathcal{A}, x^{\frac{1}{2}}) = S(\mathcal{A}, x^{\frac{1}{3}}) - \sum_{x^{1/3} \leq p < x^{1/2}} S(\mathcal{A}_p, p)$$

$$= S(\mathcal{A}, x^{\frac{1}{3}}) - \sum_{x^{1/3} \leq p < x^{1/2}} S(\mathcal{A}_p, x^{\frac{1}{3}})$$

by a simple observation. Hence we can apply the fundamental theorem requiring only $x^{1/3} < M < x^{2/3}$. The proof is completed by noting that

$$S(\mathcal{B}, x^{\frac{1}{3}}) - \sum_{x^{1/3} \leq p < x^{1/2}} S(\mathcal{B}_p, p) = S(\mathcal{B}, x^{\frac{1}{2}}).$$

This now looks exactly like our application of Vaughan's identity in Section 2.3. It is vital in the next section that however we apply the Buchstab identity to decompose $S(\mathcal{A}, x^{1/2})$, we can do the same for $S(\mathcal{B}, x^{1/2})$. So, if for example,

$$S(\mathcal{A}, x^{\frac{1}{2}}) = S(\mathcal{A}, x^{\nu}) - \sum_{x^{\nu} \leq p < x^{1/2}} S(\mathcal{A}_p, x^{\nu}) + \sum_{x^{\nu} \leq q < p < x^{1/2}} S(\mathcal{A}_{pq}, q),$$

then

$$S(\mathcal{B}, x^{\frac{1}{2}}) = S(\mathcal{B}, x^{\nu}) - \sum_{x^{\nu} \leq p < x^{1/2}} S(\mathcal{B}_p, x^{\nu}) + \sum_{x^{\nu} \leq q < p < x^{1/2}} S(\mathcal{B}_{pq}, q),$$

and similarly for other expressions.

3.5 A LOWER-BOUND SIEVE

Now we consider what happens if

$$\mathcal{A} = \{n \in \mathcal{B} : \|\xi n + \kappa\| < x^{\epsilon - \theta}\}$$

with $\theta > \frac{1}{4}$. We have added in the ϵ so that we can now ignore the powers of $\log x$ and the divisor function which is estimated at its maximum in the Type I sum. The Type I estimate remains satisfactory so long as $M < x^{1-\theta}$. It is the Type II sum estimates that become problematic since they are now only valid for

$$x^{\theta} \leq M \leq x^{1-2\theta} \quad \text{or} \quad x^{2\theta} \leq M \leq x^{1-\theta}.$$

In other words, a *hole* has opened up around $x^{1/2}$. The Fundamental Theorem now only gives a formula for

$$S(\mathcal{A}, z) \quad \text{with} \quad z = x^{1-3\theta}.$$

Write $T = x^{\theta}$, $U = x^{1-2\theta}$, $X = x^{1/2}$. By Buchstab's identity (3.1.3) we have

$$\begin{aligned}
S(\mathcal{A}, X) &= S(\mathcal{A}, z) - \sum_{z \leq p \leq X} S(\mathcal{A}_p, p) \\
&= S(\mathcal{A}, z) - \sum_{z \leq p < T} S(\mathcal{A}_p, p) - \sum_{T \leq p \leq U} S(\mathcal{A}_p, p) - \sum_{U < p < X} S(\mathcal{A}_p, p) \\
&= \sum_1 - \sum_2 - \sum_3 - \sum_4, \quad \text{say.}
\end{aligned}$$

Now we can apply the Fundamental Theorem to $\sum_1$, while $\sum_3$ is automatically a satisfactory Type II sum. Hence, writing $\lambda = 2x^{\epsilon - \theta}$ and η for a suitable error, we have

$$\sum_1 = \lambda S(\mathcal{B}, z) + \eta, \quad \sum_3 = \lambda \sum_{T \leq p \leq U} S(\mathcal{A}_p, p) + \eta.$$

Now

$$\sum\nolimits_2 = \sum_{z \le p < T} S(\mathcal{A}_p, z) - \sum_{z \le q < p < T} S(\mathcal{A}_{pq}, q) = \sum\nolimits_5 - \sum\nolimits_6, \text{ say.}$$

We can apply our Fundamental Theorem to $\sum_5$, but we cannot apply the Fundamental Theorem immediately to $\sum_6$, nor have we a Type II sum for all the ranges of p and q. When θ is not much larger than $\frac{1}{4}$, the problem region has $U < pq < x/U$, while for $\theta > \frac{2}{7}$ there is also the possibility that $pq < T$. For simplicity, at present we assume we cannot estimate any of $\sum_6$.

We also have

$$\sum\nolimits_4 = \sum_{U < p < X} S(\mathcal{A}_p, z) - \sum_{\substack{U < p < X \\ z \le q, pq^2 < x}} S(\mathcal{A}_{pq}, q) = \sum\nolimits_7 - \sum\nolimits_8, \text{ say,}$$

where we have noted that $S(\mathcal{A}_{pq}, q) = 0$ if $pq^2 > x$, and in this sum $pq^2 < x \Rightarrow q < p$. The sum $\sum_7$ can be dealt with by our Fundamental Theorem, but not all of $\sum_8$ can be treated immediately. For simplicity's sake we ignore those parts of $\sum_8$ for which Type II estimates hold. Write $\sum_j^*$ for $\sum_j$ with $\mathcal{A}$ replaced by $\mathcal{B}$. We then have

$$\begin{aligned}
S(\mathcal{A}, X) &= \sum\nolimits_1 - \sum\nolimits_3 - \sum\nolimits_5 + \sum\nolimits_6 - \sum\nolimits_7 + \sum\nolimits_8 \\
&= \lambda \left(\sum\nolimits_1^* - \sum\nolimits_3^* - \sum\nolimits_5^* - \sum\nolimits_7^* \right) + \eta + \sum\nolimits_6 + \sum\nolimits_8 \\
&= \lambda \left(\sum\nolimits_1^* - \sum\nolimits_3^* - \sum\nolimits_5^* - \sum\nolimits_7^* + \sum\nolimits_6^* + \sum\nolimits_8^* \right) + \eta \\
&\quad + \left(\sum\nolimits_6 - \lambda \sum\nolimits_6^* + \sum\nolimits_8 - \lambda \sum\nolimits_8^* \right) \\
&= \lambda S(\mathcal{B}, X) + \eta + \left(\sum\nolimits_6 - \lambda \sum\nolimits_6^* + \sum\nolimits_8 - \lambda \sum\nolimits_8^* \right) \\
&\ge \lambda S(\mathcal{B}, X) - \lambda \left(\sum\nolimits_6^* + \sum\nolimits_8^* \right) + \eta.
\end{aligned}$$

We can thus obtain a nontrivial lower bound for $S(\mathcal{A}, X)$ provided that

$$\sum_{p,q} S(\mathcal{B}_{pq}, q) < (1 - \epsilon) S(\mathcal{B}, X),$$

where the sum over p, q corresponds to the *parts we cannot do*, that is, the ranges in $\sum_6$ and $\sum_8$.

Up until now we have not used the prime number theorem, but to be efficient we must turn to the results established on sums over primes and the function $\omega(u)$ given in Section 1.4. Now $\sum_8^*$ counts all products of three primes p, q, r, with

$$pqr \in \mathcal{B}, \quad U < p < X, \quad pq^2 < x, \quad r > q > z$$

(no other numbers are counted since $pq^3 > x$ here, at least for $\theta < \frac{3}{11}$). Using our previous work (compare (1.4.15) and note that $\omega(u) = 1/u$ in this case) this is

$$\frac{x}{2 \log x} \int_{u=1-2\theta}^{1/2} \int_{v=1-3\theta}^{(1-u)/2} \frac{du \, dv}{uv(1 - u - v)} + \eta \le \frac{C_1 x}{2 \log x} \left(\theta - \frac{1}{4} \right)^2.$$

Here we estimated the integral as "area" times "maximum value of integrand." If $\theta < \frac{3}{11}$, then the integrand does not exceed $\frac{232}{7} < 34$. Similarly products of three primes counted by $\sum_6^*$ give a contribution

$$\frac{x}{2\log x} \int_{u=1-3\theta}^{\theta} \int_{v=1-3\theta}^{u} \frac{du\, dv}{uv(1-u-v)} + \eta \leq \frac{C_2 x}{2\log x} \left(\theta - \frac{1}{4}\right)^2 .$$

There are also some products of four primes counted by $\sum_6^*$, and their contribution is

$$\frac{x}{2\log x} \int_{u=1-3\theta}^{\theta} \int_{v=1-3\theta}^{\min(u,(1-u)/3)} \int_{w=v}^{(1-u-v)/2} \frac{du\, dv\, dw}{uvw(1-u-v-w)}$$

$$\leq \frac{C_3 x}{2\log x} \left(\theta - \frac{1}{4}\right)^3 .$$

We can thus conclude that for θ not too much bigger than $\frac{1}{4}$ we have

$$S(\mathcal{A}, X) \geq \lambda S(\mathcal{B}, X) \left(1 - C\left(\theta - \frac{1}{4}\right)^2\right) .$$

In fact, with a little more work (remember we discarded parts of $\sum_6$ and $\sum_8$ for which we could obtain a formula) it is possible to show that the above holds with $C = 80$ up to $\theta = \frac{2}{7}$. We will justify this in Section 5.3.

Now we have not exhausted the possibilities of what can be done here. We can make further decompositions with Buchstab's identity. Also, when we have multiple sums over primes, it is sometimes possible and more efficient to swap the roles of the variables. We shall return to this in Chapter 5. For now we briefly consider the extra terms introduced by further decompositions. After $2n$ applications of Buchstab's identity we get n integrals corresponding to the sums that cannot be further decomposed, and for which the Type II estimates do not apply. Let these be $I_2, \ldots, I_{2n}$. Write $\boldsymbol{\alpha}_r = (\alpha_1, \ldots, \alpha_r)$ and put

$$d\boldsymbol{\alpha}_r = \frac{d\alpha_r}{\alpha_1 \ldots \alpha_{r-1}\alpha_r^2}, \quad \mathcal{D}_r = \{\boldsymbol{\alpha}_r : \alpha_j < \alpha_{j-1}, 2 \leq j \leq r\}.$$

We then have

$$I_{2r} = \int_{\mathcal{R}} \omega\left(\frac{1 - \alpha_1 - \cdots - \alpha_{2r}}{\alpha_{2r}}\right) d\boldsymbol{\alpha}_{2r}.$$

Here $\mathcal{R} \subset (1 - 3\theta, \frac{1}{2})^{2r} \cap \mathcal{D}_{2r}$. For large r one can replace $\mathcal{R}$ by the whole of $(1 - 3\theta, \frac{1}{2})^{2r} \cap \mathcal{D}_{2r}$, replace $1/\alpha_{2r}^2$ by $(1/\alpha_{2r})(1/(1 - 3\theta))$, and substitute 1 for $\omega(u)$. The resulting integral can then be calculated by induction and elementary integration as

$$\frac{\left(\log\left(\frac{1}{2}/(1 - 3\theta)\right)\right)^{2r}}{(1 - 3\theta)(2r)!}. \tag{3.5.1}$$

This provides a rough bound for I_{2r} that shows that it will quickly become small. Clearly this estimate can be improved with some simple modifications. For example, one would only use this when $\mathcal{R} \subset (\nu, g)^{2r}$ for some $g < \frac{1}{2}$, and then (3.5.1) becomes substantially smaller. For example, later we will have an application where one might take $\nu = \frac{1}{20}$, $g = \frac{1}{10}$, $r = 4$. This gives $I_8 < 0.00004$.

For smaller r one would want to use numerical integration and treat the terms more carefully since I_{2r} for small r will provide the largest contribution to the discarded sums. The final part of the proof then becomes a demonstration that

$$\sum_r I_{2r} < 1.$$

It seems rather disappointing that the end of a number-theoretic proof should rest on this numerical work! This has been a feature of sieve applications in recent years, and the advent of cheap and powerful personal computers has facilitated this trend.

It is worthwhile remarking on the conditions needed to make further decompositions with the Buchstab identity. One more decomposition will not be of any use since this will lead to sums of the form

$$\sum_{p,q,r} S(\mathcal{A}_{pqr}, r),$$

which must be discarded but are counted with a negative weight. It is not impossible (even though extremely unlikely) that some of these are too big, thus canceling out the main terms. We must only discard positive terms, that is, sums over an even number of prime variables resulting from the Buchstab decomposition. Thus we can only apply this technique when we can apply it *twice*. Thus, the condition for applying Buchstab's decomposition is not $pq < x^{1-\theta}$ (required for the Fundamental Theorem) but $\min(pq^2, pq(x/pq)^{1/2}) < x^{1-\theta}$, that is, either $pq^2 < x^{1-\theta}$ or $pq < x^{1-2\theta}$. In general this becomes

$$p_1 p_2 \ldots p_{2r}^2 \le x^{1-\theta}. \tag{3.5.2}$$

Those familiar with the construction of the Rosser-Iwaniec sieve will note that the requirement for that sieve (in its one-dimensional form) is (noting that $x^{1-\theta}$ here corresponds to the parameter D in the linear sieve)

$$p_1 p_2 \ldots p_{2r}^3 \le x^{1-\theta}.$$

In the construction of the Rosser-Iwaniec sieve, the rule (3.5.2) for further decompositions only occurs in the half-dimensional case. Since the Rosser-Iwaniec construction is optimal in dimension 1 (as will be demonstrated in Chapter 4), this may at first seem puzzling, especially since, as the Type II information becomes less useful ($\beta \to 0$ in the fundamental theorem), the sieve process we have described should reduce to the Rosser-Iwaniec approach. The problem is resolved by noting that it may not always be *optimal* to decompose twice more even when it is *possible* so to do. If we work with a bounded number of decompositions, then it is possible to use a computer to check at each stage whether it is most efficient to keep on decomposing. If (3.5.2) holds and $p_{2r} \le z^e$ ($e = 2.718\ldots$ here), it certainly is more efficient to keep going, but for larger p_r the most important question is: How many of the new sums can be estimated as Type II sums?

Summing up, we can show that the lower-bound sieve we have constructed for this problem gives

$$\frac{S(\mathcal{A}, y)}{\lambda S(\mathcal{B}, y)} \ge (1 + o(1)) f(\Delta, y)$$

where Δ represents the Type I and Type II information we possess and $f(\Delta, y) = 1$ when Δ is good or y is small,

$$f(\Delta, y) \geq e^{-\gamma} \frac{f(u)}{\omega(u)} \quad \text{with} \quad u = (1 - \theta)\frac{\log x}{\log y},$$

where $f(u)$ has its standard meaning for the Rosser-Iwaniec sieve (see Chapter 4), and $\omega(u)$ is Buchstab's function. As $\theta \to \frac{1}{3}$ we should expect

$$f(\Delta, y) \to e^{-\gamma}\frac{f(u)}{\omega(u)}.$$

The reader may wonder what would happen if, instead of trying to obtain a non-trivial lower bound for $S(\mathcal{A}, X)$ at the breakdown point $\theta = \frac{1}{4}$ for the αp problem, one considers $S(\mathcal{A}, z)$ with z as large as possible. It would be natural to assume that one could obtain an asymptotic formula with z close to X so long as θ is close to $\frac{1}{4}$. Sadly this is not the case. Viewed from this angle, there is a discontinuity at $\theta = \frac{1}{4}$. For $\theta \in (\frac{1}{4}, \frac{2}{7}]$ the largest value of z one can take and still obtain an asymptotic formula is $x^{(1-\theta)/3}$, and the value worsens for larger θ. The reason for this value for z is as follows. If we have two primes p, q arising from Buchstab's decomposition, then $pq \geq x^\theta$. If $pq > x^{1-2\theta}$, then we can decompose again to get another prime variable r. We must have $pqr > x^{2\theta}$, but from our choice of z we have $pqr < z^3 = x^{1-\theta}$. Hence, either pq or pqr will be in an asymptotic formula region. The reader can see the case $\theta = \frac{2}{7}$ illustrated in the next section.

3.6 A CHANGE OF NOTATION

We now revert to the notation $\rho(n, z)$. Then the Buchstab decomposition for $S(\mathcal{A}, X)$ implies precisely the same decomposition for $\rho(n, X)$ when $n \in \mathcal{B}$. For the sake of illustration we take the value $\theta = \frac{2}{7}$. From the discussion at the end of the last section, we expect to be able to deal with $\rho(m, x^{5/21})$. Buchstab's identity gives

$$\rho(m, X) = \rho(m, x^{\frac{5}{21}}) - \sum_{x^{5/21} \leq p < X} \rho(m/p, p).$$

Hence we have $\rho(m, X) \geq \rho(m)$ if

$$\rho(m) = \rho(m, x^{\frac{5}{21}}) - \sum_{x^{5/21} < p < X} \rho(m/p, z(p)), \tag{3.6.1}$$

for any choice of $z(p) \leq p$. The appropriate choice to make here is $z(p) = p$ for $x^{2/7} \leq p \leq x^{3/7}$ and

$$z(p) = \begin{cases} (x^{\frac{5}{7}}/p)^{\frac{1}{2}} & \text{if } p < x^{\frac{2}{7}}, \\ x^{\frac{5}{7}}/p & \text{if } p > x^{\frac{5}{7}}. \end{cases}$$

We now write out all the details to prove that we can deal with $\rho(m, x^{5/21})$. We have

$$\rho(m, x^{\frac{5}{21}}) = \rho(m, x^{\frac{1}{7}}) - \sum_{x^{1/7} \leq p < x^{5/21}} \rho(m/p, x^{\frac{1}{7}}) + \sum_{x^{1/7} \leq q < p < x^{5/21}} \rho(m/(pq), q).$$

$$\tag{3.6.2}$$

We note that in the final sum in (3.6.2), $pq > x^{2/7}$. Write S^* for the part of this sum with $pq > x^{3/7}$. Then we observe that

$$S^* = \sum_{x^{3/7}/p < q < p < x^{5/21}} \rho(m/(pq), q)$$

$$= \sum_{x^{3/7}/p < q < p < x^{5/21}} \rho(m/(pq), x^{\frac{1}{7}}) - \sum_{\substack{x^{3/7}/p < q < p < x^{5/21} \\ x^{1/7} \leq r < q}} \rho(m/(pqr), r).$$

(3.6.3)

In the final sum in (3.6.3) we have $x^{4/7} < pqr < x^{5/7}$. There is an analogue of our treatment of $\rho(m, x^{5/21})$ for the final term in (3.6.1). Combining (3.6.1), (3.6.2), and (3.6.3) yields a decomposition of ρ into functions of the form

$$\sum_{\substack{rs=m \\ x^{2/7} < r < x^{3/7}}} f_r g_s \qquad (3.6.4)$$

or

$$\sum_h a_h \rho(m/h, x^{\frac{1}{7}}), \qquad (3.6.5)$$

where h does not exceed $x^{4/7}$. We also note from (3.6.1) that $\rho(m) = 0$ if $p|m$ with $p < x^{3/14}$. (Note that we have $x^{3/14}$ and not $x^{5/21}$ in view of the minimum value taken by $z(p)$.) Now we can immediately give an asymptotic formula for (3.6.4) summed over m using our Type II information. Our method converts (3.6.5) into sums of the form

$$\rho(m) = \sum_{k=1}^{K} \sum_{\substack{rs=m \\ x^{2/7} < r < x^{3/7}}} f_r^{(k)} g_s^{(k)},$$

with $K < \log x$, and the f_r, g_s are at worst divisor functions times a small power of $\log x$. Another way of viewing $\rho(m)$ would be as a sum of functions supported on numbers with no prime factor less than $x^{1/7}$. The above work is used in [64] to consider the values taken by an additive quadratic form with three prime variables (see Theorem 10.6) and in [67] to study the exceptional sets of numbers in the correct congruence classes that cannot be represented as sums of three or four squares of primes. In both of these applications our method is necessary to bound the exponential sum

$$\sum_{n \leq x} \rho(n) e\left(\alpha n^2\right)$$

for some function ρ that is a lower bound for the characteristic function of the primes. Using Vaughan's identity $x^{1/8}$ can be saved on the trivial estimate if $\rho(n)$ is replaced by $\Lambda(n)$ [45], but our approach increases this saving to $x^{1/7}$. In the application given in [67] we make use of all the different ways that one can write down a decomposition of $\rho(m)$.

Looking at the first decomposition (3.6.1) we can analyse the behaviour of $\rho(m)$ rather well. We have $\rho(m) = 1$ if $m \in \mathcal{B}$ is a prime since in this case the first term on the right-hand side is 1 while the sum over p is zero. If m has no prime factor exceeding $x^{5/21}$, then $\rho(m)$ is zero. We can therefore write the remaining m that might give nonzero $\rho(m)$ as $m = pu, p > x^{5/21}, q|u \Rightarrow q > x^{3/14}$. Hence u has at most three prime factors. It follows that $\rho(m) \geq -3$ for all m, the worst m having four prime factors all around $x^{1/4}$ in size. Note that $z(p) \approx x^{13/56} < x^{1/4}$ if $p \approx x^{1/4}$ so the sum on the right-hand side of (3.6.1) will count all four prime factors in this case. We thus see that $\rho(m) \in \{1, 0, -1, -2, -3\}$ in this situation.

3.7 THE PIATETSKI-SHAPIRO PNT

This section is by way of an extended exercise so that an interested reader can try to attack a problem with the alternative sieve for themselves. The solution can be found in [13], although the implementation of the Fundamental Theorem there rests on an earlier version of the author's sieve method [59]. We recall Exercise 1 from Chapter 2: the Piatetski-Shapiro PNT. Now we will write $\mathcal{A} = \{m = [n^c] : m \in \mathcal{B}\}$ and wish to show that

$$S(\mathcal{A}, x^{\frac{1}{2}}) > C\lambda S(\mathcal{B}, x^{\frac{1}{2}}) \tag{3.7.1}$$

for some $C > 0$ with c as large as possible. Here $\lambda = x^{\gamma-1}(1 - (\frac{1}{2})^\gamma)$. The Type I and Type II information is obtained via the following two estimates, which may be found as Lemmas 5 and 4 in [13], respectively. We write $\eta = \epsilon/100$ in the sequel.

Lemma 3.1. *For $\frac{11}{13} < \gamma < 1$ and $D \leq x^{3\gamma-2-\epsilon}$, we have*

$$\sum_{h \leq H} \left| \sum_{d \sim D} a_d \sum_{nd \in \mathcal{B}} e\left(h(nd)^\gamma\right) \right| \ll x^{1-5\eta}$$

for any complex numbers a_d with $|a_d| \leq 1$.

Lemma 3.2. *Let a_m, b_n be sequences of complex numbers with modulus at most 1. Suppose that $\frac{11}{13} < \gamma < 1$ and $M = x^\alpha$, where*

$$\alpha \in [1 - \gamma + \epsilon, 5\gamma - 4 - \epsilon] \cup [3 - 3\gamma + \epsilon, 3\gamma - 2 - \epsilon] \cup [5 - 5\gamma + \epsilon, \gamma - \epsilon].$$

Then we have

$$\sum_{h \leq H} \left| \sum_{m \sim M} \sum_{mn \in \mathcal{B}} a_m b_n e\left(h(mn)^\gamma\right) \right| \ll x^{1-5\eta}.$$

The above lemmas come into play using (2.6.4) because

$$m = [n^c] \iff [-m^\gamma] - [-(m+1)^\gamma] = 1$$

and

$$[-m^\gamma] - [-(m+1)^\gamma] = (m+1)^\gamma - m^\gamma + (\{-(m+1)^\gamma\} - \{m^\gamma\}).$$

Now let $\nu = 6\gamma - 5 - 2\epsilon$ (so, for example, $\nu = \frac{1}{10}$ when $\gamma = \frac{17}{20} + \frac{\epsilon}{3}$). The reader should be able to deduce that we can give an asymptotic formula for

$$\sum_{r \sim R} c_r S(\mathcal{A}_r, x^\nu)$$

for certain values R. Applying Buchstab's identity to $S(\mathcal{A}, x^{1/2})$ in a similar manner to Section 3.5 along with some numerical calculations should then establish (3.7.1) for $\gamma > \frac{20}{17}$. For the latest published result on this problem see [150] (a small improvement on the previous best known value in [115]), where the alternative sieve is applied with improved exponential sum estimates and quite detailed calculations. They achieve $\frac{243}{205} \approx 1.18536\ldots$ for the exponent. The reader might also like to compare the related applications of the alternative sieve in [116, 118].

3.8 HISTORICAL NOTE

We finish this chapter by commenting on the author's original derivation of a result like Theorem 3.1 in [59]. The basic idea to start with was that the Rosser-Iwaniec sieve is neutral with respect to Buchstab's identity. That is, if you apply Buchstab's identity and then the Rosser-Iwaniec sieve to some set, you will get the same answer as when applying the sieve in the first place (see Section 4.3). However, if you can give an asymptotic formula for some of the new sums after applying the sieve, then a better result than applying Rosser-Iwaniec has been obtained. I had first seen that idea in [85]. It did not take long for me to realize that the correct starting point should not be the Rosser-Iwaniec sieve but the fundamental lemma that underpins the construction of that sieve, and the Type II information for the problem under consideration then enables one to get an asymptotic formula of the form given in (3.3.4), albeit with a weaker error term resulting from the error in the fundamental lemma. Within a couple of years I had realized that the fundamental lemma could be dispensed with by working as in this chapter. However, this was not written up until [62]. The paper [59] appears to be the first that used both the ideas of working with $S(\mathcal{A}, z)$ (where z comes from the Type II information available) and carrying out parallel decompositions so that one only needs to give an upper bound for the numbers thrown away. These are the two principal features of what we have called the alternative sieve. This appellation was first given to the technique in [6], where the method was applied as an *alternative* to the Rosser-Iwaniec upper-bound sieve. For another early application of the method see [14].

Chapter Four

The Rosser-Iwaniec Sieve

4.1 INTRODUCTION

In this chapter we shall give a derivation of the Rosser-Iwaniec sieve (in its one-dimensional form) that is modelled on our treatment of the alternative sieve. This may seem strange since the author's original formulation of this sieve had its genesis in the Rosser-Iwaniec construction! However, the comparison principle used before (comparing $\mathcal{A}$ with $\mathcal{B}$) enables us to give a relatively straightforward exposition of the Rosser-Iwaniec sieve in the simplest case. The comparison set $\mathcal{B}$ is no longer the set of all integers in an interval, but the choice we make automatically demonstrates that the Rosser-Iwaniec sieve is optimal for the arithmetical information fed in. Our account thus differs from the standard one where Selberg's extremal examples are given *after* the sieve is developed (see [51, 104]). This approach is also outlined in [84]. In our work the extremal examples are our comparison sets. Our first task is therefore to introduce these sets and determine their properties. In the following section we introduce a fundamental lemma that we will require both for the construction of the Rosser-Iwaniec sieve and for the development of our own sieve in later chapters. We then give a heuristic argument why the sieve bounds must take the form stated before giving the proofs of the main theorems.

Let $\lambda(n)$ be Liouville's function given by $\lambda(n) = (-1)^{t(n)}$, where $t(n)$ is the total number of prime factors of n. Let

$$\mathcal{B}^- = \{n \leq D : \lambda(n) = 1\}, \qquad \mathcal{B}^+ = \{n \leq D : \lambda(n) = -1\}.$$

Write

$$V(z) = \prod_{p < z} \left(1 - \frac{1}{p}\right) \sim \frac{e^{-\gamma}}{\log z}$$

by Theorem 1.2. Since $\mathcal{B} = \mathcal{B}^- \cup \mathcal{B}^+$, we have

$$S(\mathcal{B}^+, z) + S(\mathcal{B}^-, z) = S(\mathcal{B}, z) \sim s\omega(s)\frac{D}{\log D} \sim De^{\gamma}\omega(s)V(z),$$

where

$$s = \frac{\log D}{\log z}.$$

Now suppose that

$$S(\mathcal{B}^+, z) \sim sG(s)\frac{D}{\log D}, \qquad S(\mathcal{B}^-, z) \sim sg(s)\frac{D}{\log D}, \qquad (4.1.1)$$

as $D \to \infty$ with s fixed. Using Buchstab's identity, if $k \leq s < k+1$, we have

$$S(\mathcal{B}^-, D^{\frac{1}{s}}) = S(\mathcal{B}^-, D^{\frac{1}{k}}) + \sum_{D^{1/s} \leq p < D^{1/k}} S(\mathcal{B}_p^-, p).$$

However,

$$S(\mathcal{B}_p^-, p) \sim \frac{D}{p \log(D/p)} \frac{\log(D/p)}{\log p} G\left(\frac{\log(D/p)}{\log p}\right).$$

Working as in our analysis of $w(u)$ in Chapter 1 (and made rigorous in the appendix), we get, for $k \leq s \leq k+1$,

$$sg(s) = kg(k) + \int_k^s G(y-1)\, dy.$$

Similarly we get

$$sG(s) = kG(k) + \int_k^s g(y-1)\, dy.$$

Thus g, G satisfy the coupled difference-differential equations

$$\begin{aligned}
(sg(s))' &= G(s-1), \\
(sG(s))' &= g(s-1).
\end{aligned} \tag{4.1.2}$$

The reader will readily verify that

$$G(s) = \frac{1}{s} \quad \text{for } 1 \leq s \leq 3,$$

$$g(s) = 0 \quad \text{for } 1 \leq s \leq 2.$$

The functions g, G can then be defined inductively for all $s \geq 1$ and the derivation of (4.1.1) made rigorous in like manner to our treatment of $w(u)$ in the appendix. The following properties quickly follow:

$$g(s) + G(s) = w(s), \quad 0 \leq g(s) \leq G(s) \leq 1 \quad \text{for all } s \geq 1,$$

$$g(s) \to \tfrac{1}{2} e^{-\gamma}, \quad G(s) \to \tfrac{1}{2} e^{-\gamma} \quad \text{as } s \to \infty.$$

Remark. The reader should note that our functions $g(s), G(s)$ are $\frac{1}{2} e^{-\gamma}$ times the functions $f(s), F(s)$ that occur in standard expositions of the theory.

In this chapter we write $\log_2 x$ for $\log \log x$ and, in general, $\log_k x = \log \log_{k-1} x$. We can now state our lower-bound result.

Theorem 4.1. *Let $x \geq D \geq z^2 \geq 2, \mathcal{A} \subseteq \mathbb{N} \cap [1, x]$. Suppose that*

$$|\mathcal{A}_d| = \frac{H}{d} + R_d. \tag{4.1.3}$$

Then we have

$$S(\mathcal{A}, z) \geq \frac{2H}{D} S(\mathcal{B}^-, z) \left(1 + O((\log_3 D)^{-1})\right) + O\left(\sum_{\substack{d < D \\ \mu(d) \neq 0}} |R_d|\right). \tag{4.1.4}$$

In particular

$$S(\mathcal{A}, z) \geq sg(s) \frac{2H}{\log D} (1 + o(1)) + O\left(\sum_{\substack{d < D \\ \mu(d) \neq 0}} |R_d|\right). \tag{4.1.5}$$

Remarks. The main term in (4.1.5) can also be written in the form

$$g(s)\frac{2H}{\log z}(1+o(1)) = f(s)HV(z)(1+o(1))$$

in the standard sieve notation. The final error term on the right-hand side of both (4.1.4) and (4.1.5) can be estimated using what we have called Type I information. To see this, note that

$$\sum_{d<D}|R_d| = \sum_{d<D}\left||A_d| - \frac{H}{d}\right|$$

$$= \sum_{d<D}\left|\sum_{dn\in A}1 - \frac{H}{x}\sum_{dn\in[1,x]}1\right| + O(D)$$

$$= \sum_{d<D}a_d\sum_{dn\in A}1 - \frac{H}{x}\sum_{d<D}a_d\sum_{dn\in[1,x]}1 + O(D),$$

for certain a_d. The final two sums are just the left- and right-hand sides of (3.3.1) with $\lambda = H/x$, and so estimating the error term in the above theorem is equivalent to obtaining Type I information in our Fundamental Theorem.

The upper-bound result is as follows. We shall only prove the lower-bound result; the reader should see that the upper bound will follow similarly.

Theorem 4.2. *Given the hypotheses of Theorem 4.1, we have*

$$S(A,z) \le \frac{2H}{D}S(B^+,z)\left(1+O\big(((\log_3 D)^{-1})\big)\right) + O\left(\sum_{\substack{d<D \\ \mu(d)\neq 0}}|R_d|\right). \qquad (4.1.6)$$

In particular,

$$S(A,z) \le sG(s)\frac{2H}{\log D}(1+o(1)) + O\left(\sum_{\substack{d<D \\ \mu(d)\neq 0}}|R_d|\right). \qquad (4.1.7)$$

Remarks. It should be clear from our proof that the above theorems are the best possible since the actual bounds obtained (see (4.4.8) and (4.4.9)) are attained for $A = B^-$ in the case of Theorem 4.1, and similar remarks apply to Theorem 4.2 with $A = B^+$. The standard form of the main term in (4.1.7) is $HF(s)V(z)$.

4.2 A FUNDAMENTAL LEMMA

The purpose of a fundamental lemma is to sieve out all the small prime factors. If one attempts to construct the sieve bounds without such a result, one runs into serious difficulties with accumulated errors from numbers with many small prime factors. The following lemma appears to have been first given by Heath-Brown ([80, Lemma 15]) and has some advantages over other fundamental lemma type results (see [51] for examples). We state the lemma in a different but equivalent way to Heath-Brown's formulation.

Lemma 4.1. *Let Q be the product of r distinct primes not exceeding w. Let $E \geq 1$. Then*

$$\left| \sum_{\substack{d|Q \\ d \geq E}} \mu(d) \right| \leq \sum_{\substack{d|Q \\ E \leq d < Ew}} 1. \tag{4.2.1}$$

Proof. Our proof is essentially the same as Heath-Brown's. Let Q have k prime factors. We proceed by induction on k.

(i) The case $k = 1$. Here $Q = p$, a prime. If $E > p$, then both sides of (4.2.1) are 0. If $E \leq p$, then the left-hand side is 0 (if $E = 1$) and otherwise equal to 1. The right-hand side is always at least 1 when $E \leq p$. Hence (4.2.1) is true for $k = 1$.

(ii) Suppose (4.2.1) is true whenever Q has k distinct prime factors. Now replace Q by Qp, where $p \nmid Q$. We have

$$\sum_{\substack{d|Qp \\ d \geq E}} \mu(d) = \sum_{\substack{d|Q \\ d \geq E}} \mu(d) - \sum_{\substack{d|Q \\ d \geq E/p}} \mu(d) = S_1 - S_2, \quad \text{say.}$$

By the inductive hypothesis we have

$$|S_1| \leq \sum_{\substack{d|Q \\ E \leq d < Ew}} 1, \quad |S_2| \leq \sum_{\substack{d|Q \\ E/p \leq d < Ew/p}} 1.$$

Now

$$\sum_{\substack{d|Q \\ E/p \leq d < Ew/p}} 1 = \sum_{\substack{p|d|pQ \\ E \leq d < Ew}} 1$$

and

$$\sum_{\substack{p|d|pQ \\ E \leq d < Ew}} 1 + \sum_{\substack{d|Q \\ E \leq d < Ew}} 1 = \sum_{\substack{d|Qp \\ E \leq d < Ew}} 1.$$

Hence (4.2.1) holds.

The proof then follows by mathematical induction. □

Lemma 4.2. *Let a, u be positive numbers, $w = x^{1/u}$, $D = x^a$. Suppose that*

$$\frac{1}{a} < u < (\log x)^{1-\epsilon}. \tag{4.2.2}$$

Then

$$\sum_{\substack{d|P(w) \\ d > D}} \frac{1}{d} \ll \exp(\log \log w + 2ua - ua \log ua). \tag{4.2.3}$$

The implied constant is absolute.

Remark. If we write $s = (\log D)/(\log w)$ then (4.2.2) becomes

$$1 < s < (\log D)(\log x)^{-\epsilon},$$

and (4.2.3) becomes

$$\sum_{\substack{d|P(w) \\ d>D}} \frac{1}{d} \ll \exp(\log \log w + 2s - s \log s).$$

Proof. Let $\rho = (u \log ua)/(\log x)$. Now, using what is commonly called Rankin's trick,

$$\sum_{\substack{d|P(w) \\ d>D}} \frac{1}{d} \leq \frac{1}{D^\rho} \sum_{d|P(w)} \frac{d^\rho}{d} = \frac{1}{D^\rho} \prod_{p<w} (1 + p^{\rho-1})$$

$$= \frac{1}{D^\rho} \exp\left(\sum_{p<w} \log\left(1 + p^{\rho-1}\right)\right) \qquad (4.2.4)$$

$$\leq \frac{1}{D^\rho} \exp\left(\sum_{p<w} \frac{p^\rho - 1}{p} + \sum_{p<w} \frac{1}{p}\right).$$

We use the simple inequality

$$\exp(cy) - 1 \leq (\exp(c) - 1)y,$$

which is valid for $c > 0$, $0 \leq y \leq 1$, with $y = (\log p)/(\log w)$, $c = \log ua$. The last expression in (4.2.4) is thus

$$\leq \frac{1}{D^\rho} \exp\left(\frac{\exp(\log ua) - 1}{\log w} \sum_{p<w} \frac{\log p}{p} + \sum_{p<w} \frac{1}{p}\right)$$

$$= \exp\left(-ua \log ua + ua\left(1 + O\left(\frac{1}{\log w}\right)\right) + \log \log w + O(1)\right)$$

by Theorem 1.2. This completes the proof. $\qquad \square$

One can insert a multiplicative function $v(d)$ into the sum on the left-hand side of (4.2.3) to bound

$$\sum_{\substack{d|P(w) \\ d>D}} \frac{v(d)}{d}.$$

If $v(p)$ is 1 on average over primes p, then essentially the same result as (4.2.3) can be obtained. We shall need a result where $v(p) \leq k$ in later chapters for k-dimensional forms of the alternative sieve in certain situations. The average of $v(p)$ over the primes, by which we mean the minimum value that k could take in (4.2.5), is called the *dimension* of the sieve and corresponds to k residue classes being removed from the set under investigation on average for each prime. For example, to find sums of two squares the dimension is $\frac{1}{2}$ since one residue class is removed for each $p \equiv 3 \pmod 4$. To consider prime twins we wish to sieve $n(n+2)$ and so remove two residue classes for each prime, leading to a two-dimensional sieve problem. This last problem can also be considered as a one-dimensional problem; see the remarks after Theorem 4.3. We state the variant of Lemma 4.2 that can be used in general and leave its proof as an exercise for the reader.

Lemma 4.3. *Given the hypotheses of Lemma 4.2, let $v(d)$ be a multiplicative function satisfying*

$$\sum_{p \leq X} \frac{v(p) \log p}{p} \leq k \log X + O(1),$$

$$\sum_{p \leq X} \frac{v(p)}{p} \leq k \log \log X + O(1), \tag{4.2.5}$$

for some $k > 0$. Then

$$\sum_{\substack{d \mid P(w) \\ d > D}} \frac{v(d)}{d} \ll \exp(k \log \log w + 2kua - ua \log ua). \tag{4.2.6}$$

Here the implied constant depends only on k and the implied constants in (4.2.5).

We can now state the result we shall use as our Fundamental Lemma to sieve out small prime factors.

Theorem 4.3. (The Fundamental Lemma) *Given a, u, w, D, as in Lemma 4.2, and suppose that $\mathcal{A}$ satisfies (4.1.3). Then*

$$S(\mathcal{A}, w) = HV(w)(1 + O(\epsilon)) + O\left(\sum_{d \leq Dw} |R_d|\right), \tag{4.2.7}$$

where $\epsilon = \exp(2 \log \log w + 2ua - ua \log(ua))$.

Remarks. The reader should compare and contrast this result with Theorem 3.1. Here we have only Type I information and are restricted to quite small values of w. Theorem 3.1 requires greater arithmetical input in terms of Type II information but delivers a much stronger output.

The condition (4.1.3) can be replaced with

$$|\mathcal{A}_d| = \frac{Hv(d)}{d} + R_d,$$

with $v(d)$ a multiplicative function averaging 1 over the primes, by modifying $V(w)$ to

$$V(w) = \prod_{p < w} \left(1 - \frac{v(p)}{p}\right).$$

For example, one can take, for $2 \nmid d$,

$$|\mathcal{A}_d| = \frac{H}{\phi(d)} + R_d, \quad \text{with } V(w) = \prod_{3 \leq p < w} \left(1 - \frac{1}{p - 1}\right).$$

This case is relevant for the twin-prime problem approached using a one-dimensional sieve by taking $\mathcal{A} = \{n : n = p + 2\}$ and $H = \pi(x)$.

Proof. By the sieve of Eratosthenes-Legendre

$$S(\mathcal{A}, w) = \sum_{d|P(w)} \mu(d)|\mathcal{A}_d|$$

$$= \sum_{\substack{d|P(w) \\ d \leq D}} \mu(d)|\mathcal{A}_d| + \sum_{\substack{d|P(w) \\ d > D}} \mu(d)|\mathcal{A}_d| = S_1 + S_2, \quad \text{say.}$$

Let $E = Dw$. Henceforth we simply write P for $P(w)$. Now, using Lemma 4.1 we have

$$|S_2| = \left| \sum_{n \in \mathcal{A}} \sum_{\substack{d|(P,n) \\ d > D}} \mu(d) \right|$$

$$\leq \sum_{n \in \mathcal{A}} \sum_{\substack{d|(P,n) \\ D < d \leq E}} 1 = \sum_{\substack{d|P \\ D < d \leq E}} |\mathcal{A}_d|$$

$$= \sum_{\substack{d|P \\ D < d \leq E}} \frac{H}{d} + \sum_{\substack{d|P \\ D < d \leq E}} R_d$$

$$\ll H \exp(\log \log w + 2ua - ua \log(ua)) + \sum_{d \leq E} |R_d| \quad \text{(by Lemma 4.2)}$$

$$= \frac{H}{\log w} \exp(2 \log \log w + 2ua - ua \log(ua)) + \sum_{d \leq E} |R_d|.$$

Also,

$$S_1 = \sum_{\substack{d|P \\ d \leq D}} \mu(d)|\mathcal{A}_d| = \sum_{\substack{d|P \\ d \leq D}} \frac{\mu(d)H}{d} + \sum_{\substack{d|P \\ d \leq D}} \mu(d)R_d$$

$$= H \sum_{d|P} \frac{\mu(d)}{d} - H \sum_{\substack{d|P \\ d > D}} \frac{\mu(d)}{d} + \sum_{\substack{d|P \\ d \leq D}} \mu(d)R_d$$

$$= S_3 - S_4 + S_5, \quad \text{say.}$$

Now

$$S_3 = HV(w), \qquad |S_5| \leq \sum_{\substack{d|P \\ d \leq D}} |R_d|,$$

and

$$S_4 \ll H \sum_{\substack{d|P \\ d > D}} \frac{1}{d} \ll \frac{H}{\log w} \epsilon$$

by Lemma 4.2. This completes the proof. $\qquad \square$

Before returning to the main subject matter of this chapter we pause to consider this fundamental lemma further. We first note that the error terms are actually

$$O\left(\sum_{\substack{d|P \\ d\leq D}} \mu(d)R_d\right) + O\left(\sum_{\substack{d|P \\ D<d\leq E}} R_d\right),$$

which leaves the door open for cancellation. We see from the above that the "worst" sets should have $|R_d|$ consistently taking either the same or opposite sign to $\mu(d)$. This is exactly the case for $\mathcal{B}^{\pm}$, of course.

Now to get a nontrivial result from the fundamental lemma we need to have $ua \log(ua)$ growing faster than $\log \log w$ in order to get $\epsilon \to 0$. We can arrange this by taking

$$a = \delta > 0 \text{ (fixed)}, \quad u = \log \log x, \quad w = \exp\left(\frac{\log x}{\log \log x}\right).$$

The reader should have no problem in using the above to prove the following result.

Theorem 4.4. *Let* $\epsilon > 0, x > 1, y = x^{\epsilon}$. *Write*

$$\mathcal{A} = \mathbb{N} \cap [x, x+y].$$

Then, for

$$w \leq \exp\left(\frac{\log x}{\log \log x}\right),$$

we have

$$S(\mathcal{A}, w) \sim yV(w) \sim y\frac{e^{-\gamma}}{\log w}.$$

Remarks. The reader should note how much more powerful this is than our original application of Eratosthenes-Legendre in Chapter 1. However, the deepest result fed into the proof is Mertens' PNT. The reader should compare the remarks in [93, p. 3]. Indeed, the reader should have no trouble proving that the Fundamental Lemma gives a quick and simple proof of Brun's result that

$$\sum_{p, p+2 \text{ both prime}} \frac{1}{p}$$

converges.

4.3 A HEURISTIC ARGUMENT

Before proving our main results we give here a brief justification why one should expect the inequalities to have the form stated. We would like to prove that

$$S(\mathcal{A}, z) \sim \frac{H}{\log z}w(u) \quad \text{with } u = \frac{\log x}{\log z},$$

and the result of the previous section shows that this is indeed true for small z. Since the arithmetical information fed into the sieve is an upper bound on

$$\sum_{d \leq D} |R_d|,$$

it seems reasonable to suppose that upper and lower bounds would have the form

$$\frac{H}{\log z} j(s) \leq S(\mathcal{A}, z) \leq \frac{H}{\log z} J(s), \tag{4.3.1}$$

with $s = (\log D)/(\log z)$. If the result is optimal, it should be neutral under Buchstab's identity. That is, if we write

$$S(\mathcal{A}, z) = S(\mathcal{A}, w) - \sum_{w \leq p < z} S(\mathcal{A}_p, p), \tag{4.3.2}$$

then applying (4.3.1) to the right-hand side should give the same result as applying it to the left-hand side. Well, to tackle the second sum on the right-hand side we need to deal with a remainder term

$$\sum_{w \leq p < z} \sum_{d < D'} |R_{dp}|.$$

We will therefore be restricted to taking $D' = D/p$. Using (4.3.2) to obtain a lower bound, we therefore expect

$$\frac{H}{\log z} j(s) = \frac{H}{\log w} j\left(\frac{\log D}{\log w}\right) - \sum_{w \leq p < z} \frac{H}{p \log p} J\left(\frac{\log(D/p)}{\log p}\right).$$

Now we remove the common factor H throughout, convert sums into integrals in our usual fashion, and change variables with $D = p^t = z^s = w^\theta$ and so obtain (multiplying throughout now by $\log D$)

$$sj(s) = \theta j(\theta) - \int_s^\theta J(t-1)\, dt.$$

Similarly we must obtain the same formula with the roles of $j(s), J(s)$ reversed. These functions must therefore satisfy the same linked pair of differential delay equations as $g(s), G(s)$ do, namely, (4.1.2). If we show that the upper and lower bounds are attained for $\mathcal{B}^\pm$, we have thus demonstrated that the results are optimal for the quality of information fed into the method.

4.4 PROOF OF THE LOWER-BOUND SIEVE

We first need some more technical results.

Lemma 4.4. *Let $\alpha_1, \ldots$ be a sequence of positive reals with $\alpha_1 \leq \frac{1}{2}$ satisfying*

$$\alpha_j \leq \alpha_{j-1} \text{ for } j > 1 \text{ and } \alpha_{2j} \leq \frac{1 - \alpha_1 - \cdots - \alpha_{2j-1}}{3} \text{ for } j \geq 1.$$

Then, for all $j \geq 1$, we have

$$\sum_{\ell=1}^{2j+1} \alpha_\ell \leq 1 - \frac{1}{2 \times 3^j}. \tag{4.4.1}$$

Proof. We have

$$\alpha_1 + \alpha_2 + \alpha_3 \le \alpha_1 + 2\frac{1 - \alpha_1}{3} = \frac{2 + \alpha_1}{3} \le 1 - \frac{1}{6},$$

so the result is true for $j = 1$. Now suppose it is true for j and consider $j + 1$:

$$\sum_{\ell=1}^{2j+3} \alpha_\ell = \alpha_{2j+3} + \alpha_{2j+2} + \sum_{\ell=1}^{2j+1} \alpha_\ell$$

$$\le 2\frac{1 - \alpha_1 - \cdots - \alpha_{2j+1}}{3} + \sum_{\ell=1}^{2j+1} \alpha_\ell$$

$$= \frac{1}{3}\left(2 + \sum_{\ell=1}^{2j+1} \alpha_\ell\right)$$

$$\le 1 - \frac{1}{2 \times 3^{j+1}},$$

as required. The proof is then completed by induction. $\qquad\square$

Now we need to borrow some terminology from standard treatments of the Rosser-Iwaniec sieve ([51, p. 110]). For $n \ge 2$ let

$$\mathfrak{I}_n(s) = \{(t_1, \ldots, t_n) \in \mathbb{R}^n : 0 < t_n < \cdots < t_1 < \tfrac{1}{s}, t_1 + \cdots + t_{n-1} + 3t_n > 1,$$
$$t_1 + \cdots + t_{m-1} + 3t_m < 1 \text{ if } m < n, m \equiv n \bmod 2\}.$$

Write $\mathbf{dt}_n = dt_n \ldots dt_1$ and put (again for $n \ge 2$)

$$f_n(s) = \frac{1}{s} \int_{\mathfrak{I}_n(s)} \frac{1}{t_1 \ldots t_{n-1} t_n^2} \, \mathbf{dt}_n. \tag{4.4.2}$$

We note that $\mathfrak{I}_n(s) = \emptyset$, and so $f_n(s) = 0$ if $s \ge n + 2$. We write

$$f_1(s) = \frac{3}{s} - 1, \quad \text{for } 1 \le s < 3 \tag{4.4.3}$$

and put

$$b_n = \begin{cases} 3 & \text{if } n \text{ is odd,} \\ 2 & \text{if } n \text{ is even.} \end{cases}$$

Lemma 4.5. *The functions $f_n(s)$ satisfy the recurrence relation*

$$f_n(s) = \frac{1}{s} \int_{\max(s,b_n)}^{2+n} f_{n-1}(t - 1) \, dt. \tag{4.4.4}$$

It follows that, for $s \ge 3$,

$$f_n(s) \le A\alpha^n e^{-s}, \tag{4.4.5}$$

where A, α are positive constants with $\alpha < 1$.

Proof. We have

$$\int_{\max(s,b_n)}^{2+n} f_{n-1}(t-1)\, dt$$

$$= \int_{\max(s,b_n)}^{2+n} \frac{1}{t-1} \int_{\Im_{n-1}(t-1)} \frac{1}{t_1\dots t_{n-2}t_{n-1}^2}\, \mathbf{dt}_{n-1}\, dt. \tag{4.4.6}$$

We first make the change of variables $tu_j = (t-1)t_j$ to the inner integral on the right-hand side of (4.4.6). Then we put

$$t_1 = \frac{1}{t}, \quad t_j = u_{j-1} \text{ for } j \geq 2.$$

This transforms the integral into the integral on the right hand side of (4.4.2) and so establishes (4.4.4).

To prove (4.4.5) it suffices to show that, for some absolute constant B, we have

$$f_{2n+1}(s) \leq B e^{-s}\alpha^{2n+2}. \tag{4.4.7}$$

We shall prove this by induction; the case $n = 0$ follows from (4.4.3) (note that $f_1(s) = 0$ for $s > 3$). We define α by

$$\alpha^2 = \frac{e^5}{3} \int_3^\infty \frac{e^{-t}}{t-1}\, dt.$$

A numerical calculation shows that $\alpha = 0.94\dots < 1$. Let $q(s) = \max(3,s)$ and assume that (4.4.7) holds for $n-1$. Then

$$f_{2n+1}(s) = \frac{1}{s} \int_{q(s)}^{2n+3} \frac{1}{t-1} \int_{t-1}^{2n+2} f_{2n-1}(u-1)\, du\, dt$$

$$< \frac{B\alpha^{2n}}{s} \int_{q(s)}^\infty \frac{1}{t-1} \int_{t-1}^\infty e^{1-u}\, du\, dt$$

$$= \frac{Be^2\alpha^{2n}}{s} \int_{q(s)}^\infty \frac{e^{-t}}{t-1}\, dt.$$

First suppose that $s \leq 3$. Then, from the definition of α,

$$f_{2n+1}(s) < \frac{3B\alpha^{2n+2}}{se^3}.$$

Since

$$e^{-s} \geq \frac{3}{se^3} \quad \text{for } 1 \leq s \leq 3,$$

this completes the proof in this case.

Now suppose that $s > 3$. We note that

$$\int_s^\infty \frac{e^{-t}}{t-1}\, dt < e^{3-s} \int_3^\infty \frac{e^{-t}}{t-1}\, dt = 3\alpha^2 e^{-2-s},$$

and this establishes (4.4.7) in this case to complete the proof.

Remark. One can find a much more precise treatment of the functions $f_n(s)$, and in particular their sum over n, in [51, Chapter 4.3].

☐

Proof. (Theorem 4.1) Put

$$w = \exp\left(\frac{\log D}{(\log_2 D)^2}\right), \quad W = \exp\left(\frac{\log D}{\log_2 D}\right), \quad E = D\exp\left(-\frac{2\log D}{\log_2 D}\right).$$

We also write

$$P(w, z) = \prod_{w \leq p < z} p.$$

Two applications of Buchstab's identity yield

$$S(\mathcal{A}, z) = S(\mathcal{A}, w) - \sum_{w \leq p < z} S(\mathcal{A}_p, w) + \sum_{w \leq p_2 < p < z} S(\mathcal{A}_{p_1 p_2}, p_2).$$

We apply Buchstab's identity twice more to that part of the final sum with $p_1 p_2^3 < E$. There is no point in applying Buchstab's identity to the remainder of the sum since E is not much smaller than D,

$$p_1 p_2^3 \geq D \Rightarrow \frac{\log(D/p_1 p_2)}{\log p_2} \leq 2,$$

and $g(s) = 0$ for $s < 2$. Put another way, if the method works for $\mathcal{A} = \mathcal{B}^-$, we note that $S(\mathcal{B}_{p_1 p_2}^-, p_2) = 0$, so we are not throwing anything away in this case. Later in the argument we show that even though E is slightly smaller than D, the terms discarded have smaller order than the main term. Similar remarks apply to the following discussion.

In general, after r steps, we have

$$S(\mathcal{A}, z) = \sum_{d | P(w,z)}^{(1)} \mu(d) S(\mathcal{A}_d, w) + \sum_{p_1,\dots,p_{2r}}^{(2)} S(\mathcal{A}_{p_1\dots p_{2r}}, p_{2r}) + \sum{}^*,$$

where $^{(1)}$ denotes the conditions:

$$d = p_1 \dots p_\ell, \ \ell \leq 2r - 1, \ \text{and} \ p_1 \dots p_{h-1} p_h^3 < E \ \text{ for even h},$$

while $^{(2)}$ indicates the following:

$$p_{2r} < p_{2r-1} < \cdots < p_1, \quad p_1 \dots p_{2k}^3 < E \ \text{ if } k < 2r.$$

Finally, the term $\sum^*$ counts the terms with $p_1 \dots p_{2k}^3 > E$ for which we did not iterate Buchstab's identity at the kth stage. The above iteration must stop after $\ll (\log\log x)^2$ steps, and we thereby obtain

$$S(\mathcal{A}, z) = \sum_{d \in \mathcal{D}} \mu(d) S(\mathcal{A}_d, w) + \sum_r \sum_{p_1,\dots,p_{2r}}{}^* S(\mathcal{A}_{p_1\dots p_{2r}}, p_{2r}),$$

where

$$\mathcal{D} = \{d \leq E : d = p_1 \dots p_\ell, w \leq p_\ell < p_{\ell-1} < \cdots < p_1 < z,$$
$$\text{and } p_1 \dots p_{2r}^3 < E \text{ for } 2r < \ell\}$$

and $\sum^*$ indicates that $p_1 \ldots p_{2r}^3 \geq E$, $p_1 \ldots p_{2j}^3 < E$ for $j < r$. Now, by Theorem 4.3 with the value of D there taken to be W, we have

$$S(\mathcal{A}_d, w) = \frac{H}{d} V(w) \left(1 + O(\exp\left(-(\log_2 D)(\log_3 D - 3)\right))\right)$$

$$+ O\left(\sum_{f < Ww} |R_{df}|\right).$$

Here we tacitly assume that $p | f \Rightarrow p < w$ and $\mu(f) \neq 0$. Thus (with the variable d in the following only taking on square-free values)

$$S(\mathcal{A}, z) = \sum_{d \in \mathcal{D}} \mu(d) \frac{H}{d} V(w) + O\left(\sum_{d \leq EWw} |R_d|\right)$$

$$+ O\left(H (\log D)^{-2}\right) + \sum_r \sideset{}{^*}\sum_{p_1, \ldots, p_{2r}} S(\mathcal{A}_{p_1 \ldots p_{2r}}, p_{2r}). \tag{4.4.8}$$

In particular, taking $\mathcal{A} = \mathcal{B}^-$ gives

$$S(\mathcal{B}^-, z) = \sum_{d \in \mathcal{D}} \mu(d) \frac{D}{2d} V(w) + O\left(\sum_{d \leq EWw} |R_d|\right)$$

$$+ O\left(D (\log D)^{-2}\right) + \sum_r \sideset{}{^*}\sum_{p_1, \ldots, p_{2r}} S(\mathcal{B}^-_{p_1 \ldots p_{2r}}, p_{2r}). \tag{4.4.9}$$

Now we know that the left-hand side of (4.4.9) is $\sim sg(s)D(\log D)^{-1}$. Also,

$$R_d = \sum_{\substack{n \in \mathcal{B}^- \\ d | n}} 1 - \frac{D}{2d} = \sum_{n \leq D/d} \frac{1 + \lambda(dn)}{2} - \frac{D}{2d},$$

while $\lambda(dn) = \lambda(d)\lambda(n)$ and

$$\sum_{n \leq X} \lambda(n) \ll X \exp\left(-(\log X)^{\frac{1}{2}}\right),$$

as shown in Exercise 3 in Chapter 1. It quickly follows that

$$\sum_{d \leq EWw} |R_d| \ll EWw + D(\log(EWw)) \exp\left(-\left(\log\left(\frac{D}{EWw}\right)\right)^{\frac{1}{2}}\right)$$

$$\ll \frac{D}{(\log D)^A}$$

for any $A > 0$.

The difficult term to consider is

$$\sum_r \sideset{}{^*}\sum_{p_1, \ldots, p_{2r}} S(\mathcal{B}^-_{p_1 \ldots p_{2r}}, p_{2r}).$$

In our approach we make use of the special nature of the set $\mathcal{B}^-$ for small r, while our treatment of larger values of r consists of a slimmed-down version of the standard linear sieve working. Converting sums into integrals in a familiar way (except that we put $p_j = E^{\alpha_j}$ rather than D^{α_j} to make the conditions on the variables easier to handle) gives a typical term

$$\leq \frac{D}{\log E} \int \cdots \int \frac{1}{\alpha_1 \ldots \alpha_{2r-1}\alpha_{2r}} U(\boldsymbol{\alpha}) \, d\boldsymbol{\alpha} = I_{2r}, \text{ say.}$$

Here we have written $U(\boldsymbol{\alpha})$ for a function approximating

$$\frac{p_1 \ldots p_{2r} \log E}{D} S(\mathcal{B}^-_{p_1 \ldots p_{2r}}, p_{2r}),$$

so $U(\boldsymbol{\alpha}) \leq (\alpha_{2r})^{-1}$ would be suitable. We note that the error involved in doing this introduces a factor

$$\leq \left(1 + O\left(\frac{(\log_2 D)^2}{\log D}\right)\right)^{2r+1} \leq 2$$

for all large D since $2r \leq (\log_2 D)^2$.

However, for small r, one can do substantially better. To see this, note that

$$p_1 \ldots p_{2r-1} p_{2r}^3 > E, \text{ which gives } \alpha_1 + \cdots + \alpha_{2r-1} + 3\alpha_{2r} > 1. \quad (4.4.10)$$

Hence $S(\mathcal{B}^-_{p_1 \ldots p_{2r}}, p_{2r})$ counts numbers n with an even number of prime factors such that

$$p|n \Rightarrow p > p_{2r}, \quad n \leq \frac{D}{p_1 \ldots p_{2r}}.$$

If $p_{2r} > D/E$, it follows from (4.4.10) that $p_1 \ldots p_{2r-1} p_{2r}^4 > D$, and so only products of two primes are counted. We note that, from Lemma 4.4,

$$\alpha_1 + \cdots + \alpha_{2r-1} \leq 1 - \frac{1}{2 \times 3^{r-1}},$$

so (4.4.10) gives

$$\alpha_{2r} \geq \frac{1}{2 \times 3^r}.$$

Hence $p_{2r} > D/E$ certainly holds if $r < \frac{1}{2}\log_3 D$. Now suppose that $n = q_1 q_2$ with $q_1 < q_2$. Let $q_j = E^{\beta_j}$ and put $\nu = (\log_2 D)^{-1}$. Then

$$E < p_1 \ldots p_{2r} q_1 q_2 \leq D$$

becomes

$$1 < \beta_1 + \beta_2 + \alpha_1 + \cdots + \alpha_{2r} \leq 1 + 2\nu + O(\nu^2).$$

Converting a sum over primes to an integral (whose length is $\leq \nu + O(\nu^2)$) we then have

$$U(\boldsymbol{\alpha}) \leq \frac{\nu + O(\nu^2)}{\alpha_{2r}^2}.$$

If we approximate all those terms with $r \leq K \log_4 D$ this way and bound the integral over $\boldsymbol{\alpha}$ crudely, we then get an estimate

$$\ll \frac{D\nu}{\log E} \left(3^{K \log_4 D}\right)^{2+K \log_4 D} \ll \frac{D}{\log D} (\log_2 D)^{-\frac{1}{2}}.$$

For larger r we note that, with $U(\alpha) = (\alpha_{2r})^{-1}$,

$$I_{2r} = \frac{D}{\log E} f_{2r}(s) \quad \text{with} \quad s = \frac{\log z}{\log D},$$

and hence, by (4.4.5),

$$\sum_{r > K \log_4 D} I_{2r} \ll \frac{D}{\log E} \alpha^{2K \log_4 D} \ll \frac{D}{\log D} (\log_3 D)^{-1}$$

by choosing K sufficiently large.

We now finish the proof by bounding the last term on the right-hand side of (4.4.8) by zero. Comparing (4.4.8) and (4.4.9) with the bounds we have obtained for the second two terms on the right-hand side of (4.4.9) gives (4.1.4) and completes the proof. $\qquad\square$

4.5 DEVELOPMENTS OF THE ROSSER-IWANIEC SIEVE

Much progress has been made in applications of the Rosser-Iwaniec sieve by noting that the remainder term can be expressed in the form

$$\sum_{m \le M} \sum_{n \le N} a_m b_n R_{mn},$$

for any given $M, N \ge 1$ with $MN = D$. The reader should note that the expression of the remainder term in this bilinear form is not immediate but requires some work (see [103] or [51, Chapter 6]). The flexible nature of this error term has facilitated its treatment by Fourier analysis or the use of mean-value results on Dirichlet polynomials or character sums. If one is using a lower-bound sieve to detect almost-primes in a sequence, then it is usually beneficial to attach weights to the sieve; see [51, Chapter 5], for example.

It should be noted that we assumed in the previous section for ease of exposition that

$$|\mathcal{A}_d| = \frac{H}{d} + R_d.$$

In many applications one instead has

$$|\mathcal{A}_d| = \frac{H}{d} v(d) + R_d,$$

where $v(d)$ is a multiplicative function that is 1 on average over the primes. We have already explained that this can be handled as far as the fundamental lemma is concerned, and the function $V(z)$ then becomes

$$\prod_{p < z} \left(1 - \frac{v(p)}{p}\right).$$

For some common applications we have

$$v(p) = \frac{p}{p - 1} \qquad (4.5.1)$$

and this can be easily accommodated within the present framework. To see this, simply note that we obtain (assuming $p|d \Rightarrow p > w$ and writing η for an error whose treatment is essentially unchanged from previous discussion)

$$S(\mathcal{A}_d, w) = v(d) \left(\frac{H}{d} V(w) + \eta \right),$$

but

$$\prod_{p|d} \left(1 - \frac{v(p)}{p} \right) = \prod_{p|d} \left(1 - \frac{1}{p-1} \right)$$

$$= Y_d \prod_{p|d} \left(1 - \frac{1}{p} \right),$$

where

$$Y_d = \prod_{p|d} \left(1 - \frac{1}{(p-1)^2} \right) = 1 + O\left(\frac{1}{w} \right).$$

Also,

$$v(d) = 1 + O\left(\frac{(\log_2 D)^2}{w} \right).$$

We are thus able to use our work on (4.4.8) and (4.4.9) with only minor alterations.

Another common variant involves missing out certain primes from the sieving process because the elements of the set to be sieved are automatically free of these prime factors. One example of this situation is

$$\mathcal{A} = \{ n : n \equiv a \, (\mathrm{mod} \, q), \, n \leq X \},$$

where $(a, q) = 1$. This is the situation we shall meet in Chapter 8 when discussing the Brun-Titchmarsh inequality; it involves simply changing

$$V(z) \quad \text{to} \quad \frac{q}{\phi(q)} V(z).$$

For the completely general case of the Rosser-Iwaniec sieve in dimension 1 we refer the reader to [51] or [102]. The Rosser-Iwaniec sieve is only optimal in dimensions $\frac{1}{2}$ and 1. For other dimensions it is not even known what the best possible results are since we do not have extremal examples as used above. This is a big open question in sieve theory.

Exercise
Let $1 < q < y/e < x/e$ and suppose $(a, q) = 1$. Prove the Brun-Titchmarsh inequality in the form

$$\pi(x, q, a) - \pi(x - y, q, a) \leq \frac{2y \left(1 + O((\log_3(y/q))^{-1} \right)}{\phi(q) \log(y/q)}.$$

Hint: Take

$$D = \frac{y}{q(\log(y/q))^2}.$$

The reader should note that the inequality

$$0 \le \pi(x,q,a) \le (2+o(1))\frac{x}{\phi(q)\log x},$$

which follows from the above, is essentially no worse than the best that can be obtained by analytic means using L-functions for q in certain ranges, in view of the possible existence of Siegel zeros. In that case we could have

$$\pi(x,q,a) = \frac{x}{\phi(q)\log x}(1+o(1)) - \frac{\chi(a)x^\beta}{\beta\phi(q)\log x},$$

where β is the potential Siegel zero for $L(s,\chi)$ and χ is a real character $(\bmod\, q)$. Since β could be very close to 1, this gives very few (or no) primes in half the reduced residue classes and correspondingly twice as many as expected in the others. It is intriguing that we know this is the limit of what can be obtained by the Rosser-Iwaniec sieve because of the extremal examples, yet if it had been possible to reduce the upper-bound constant below 2, we would have disproved the existence of Siegel zeros!

Chapter Five

Developing the Alternative Sieve

5.1 INTRODUCTION

In this chapter we shall give different forms of the Fundamental Theorem from Chapter 3 (first given in [62]) and consider how to reverse the roles of the variables appearing implicitly and explicitly in our sums. We shall then go on to give an improvement in our result for small values of $\|\xi p + \kappa\|$. Finally we shall give another application: finding integers in short intervals that have a large prime factor. This will enable us to introduce the techniques we shall require in later chapters to study primes in short intervals.

5.2 NEW FORMS OF THE FUNDAMENTAL THEOREM

The following is an alternative version of the Fundamental Theorem that makes use of an extra summation range.

Theorem 5.1. *Suppose we have* (3.3.1), (3.3.2), *and* (3.3.3), *but now* $M \geq x^{1-\beta}$, $S < x^{\alpha}, R < x^{1-\alpha-\beta}$. *Let* d_s *be coefficients with* $|d_s| \leq 1, d_s = 0$ *unless* $p|d_s \Rightarrow p > x^{\delta}$ *for some* $\delta > 0$. *Then we have*

$$\sum_{\substack{r \sim R \\ s \sim S}} c_r d_s S(\mathcal{A}_{rs}, x^{\beta}) = \lambda \sum_{\substack{r \sim R \\ s \sim S}} c_r d_s S(\mathcal{B}_{rs}, x^{\beta}) + O(Y \log^3 x). \tag{5.2.1}$$

Remark. The advantage in the above result is that RS can be as big as $x^{1-\beta}$, not $x^{1-\alpha}$ as in the original form of the Fundamental Theorem.

Proof. We need to consider sums

$$\sum_{r,s} c_r d_s \sum_{\substack{d|P(z) \\ drsn \in \mathcal{A}}} \mu(d).$$

We can deal with that part of the sum with $n \geq x^{\beta}$ by (3.3.1). For the rest we have

$$nsd \geq x^{\alpha}, \quad ns < x^{\alpha+\beta}, \quad p|d \Rightarrow p \leq x^{\beta}.$$

We can thus apply (3.3.2) to that part of the sum with $ns \geq x^{\alpha}$ and apply our iterative procedure to the rest of the sum and thus eventually obtain a variable in the range $[x^{\alpha}, x^{\alpha+\beta}]$ as required. $\qquad\square$

Now suppose we have a situation where we can estimate Type I sums with $M \leq x^{4/5}$ and Type II sums with $\alpha = \frac{2}{5}, \beta = \frac{1}{5}$. At first this gives only a formula for

$$\sum_{r \sim R} c_r S(\mathcal{A}_r, z) \quad \text{for} \quad R < z^3,$$

where $z = x^{1/5}$, or, using (5.2.1), we could deal with

$$\sum_{\substack{r \sim R \\ s \sim S}} c_r d_s S(\mathcal{A}_{rs}, z) \quad \text{for} \quad S < z^2, R < z^2.$$

Writing $X = x^{1/2}$ as previously, we have

$$S(\mathcal{A}, X) = S(\mathcal{A}, z) - \sum_{z \leq p < X} S(\mathcal{A}_p, p)$$

$$= S(\mathcal{A}, z) - \sum_{z^2 \leq p < X} S(\mathcal{A}_p, p) - \sum_{z \leq p < z^2} S(\mathcal{A}_p, z)$$

$$+ \sum_{\substack{z < p < z^2 \\ z \leq q < \min(p, (x/p)^{1/2})}} S(\mathcal{A}_{pq}, q)$$

$$= \sum\nolimits_1 - \sum\nolimits_2 - \sum\nolimits_3 + \sum\nolimits_4, \quad \text{say.}$$

We can give asymptotic formulae for $\sum_1$ and $\sum_3$ by the Fundamental Theorem, while $\sum_2$ is automatically a Type II sum. In $\sum_4$ we always have $pq > z^2$, so we have a type II sum unless $pq > z^3$. If $pq > z^3$, we note that

$$S(\mathcal{A}_{pq}, q) = S(\mathcal{A}_{pq}, z),$$

and hence (5.2.1) is applicable to the remainder of $\sum_4$. We have thus given an asymptotic formula for every sum in our decomposition giving

$$S(\mathcal{A}, X) = \lambda S(\mathcal{B}, X) + O(Y \log^3 x).$$

Now, since we have an asymptotic formula, we can replace $S(\mathcal{A}, X)$ in the above by

$$\sum_{p \sim x} e(\gamma p)$$

(this entails replacing $S(\mathcal{B}, X)$ by zero at the same time) and so obtain a bound for this sum that is $\ll \log^3 x$ times

$$\left(\max_M \left(\frac{x}{q} + M + q \right) \log x + \max_H x^{\frac{1}{2}} \left(\frac{x}{q} + \frac{x}{H} + q + H \right)^{\frac{1}{2}} (\log x)^{\frac{7}{2}} \right),$$

where the maxima are taken over $M \leq x^{4/5}, x^{2/5} \leq H \leq x^{3/5}$, using our earlier working and assuming that

$$\left| \gamma - \frac{a}{q} \right| < \frac{1}{q^2}, \quad (a, q) = 1.$$

We thus obtain (apart from a worse power of $\log x$) the Vinogradov-Vaughan bound for the simplest exponential sum over primes:

$$\sum_{p \sim x} e(\gamma p) \ll (\log x)^{\frac{13}{2}} \left(x^{\frac{4}{5}} + \frac{x}{q^{\frac{1}{2}}} + (xq)^{\frac{1}{2}} \right). \tag{5.2.2}$$

The example we have given above of an asymptotic formula when $\alpha = \frac{2}{5}, \beta = \frac{1}{5}, M = x^{4/5}$ can be generalized to

$$0 < \alpha < \tfrac{1}{2}, \ \beta = 1 - 2\alpha, \ M = \max(x^{1-\alpha}, x^{2\alpha}).$$

In Chapter 2 we saw that these were the conditions required by Vaughan's identity with $U = V$.

We have shown that use can be made of a double sum in the Fundamental Theorem. Clearly quite complicated multiple sums arise in our method, and this suggests that there may be problems when more arithmetical information is available, for example, to deal with a sum like

$$\sum_{mnrs \in \mathcal{A}} a_m b_n c_r d_s \quad \text{or} \quad \sum_{klm \in \mathcal{A}} a_m,$$

that is, a quadruple general sum or a sum with two Type I ranges. For the αp problem it seemed that no use could be made of these extra types of sums, but Heath-Brown and Jia introduced some new ideas in their paper [87] that we shall mention below. We shall see later how different types of multiple sums can be used for other questions.

Finally, we mention that for certain problems it may be impossible to estimate general Type II sums, and we need to use the fact that the coefficients a_m, b_n that arise have a special form, namely, that they will always be the characteristic functions of integers built up from primes in certain ranges. For some of these problems (primes in short intervals, the Bombieri-Vinogradov theorem in an extended range), the following version of the Fundamental Theorem is more useful. We shall distinguish it from its progenitor by the addition of an asterisk to its appellation.

Theorem 5.2. (Fundamental Theorem*) *Let $\epsilon > 0$ and $u > 0$ such that*

$$\frac{\log \log x}{\log \log \log x} < u < (\log x)^{1-\epsilon}.$$

Suppose that $R < Mx^{-\epsilon}$ and let $w = x^{1/u}$. Suppose that (3.3.1) holds and that (3.3.2) holds whenever a_m has the form

$$a_m = \sum_{\substack{w \leq p < x^\beta \\ ph=m}} d_h p^{it}, \ |t| \leq x^2 \lambda^{-1}, \ |d_h| \ll \tau(h). \tag{5.2.3}$$

Suppose also that $x^\alpha < M$, $R < x^{1-\alpha}$, and $M \geq x^{1-\alpha}$ if $R > x^{\alpha+\beta}$. Then

$$\sum_{r \sim R} c_r S(\mathcal{A}_r, x^\beta) = \lambda \sum_{r \sim R} c_r S(\mathcal{B}_r, x^\beta)(1 + E) + O(Y \log^3 x), \tag{5.2.4}$$

where

$$E \ll (\log x) \exp \left(\log \log w + 2u\epsilon - u\epsilon \log(u\epsilon) \right).$$

In particular, if $u \geq \log \log x$, we have $E \ll_A (\log x)^{-A}$ for any $A > 0$.

Remark. Clearly, in general u needs to be chosen to balance the various errors in the above. Smaller u leads to larger E, while larger u makes the coefficients a_m more difficult to deal with in certain situations. The presence of the term E makes the above result more closely resemble the classical fundamental lemma of sieve theory. Indeed, our proof uses such a result from Chapter 4.

Proof. Let $z = x^\beta$. We begin by writing

$$S(\mathcal{A}_r, z) = S(\mathcal{A}_r, w) - \sum_{w \le p < z} S(\mathcal{A}_{rp}, p) = \Sigma_A - \Sigma_B, \quad \text{say.}$$

We can treat Σ_B by the method used to prove Theorem 3.1 since the coefficients a_m appearing there have the form (5.2.3). This gives

$$\sum_{r \sim R} c_r \Sigma_B = \lambda \sum_{r \sim R} c_r \sum_{w \le p < z} S(\mathcal{B}_{pr}, p) + O\big(Y(\log x)^3\big).$$

We can apply Theorem 4.3, with $D = wM/R > x^\epsilon$, to $\sum_r c_r \Sigma_A$ to show that

$$\sum_{r \sim R} S(\mathcal{A}_r, w)$$
$$= \lambda \sum_{r \sim R} S(\mathcal{B}_r, w)(1 + O(\exp(2 \log \log w + 2u\epsilon - u\epsilon \log(u\epsilon)))) + O(Y).$$

We quickly obtain (5.2.4) since

$$S(\mathcal{B}_r, w) \exp(2 \log \log w) \ll S(\mathcal{B}_r, z)(\log x) \exp(\log \log w).$$

$\square$

5.3 REVERSING ROLES

We now return to the problem of the distribution of αp modulo 1. In Chapter 3 we let

$$\mathcal{A} = \{n \le x : ||\xi p + \kappa|| < x^{\epsilon - \theta}\}$$

and used the two-stage Buchstab decomposition:

$$S(\mathcal{A}, X) = S(\mathcal{A}, z) - \sum_p S(\mathcal{A}_p, z) + \sum_{p, q} S(\mathcal{A}_{pq}, q).$$

We first split off those sums that were automatically of Type II form. If we try to increase θ, we run into problems because the size of the final sum discarded becomes larger than the main term. We can perform further decompositions for parts of this sum, but there is an advantage in reversing the roles of variables for part of this sum, in particular when p is large. We note that $S(\mathcal{A}_{pq}, q)$ counts numbers pqt, where $r|t$ (r here denoting a prime) $\Rightarrow r \ge q$. Suppose that the point (p, q) belongs to a set $\mathcal{F}$. Then we can rewrite

$$\sum_{(p,q) \in \mathcal{F}} S(\mathcal{A}_{pq}, q) \quad \text{as} \quad \sum_{p|t \Rightarrow p \ge q} S(\mathcal{A}_{tq}^*, X/(tq)^{\frac{1}{2}}).$$

Here $\mathcal{A}_{tq}^* = \{n \in \mathcal{A}_{tq} : (n,q) \in \mathcal{F}\}$. Since the set $\mathcal{F}$ will be of the form

$$\mathcal{F} = \{(x^\eta, x^\varsigma) : \eta \in [a,b], f(\eta) \le \varsigma \le g(\eta)\},$$

where f, g are piecewise-linear polynomials (for example, $g(\eta) = (1-\eta)/2$ for part of the range), we can obtain the required Type I and Type II information for this set with just one additional application of Perron's formula. The extra logarithmic factor included in the error terms is absorbed in the x^ϵ factor here. We shall illustrate this technique by giving the proof of the result in [62].

Write $\tau = 1 - 3\theta$ and put

$$\mathcal{C} = \{(\eta, \varsigma) : \eta \in [\tau, \theta) \cup (1 - 2\theta, \tfrac{1}{2}), \tau \le \varsigma \le \min(\eta, \tfrac{1}{2}(1-\eta)),$$
$$\{\varsigma, \eta + \varsigma\} \cap ([\theta, 1-2\theta] \cup [2\theta, 1-\theta]) = \emptyset\}$$

and

$$\nabla = \{(m,n) : m = x^\eta, n = x^\varsigma, (\eta, \varsigma) \in \mathcal{C}\}.$$

In our previous treatment of the problem we discarded the whole of ∇ since we are unable to express these terms as Type II sums. We now divide this region into three subregions ∇_j, $j = 1, 2, 3$. We write ∇_1 for the part with $mn^2 \le x^{1-\theta}$ and ∇_2 for the subregion with $mn \le x^{1-\theta}$, $mn^2 > x^{1-\theta}$, $n > x^{1-2\theta}$. The remainder of ∇ is denoted by ∇_3. Unless otherwise stated, ∇_j is an abbreviation for $(p,q) \in \nabla_j$. We then have

$$S(\mathcal{A}, X) = \lambda \left(S(\mathcal{B}, X) - \sum_\nabla S(\mathcal{B}_{pq}, q) \right) + \sum_{j=1}^{3} \sum_{\nabla_j} S(\mathcal{A}_{pq}, q). \quad (5.3.1)$$

Now we can apply Buchstab's identity twice more to the sum over ∇_1 to obtain

$$\sum_{\nabla_1} S(\mathcal{A}_{pq}, q) = \sum_{\nabla_1} S(\mathcal{A}_{pq}, z) - \sum_{p,q,r} S(\mathcal{A}_{pqr}, z) + \sum_{p,q,r,s} S(\mathcal{A}_{pqrs}, s). \quad (5.3.2)$$

Here we have suppressed the ranges of summation on the final two sums in the interest of clarity. The first two sums on the right-hand side of (5.3.2) can be estimated by Theorem 3.1 since $pqr < pq^2 \le x^{1-\theta}$. The final sum is empty for $\theta < \frac{4}{15}$, while for $\theta \le \frac{2}{7}$ we note that the numbers counted have the form $pqrst$ with

$$x^\theta \le x^{2-6\theta} \le \min(rs, st) \le x^{\frac{2}{5}} < x^{1-2\theta},$$

and so this corresponds to a Type II sum with ranges of a suitable size. Even when $\theta > \frac{2}{7}$ it seems hopeful that the errors introduced by discarding sums that cannot be estimated will not be too great. We shall return to this point below.

We now turn our attention to ∇_2 and it is here that we use the role reversal device. Thus

$$\sum_{\nabla_2} S(\mathcal{A}_{pq}, q) = \sum_\Xi S(\mathcal{A}_{qt}^*, X/(qt)^{\frac{1}{2}}).$$

Here Ξ stands for the conditions

$$qx^\theta > t > \tfrac{1}{2}x^\theta, \quad x^{\frac{1}{2}} < qt < x^{2\theta}, \quad x^\tau \le q < x^\theta, \quad p \mid t \Rightarrow p \ge q.$$

Of course, $\mathcal{A}^*_{qt} = \{n \in \mathcal{A}_{qt} : (n, q) \in \nabla_2\}$. An application of Buchstab's identity then yields

$$\sum_{\nabla_2} S(\mathcal{A}_{pq}, q) = \sum_{\Xi} S(\mathcal{A}^*_{qt}, z) - \sum_{\substack{\Xi \\ z \le s < X/(qt)^{1/2}}} S(\mathcal{A}^*_{qts}, s)$$

$$= S_1 - \sum_{\substack{(us,q) \in \nabla_2, u > 1 \\ p|u \Rightarrow p \ge s > z}} S(\mathcal{A}_{qus}, q), \quad \text{say,}$$

$$= S_1 - \sum S(\mathcal{A}_{qus}, z) + \sum S(\mathcal{A}_{qusr}, r)$$

$$= S_1 - S_2 + S_3, \quad \text{say,}$$

after another application of Buchstab's identity. The reader should note how in the second line we reversed roles back again. One could picture this as turning the sum inside out, performing a decomposition, and then turning it inside out again (or should that be "outside in"?). We are able to perform the last decomposition because $pq \le x^{1-\theta} \Rightarrow qus \le x^{1-\theta}$. We can obtain asymptotic formulae for S_1 and S_2 by Theorem 3.1. The sum S_3 can be dealt with in the same way as the final term in (5.3.2) for $\theta < \frac{2}{7}$.

We now consider ∇_3, the region for which no asymptotic formula has been shown to hold when $\theta < \frac{2}{7}$ and which corresponds to the condition

$$pq > x^{1-\theta} \quad \text{or} \quad pq^2 > x^{1-\theta}.$$

We note that, for $\theta \le \frac{2}{7}$, $p > x^{1-2\theta} \Rightarrow pq^2 > x^{1-2\theta}z^2 \ge x^{1-\theta}$. Hence, if $pq^2 < x^{1-\theta}$, we can assume that $p < x^\theta$. We also have $p < x^{(1-\theta)/3} \Rightarrow pq^2 < x^{1-\theta}$. Hence, for $\frac{1}{4} \le \theta \le \frac{2}{7}$ we obtain

$$S(\mathcal{A}, X) \ge \lambda S(\mathcal{B}, X)(1 - C(\theta))(1 + o(1)),$$

where

$$C(\theta) = I_1(\theta) + I_2(\theta)$$

with

$$I_1(\theta) = \int_{1-2\theta}^{1/2} \int_{1-\theta-\eta}^{(1-\eta)/2} \frac{d\zeta\, d\eta}{\eta\zeta(1 - \eta - \zeta)},$$

and

$$I_2(\theta) = \int_{(1-\theta)/3}^{\theta} \int_{(1-\theta-\eta)/2}^{\eta} \omega\left(\frac{1-\eta-\zeta}{\zeta}\right) \frac{d\zeta\, d\eta}{\eta\zeta^2}.$$

Simple bounds lead to the inequality $C(\theta) < 80(\theta - \frac{1}{4})^2$ mentioned in Section 3.5. There appears to be no way we can reduce this term with our present arithmetical information since there is no way the variables can be grouped together to give a satisfactory Type II sum.

For $\theta > \frac{2}{7}$ further losses are introduced from ∇_1 and ∇_2. We consider the errors from ∇_1 first. Write $\boldsymbol{\alpha}_n = (\alpha_1, \ldots, \alpha_n) \in (\tau, 1)^n$ and $d\boldsymbol{\alpha} = d\alpha_n \ldots d\alpha_1$. The integral corresponding to the region of ∇_1 for which an asymptotic formula has not been given is

$$\int_{\mathcal{D}} \omega \left(\frac{1 - \alpha_1 - cldots - \alpha_4}{\alpha_4} \right) \frac{d\alpha_4}{\alpha_1 \alpha_2 \alpha_3 \alpha_4^2}$$

for some region $\mathcal{D}$. Comparing the conditions for ∇_1 with the requirements for the Type II estimates (that is, a variable must lie between x^θ and $x^{1-2\theta}$) shows that $\mathcal{D} = (\mathcal{D}_1 \cup \mathcal{D}_2 \cup \mathcal{D}_3) \cap \mathcal{G}_4$, where

$$\mathcal{D}_1 = \{\alpha_4 : 1 - 2\theta < \alpha_1 \le \tfrac{1}{2}, \alpha_2 < \tfrac{1}{2}(1 - \theta - \alpha_1),$$
$$\alpha_3 < \min(\alpha_2, \tfrac{1}{2}(1 - \alpha_1 - \alpha_2)), \alpha_4 < \min(\alpha_3, \tfrac{1}{2}(1 - \alpha_1 - \alpha_2 - \alpha_3))\},$$
$$\mathcal{D}_2 = \{\alpha_4 : \alpha_1 < 4\theta - 1, \alpha_4 < \alpha_3 < \alpha_2 < \theta - \alpha_1\},$$
$$\mathcal{D}_3 = \{\alpha_4 : \tfrac{1}{2} - \theta < \alpha_1 < \theta, 1 - 2\theta - \alpha_1 < \alpha_2 < \tfrac{1}{2}(1 - \theta - \alpha_1),$$
$$\alpha_3 < \min(\alpha_2, \tfrac{1}{2}(1 - \alpha_1 - \alpha_2)), \alpha_4 < \min(\alpha_3, \tfrac{1}{2}(1 - \alpha_1 - \alpha_2 - \alpha_3))\},$$

and $\mathcal{G}_n$ consists of those α_n for which no combination of the variables lies in the interval $[\theta, 1 - 2\theta]$. Further decompositions are possible, of course, if $\alpha_1 + \alpha_2 + \alpha_3 + 2\alpha_4 < 1$. Indeed, when $\alpha_1 + \alpha_2 + \alpha_3 + 4\alpha_4 < 1$, we can decompose further still and use the bounds we obtained for the expressions I_{2r} in Chapter 3 (see (3.5.1)). As mentioned in Chapter 3, when using a computer to obtain approximate values for the integrals, it is easy to check whether it is more efficient to perform further decompositions. In other words, one actually calculates expressions like

$$\int_{\alpha_2} \frac{d\alpha_1 \, d\alpha_2}{\alpha_1 \alpha_2} \min \left(\frac{1}{\alpha_2} \omega \left(\frac{1 - \alpha_1 - \alpha_2}{\alpha_2} \right), \int_{\alpha_4 \in \mathcal{D}} \frac{d\alpha_3 \, d\alpha_4}{\alpha_3 \alpha_4} \right.$$
$$\left. \min \left(\frac{1}{\alpha_4} \omega \left(\frac{1 - \alpha_1 - \alpha_2 - \alpha_3 - \alpha_4}{\alpha_4} \right), \int (\cdots) \right) \right).$$

Still further progress is possible by making visible the almost-primes counted by $S(\mathcal{A}_n, z)$. For example, if $pqrs^3 \le x$, we may write

$$S(\mathcal{A}_{pqrs}, s) = S(\mathcal{A}_{pqrs}, X/(pqrs)^{\frac{1}{2}}) + \sum_{s \le t < X/(pqrs)^{1/2}} S(\mathcal{A}_{pqrst}, t). \quad (5.3.3)$$

This is just Buchstab's identity rearranged, of course. Although we may not be able to give an asymptotic formula for a sum over p, q, r, s of the left-hand side of (5.3.3), and *a fortiori* neither for the first term on the right-hand side, nevertheless there may be parts of the final sum that are amenable to the Type II estimates.

Now we turn to the sums discarded from ∇_2. Write

$$\mathcal{E} = \{\alpha_4 : 1 - 2\theta \le \alpha_1 \le \tfrac{1}{2}, \alpha_3 \le \tfrac{1}{2}\alpha_1, (\alpha_1 - \alpha_3, \alpha_2, \alpha_3, \alpha_4) \in \mathcal{G},$$
$$1 - \theta - \alpha_1 \le \alpha_2 \le \tfrac{1}{2}(1 - \alpha_1), \alpha_4 \le \tfrac{1}{2}(1 - \alpha_1 - \alpha_2)\}.$$

When converting the sum corresponding to ∇_2 to an integral, we meet a new phenomenon: There are *two* variables that instead of being prime have only large prime factors. To be precise, we have a variable $u = x^{\alpha_1 - \alpha_3}$ with $p|u \Rightarrow p > s = x^{\alpha_3}$, as well as a variable $t \sim \tfrac{1}{2} x^{1 - \alpha_1 - \alpha_2 - \alpha_4}$ with $p|t \Rightarrow p > r = x^{\alpha_4}$. Using our standard procedures for converting sums to integrals (see below for a little more justification), we have the expression

$$\int_{\mathcal{E}} \omega\left(\frac{\alpha_1 - \alpha_3}{\alpha_3}\right) \omega\left(\frac{1 - \alpha_1 - \alpha_2 - \alpha_4}{\alpha_4}\right) \frac{d\boldsymbol{\alpha}_4}{\alpha_2 \alpha_3^2 \alpha_4^2}. \qquad (5.3.4)$$

As in the case of ∇_1 there may be the possibility of further decompositions, or making almost-primes visible. Now, of course, we have two almost-prime variables.

We pause briefly to give more details for the derivation of (5.3.4). We here write $\eta = (\log x)^{-1}$. We have, by (1.4.14) (as established for all u in the appendix),

$$\sum S(\mathcal{B}_{qrsu}, r) = \frac{x(1 + O(\eta))}{2 \log x} \sum_{q,r,s,u} \frac{1}{qrsu\alpha_4} \omega\left(\frac{1 - \alpha_1 - \alpha_2 - \alpha_4}{\alpha_4}\right).$$

Let $\mathcal{I}(V) = [V, V + V\eta)$. Now we restrict q, r, s, u to ranges of the form

$$q \in \mathcal{I}(Q), \quad r \in \mathcal{I}(R), \quad s \in \mathcal{I}(S), \quad u \in \mathcal{I}(U), \qquad (5.3.5)$$

with

$$Q = x^{\beta_2}, \quad R = x^{\beta_4}, \quad S = x^{\beta_3}, \quad U = x^{\beta_1 - \beta_3}.$$

In this way the region of summation is split into $O(\eta^{-8})$ subranges. We only need an *upper* bound for the sum, so we can include every range for which *at least one* quadruple (q, r, s, u) satisfies the restrictions of ∇_2. Of course, a lower bound, differing from this by only a factor $(1 + O(\eta))$, can be obtained by including merely those ranges for which the ∇_2 conditions are satisfied for *every* quadruple. We first consider the sum over u. We have

$$\sum_{\substack{u \in \mathcal{I}(U) \\ p|u \Rightarrow p \geq s}} \frac{1}{u} = (1 + O(\eta)) \frac{1}{U} \sum_{\substack{u \in \mathcal{I}(U) \\ p|u \Rightarrow p \geq s}} 1$$

$$= (1 + O(\eta)) \, \eta^2 \beta_3^{-1} \omega\left(\frac{\beta_1}{\beta_3} - 1\right).$$

Here we have observed that $\omega(\alpha_1/\alpha_3 - 1)$ varies by $O(\eta)$ as u varies over $\mathcal{I}(U)$. We note that $x^{\eta^2} = (1 + \eta + O(\eta^2))$. From the PNT applied to the summation ranges over q, s, r, we then obtain

$$\sum_{(5.3.5)} S(\mathcal{B}_{qrsu}, r) \leq \eta^9 \frac{1 + O(\eta)}{2\beta_2 \beta_3^2 \beta_4^2} \omega\left(\frac{\beta_1}{\beta_3} - 1\right) \omega\left(\frac{1 - \beta_1 - \beta_2 - \beta_4}{\beta_4}\right)$$

$$= \frac{\eta(1 + O(\eta))}{2} \int_{\beta_1}^{\beta_1 + \eta^2} \cdots \int_{\beta_4}^{\beta_4 + \eta^2} w(\boldsymbol{\alpha}_4) \frac{d\boldsymbol{\alpha}_4}{\alpha_2 \alpha_3^2 \alpha_4^2},$$

with

$$w(\boldsymbol{\alpha}_4) = \omega\left(\frac{\alpha_1 - \alpha_3}{\alpha_3}\right) \omega\left(\frac{1 - \alpha_1 - \alpha_2 - \alpha_4}{\alpha_4}\right).$$

Combining all the ranges then gives

$$\sum S(\mathcal{B}_{qrsu}, r) \leq \mathcal{I} \frac{x(1 + O(\eta))}{2 \log x},$$

where $\mathcal{J}$ is the integral given in (5.3.4).

To sum up, we now have a more efficient method to prove that

$$S(\mathcal{A}, X) \geq \lambda S(\mathcal{B}, X) C'(\theta),$$

where $C'(\theta)$ is a continuous decreasing function on the interval $[\frac{1}{4}, \frac{1}{3}]$. The determination of the value $\theta_0 = \sup\{\theta : C'(\theta) > 0\}$ apparently can only be approximated using numerical integration. We list in Table 5.1 the contribution from each of the cases we have discussed above for the value $\theta = 0.3182$ (from [62]).

Integral	I_1	I_2	$\mathcal{D}_1$	$\mathcal{D}_2$	$\mathcal{D}_3$	$\mathcal{E}$
Bound	0.141	0.217	0.091	0.013	0.1	0.33

Table 5.1 Integral losses at $\theta = 0.3182$

Calculations indicate that it is the integral over region $\mathcal{E}$ that is increasing most quickly at this point and prevents much further progress without new ideas.

5.4 A NEW IDEA

Here we briefly present the new ideas that enabled Heath-Brown and Jia to increase θ to $\frac{16}{49} (= 0.3265\ldots)$. These are innovations concerning the arithmetical information fed into the sieve method, rather than improvements on the sieve technique itself. First, they replace ξ by its approximation a/q at the outset. By itself this does not affect things substantially. The Type I and Type II information we have used is still valid, and the error introduced, namely,

$$\left\| p \left(\xi - \frac{a}{q} \right) \right\|,$$

is of no real consequence. To make further progress they suppose that $\kappa = 0$ and so take

$$\mathcal{A} = \{n : x < n \leq 2x, (n, q) = 1, an \equiv r \,(\mathrm{mod}\, q), r \leq R\},$$

where

$$R = \left[x^{(1-\theta)/2} \right], \quad x = q^{2/(1+\theta)}.$$

The new arithmetical information they derive is for the triple sum

$$\Omega = \sum_{\substack{n,y,z \\ nyz \in \mathcal{A}}} b(n) c(y) h(z),$$

where $c(y), h(z)$ are real coefficients of modulus at most 1, supported on integers $Y < y \leq 2Y, Z < z \leq 2Z$ with $(yz, q) = 1$. They relate Ω to

$$\Omega_4 = \sum_{\substack{u \sim U \\ (u,q)=1}} \sum_{|k| \sim K} \left| \sum_{s,t} \alpha(s) \beta(t) e \left(\frac{qk\overline{st}}{u} \right) \right|.$$

Here $\bar{a}$ indicates multiplicative inverse $\pmod{q}$. This sum is bounded using a familiar short Kloosterman sum estimate [93, p. 36] (we quote this explicitly below as (8.5.2)) or by techniques from the geometry of numbers. Lemmas 4, 6, and 7 of [87] then provide acceptable bounds under the conditions

$$h(z) = \begin{cases} 1 & \text{for } (z,q) = 1, z \in \mathfrak{I} \subset (Z, 2Z], \\ 0 & \text{otherwise}, \end{cases} \qquad Y^5 Z^3 \ll x^{2-2\theta};$$

or

$$\max(Y^{10}Z^{12}, Y^{12}Z^{10}) \ll x^{7-9\theta};$$

or

$$Y^7 Z^8 \ll x^{4-4\theta}, \quad Y^4 Z^3 \ll x^{3-5\theta}, \quad Y^{12} Z^8 \ll x^{7-9\theta}.$$

The writers then deduce that for the types of coefficients that arise from the sieve method, Type II sums can be given an acceptable treatment provided that one range has length M with

$$x^\theta \ll M \ll x^{\frac{15-17\theta}{26}}. \tag{5.4.1}$$

The upper bound in (5.4.1) is larger than ours for $\theta > \frac{11}{35} = 0.314\ldots$. For example, when $\theta = 0.3265$, we need $M = x^\alpha$ with $0.3265 \le \alpha \le 0.347$, whereas Heath-Brown and Jia only require $0.3265 \le \alpha \le 0.3634\ldots$. In particular, this Type II information does not become trivial as $\theta \to \frac{1}{3}$, although the value $\frac{1}{3}$ cannot be exceeded in view of other terms that come into play at that point. It is clear that this extra information must lead to superior results, but the value obtained requires some onerous and detailed calculations!

5.5 HIGHER-DIMENSIONAL VERSIONS

Now we consider a two-dimensional version of the method, and the reader will grasp that the method can be similarly extended to any higher dimension. By *two-dimensional* we indicate that two variables are sieved (by one residue class for each prime) simultaneously, and this corresponds exactly to the definition of sieve dimension given in Chapter 4. We suppose that $\mathcal{A} = \{mn \in \mathcal{B} : U \le m \le V\}$ and consider $S(\mathcal{A}, z)$, which equals

$$\sum_{\substack{mn \in \mathcal{A} \\ (mn, P(z))=1 \\ U \le m \le V}} 1 = \sum_{\substack{d|P(z) \\ e|P(z) \\ demn \in \mathcal{A} \\ U \le dm \le V}} \mu(d)\mu(e)$$

$$= \lambda \sum_{\substack{d|P(z) \\ e|P(z) \\ demn \in \mathcal{B} \\ U \le dm \le V}} \mu(d)\mu(e) + \sum_{\substack{d|P(z) \\ e|P(z)}} \mu(d)\mu(e)\phi(d,e). \tag{5.5.1}$$

Here

$$\phi(d,e) = \sum_{\substack{demn \in \mathcal{A} \\ U \le md \le V}} 1 - \lambda \sum_{\substack{demn \in \mathcal{B} \\ U \le md \le V}} 1.$$

Now, the first term on the right-hand side of (5.5.1) is

$$\lambda \sum_{\substack{mn\in\mathcal{B} \\ (mn,P(z))=1 \\ U\leq m\leq V}} 1 = \lambda S(\mathcal{B}, Z),$$

while the second term could either be split into Type I and Type II sums as before (except that there is now greater flexibility with more variables) or we could use any extra information that might be available for more complicated sums. In particular, we now have sums that have two Type I ranges:

$$\sum_{\substack{\ell mn\in\mathcal{A} \\ \ell\sim L}} c_\ell,$$

and in the next section we shall see that this can be used when the problem is reduced to Dirichlet series.

5.6 GREATEST PRIME FACTORS

If we are unable to prove there are primes in short intervals, it is natural to try to detect prime-like numbers. We thus consider the problem of the greatest prime factor of the integers in an interval. As the reader will have gathered by now, our first task is to assemble appropriate Type I and II information. In fact we will use other shapes of multiple sums as well. Let $\mathfrak{I}(x) = \mathbb{Z}\cap [x-y,x], y = x^{1/2+\epsilon}$. We want to find numbers in $\mathfrak{I}(x)$ with large prime factors. Jutila [113] had the clever idea to restrict numbers to those of the form $nm \in \mathfrak{I}(x)$ where m is *smooth*, that is, composed of only small prime factors, which thus factorize easily. Our aim will be to find such products of numbers with n a prime. Heath-Brown and Jia [86] have revived this method to great effect to prove that $\mathfrak{I}(x)$ contains numbers with their greatest prime factor exceeding $x^{17/18}$. Haugland improved the exponent to $\frac{24}{25}$ [71], which was further improved to $\frac{25}{26}$ by Jia and Liu [112]. We will outline here a method to give $\frac{19}{20}$, which can be further refined to get a better exponent, although it becomes an exercise in computation rather then number theory at this stage! From now on we shall assume some familiarity with the theory of the Riemann zeta-function. Most of the material we need can be found in [157, Chapters 4 and 5].

For $0 < \gamma < \frac{1}{3}$ write

$$k = [1/\epsilon], \qquad X = x^{\frac{\gamma}{k}},$$

$$\mathcal{D} = \{m = n_1\cdots n_k : n_j \sim X, j < k, Y < n_k \leq Z\}, \quad Y = 2^{-2k}X, \ Z = 2^{4k}X,$$

$$\mathcal{B} = \{n : mn \sim x, m \in \mathcal{D}\}, \quad \mathcal{A} = \{n : mn \in \mathfrak{I}(x), m \in \mathcal{D}\}.$$

We will follow our usual philosophy of trying to relate $S(\mathcal{A}, z)$ to $\lambda S(\mathcal{B}, z)$. Here $\lambda = y/(x/2) = 2x^{\epsilon-1/2}$. Our goal will be to establish that

$$S(\mathcal{A}, x^{\frac{1-\gamma}{2}}) > 0,$$

which indicates that A contains primes and hence that $\mathfrak{I}(x)$ contains integers with their greatest prime factor exceeding $x^{1-\gamma}$. In this problem we therefore have *decreasing* γ corresponding to *increasing* difficulty. We will find that for $\gamma \geq \frac{1}{6}$ we can obtain an asymptotic formula that goes smoothly to a lower bound for $\gamma < \frac{1}{6}$.

To obtain our Type I and Type II estimates we first need the following result. We will write $\eta = \eta(\epsilon)$ for a small positive number that depends only on ϵ and which may vary from line to line. We may thus write $\eta^2 = \eta$, for example.

Lemma 5.1. *Let* $|a_u|, |b_v| \leq 1$. *Then*

$$\sum_{\substack{u\sim U, v\sim V \\ uv\in\mathcal{A}}} a_u b_v = \lambda \sum_{\substack{u\sim U, v\sim V \\ uv\in\mathcal{B}}} a_u b_v + O\left(yx^{-\eta}\right) + O(E), \tag{5.6.1}$$

where

$$E = O\left(yX^{\frac{1}{2}}x^{-\frac{1}{2}-\eta}\int_{1/2-iT}^{1/2+iT} |A(s)B(s)||L(s)|^{k-1}\min\left(1, \frac{x}{y|s|}\right)|ds|\right).$$

Here

$$A(s) = \sum_{u\sim U} a_u u^{-s}, \quad B(s) = \sum_{v\sim V} b_v v^{-s}, \quad L(s) = \sum_{n\sim X} n^{-s},$$

and $T = x^{1+\epsilon/2}/y = x^{(1-\epsilon)/2}$.

Proof. We may use Lemma 1.1 to establish that

$$\sum_{\substack{u\sim U, v\sim V \\ uv\in\mathcal{A}}} a_u b_v \tag{5.6.2}$$
$$= \frac{1}{2\pi i}\int_{c-iT}^{c+iT} M(s)A(s)B(s)L(s)^{k-1}\frac{x^s - (x-y)^s}{s}ds + O\left(\frac{x^{1+\frac{\epsilon}{3}}}{T}\right),$$

where

$$c = 1 + \frac{1}{\log x} \quad \text{and} \quad M(s) = \sum_{Y<n\leq Z} n^{-s}.$$

We shift the contour to the line $\sigma = \frac{1}{2}$ and encounter no poles. Now we know that

$$|L(s)|, |M(s)| \ll X^{1-\sigma}x^{-\eta} \tag{5.6.3}$$

since, for example,

$$\sum_{n\sim G} n^{-s} = \sum_{n\sim G} n^{-\sigma}\exp(-it\log n),$$

and we can get a saving of x^η over the trivial bound for such a sum provided that $G > x^{\epsilon^2}, Y < |t| < x$. This can be seen from [157, Section 5.9]. It follows that the errors introduced in shifting the integral in (5.6.2) (the horizontal line segments as I like to think of them) are of size

$$\ll x^{1-k\eta-\sigma}\frac{1}{T}x^\sigma = O\left(yx^{-\eta}\right)$$

as required.

We consider the integral on $\sigma = \frac{1}{2}$ in two parts: $|t| > Y, |t| \leq Y$. We obtain the main term in (5.6.1) from the region $|t| \leq Y$. We have

$$\frac{x^s - (x-y)^s}{s} = yx^{s-1} + O\left(y^2 Y x^{-\frac{3}{2}}\right).$$

Using $M(s) \ll X^{1/2}$ on $\sigma = \frac{1}{2}$ we can therefore replace $(x^s - (x-y)^s)/s$ with yx^{s-1} and incur a loss

$$\ll yx^{-\eta - \frac{1}{2}} \int_{1/2-iY}^{1/2+iY} |A(s)B(s)||L(s)|^{k-1}|ds|. \tag{5.6.4}$$

We also note that, since $|t| \leq Y$, we have

$$M(s) = \frac{Z^{1-s} - Y^{1-s}}{1-s} + O\left(X^{-\sigma}\right) \tag{5.6.5}$$

by [157, Theorem 4.11] applied with $x = Z$ and $x = Y$. We use (5.6.5) to substitute for $M(s)$ and thus incur a further loss no worse than (5.6.4). We shift the integral back to $\sigma = c$ to obtain errors from horizontal line segments that can be bounded in a satisfactory manner plus a main term

$$y \int_{c-iY}^{c+iY} L(s)^{k-1} A(s) B(s) \frac{(x/Y)^{s-1} - (x/Z)^{s-1}}{s-1} ds$$

$$= y \int_{c-iY-1}^{c+iY-1} L(s+1)^{k-1} A(s+1) B(s+1) \frac{(x/Y)^s - (x/Z)^s}{s} ds$$

$$= y \sum_{\substack{u \sim U, v \sim V \\ n_j \sim X}} \frac{a_u b_v}{uvn_1 \ldots n_{k-1}} + O\left(\frac{yx^\eta}{Y}\right) \tag{5.6.6}$$

using Perron's formula again. The error in (5.6.6) is $O(yx^{-\eta})$. Now let $R = uvn_1 \ldots n_{k-1}$. Then

$$\sum_{\substack{u \sim U, v \sim V \\ n_j \sim X}} \frac{a_u b_v}{R} = \frac{2}{x} \sum_{\substack{u \sim U, v \sim V \\ n_j \sim X}} a_u b_v \left(\left[\frac{x}{R}\right] - \left[\frac{x}{2R}\right] + O(1)\right)$$

$$= \frac{2}{x} \sum_{\substack{u \sim U, v \sim V \\ uv \in \mathcal{B}}} a_u b_v + O\left(\frac{1}{X}\right). \tag{5.6.7}$$

From (5.6.6) and (5.6.7) we get our main term in (5.6.1) with error $O(yx^{-\eta})$ plus the integral in (5.6.4). The rest of the integral on $\sigma = \frac{1}{2}$ is easily absorbed into the integral in (5.6.1) since $M(s) \ll X^{1/2}x^{-\eta}$ for $Y < |\operatorname{Im} s| \leq T$. This completes the proof of the lemma. □

It therefore remains to establish that

$$\int_{1/2-iT}^{1/2+iT} |A(s)B(s)||L(s)|^{k-1}|ds| \ll \left(\frac{x}{X}\right)^{\frac{1}{2}} \log^A x \tag{5.6.8}$$

for some A, to obtain an asymptotic formula for

$$\sum_{\substack{u \sim U, v \sim V \\ uv \in \mathcal{A}}} a_u b_v. \tag{5.6.9}$$

We recall the familiar mean value theorem for Dirichlet polynomials (for example, see [157, (7.20.1)]), which will be used frequently in this and subsequent chapters. The result as stated below is not the best known, but we are able to supply a quick proof of this form in the appendix.

Lemma 5.2. *Suppose that* $T > 0, U \geq 1$ *and let* $f_u \in \mathbb{C}, u \sim U$. *Then*

$$\int_{1/2-iT}^{1/2+iT} \left| \sum_{u \sim U} f_u u^{-s} \right|^2 |ds| \ll (T+U) \sum_{u \sim U} \frac{|f_u|^2}{u}. \tag{5.6.10}$$

Remark. The best form of this result (see [135]) has

$$\int_{1/2-iT}^{1/2+iT} \left| \sum_{u \leq U} f_u u^{-s} \right|^2 |ds| = \sum_{u \leq U} \frac{|f_u|^2}{u} (2T + O(u)). \tag{5.6.11}$$

From now on we shall write all integrals as $\int dt$ for convenience since they will all be over the range indicated on the left side of (5.6.10). Suppose that $\max(U, V) \ll x^{1/2}, UVX^k \leq x$. Then we can find non-negative integers r, w with $r + w = k - 1$, and

$$\max(UX^r, VX^w) \leq (xX)^{\frac{1}{2}}.$$

Note that

$$(UX^r)(VX^w) \ll \frac{x}{X} \quad \text{and} \quad T\max(UX^r, VX^w) \ll \frac{x}{X}.$$

We can then establish (5.6.9) by applying Cauchy's inequality to the left-hand side of (5.6.8) to obtain a bound

$$\leq \left(\int |A(s)|^2 |L(s)|^{2r} dt \right)^{\frac{1}{2}} \left(\int |B(s)|^2 |L(s)|^{2w} dt \right)^{\frac{1}{2}}$$

$$\ll \log^A x \, (T + UX^r)^{\frac{1}{2}} (T + VX^w)^{\frac{1}{2}} \ll \left(\frac{x}{X} \right)^{\frac{1}{2}} \log^A x.$$

We can thus estimate a Type II sum if $\max(U, V) \ll x^{1/2}$, that is, both ranges between constant multiples of $x^{1/2-\gamma}$ and $x^{1/2}$. This gives a range $x^{1/3}$ to $x^{1/2}$ if $\gamma = \frac{1}{6}$. Note, though, that this corresponds to $x^{2/5}$ to $x^{3/5}$ in our previous work since the numbers we are studying have size $x^{5/6}$ with this value of γ.

To tackle the Type I sum we actually consider the more general expression that we shall refer to as a Type I/II sum, namely,

$$\sum_{\substack{u \sim U, v \sim V \\ uvn \in \mathcal{A}}} a_u b_v$$

with $V \leq x^{1/2}, VU^2 \leq x^{1-\gamma}$. By a well-known principle in analytic number theory (the reflection principle) we can replace

$$\sum_{N \leq n \leq 2N} n^{-s} \quad \text{with} \quad \sum_{\frac{|t|}{4\pi N} \leq n \leq \frac{|t|}{2\pi N}} n^{-s} \tag{5.6.12}$$

when $|t| > N$ with an error of size

$$N^{-\frac{1}{2}} + \left(\frac{|t|}{N}\right)^{-\frac{1}{2}}.$$

This is demonstrated using the approximate functional equation for the Riemann zeta-function [157, (4.12.4)], which with $\operatorname{Re} s = \frac{1}{2}$ gives

$$\zeta(s) = \sum_{n \leq N} n^{-s} + \chi(s) \sum_{m \leq M} m^{s-1} + O\left(N^{-\frac{1}{2}} + M^{-\frac{1}{2}}\right),$$

where $2\pi M N = |t|, s = \frac{1}{2} + it, |\chi(s)| = 1$. Of course, we then need to apply Perron's formula to the sum to remove the summation range dependence on t, but this only introduces a further $\log^2 x$ factor into the error term.

On the other hand, for $|t| < N$ the first sum in (5.6.12) equals

$$\frac{(2N)^{1-s} - N^{1-s}}{1-s} + O\left(N^{-\frac{1}{2}}\right)$$

by the same working as (5.6.5).

Now the size of N will be around $x^{1-\gamma}/UV$, so that

$$U\left(\frac{T}{N}\right) \approx \frac{U^2 V}{x^{1-\gamma}} T \leq T.$$

This enables us to give a suitable estimate for the error terms using Lemma 5.2.

We note that this does not give Type I information as nice as we had before. The value $\gamma = \frac{1}{6}$ gives Type I information up to $x^{3/5}$ in our previous notation, but we need $x^{4/5}$ to work as we did to obtain (5.2), say. To compensate for this, we can include the extra range of summation with U up to $x^{1/6}$, that is, $x^{1/5}$ in our previous notation. The following lemma incorporates this information.

Lemma 5.3. *For $M \leq x^{1/2}, N \leq x^{1/4}, |a_m|, |b_n| \ll 1$, we can obtain an asymptotic formula for*

$$\sum_{m \sim M, n \sim N} a_m b_n S(\mathcal{A}_{mn}, x^\gamma). \tag{5.6.13}$$

Proof. If $x^{1/2-\gamma} \leq M \leq x^{1/2}$ then we automatically have a Type II sum. On the other hand, $M \leq x^{1/2-\gamma} \Rightarrow MN^2 \leq x^{1-\gamma}$. We may thus work as in the Fundamental Theorem, taking out successive prime factors $p_1, p_2, \ldots$. At each stage we either have a Type II sum when

$$x^{\frac{1}{2}-\gamma} \leq mp_1 \ldots p_h < x^{\frac{1}{2}}$$

or

$$mp_1 \ldots p_h < x^{\frac{1}{2}-\gamma} \Rightarrow mp_1 \ldots p_h n^2 \leq x^{1-\gamma},$$

which gives the Type I sum estimate needed to take out the next prime factor. The proof is then completed in the usual way. □

We will find we need even more arithmetical information than we have garnered to date. In particular, we require estimates for what we might call a Type I_2 sum (compare the discussion in Section 5.5):

$$\sum_{\substack{unm\in\mathcal{A} \\ n\sim N, m\sim M}} a_u$$

with $MN > x^{1/2-\gamma}$. The new feature of the above sum is the presence of two ranges of consecutive integers. If $MN < x^{1/2}$ we have a Type II sum, so we can assume that $MN > x^{1/2}$. It follows that at least one of M, N must exceed $x^{1/4}$. Without loss of generality $N > M$. We then use the reflection principle as above to change N to T/N in size. If $M > T^{1/2}$, we reflect M as well. In this way Cauchy's inequality and the mean value theorem for Dirichlet polynomials give an upper bound for the error

$$\ll \log^A x \left(\frac{x}{MN} + T\right)^{\frac{1}{2}} \left(\min\left(\frac{MT}{N}, \frac{T^2}{MN}\right) + T\right)^{\frac{1}{2}} \ll \left(\frac{x}{X}\right)^{\frac{1}{2}} \log^A x$$

as required.

We are now in a position to start our Buchstab decomposition. We will begin with the value $\gamma = \frac{1}{6}$. Write $z = x^{1/6}$. We then want to consider $S(\mathcal{A}, x^{5/12})$ which equals

$$S(\mathcal{A}, z) - \sum_{z\leq p<x^{5/12}} S(\mathcal{A}_p, p).$$

The Fundamental Theorem gives an asymptotic formula for the first term above, while the parts of the sum with $x^{1/3} \leq p < x^{5/12}$ can be estimated by our Type II sums. The remainder we express as

$$- \sum_{z\leq p<z^2} S(\mathcal{A}_p, z) + \sum_{\substack{z\leq q<p<z^2 \\ pq^2<z^5}} S(\mathcal{A}_{pq}, q).$$

Now the first term can be evaluated by the Fundamental Theorem, while those parts of the second term with $qp \leq x^{1/2}$ are covered by the Type II estimates since $pq > z^2$. For the rest of this sum we note that $S(\mathcal{A}_{pq}, q)$ counts numbers r in size less than z^2 and hence

$$S(\mathcal{A}_{pq}, q) = S(\mathcal{A}_{pq}, z).$$

Thus those parts of the sum with $q \leq x^{1/4}$ are covered by (5.6.13). For the case $q > x^{1/4}$ we use a two-dimensional version of the method as outlined in Section 5.5. Given Q, write

$$\mathcal{A}_p^* = \{mn : mnp \in \mathcal{A}, m \sim Q\}$$

and define $\mathcal{B}_p^*$ analogously. Then, since $q < z^2$,

$$\sum_{q\sim Q} S(\mathcal{A}_{pq}, z) = S(\mathcal{A}_p^*, z).$$

As in (5.5.1), we have

$$\sum_p S(\mathcal{A}_p^*, z) = \lambda \sum_p S(\mathcal{B}_p^*, z) + \sum_p \sum_{d|P,e|P} \mu(d)\mu(e)\phi_p(d,e)$$

with $P = P(z)$ and

$$\phi_p(d,e) = \sum_{\substack{demn\in\mathcal{A}_p^* \\ dm\sim Q}} 1 - \lambda \sum_{\substack{demn\in\mathcal{B}_p^* \\ dm\sim Q}} 1.$$

We can deal with those parts of the remainder with $z^2 \le de \le x^{1/2}$ as a Type II sum. If $de > x^{\frac{1}{2}}$, we can take out the largest prime factor of de and repeat until we get a product lying between z^2 and $x^{\frac{1}{2}}$. This is possible, of course, since all the prime factors of de are less than z and $x^{\frac{1}{2}} = z^3$. We can thus suppose that $de < z^2$. Now if $pde > x^{\frac{1}{2}}$ we can again take out the largest prime factor and repeat until at last $z^2 \le pde \le x^{1/2}$. We may thus suppose that $pde < z^2$, in which case $mn > x^{1/2}$ and our Type I_2 estimates are applicable.

We have thus obtained an asymptotic formula for $S(\mathcal{A}, x^{5/12})$ and so proved the following.

Theorem 5.3. *For all large x the interval $[x - x^{1/2+\epsilon}, x]$ contains integers with prime factors exceeding $x^{5/6}$.*

Now we consider what happens for $\gamma < \frac{1}{6}$. Let

$$\mathbf{a} = (\alpha_1, \alpha_2), \quad \alpha_j = \frac{\log p_j}{\log x}.$$

Write (the sets are illustrated below for $\gamma = \frac{1}{20}$)

$$A = \left\{ \mathbf{a} : \gamma \le \alpha_1 \le \frac{1}{2} - 2\gamma, \gamma \le \alpha_2 \le \min\left(\alpha_1, \frac{1}{2} - \gamma - \alpha_1\right) \right\},$$

$$X = \left\{ \mathbf{a} : \frac{1}{2}\left(\frac{1}{2} - \gamma\right) \le \alpha_1 \le \frac{1}{2} - \gamma, \right.$$

$$\left. \max\left(\frac{1}{2} - \gamma - \alpha_1, \gamma\right) \le \alpha_2 < \min\left(\alpha_1, \frac{1}{2} - \alpha_1\right) \right\},$$

$$B = \left\{ \mathbf{a} : \frac{1}{4} \le \alpha_1 \le \frac{1}{2} - \gamma, \frac{1}{2} - \alpha_1 \le \alpha_2 \le \frac{1}{4} \right\},$$

$$C = \left\{ \mathbf{a} : \frac{1}{4} \le \alpha_1 \le \frac{1}{2} - \gamma, \frac{1}{4} \le \alpha_2 \le \min\left(\alpha_1, \frac{1 - \alpha_1 - \gamma}{2}\right) \right\}.$$

Then, with z now equal to x^γ and putting $W = x^{1/2}/z$, $R = (x/z)^{1/2}$, we have

$S(\mathcal{A}, R)$

$$= S(\mathcal{A}, z) - \sum_{W \le p \le R} S(\mathcal{A}_p, p) - \sum_{z \le p \le W} S(\mathcal{A}_p, z) + \sum_{\substack{z \le p_2 < p_1 < W \\ p_1 p_2^2 < x/z}} S(\mathcal{A}_{p_1 p_2}, p_2)$$

$$= S_1 - S_2 - S_3 + S_4 \quad \text{say.}$$

We can evaluate S_1 and S_3 by the Fundamental Theorem, and we note that S_2 is of the right shape for our Type II estimates. We split up S_4 thus:

$$S_4 = S_5 + S_6 + S_7 + S_8,$$

where the summation ranges correspond to A, X, B, C, respectively. We can evaluate S_6 immediately as a Type II sum. We can apply Buchstab's decomposition twice more to the sum S_5: indeed, for small γ, some parts of this sum can be decomposed further. For S_7 we can perform two more decompositions, reversing the roles of the variables if necessary. It is not always more efficient to do this, however, and this needs to be taken into account when performing the calculations. The final term S_8 corresponds to the awkward sum we treated with a two-dimensional sieve when $\gamma = \frac{1}{6}$. We can treat some parts of S_8 in this way for $\gamma < \frac{1}{6}$.

In conclusion we get the familiar situation

$$S(\mathcal{A}, R) \geq \lambda C(\lambda) S(\mathcal{B}, R)$$

where

$$C\left(\tfrac{1}{6}\right) = 1, \quad C(\gamma) > 0 \ \text{ for } \ \gamma > \gamma_0.$$

As usual with our method, γ_0 is hard to pin down precisely, but certainly $\gamma_0 < \frac{1}{20}$. With more work and more sophisticated mean value estimates, however, one can get $C\left(\frac{1}{26}\right) > 0$ ([112]).

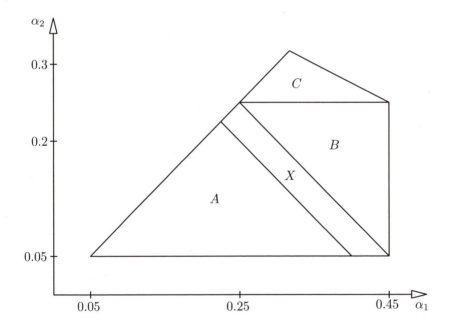

Diagram 5.1 The sets A, B, C, X at $\gamma = 0.05$

Exercises

1. There are other asymptotic regions that can be included in Diagram 5.1. Can you find them? Carry out the detailed calculations for $\gamma = 0.05$ in the above. Reduce the value of γ as much as you can. If you are feeling brave, try to bring in Watt's mean value theorem (see Lemma 7.8).

2. In Section 5.3 we said that $C'(\theta)$ is a continuous decreasing function on $[\frac{1}{4}, \frac{1}{3}]$. Hence its inverse $g(x)$ exits. Have you any ideas how to express $g(x)$? Do you know any techniques that could determine $g(0)$?

Chapter Six

An Upper-Bound Sieve

6.1 THE METHOD DESCRIBED

For an upper-bound sieve we can work in a similar way to Chapters 3 and 5, but we now must only discard negative sums. As the simplest example, we consider giving upper bounds to the number of primes p with $||\xi p + \kappa|| < x^{\epsilon-\theta}$. As before, we write $T = x^\theta, U = x^{1-2\theta}, X = x^{1/2}, z = x^{1-3\theta}$ and put

$$\mathcal{A} = \left\{ n \in \mathcal{B} : ||n\xi + \kappa|| < x^{\epsilon-\theta} \right\}.$$

Our fundamental theorem instantly gives the upper bound

$$S(\mathcal{A}, X) \leq S(\mathcal{A}, z) = \lambda S(\mathcal{B}, z)(1 + o(1)),$$

where $\lambda = 2x^{\epsilon-\theta}$. For example, with $\theta = \frac{2}{7}$ we obtain the rather poor upper bound

$$S(\mathcal{A}, X) \leq \lambda K S(\mathcal{B}, X)(1 + o(1)), \tag{6.1.1}$$

where

$$K = 7\omega(7) \approx 3.93.$$

One could compare a simple application of the Rosser-Iwaniec sieve. From our work in Chapter 4 we can take $D = x^{1-\theta}$ and immediately obtain (6.1.1) with $K = 2/(1 - \theta) = \frac{14}{5}$ for $\theta = \frac{2}{7}$. However, the inclusion of Type II information with the Rosser-Iwaniec sieve enables one to increase D to $x^{2-4\theta}$, thus improving K to $\frac{7}{3}$ at $\theta = \frac{2}{7}$. To see this, express the remainder in a bilinear form:

$$\sum_{m \leq M, n \leq N} |R_{mn}|.$$

We can take $M = N = U$ since if either m or n exceeded T, we would have a Type II sum. Otherwise $mn < T^2 < x^{1-\theta}$ for $\theta < \frac{1}{3}$ (where the Type II information disappears in any case). We could then get a still better bound by writing $z' = \min(D^{1/3}, T)$ and, using our Type II information, estimate

$$S(\mathcal{A}, X) \leq S(\mathcal{A}, z') - \sum_{T \leq p \leq U} S(\mathcal{A}_p, p)$$

$$= S(\mathcal{A}, z) - (S(\mathcal{A}, T) - S(\mathcal{A}, U))$$

$$\leq \lambda K S(\mathcal{B}, X)(1 + o(1)),$$

where, writing $\zeta = \min((2 - 4\theta)/3, \theta)$,

$$K = e^{-\gamma} \frac{F((2 - 4\theta)/\zeta)}{\zeta} - \frac{\omega(1/\theta)}{\theta} + \frac{\omega(1/(1 - 2\theta))}{1 - 2\theta}.$$

We note that for $\theta = \frac{1}{4}$ this gives $4(e^{-\gamma}F(4) - w(4)) + 1$. Since $e^{-\gamma}F(4) > w(4)$, this shows that we cannot get the "correct" upper bound for θ near $\frac{1}{4}$ by this method.

However, we can do much better with our sieve method, as we now show. We begin as before:

$$S(\mathcal{A}, X) = S(\mathcal{A}, z) - \left(\sum_{z \leq p < T} + \sum_{T \leq p \leq U} + \sum_{U < p < X} \right) S(\mathcal{A}_p, p).$$

Now the range $T \leq p \leq U$ gives a Type II sum as before. In the range $z \leq p < T$ there are no problems in making two further decompositions (we assume here, as we must, that $\theta < \frac{1}{3}$). For the range $U < p < x$ we run into problems, since if q is the next prime variable in the decomposition, pq can be as large as $(xp)^{1/2} > x^{1-\theta}$. However, we can swap roles in this last case as follows.

$$\sum_{*p,q*} S(\mathcal{A}_{pq}, q) = \sum_{q,r} S(\mathcal{A}_{qr}, (x/qr)^{\frac{1}{2}})$$

$$= \sum_{q,r} S(\mathcal{A}_{qr}, z) - \sum_{q,r,s} S(\mathcal{A}_{qrs}, s).$$

(We leave it to the reader to fill in the summation ranges). It follows that all the sums that are discarded have the form

$$\sum_{p,q,r} S(\mathcal{A}_{pqr}, r) \quad \text{or} \quad \sum_{q,r,s} S(\mathcal{A}_{qrs}, s).$$

Thus, using the methods of Chapters 3 and 5 the *loss* we make in the sieve construction has the form $C(\theta - \frac{1}{4})^3$. Hence we get

$$S(\mathcal{A}, X) \leq \lambda S(\mathcal{B}, X) \left(1 + C\left(\theta - \frac{1}{4}\right)^3 \right),$$

at least for θ not too much greater that $\frac{1}{4}$.

Combining this with Chapters 3 and 5, we see that we have used the sieve of Eratosthenes-Legendre with Buchstab's identity to produce a sieve method that goes smoothly from an asymptotic formula to upper and lower bounds at the breakdown point $\theta = \frac{1}{4}$. Clearly this technique should work for a wide variety of problems for which we can obtain Type I and Type II information. As the Type II information disappears (that is, increasing θ) the method should go smoothly to the standard Rosser-Iwaniec upper and lower bounds for the linear sieve problems we obtained in Chapter 4. Updating our statement for the lower bound in Chapter 3, we must have

$$(1 + o(1))f(\Delta, z) \leq \frac{S(\mathcal{A}, z)}{\lambda S(\mathcal{B}, z)} \leq F(\Delta, z)(1 + o(1)),$$

where Δ represents the arithmetical information fed into the sieve method and

$$f(\Delta, z) \to e^{-\gamma} \frac{f(u)}{w(v)}, \quad F(\Delta, z) \to e^{-\gamma} \frac{F(u)}{w(v)},$$

as $\Delta \to$ "Type I information only available," with

$$u = \frac{\log M}{\log z}, \quad v = \frac{\log x}{\log z},$$

where M is the parameter in Theorem 3.1. As $\Delta \to$ "very good information" we have

$$f(\Delta, z) \to 1, \qquad F(\Delta, z) \to 1.$$

In the case $z = X$ one would expect Vaughan's identity or one of the related formulae given in Chapter 2 to be applicable once $f(\Delta, z) = 1$. This may not always be the case, as we will see in later chapters.

The reader may at this point wonder if one can set up difference-differential equations for $f(\Delta, z), F(\Delta, z)$ just as we did in Chapter 4 for functions proportional to $f(u), F(u)$. Surely, if we have obtained the best possible decomposition with the given arithmetical information, it must be neutral under Buchstab's identity. Redefining $f(\Delta, z), F(\Delta, z)$ to be functions of u or v rather than z, this seems plausible, but the problem arises that Δ will also vary as $\mathcal{A}$ is replaced by $\mathcal{A}_p$, and this creates a more complicated situation.

We shall return to the application of our upper-bound sieve in Chapter 8, contenting ourselves with the following result (based on [6]) as a relatively simple example now. Here we shall use the alternative sieve and the Rosser-Iwaniec sieve in tandem. We write $P(n)$ for the greatest prime factor of an integer n.

Theorem 6.1. *For all sufficiently large x there are integers n in the interval $[x, x + x^{1/2}]$ with*

$$P(n) > x^{0.74}. \tag{6.1.2}$$

Remark. This result is stronger than any previously published. The current sieve method was first applied to this problem in [6] to obtain the exponent 0.732. An improved exponential sum estimate was then used in [120] to get 0.738.

6.2 A DEVICE BY CHEBYCHEV

There is an argument that can be traced back to Chebychev in the 19th century that establishes the existence of numbers with large prime factors in certain sets provided that we can give sufficiently tight upper bounds for the numbers with smaller prime factors in the set.

Let $y = x^{1/2}, \mathcal{J} = (x, x + y]$ and write

$$N(d) = \sum_{\substack{n \in \mathcal{J} \\ d|n}} 1.$$

Put $L = \log x, U = x^{3/5-\epsilon}$. We then have

$$\sum_{d \leq x} \Lambda(d) N(d) = \sum_{n \in \mathcal{J}} \sum_{d|n} \Lambda(d)$$

$$= \sum_{n \in \mathcal{J}} \log n$$

$$= yL + O(y).$$

We then need to establish the three following statements:

$$\sum_{d<U} \Lambda(d)N(d) = \left(\frac{3}{5} - \epsilon\right) yL + O(y); \tag{6.2.1}$$

$$\sum_{\substack{U \leq d \leq x \\ d \text{ not prime}}} \Lambda(d)N(d) = O(y); \tag{6.2.2}$$

$$\sum_{U \leq p \leq x^{0.74}} (\log p)N(p) < \frac{2}{5} yL. \tag{6.2.3}$$

Now (6.2.2) is elementary. Since $N(d) \leq 1$ for $d > y$, we have

$$\sum_{\substack{U \leq d \leq x \\ d \text{ not prime}}} \Lambda(d)N(d) = \sum_{r \geq 2} \sum_{U \leq p^r \leq x} (\log p)N(p^r)$$

$$\leq \sum_{p^2 \leq y} \log p + \sum_{r \geq 3} \sum_{p^r \leq x} \log p$$

$$\leq y + x^{\frac{1}{3}} \log x \leq 2y$$

for all large x, where we have used nothing more than Chebychev's upper bound $\sum_{p \leq h} \log p \leq 2h$ for all large h.

Now let $v \in (y, x^{3/4}]$. Define θ by $v = x^\theta$. We note that at most one integer in $\mathcal{I}$ can be divisible by an integer in the interval $\mathcal{K} = (v, ev]$. We write

$$\mathcal{A} = \{n : n \in \mathcal{K}, N(n) = 1\}.$$

This is our set to be sieved, with comparison set $\mathcal{B} = \{n : n \in \mathcal{K}\}$, where each n in $\mathcal{B}$ is to be counted

$$h(n) = \left[\frac{2x}{n}\right] - \left[\frac{x}{n}\right]$$

times. Thus

$$\sum_{m \in \mathcal{B}} a_m = \sum_{m \in \mathcal{K}} a_m \sum_{\substack{x < n \leq 2x \\ m|n}} 1.$$

Since $h(d) = x/d + O(1)$ there is no problem using partial summation to give an asymptotic formula for $S(\mathcal{B}, (ev)^{1/2}) \sim x/(\log v)$ or sums involving $S(\mathcal{B}_r, r)$.

We note that we now have to deal with a whole range of sets as v or, equivalently, θ varies. Let

$$S(\theta) = S(\mathcal{A}, (ev)^{\frac{1}{2}}).$$

Then (6.2.3) will follow from the bound

$$\int_{0.6-\epsilon}^{0.74} \theta S(\theta)\, d\theta < \left(\frac{2}{5} - \epsilon\right) \frac{y}{L}. \tag{6.2.4}$$

6.3 THE ARITHMETICAL INFORMATION

We write

$$\psi(\alpha) = \alpha - [\alpha] - \tfrac{1}{2}.$$

This function enters our proof via the following simple observation:

$$N(d) = \left[\frac{x+y}{d}\right] - \left[\frac{x}{d}\right]$$

$$= \frac{y}{d} - \psi\left(\frac{x+y}{d}\right) + \psi\left(\frac{x}{d}\right).$$

Thus

$$\sum_{mn\in\mathcal{A}} a_m b_n = \sum_{mn\in\mathcal{K}} a_m b_n N(mn)$$

$$= \sum_{mn\in\mathcal{K}} a_m b_n \frac{y}{mn} - \sum_{mn\in\mathcal{K}} a_m b_n \left(\psi\left(\frac{x+y}{mn}\right) - \psi\left(\frac{x}{mn}\right)\right)$$

$$= S_1 - S_2, \quad \text{say}.$$

Now write $\lambda = y/x$, then

$$S_1 = \lambda \sum_{mn\in\mathcal{K}} a_m b_n \frac{x}{mn}$$

$$= \lambda \sum_{mn\in\mathcal{B}} a_m b_n + O\left(\lambda \sum_{mn\in\mathcal{K}} |a_m b_n|\right)$$

$$= \lambda \sum_{mn\in\mathcal{B}} a_m b_n + O\left(\lambda v(\log x)^C\right),$$

assuming that a_m, b_n are bounded by divisor functions. To obtain our required Type I and Type II information we must therefore show that $S_2 \ll yx^{-\eta}$ for some $\eta > 0$. Such a bound is immediate for $v < x^{1/2-2\eta}$ using the trivial bound $|\psi(x)| \le 1$. We thus get

$$\sum_{d < x^{\frac{1}{2}-\epsilon}} \Lambda(d) N(d) = (1/2 - \epsilon) yL + O(y)$$

straightaway. We therefore only need concern ourselves with $v > x^{1/2-\epsilon}$ in the following, and what remains to prove for (6.2.1) is that

$$\sum_{x^{1/2-\epsilon} < d < U} \Lambda(d) N(d) = \frac{1}{10} yL + O(y). \tag{6.3.1}$$

The first step is to approximate $\psi(x)$ with a finite Fourier series. We effectively did the same thing in our consideration of the Piatetski-Shapiro prime number theorem at the end of Chapters 2 and 3. We first note that

$$\psi(x) = \frac{1}{2\pi i} \sum_{h\neq 0} \frac{1}{h} e(hx). \tag{6.3.2}$$

Now this series is only slowly converging and is not uniformly convergent. One way around this problem would be to use the following analogue of Lemma 2.1.

Lemma 6.1. *Given $N \geq 1$, there is a trigonometric polynomial $\psi^*(x)$ such that*

$$|\psi(x) - \psi^*(x)| \leq \frac{1}{2N+2} \sum_{|n| \leq N} \left(1 - \frac{|n|}{N+1}\right) e(nx),$$

with

$$\psi^*(x) = \sum_{1 \leq |n| \leq N} c_n e(nx)$$

where

$$c_n \ll \frac{1}{n}.$$

Proof. See the appendix in [49]   ☐

However, it suffices to use a simpler approximation, which has the advantage that the coefficient of h remains $1/h$. If we cut off the series (6.3.2) at H and apply partial summation to the tail of the series (that is, $|h| > H$), we obtain

$$\psi(x) = \frac{1}{2\pi i} \sum_{0 < |h| \leq H} \frac{1}{h} e(hx) + O\left(\min\left(1, \frac{1}{H\|x\|}\right)\right). \tag{6.3.3}$$

Now

$$\frac{1}{2\pi i h}\left(e\left(\frac{h(x+y)}{mn}\right) - e\left(\frac{hx}{mn}\right)\right) = \int_x^{x+y} \frac{1}{mn} e\left(\frac{h\xi}{mn}\right) d\xi.$$

Hence

$$S_2 = \int_x^{x+y} \sum_{0 < |h| \leq H} \sum_{mn \in \mathcal{K}} \frac{a_m b_n}{mn} e\left(\frac{h\xi}{mn}\right) d\xi$$

$$+ O\left(\max_{x \leq \xi \leq x+y} \sum_{mn \in \mathcal{K}} a_m b_n \min\left(1, \frac{1}{H\|\xi/mn\|}\right)\right)$$

$$= S_3 + S_4, \quad \text{say.}$$

We have, for any $\eta > 0$,

$$S_4 \ll \max_{x \leq \xi \leq x+y} x^\eta \sum_{n \in \mathcal{K}} \min\left(1, \frac{1}{H\|\xi/n\|}\right)$$

$$\ll x^{2\eta} \frac{v}{H} \ll y x^{-\eta},$$

for $x^{2/5} < v < x^{3/4}$ if we pick $H = vx^{3\eta}/y$. To see this, note that, by Lemma (2.1)

$$\sum_{\substack{n \in \mathcal{K} \\ \|\xi/n\| < 1/Q}} 1 \ll \frac{v}{Q} + \frac{1}{Q} \sum_{\ell=1}^{Q} \left|\sum_{n \in \mathcal{K}} e\left(\frac{\ell\xi}{n}\right)\right|.$$

If we pick $Q = H2^{-j}$, it remains to show that

$$\sum_{\ell=1}^{Q} \left|\sum_{n \in \mathcal{K}} e\left(\frac{\ell\xi}{n}\right)\right| \ll v.$$

This follows from the Kusmin-Landau bound ([49, Theorem 2.1]) if $v \geq 2(Qx)^{1/2}$, and from the simplest van der Corput estimate ([49, Theorem 2.2]) for $x^{2/5} < v < 2(Qx)^{1/2}$,

To tackle S_3 we can apply partial summation to the m and n variables to reduce the problem to a demonstration that

$$\sum_{m,n,\ell} a_m b_n\, e\left(\frac{\xi\ell}{mn}\right) \ll vx^{-2\eta}, \tag{6.3.4}$$

where $x \leq \xi \leq x + y, mn \in \mathcal{K}, \ell \leq vx^{1/2+3\eta}$. Such bounds are available via Van der Corput's method of exponent pairs with various technical refinements wedded to work of Fouvry and Iwaniec [35] on certain types of exponential sums. We include all the bounds we require in the following two lemmas. The proofs are fairly involved and technical, and most of the details can be found in [120].

Lemma 6.2. *Suppose that* $m \sim M \leq x^{2/5-\epsilon}$ *and* $b_n \equiv 1$. *Then* (6.3.4) *holds. In particular* $S_2 \ll yx^{-\eta}$ *under these conditions.*

Before stating the next lemma we require some notation. We define ϕ_j as in Table 6.1.

	ϕ_1	ϕ_2	ϕ_3	ϕ_4	ϕ_5	ϕ_6	ϕ_7	ϕ_8	ϕ_9
$\frac{a}{b}$	$\frac{3}{5}$	$\frac{11}{18}$	$\frac{35}{54}$	$\frac{2}{3}$	$\frac{90}{131}$	$\frac{226}{323}$	$\frac{547}{771}$	$\frac{23}{32}$	$\frac{37}{50}$
$\approx$	0.6	0.611	0.648	0.666	0.687	0.699	0.709	0.718	0.74

Table 6.1 Definition of ϕ_j

In the above, $\approx$ gives the first three places of the decimal whether the decimal is terminating or not. Put $\mathfrak{I}_j = [\phi_j, \phi_{j+1}), j = 1, \dots 8$. We then write

$$\mathfrak{I}(\theta) = [\theta - \tfrac{1}{2} + \epsilon, \tau(\theta) - \epsilon]$$

where $\tau(\theta)$ is in Table 6.2 (we do not go beyond ϕ_8 for the next result).

Interval	$\mathfrak{I}_1$	$\mathfrak{I}_2$	$\mathfrak{I}_3$	$\mathfrak{I}_4$	$\mathfrak{I}_5$	$\mathfrak{I}_6$	$\mathfrak{I}_7$
$\tau(\theta)$	$2 - 3\theta$	$\frac{1}{6}$	$\frac{9\theta - 3}{17}$	$\frac{12\theta - 5}{17}$	$\frac{55\theta - 25}{67}$	$\frac{59\theta - 28}{66}$	$\frac{245\theta - 119}{261}$

Table 6.2 Definition of $\tau(\theta)$

Lemma 6.3. *Suppose that* $m \sim M = x^\beta$ *and* $\phi_1 \leq \theta \leq \phi_8$. *Then* (6.3.4) *holds provided that* $\beta \in \mathfrak{I}(\theta)$. *In particular,* $S_2 \ll yx^{-\eta}$ *under these conditions.*

We illustrate the deterioration that occurs with both the Type I and Type II information as θ increases in the following diagram.

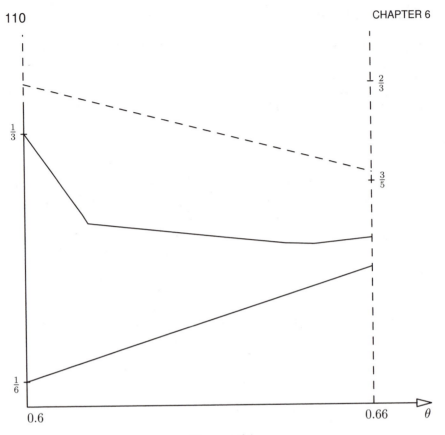

Diagram 6.1

Here we have used the vertical scale $(\log N)/(\theta \log x)$. This makes the graphs consist of piecewise-linear functions rather than sections of hyperbolae. The solid lines represent the Type II information with the vertical scale on the left. The gap between these lines represents the "width" of the Type II information. The lower line represents $y = 1 - 1/(2\theta)$. The upper line represents

$$y = \begin{cases} 2/\theta - 3 & \text{if } \theta < \frac{11}{18}, \\ 1/(6\theta) & \text{if } \frac{11}{18} < \theta < \frac{35}{54}, \\ (9\theta - 3)/(17\theta) & \text{if } \theta > \frac{35}{54}. \end{cases}$$

The broken lines represent the Type I information $y = 2/(5\theta)$ with the vertical scale on the right. The scale is again $(\log N)/(\theta \log x)$.

6.4 APPLYING THE ROSSER-IWANIEC SIEVE

From Lemma 6.2 we can see that it is possible to apply Theorem 4.2 with $z^3 = D = D_0 = x^{2/5-\epsilon}$. From (4.1.7) this delivers a bound

$$S(\theta) \leq y \frac{5}{L(1-\epsilon)}$$

that is valid for all $\theta < \frac{3}{4}$. We shall want to use this bound for as short a range as is possible since the expected value is

$$S(\theta) \sim y\frac{1}{\theta L} \leq y\frac{2}{L}$$

throughout our range of consideration ($\theta \geq \frac{1}{2}$). When $\theta < \phi_8$, it is possible to use Lemma 6.3 to increase the permissible size of D in a similar way to our use of Type II information in Section 6.1 for the αp problem. We recall our statement that the sum over d could be broken down to a double sum $d = mn, m \leq M, n \leq N$ with $MN = D$. Now, if we write $a = v/y, b = x^{\tau(\theta)}$, then we can take

$$M = b, \quad N = D_0 a^{-1}.$$

To see this, note that if $m \leq a$, then $mn \leq D_0$, which we have already noted is satisfactory. On the other hand, if $m > a$, then $\log m \in \mathcal{J}(\theta)L$ and so Lemma 6.3 supplies the required bound. We may thus apply Theorem 4.2 with $D = x^{\rho(\theta)-\epsilon}$ where ρ is given in Table 6.3 for $\theta < \phi_7$.

Interval	$\mathcal{J}_1$	$\mathcal{J}_2$	$\mathcal{J}_3$	$\mathcal{J}_4$	$\mathcal{J}_5$	$\mathcal{J}_6$
$\rho(\theta)$	$\dfrac{29-4\theta}{10}$	$\dfrac{16-15\theta}{15}$	$\dfrac{123-80\theta}{170}$	$\dfrac{103-50\theta}{170}$	$\dfrac{353-120\theta}{670}$	$\dfrac{157-35\theta}{330}$

Table 6.3 Definition of $\rho(\theta)$

For $\theta \in \mathcal{J}_7$ we have $\rho(\theta) = (1159 - 160\theta)/(2610)$ and $\rho(\theta) = \frac{2}{5}$ for $\theta \geq \phi_8$. We note that the improvement becomes much less significant as θ increases. With these values we then obtain

$$S(\theta) \leq y\frac{2}{(\rho - \epsilon)L}.$$

This can be immediately improved for $\theta \leq \phi_8$ by noting that

$$S(\theta) \leq S(\mathcal{A}, z) - \sum_{z \leq p < (ev)^{1/2}} S(\mathcal{A}_p, p), \tag{6.4.1}$$

and parts of the final sum on the right-hand side of (6.4.1) can be estimated by Lemma 6.3. For obvious reasons, these terms were called "deductibles" in [6]. Further progress can be made by analysing the sums thrown away in the construction of the upper-bound sieve and determining which parts can be given an asymptotic formula. All these asymptotic formulae regions shrink to zero as $\theta \to \phi_8$. It is possible (by extending the sequence of exponent pairs considered in [120, Corollary 3]) to obtain asymptotic formulae regions right up to $\theta = 0.75 - \epsilon$, but they are extremely small.

As an example to compare with the results of the alternative sieve, we note that for $\theta = 0.6$ (where $D = x^{1/2-\epsilon}, z \approx x^{\frac{1}{6}}$) we have a deductible

$$\sum_{x^{1/6} < p < x^{1/5}} S(\mathcal{A}_p, p) \sim \frac{y}{L} \int_{1/6}^{1/5} \frac{1 + \log(3/(5\alpha) - 2)}{\alpha(3/5 - \alpha)} d\alpha.$$

Numerical integration gives a value > 0.5435 for the integral above, so at $\theta = 0.6$ this approach yields an upper bound

$$< \frac{3.4565}{L} y.$$

The bound at this value could be further improved by taking $z = x^{1/10}$ in the application of the upper-bound sieve and using the whole range for which an asymptotic formula can be obtained.

6.5 AN ASYMPTOTIC FORMULA

We now turn our attention to (6.3.1). Using Lemmas 6.2 and 6.3 our Fundamental Theorem gives an asymptotic formula for

$$S(\mathcal{A}, ba^{-1}) \quad \text{or} \quad \sum_{r \sim R} c_r S(\mathcal{A}_r, ba^{-1}), \tag{6.5.1}$$

where $R < x^{9/10 - \theta - \epsilon}$. Now for $v < U$ we have $ba^{-1} = x^{5/2 - 4\theta} \in (a, b)$. Hence the occurrences of ba^{-1} in (6.5.1) can be replaced with b. We thus have, by Buchstab's identity,

$$S(\mathcal{A}, (ev)^{\frac{1}{2}}) = S(\mathcal{A}, b) - \sum_{b \leq p < (ev)^{1/2}} S(\mathcal{A}_p, p). \tag{6.5.2}$$

If we write $v/p = x^{\alpha}$, we have

$$\alpha \leq \theta - (2 - 3\theta) \leq 2(2 - 3\theta).$$

In other words, $v/p \leq b^2$. Hence we can replace p by b in the second sum on the right-hand side of (6.5.2) to give an asymptotic formula for the whole of the equation. This establishes (6.3.1).

6.6 THE ALTERNATIVE SIEVE APPLIED

We now consider what happens as θ increases above $0.6 - \epsilon$. As might be expected, so long as we have sufficiently wide asymptotic formula regions (furnished by Lemma 6.3), the alternative sieve will produce much better results than the Rosser-Iwaniec sieve. We henceforth will write $w = ba^{-1}$. Our starting point is (6.5.1). One drawback for further work is the restriction $R < x^{9/10 - \theta - \epsilon}$. This can be relaxed by using double sums. Working as in Theorem 5.1, we obtain the following result.

Lemma 6.4. *Let $R \leq a, S \leq x^{2/5 - \epsilon} a^{-1} = x^{0.9 - \theta - \epsilon}$. Suppose that $|c_r|, |d_s| \leq 1$. Then*

$$\sum_{r \sim R} c_r \sum_{s \sim S} d_s S(\mathcal{A}_{rs}, w) = \lambda \sum_{r \sim r} c_r \sum_{s \sim S} d_s S(\mathcal{B}_{rs}, w). \tag{6.6.1}$$

When θ is not too large, we can obtain other useful results as follows.

Lemma 6.5. *Let $\theta \in \mathcal{I}_1, evb^{-2} < P \le x^{-\epsilon}va^{-3}$, and $w < Q < a$. Then*

$$\sum_{p \sim P} \sum_{q \sim Q} S(\mathcal{A}_{pq}, q) = \lambda \left(1 + O\left(L^{-1}\right)\right) \sum_{p \sim P} \sum_{q \sim Q} S(\mathcal{B}_{pq}, q). \tag{6.6.2}$$

Proof. By Buchstab's identity

$$\sum_{p \sim P} \sum_{q \sim Q} S(\mathcal{A}_{pq}, q) = \sum_{p \sim P} \sum_{q \sim Q} S(\mathcal{A}_{pq}, w) - \sum_{p \sim P} \sum_{q \sim Q} \sum_{w \le r < q} S(\mathcal{A}_{pqr}, r). \tag{6.6.3}$$

Since $P \le x^{-\epsilon}va^{-3} \le x^{2/5-\epsilon}a^{-1}$ and $Q < a$, the first sum on the righthand side of (6.6.3) is covered by Lemma 6.4. The part of the second sum with $rq < b$ can be dealt with by Lemma 6.3 since $rq > (ba^{-1})^2 > a$ for $\theta \in \mathcal{I}_1$. The remaining part of the sum counts numbers $pqrs \in \mathcal{A}$, where

$$s < \frac{ev}{pqr} \le \frac{ev}{(ev/b^2)b} = b,$$

and

$$s > \frac{v}{8PQ^2} \ge \frac{v}{8(x^{-\epsilon}va^{-3})a^2} = \frac{x^{\epsilon}a}{8} \ge a.$$

Thus an appeal to Lemma 6.3 completes the proof. $\qquad\square$

Lemma 6.6. *For $\theta \in \mathcal{I}_1$ we have*

$$\sum_{evb^{-2} < p < x^{-\epsilon}va^{-3}} S(\mathcal{A}_p, p) = \lambda \left(1 + O\left(L^{-1}\right)\right) \sum_{evb^{-2} < p < x^{-\epsilon}va^{-3}} S(\mathcal{B}_p, p).$$

Proof. By Buchstab's identity

$$\sum_{evb^{-2} < p < x^{-\epsilon}va^{-3}} S(\mathcal{A}_p, p) = \sum_{evb^{-2} < p < x^{-\epsilon}va^{-3}} S(\mathcal{A}_p, w)$$

$$- \sum_{\substack{evb^{-2} < p < x^{-\epsilon}va^{-3} \\ w \le q < \min(p, (ev/p)^{1/2})}} S(\mathcal{A}_{pq}, q).$$

The first term on the right-hand side above is covered by (6.5.1) since $x^{-\epsilon}va^{-3} \le x^{2/5-\epsilon}(2a)^{-1}$. For the second term we note that $p > evb^{-2}$ forces $q < (ev/p)^{1/2} \le b$. For the part with $q \ge a$ we obtain an asymptotic formula via Lemma 6.3. For the remainder we can use (6.6.2). $\qquad\square$

The above results were used in [120], and the next lemma helps to extend things further. To be precise, for larger p we can show that even if an asymptotic formula cannot be established, nevertheless we can perform two more Buchstab iterations and thus greatly reduce the size of the sums thrown away.

Lemma 6.7. *Let $\theta \le 0.65 - \epsilon, P < b^2$. Then we have*

$$\sum_{p \sim P} S(\mathcal{A}_p, w) = \lambda \left(1 + O\left(L^{-1}\right)\right) \sum_{p \sim P} S(\mathcal{B}_p, w). \tag{6.6.4}$$

Proof. For the sake of simplicity we appeal to Vaughan's identity in place of mounting a two-dimensional alternative sieve attack on this problem. By partial summation it suffices to obtain an asymptotic formula for

$$\sum_{n \sim N} \Lambda(n) S(\mathcal{A}_n, w).$$

Splitting the sum over n into Type I and Type II sums, we can assume that for the Type II sums the larger variable does not exceed $P^{2/3} < x^{0.9-\theta-\epsilon}$. The smaller variable will not exceed $P^{1/2} \leq b$. Well, either the smaller variable exceeds a, in which case Lemma 6.3 is applicable, or the variable is less than a and the hypotheses for Lemma 6.4 are satisfied.

Turning to the Type I sums that arise, we can suppose that the sum over consecutive integers has length

$$> x^{0.9-\theta-\epsilon} > x^{\theta-0.4+\epsilon}$$

since $\theta < 0.65 - \epsilon$. Lemma 6.2 can therefore be applied to complete the proof. $\square$

Now we extend Lemma 6.6 by allowing $\theta > \phi_2$ and increasing the range for p. In fact, the additional errors introduced for $\theta > \phi_2$ mean that this result is only used to increase the range of p for smaller θ.

Lemma 6.8. *Suppose that $\theta \in \left[\frac{3}{5}, \frac{47}{75}\right]$ and $evb^{-2} < P < (ev)^{1/2}$. Then*

$$\sum_{p \sim P} S(\mathcal{A}_p, p) \geq \lambda(1 + O(L^{-1})) \sum_{p \sim P} S(\mathcal{B}_p, p) - \lambda S_\nabla, \qquad (6.6.5)$$

where, for $P < x^{2/5}a^{-1}$,

$$S_\nabla = \sum_\nabla S(\mathcal{B}_{pqr}, r).$$

Here ∇ is the region

$$p \sim P, w \leq r < q < a, r < (ev/pq)^{\frac{1}{2}}, rq \notin [a, b], pqr \notin [vb^{-1}, va^{-1}].$$

For $P > x^{2/5}a^{-1}$ we have

$$S_\nabla = \sum_\nabla S(\mathcal{B}_{tqu}, u),$$

with ∇ as the set of conditions

$$tq \sim P, w \leq q < a, s|t \Rightarrow s > q, w \leq u < (ev/P)^{\frac{1}{2}},$$

where no combination of the variables satisfies the requirements of Lemma 6.3.

Proof. In view of Lemma 6.6 we may assume that $P > x^{-\epsilon}va^{-3}$. When $P < x^{0.9-\theta-\epsilon}$, we have

$$\sum_{p \sim P} S(\mathcal{A}_p, p) = \sum_{p \sim P} S(\mathcal{A}_p, w) - \sum_{\substack{p \sim P \\ w \leq q < (ev/p)^{1/2}}} S(\mathcal{A}_{pq}, q).$$

The first sum on the right-hand side can be handled by (6.5.1). Since $(ev/p)^{1/2} < b$, that part of the second sum with $q > a$ is covered by Lemma 6.3. We may apply Buchstab's identity to the remaining sum to obtain

$$\sum_{\substack{p \sim P \\ w \leq q < a}} S(\mathcal{A}_{pq}, q) = \sum_{\substack{p \sim P \\ w \leq q < a}} S(\mathcal{A}_{pq}, w) - \sum_{\substack{p \sim P \\ w \leq r < q < a}} S(\mathcal{A}_{pqr}, r).$$

The first sum on the right-hand side above can be estimated via Lemma 6.4. Part of the second sum can be dealt with by Lemma 6.3. The remainder of the sum must be discarded, corresponding to the region ∇.

In the following we can now assume that $\max(evb^{-2}, x^{2/5}a^{-1}) \leq P \leq v^{1/2}$. The condition $P > evb^{-2}$ ensures that the numbers we count have the form pr, where $r < b^2$, and hence (6.6.4) can be applied after a role reversal:

$$\sum_{p \sim P} = \sum_{r} S(\mathcal{A}_r^*, (ev/r)^{\frac{1}{2}})$$

$$= \sum_{r} S(\mathcal{A}_r^*, w) - \sum_{\substack{r \\ w \leq q < (ev/r)^{1/2}}} S(\mathcal{A}_{rq}^*, q).$$

Here

$$\mathcal{A}^* = \mathcal{A}^*(r) = \{n \in \mathcal{A} : n = mr, m \sim P\}.$$

Since $r < b^2$ the first sum on the right can be evaluated. For the second sum we note that $(ev/r)^{1/2} < b$, so we can give an asymptotic formula for that part with $q \geq a$. For the remainder of that sum we reverse roles again and apply Buchstab's identity once more:

$$\sum_{\substack{r \\ w \leq q < a}} S(\mathcal{A}_{rq}^*, q) = \sum_{\substack{tq \sim P \\ w \leq q < a \\ s|t \Rightarrow s \geq q}} S(\mathcal{A}_{tq}, (ev/P)^{\frac{1}{2}})$$

$$= \sum_{\substack{tq \sim P \\ w \leq q < a \\ s|t \Rightarrow s \geq q}} S(\mathcal{A}_{tq}, w) - \sum_{\substack{tq \sim P \\ w \leq q < a \\ s|t \Rightarrow s \geq q \\ w \leq u < (ev/P)^{1/2}}} S(\mathcal{A}_{tqu}, u).$$

The first sum on the right-hand side satisfies the requirements of Lemma 6.4 for $\theta < \frac{47}{75}$. We discard those parts of the second sum for which we cannot give an asymptotic formula from Lemma 6.3. This completes the proof. $\qquad \square$

6.7 UPPER-BOUNDS: REGION BY REGION

We start off for all regions with Buchstab's identity:

$$S(\theta) = S(\mathcal{A}, w) - \sum_{w \leq p < a} S(\mathcal{A}_p, p) - \sum_{a \leq p \leq b} S(\mathcal{A}_p, p) - \sum_{b < p < (ev)^{1/2}} S(\mathcal{A}_p, p)$$

$$= S_1 - S_2 - S_3 - S_4, \quad \text{say.}$$

$$(6.7.1)$$

We can evaluate S_1 and S_3 above. The challenge is to recover as much as possible from the sums S_2, S_4. In the following we write

$$T(\theta) = \frac{L}{y} S(\theta)$$

and ignore any terms in the evaluation of $T(\theta)$ that are $O(L^{-1})$.

1. $\theta \in \mathcal{I}_1$. We first observe that $w^2 \geq a$ in this region (equality occurring at $\theta = \frac{11}{18}$). It follows that we have an asymptotic formula for any sum where two variables belong to $[w, b^{1/2}]$. Hence we can obtain an asymptotic formula for that part of S_2 with $p < b^{1/2}$. We can apply Buchstab's identity twice more to the rest of S_2, and the work in the previous section did the same for all of S_4 with the exception of the term

$$\sum_{b < p \leq evb^{-2}} S(\mathcal{A}_p, p),$$

which must be discarded. Hence

$$T(\theta) \leq \frac{1}{\theta} + \int_{2-3\theta}^{7\theta-4} w\left(\frac{\theta - \alpha}{\alpha}\right) \frac{d\alpha}{\alpha^2} + R(\theta), \qquad (6.7.2)$$

where $R(\theta)$ comes from three- or higher-dimensional integrals. To be precise, $R(\theta) = \sum_{j=1}^{4} R_j(\theta)$, where:

$$R_1(\theta) = \int_{3/2-2\theta}^{9/10-\theta} \int_{5/2-4\theta}^{\theta-1/2} \int_{5/2-4\theta}^{(\theta-\alpha-\beta)/2} w\left(\frac{\theta - \alpha - \beta - \gamma}{\gamma}\right) \frac{d\gamma \, d\beta \, d\alpha}{\alpha\beta\gamma^2};$$

$$R_2(\theta) = \int_{5/2-4\theta}^{\theta-1/2} \int_{9/10-\theta-\alpha}^{\theta/2-\alpha} w\left(\frac{\beta}{\alpha}\right) \int_{5/2-4\theta}^{(\theta-\alpha-\beta)/2} w\left(\frac{\theta - \alpha - \beta - \gamma}{\gamma}\right) \frac{d\gamma \, d\beta \, d\alpha}{\alpha^2\gamma^2};$$

$$R_3(\theta) = \int_{1-3\theta/2}^{\theta-1/2} \int_{1-3\theta/2}^{\alpha} \int_{5/2-4\theta}^{\beta} w\left(\frac{\theta - \alpha - \beta - \gamma}{\gamma}\right) \frac{d\gamma \, d\beta \, d\alpha}{\alpha\beta\gamma^2};$$

$$R_4(\theta) = \int_{\mathcal{C}_5} w\left(\frac{\theta - \alpha_1 - \cdots - \alpha_5}{\alpha_5}\right) \frac{d\alpha_1 \ldots d\alpha_5}{\alpha_1 \ldots \alpha_4 \alpha_5^2}.$$

Here we assume that no combination of the variables lies in either of the two intervals

$$[\theta - \tfrac{1}{2}, 2 - 3\theta], \qquad [4\theta - 2, \tfrac{1}{2}], \qquad (6.7.3)$$

for which asymptotic formulae hold. Also, in R_3, we suppose that there is no way of combining $\alpha, \beta, \gamma, \gamma$ into two variables f, g with $f \leq \frac{9}{10} - \theta, g \leq 2 - 3\theta$, which corresponds to there being no means of performing two more Buchstab decompositions. In practice this means $\beta + 2\gamma > 0.9 - \theta$. The region $\mathcal{C}_5$, however, corresponds to the region where such a partition is possible for $\alpha_1, \alpha_2, \alpha_3, \alpha_3$, and then

$$\tfrac{5}{2} - 4\theta \leq \alpha_5 \leq \alpha_4 \leq \alpha_3.$$

We further assume for R_4 that no combination of the variables lies in the intervals (6.7.3).

It should be noted that the quality of the upper bound obtained in this region rapidly deteriorates. For θ near to ϕ_1 we can obtain an upper bound close to $S(\mathcal{A}, (ev)^{1/2})$. However, for θ near to ϕ_2 our upper bound is closer to $S(\mathcal{A}, b)$. We thus have $T(\theta)$ changing from around

$$\frac{1}{\theta} \quad \text{to approximately} \quad \frac{\omega(\theta/(2-3\theta))}{2-3\theta}.$$

Numerical calculations yield the bound

$$\int_{\phi_1}^{\phi_2} \theta T(\theta)\, d\theta < 0.01607.$$

Here the main error (size ≈ 0.004766) comes from the second term on the right-hand side of (6.7.2). The contribution from $R(\theta)$ amounts to less than 2×10^{-4}. Some remarks on the accuracy of our calculations are needed here. On the right-hand side of (6.7.2) the first term trivially poses no problems. The second term, which produces the main error, leads to a relatively simple multidimensional integral that can be estimated to several significant figures without too much difficulty. The terms involving $R_j(\theta)$ are more difficult to estimate owing to the checks that must be built in that no combination of variables lies in an asymptotic formula region. However, these integrals are already quite small, and the higher-dimensional integral (R_4) is smaller still. Building in a generous margin of error for these terms gives confidence in the overall upper bound yet does not compromise the efficiency of the method. In the next region this principle begins to break down at $\theta = 0.665$ since the errors from higher-dimensional integrals are no longer small in comparison with the other terms.

2. $\phi_2 \leq \theta \leq \phi_4$. We now discard the whole of S_4 (the small amount that can be saved from θ just over ϕ_2 is not worth considering). We thus have

$$T(\theta) \leq \frac{1}{\tau(\theta)} \omega\left(\frac{\theta}{\tau(\theta)}\right) + R_3(\theta) + R_4(\theta) + R_5(\theta),$$

where R_3 and R_4 are the integrals corresponding to those for $\theta \in \mathcal{I}_1$, and R_5 is a sevenfold integral to match those regions when further decompositions are possible. As θ approaches ϕ_4, the terms R_4 and R_5 increase in size greatly and the alternative sieve loses its efficiency. Something more could be done here if it were practicable to consider further decompositions to 9- and 11-dimensional integrals (higher-dimensional integrals can be shown to be automatically small working as we did to obtain (3.14)). The time required to run computer programs even on fast modern PCs is inordinate in this situation, however. Someone with programming experience and a laboratory full of modern PCs should be able to attack the problem though. Calculations show that the above decomposition ceases to be more efficient than the approach in Section 6.4 between 0.665 and 0.666. Very little is then lost by using the alternative sieve up to $\frac{2}{3}$. It should be noted that $\tau(\theta)$ here is less than 0.01, so $\tau/\theta < 0.016$. This means that we have very little Type II arithmetical information. Numerical integration provides the bound

$$\int_{\phi_2}^{\phi_4} \theta T(\theta)\, d\theta < 0.12935.$$

Here the average value of $\theta T(\theta)$ is ≈ 2.3. Most of the excess over 1 comes from the fact that we are giving an upper bound for $S(\mathcal{A}, b)$ rather than $S(\mathcal{A}, (ev)^{1/2})$. The contribution from the 3-, 5-, and 7-dimensional integrals is still less than 0.009 here.

3. $\phi_4 \leq \theta \leq \phi_8$. We now work as in Section 6.4. Since $z < a$ in this region, the deductible terms are made up of sums of the form

$$\sum_{a<p<b} S(\mathcal{A}_p, p) + \sum_{\substack{z<p<a \\ a<q<b}} S(\mathcal{A}_{pq}, q) + \sum_{\substack{z<p<q<a \\ a<r<b}} S(\mathcal{A}_{pqr}, r).$$

We obtain, after some simple numerical calculations,

$$\int_{\phi_4}^{\phi_8} \theta T(\theta) \, d\theta < 0.17597.$$

4. $\phi_8 < \theta \leq \phi_9$. In this region we take $D = D_0$ in the Rosser-Iwaniec sieve and thus obtain

$$\int_{\phi_8}^{\phi_9} \theta T(\theta) < \frac{5}{2}(\phi_9^2 - \phi_8^2).$$

Since

$$\int_{\phi_1}^{\phi_8} \theta T(\theta) \, d\theta < 0.32139,$$

we require

$$\phi_9^2 < \phi_8^2 + 0.4 \times 0.07861.$$

This allows us to take any $\phi_9 < 0.7403\ldots$ and so completes the proof. □

6.8 WHY A PREVIOUS IDEA FAILS

The curious reader is probably wondering whether the idea of Jutila used for the greatest prime factor in the larger interval $[x, x + x^{1/2+\epsilon}]$ can be used here. That is, What happens if we start off by counting $mn \in \mathcal{I}$ where m is smooth and of size x^α? There is no problem in adapting Chebychev's trick to accommodate this situation. Suppose we take $\alpha = 0.03$, for example. We are then able to apply the alternative sieve up to $\theta = 0.72$ since Lemma 6.3 can be amended to provide Type II information up to $0.72 = 0.75 - 0.03$. However, the argument now becomes weaker at several points. First, instead of 1 as our ultimate target for a sum of integrals, we only have 0.97. Second, Lemma 6.2 becomes weaker: 0.40 must now be replaced by 0.37. Third, we can no longer get an asymptotic formula up to 0.6. Certainly we can obtain one beyond $0.6 - 0.03$, but the alternative sieve must still be applied for smaller θ than before. Some rough calculations show that the amount gained "on the roundabouts" is more than lost "on the swings" using this approach. Further progress on this problem requires either a completely new idea or improved exponential sums estimates it seems. Improved exponential sum estimates relevant to this problem have been provided in [151] and [168], and we hope to consider this problem again elsewhere.

Chapter Seven

Primes in Short Intervals

7.1 THE ZERO-DENSITY APPROACH

In this section we give a quick résumé of the zero-density approach to finding primes in short intervals, which the reader might find helpful in comparison with the sieve approach taken in the rest of this chapter. By Perron's formula (1.4.7) we have

$$\sum_{x-y<n\leq x} \Lambda(n) = -\frac{1}{2\pi i} \int_{c-iT}^{c+iT} \frac{\zeta'}{\zeta}(s) \frac{x^s - (x-y)^s}{s}\, ds + O\left(\frac{x \log^2 x}{T}\right). \quad (7.1.1)$$

Here $c = 1 + (\log x)^{-1}$. We shift the line of integration to $\mathrm{Re}\, s = -1$. The pole of $\zeta'(s)/\zeta(s)$ at $s = 1$ contributes y to the right-hand side of (7.1.1). On the other hand, the zeros $\rho = \beta + i\gamma$ of $\zeta(s)$ lead to a term

$$-\sum_{\substack{\rho=\beta+i\gamma \\ |\gamma|<T}} \frac{x^\rho - (x-y)^\rho}{\rho}.$$

Here we only count zeros of $\zeta(s)$ with $0 < \beta < 1$ since the "trivial zeros" are at $s = -2, -4, \ldots$. On the line $\mathrm{Re}\, s = -1$ we have

$$\frac{\zeta'}{\zeta}(s) \ll \log T$$

by (4) in [27, p. 99] (reproduced as (1.4.8) here). Hence the contribution from the shifted integral is

$$\ll \frac{(\log T)^2}{x},$$

and so is negligible.

For the "horizontal" line integrals ($\mathrm{Im}\, s = \pm T$) we assume, as we may, that T has been chosen so that no zero $\beta + i\gamma$ has $|\gamma - T| < c(\log T)^{-1}$ for some $c > 0$. This is permissible since the number of zeros between t and $t+1$ is $\ll \log t$ by [27, p. 98]. By Lemma 1.3 we have

$$\frac{\zeta'}{\zeta}(s) \ll (\log T)^2$$

on these lines. The contribution from these integrals is thus

$$\ll \frac{x(\log T)^2}{T},$$

which is precisely the error arising from truncating Perron's formula (the last term in (7.1.1)). We hence obtain

$$\sum_{x-y<n\leq x} \Lambda(n) = y - \sum_{\substack{\rho=\beta+i\gamma \\ |\gamma|<T}} \frac{x^\rho - (x-y)^\rho}{\rho} + O\left(\frac{x\log^2 x}{T}\right). \qquad (7.1.2)$$

It is then clear that, if we wish to consider primes in an interval of length $y = x^\theta$, we shall need to take T of larger order than $x^{1-\theta}(\log x)^2$. We shall therefore put $T = x^{1-\theta}(\log x)^3$.

We write $E(\sigma)$ for that part of the sum over zeros in (7.1.2) with $\sigma \leq \beta < \sigma + (1/\log x)$. It then suffices to show that

$$E(\sigma) = o\left(\frac{y}{\log x}\right). \qquad (7.1.3)$$

Now

$$\frac{x^\rho - (x-y)^\rho}{\rho} = \int_{x-y}^x u^{\rho-1}\, du.$$

Thus

$$E(\sigma) \leq 2yx^{\sigma-1}N(\sigma, T),$$

where

$$N(\sigma, T) = \sum_{\substack{\rho \\ |\gamma|<T, \beta\geq\sigma}} 1.$$

By the results of Ingham [99] and Huxley [97] we have

$$N(\sigma, T) < T^{A(1-\sigma)}(\log T)^B$$

for some B with $A = \frac{12}{5}$. We note that with current methods the critical value of σ is $\frac{3}{4}$. It is at this point that the methods of Ingham and Huxley both give $A = \frac{12}{5}$. The value of A can be made smaller for other values of σ. By the zero-free region (Lemma 1.4) we have $N(\sigma, T) = 0$ for

$$\sigma > 1 - \frac{C}{(\log|T|)^{\frac{2}{3}}(\log\log|T|)^{\frac{1}{3}}}.$$

We therefore require, for all smaller σ,

$$x^{\sigma-1}\left(x^{1-\theta}(\log x)^3\right)^{A(1-\sigma)}(\log x)^{B+1} = o(1)$$

in order to establish (7.1.3). The power of x in this expression is

$$(1-\sigma)(A(1-\theta)-1) < 0,$$

so long as $\theta > \frac{7}{12}$. We thus obtain Huxley's result [97].

Theorem 7.1. *The interval $[x-y, x]$ contains*

$$\frac{y(1+o(1))}{\log x}$$

primes when $y = x^\theta$, $\frac{7}{12} < \theta < 1$.

7.2 PRELIMINARY RESULTS

In modern approaches to problems in multiplicative number theory, such as finding primes in short intervals, we use Dirichlet polynomials throughout. We then employ the mean- and large-value results that imply the best known zero-density theorems for the Riemann zeta-function. For the problem of the greatest prime factor of the integers in an interval $[x, x + x^{1/2+\epsilon})$ we only required the mean value theorem for Dirichlet polynomials and the reflection principle. Now, when we come to investigate primes in short intervals of the form $[x - y, x]$ with y somewhat larger than $x^{1/2}$, a major tool will be the Halász-Montgomery-Huxley large-values theorem, which we state as follows. This was the important new tool that enabled Huxley to obtain the value $A = \frac{12}{5}$ that we used in the last section. In a nutshell, this result tells us that a Dirichlet polynomial cannot be too large very often.

Lemma 7.1. *Let* $\mathcal{S} = \{t_1, t_2 \ldots, t_k\} \subseteq [-T, T]$ *be a set of real numbers such that* $|t_j - t_h| \geq 1$ *when* $j \neq h$. *Let* $V > 0, F \geq 1$. *Let* $a_n \in \mathbb{C}$ *for* $n \sim F$ *and put*

$$F(s) = \sum_{n \sim F} a_n n^{-s}, \quad G = \sum_{n \sim F} \frac{|a_n|^2}{n}.$$

Suppose that

$$|F(\tfrac{1}{2} + it_j)| \geq V$$

for $t_j \in \mathcal{S}$. *Then*

$$k \ll (\log(2F))^2 \left(\frac{GF}{V^2} + \frac{G^3 FT}{V^6} \right). \tag{7.2.1}$$

Proof. This follows from the work of Huxley [97]. □

Our sieve method is employed to reduce the problem to obtaining asymptotic formulae for sums such as

$$\sum_{\substack{mnr \in \mathcal{A} \\ m \sim m, n \sim N}} a_m b_n c_r, \tag{7.2.2}$$

where $\mathcal{A} = [x - y, x]$, $x^{0.52} < y < x^{3/4}$. Here the coefficients a_m, b_n, c_r are arithmetic functions that may be no easier to deal with than the characteristic function of the set of primes, but we have already seen the advantage of having a multiple sum. These coefficients will be bounded as previously by a divisor function. In fact, one of the coefficients, say c_r, will be the characteristic function of the set of primes or of a set of numbers with a bounded number of prime factors restricted to certain ranges as arises out of the application of Theorem 5.2. This means that, writing $s = 1/2 + it$, we have, by Lemma 1.5,

$$\left| \sum_{r \sim R} c_r r^{-s} \right| \ll R^{\frac{1}{2}} \mathcal{L}^{-A} \tag{7.2.3}$$

for any $A > 0$ when

$$T_0 < |t| < T = x^{1+\epsilon}/y, \quad T_0 = \exp(\mathcal{L}^{\frac{1}{3}}), \quad \mathcal{L} = \log x,$$

and

$$c_r = \sum_{\substack{p_j \sim P_j \\ r = p_1 \dots p_k}} 1 \quad \text{with} \quad \min_{1 \le j \le k} P_j \gg \exp(\mathcal{L}^{\frac{4}{5}}).$$

We shall say that a Dirichlet polynomial $\sum a_n n^{-s}$ is *divisor-bounded* if, for every $k \in \mathbb{N}$, there exists c_k such that

$$\sum_{n \le X} |a_n|^{2k} \ll X (\log X)^{c_k} \tag{7.2.4}$$

for all large X, and $|a_n| \ll n^\eta$ for any $\eta > 0$. We shall call a Dirichlet polynomial *prime-factored* if it satisfies (7.2.3). The reader may note that a Dirichlet polynomial with coefficients identically equal to 1 is by our definition also prime-factored.

We now demonstrate that an asymptotic formula can be established for (7.2.2) using Perron's formula if

$$\int_{T_0}^{T} \left| \sum_{\substack{m \sim M, n \sim N \\ r \sim R}} a_m b_n c_r (mnr)^{-s} \right| dt \ll x^{\frac{1}{2}} \mathcal{L}^{-A} \tag{7.2.5}$$

where $x \ll MNR \ll x$.

Lemma 7.2. *Let*

$$F(s) = \sum_{k \sim x} c_k k^{-s}$$

be a divisor-bounded Dirichlet polynomial. Suppose that

$$\int_{T_0}^{T} \left| F\left(\frac{1}{2} + it\right) \right| dt \ll x^{\frac{1}{2}} \mathcal{L}^{-A} \quad (all \ A > 0). \tag{7.2.6}$$

Then

$$\sum_{k \in \mathcal{A}} c_k = \frac{y}{y_1} \sum_{k \in I_1} c_k + O(y \mathcal{L}^{-A}) \quad (all \ A > 0), \tag{7.2.7}$$

where $y_1 = x \exp(-3(\log x)^{1/3})$ and $I_1 = (x - y_1, x]$.

Proof. Our proof is based on the work of Heath-Brown [73]. Using Perron's formula we have, for $Y = y$ or y_1,

$$\frac{1}{2\pi i} \int_{1/2 - iT}^{1/2 + iT} F(s) \frac{x^s - (x - Y)^s}{s} ds = \sum_{x - Y < n < x} c_n + O\left(x^{\frac{\epsilon}{4}}\left(1 + \frac{x}{T}\right)\right) \tag{7.2.8}$$

(compare [73, p. 1372]). Since $F(s)$ is divisor-bounded, on the vertical line in question we have

$$|F(s)| \ll x^{\frac{1}{2}} \mathcal{L}^B,$$

while

$$\frac{x^s - (x - Y)^s}{s} = \begin{cases} x^{s-1} Y + O(|s| x^{-\frac{3}{2}} Y^2) & \text{if } |\operatorname{Im} s| \le T_0, \\ O(x^{-\frac{1}{2}} Y) & \text{if } |\operatorname{Im} s| \ge T_0. \end{cases}$$

Hence we obtain

$$\frac{1}{2\pi i} \int_{1/2-iT_0}^{1/2+iT_0} F(s) \frac{x^s - (x-Y)^s}{s} \, ds = YE(x) + O(T_0^2 x^{-\frac{3}{2}} Y^2 x^{\frac{1}{2}} \mathcal{L}^B)$$

$$= YE(x) + O(Y\mathcal{L}^{-A}),$$

(7.2.9)

where

$$E(x) = \frac{1}{2\pi i} \int_{1/2-iT_0}^{1/2+iT_0} F(s) x^{s-1} \, ds.$$

Combining (7.2.6), (7.2.8), and (7.2.9), we obtain

$$Y^{-1} \sum_{x-Y<n\leq x} c_n = E(x) + O(\mathcal{L}^{-A}). \qquad (7.2.10)$$

The result follows by combining the cases $Y = y$ and $Y = y_1$ of (7.2.10). $\qquad \square$

Remark. The reader will note that the above lemma is a form of comparison principle — we are comparing the number of integers in a short interval with those in a longer one. Note that if c_k picks out primes, then the right-hand side of (7.2.7) is $y(\log x)^{-1} + y(\mathcal{L})^{-A}$. Because we are dealing with short intervals we do not have to involve $\mathrm{Li}(x)$, but have a "clean" $(\log x)^{-1}$ factor.

In the following we shall usually be applying the above result when $F(s)$ has the form

$$\sum_{\substack{mnr\in B \\ m\sim M, n\sim N}} a_m b_n c_r (mnr)^{-s}$$

to obtain an asymptotic formula for (7.2.2). It is then very easy to obtain (7.2.5) under the right conditions. We recall the mean value theorem for Dirichlet polynomials (Lemma 5.2),

$$\int_{-T}^{T} \left| \sum_{n\sim N} a_n n^{-s} \right|^2 dt \ll (T+N) \sum_{n\sim N} |a_n|^2 n^{-1} \ll (T+N)\mathcal{L}^B, \quad (7.2.11)$$

assuming that we have divisor-bounded Dirichlet polynomials. Hence, an application of Cauchy's inequality and (7.2.3) to the left-hand side of (7.2.5) gives an upper bound:

$$\ll R^{\frac{1}{2}} (T+M)^{\frac{1}{2}} (T+N)^{\frac{1}{2}} \mathcal{L}^{B-A}. \qquad (7.2.12)$$

This gives (7.2.5) if

$$M > \frac{x^{1+\epsilon}}{y}, \quad N > \frac{x^{1+\epsilon}}{y}, \quad MNT_0 < x.$$

We can clearly see how decreasing y reduces the allowable ranges of values for M and N, the ranges disappearing as $y \to x^{1/2}$. If we let $N = x^\alpha$, $M = x^\beta$, $R = x^\gamma$, $y = x^{\theta+\epsilon}$ and neglect ϵ for clarity, this can be represented graphically as the triangle

$$\alpha > 1 - \theta, \quad \beta > 1 - \theta, \quad \alpha + \beta < 1.$$

It will actually be better to consider this as the triangle (T in Diagram 7.1)

$$\alpha > 1 - \theta, \ \ 0 < \gamma < \theta - \alpha$$

since we shall see that we can obtain an asymptotic formula for

$$\sum_{n \sim N} a_n S(\mathcal{A}_n, x^\gamma)$$

in this region.

The inefficiency in the above simple argument is that we only took into account the maximum value of the sum over r. We need to take into account its mean value as well as use information on the infrequency of large values of Dirichlet polynomials to make an improvement. Graphically the next result gives an asymptotic formula in a parallelogram-shaped region rather than a triangular region. There is a parameter $g \in \mathbb{N}$ to be chosen optimally here that determines the height of this parallelogram. The "width" is always $2\theta - 1$, the same as the base of the triangle T. As θ decreases, g must increase. For $\theta = \frac{7}{12}$ the optimal vlaue is $g = 1$, but this parallelogram disappears when $\theta = \frac{13}{24} = 0.541\ddot{6}$. It is already more efficient to take $g = 2$ for $\theta \le \frac{19}{33}$. Diagram 7.1 illustrates this for $\theta = \frac{7}{12} + \epsilon$. The parallelogram P contains the much smaller triangle T. Lemma 7.3 is a variant of Theorem 4 in [11], which in turn is a descendant of Lemma 2 in [85] (see also the argument in [73]). This result, and its generalizations to include characters, is crucial for the work of the rest of this chapter as well as Chapters 9 – 11.

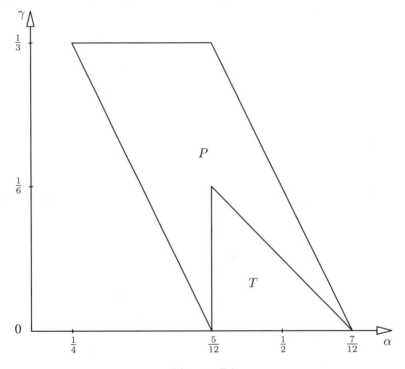

Diagram 7.1

Lemma 7.3. *Suppose that all the Dirichlet polynomials*

$$\sum_{m\sim M} a_m m^{-s}, \qquad \sum_{n\sim N} b_n n^{-s}, \qquad \sum_{r\sim R} c_r r^{-s}$$

are divisor-bounded and that $\sum_r c_r r^{-s}$ satisfies (7.2.3). Then the bound (7.2.5) holds if $M = x^{\alpha_1}, N = x^{\alpha_2}$ with

$$|\alpha_1 - \alpha_2| < 2\theta - 1, \qquad \alpha_1 + \alpha_2 > 1 - \gamma(\theta),$$

where

$$\gamma(\theta) = \max_{g \in \mathbb{N}} \gamma_g(\theta).$$

Here

$$\gamma_g = \min\left(4\theta - 2, \frac{(8g-4)\theta - (4g-3)}{4g-1}, \frac{24g\theta - (12g+1)}{4g-1}\right). \qquad (7.2.13)$$

Proof. It suffices to show that for any set $\mathcal{S}$ such that

$$t_1, t_2 \in \mathcal{S} \Rightarrow T_0 < t_1, t_2 < T, \ |t_1 - t_2| \geq 1$$

and, for $t \in \mathcal{S}$,

$$\left|\sum_{m\sim M} a_m m^{-s}\right| \sim M^{\sigma_1 - \frac{1}{2}}, \quad \left|\sum_{n\sim N} b_n n^{-s}\right| \sim N^{\sigma_2 - \frac{1}{2}}, \quad \left|\sum_{r\sim R} c_r r^{-s}\right| \sim R^{\sigma_3 - \frac{1}{2}},$$

we have

$$|\mathcal{S}| M^{\sigma_1} N^{\sigma_2} R^{\sigma_3} \ll x\mathcal{L}^{-A}. \qquad (7.2.14)$$

Note that $\sigma_j = \frac{1}{2}$ corresponds to the expected mean value — this normalization (which goes back at least as far as [85, Lemma 2]) brings out the analogy with zero-density methods. We write V for the left-hand side of (7.2.14). The number A that occurs may vary from line to line — the reader will note that we can make the A in the final result (7.2.14) as large as we like by making all the other As along the way sufficiently large. Now let $S = |\mathcal{S}|\mathcal{L}^{-B}$ for a fixed number B (which depends on which divisor function bounds the coefficients of the Dirichlet polynomials). Then the mean value theorem gives

$$SM^{2\sigma_1 - 1} \ll (M + T),$$

that is,

$$S \ll M^{2 - 2\sigma_1} + TM^{1 - 2\sigma_1}.$$

On the other hand, by Lemma 7.1 we have

$$S \ll M^{2 - 2\sigma_1} + TM^{4 - 6\sigma_1}.$$

Let

$$f(\sigma) = \min(1 - 2\sigma, 4 - 6\sigma). \qquad (7.2.15)$$

We note that equality occurs when $\sigma = \frac{3}{4}$, which turns out to be a critical value in all these proofs. Note also the connection with zero-density-type proofs. So

$$S \ll M^{2 - 2\sigma_1} + TM^{f(\sigma_1)}.$$

Similarly

$$S \ll N^{2-2\sigma_2} + TN^{f(\sigma_2)},$$

and, for any positive integer g,

$$S \ll R^{g(2-2\sigma_3)} + TR^{gf(\sigma_3)}. \tag{7.2.16}$$

The idea here is that the optimal value for g would have R^g near T in size. We now assume that g has been chosen to maximize $\gamma_g(\theta)$. We note that if we obtain three different bounds for S by the above, say S_1, S_2, S_3, then $S \ll S_1^a S_2^b S_3^c$ for any nonnegative a, b, c satisfying $a + b + c = 1$. Let $c(\sigma) = 1$ if $\sigma \le \frac{3}{4}$, $c(\sigma) = 4\sigma - 2$ if $\sigma > \frac{3}{4}$. Thus

$$M^{c(\sigma)} \ge T \Rightarrow TM^{f(\sigma)} \le M^{2-2\sigma}.$$

We now need to break the proof into different cases depending on the sizes of the σ_j. The reader should note that it is Case 2 that is most crucial and produces the expression γ_q.

Case 1. $M^{c(\sigma_1)} \ge T$, $N^{c(\sigma_2)} \ge T$. Then

$$V \ll (M^{2-2\sigma_1} N^{2-2\sigma_2})^{\frac{1}{2}} M^{\sigma_1} N^{\sigma_2} R\mathcal{L}^{-A} \ll x\mathcal{L}^{-A}$$

as required.

Case 2(i). $M^{c(\sigma_1)} \le T$, $N^{c(\sigma_2)} \le T$, $\sigma_3 \le \frac{3}{4}$. Now

$$V \ll (TM^{1-2\sigma_1} TN^{1-2\sigma_2})^{\frac{1}{2}} M^{\sigma_1} N^{\sigma_2} R^{\frac{3}{4}} = Tx^{\frac{1}{2}} R^{\frac{1}{4}} \ll x\mathcal{L}^{-A}$$

since $1 - (\alpha_1 + \alpha_2) \le 4\theta - 2$.

Case 2(ii). $M^{c(\sigma_1)} \le T$, $N^{c(\sigma_2)} \le T$, $\sigma_3 > \frac{3}{4}$. We apply (7.2.16) and take $c_1 = (2g)^{-1}, a_1 = b_1 = \frac{1}{2}(1 - c_1)$, and $c_2 = (6g)^{-1}, a_2 = b_2 = \frac{1}{2}(1 - c_2)$. Thus

$$\begin{aligned}
V &\ll (TM^{f(\sigma_1)})^{a_1}(TN^{f(\sigma_2)})^{b_1}(R^{g(2-2\sigma_3)})^{c_1} M^{\sigma_1} N^{\sigma_2} R^{\sigma_3} \\
&\quad + (TM^{f(\sigma_1)})^{a_2}(TN^{f(\sigma)})^{b_2}(TR^{g(4-6\sigma_3)})^{c_2} M^{\sigma_1} N^{\sigma_2} R^{\sigma_3} \\
&= (TM^{f(\sigma_1)})^{a_1}(TN^{f(\sigma_2)})^{b_1} M^{\sigma_1} N^{\sigma_2} R \\
&\quad + (M^{f(\sigma_1)})^{a_2}(N^{f(\sigma)})^{b_2} TR^{\frac{2}{3}} M^{\sigma_1} N^{\sigma_2} \\
&= V_1 + V_2, \quad \text{say.}
\end{aligned}$$

The reader may now see the reason for our choice of c in both cases: It removes any dependency on σ_3 in the expressions V_1, V_2.

For any $\alpha \in \left[\frac{1}{6}, \frac{1}{2}\right]$ and any real σ we have

$$\alpha f(\sigma) + \sigma \le \alpha f\left(\tfrac{3}{4}\right) + \tfrac{3}{4} = \tfrac{1}{4}(3 - 2\alpha). \tag{7.2.17}$$

This leads to a bound independent of σ_1 and σ_2. So

$$V_1 \ll T^{2a_1} R(MN)^{\frac{1}{4}(3-2a_1)} \ll T^{2a_1} x^{\frac{1}{4}(3-2a_1)} R^{\frac{1}{4}(2a_1+1)}.$$

Hence,

$$(V_1)^{8g} \ll T^{8g-4} x^{4g+1} R^{4g-1} \ll \left(x\mathcal{L}^{-A}\right)^{8g}$$

if

$$R \ll \left(x^{4g-1} T^{4-8g} \mathcal{L}^{-8g} \right)^{\frac{1}{4g-1}}.$$

This is guaranteed by

$$\gamma(\theta) \leq \frac{(8g-4)\theta - (4g-3)}{4g-1}.$$

Similarly

$$V_2 \ll T R^{\frac{2}{3}} (MN)^{\frac{1}{4}(3-2a_2)} \ll T x^{\frac{1}{4}(3-2a_2)} R^{\frac{2}{3} - \frac{1}{4}(3-2a_2)}.$$

Thus

$$(V_2)^{24g} \ll T^{24g} x^{12g+1} R^{4g-1} \ll (x\mathcal{L}^{-A})^{24g}$$

in view of the condition

$$\gamma_g(\theta) \leq \frac{24g\theta - (12g+1)}{4g-1}.$$

Case 3(i). $M^{c(\sigma_1)} \geq T > N^{c(\sigma_2)}, \sigma_3 \leq \frac{3}{4}$. In this case we have

$$V \ll M^{1-\sigma_1} (T N^{f(\sigma_2)})^{\frac{1}{2}} M^{\sigma_1} N^{\sigma_2} R^{\frac{3}{4}} \ll M(NT)^{\frac{1}{2}} R^{\frac{3}{4}} \ll x\mathcal{L}^{-A}$$

since $\alpha_1 - \alpha_2 < 2\theta - 1$.

Case 3(ii). $M^{c(\sigma_1)} \geq T > N^{c(\sigma_2)}, c(\sigma_3) > \frac{3}{4}$. We now work as in Case 2(ii), but we must let the parameter g be more general (note that we have lost the symmetry in N and M in Case 2(ii)). We obtain, with $a = \frac{1}{2}, b = \frac{1}{2} - \frac{1}{2g}, c = \frac{1}{2g}$ (missing out one horrendous-looking line!),

$$V \ll M R T^{\frac{1}{2}(1-\frac{1}{g})} N^{\frac{1}{2}+\frac{1}{4g}} + M R^{\frac{2}{3}} T^{\frac{1}{2}} N^{\frac{1}{2}+\frac{1}{12g}} = V_3 + V_4 \text{ say.}$$

To get $V_3 \ll x\mathcal{L}^{-A}$ we need

$$N \gg T^{\frac{2g-2}{2g-1}} \mathcal{L}^A.$$

We choose g so that

$$T^{\frac{2g-2}{2g-1}} \mathcal{L}^A \ll N \ll T^{\frac{2g}{2g+1}} \mathcal{L}^A.$$

This is possible since $c(\sigma) \geq 1$, so $N < T$. We further note that, if we write $3\delta = 2\theta - 1 + \alpha_2 - \alpha_1$, then $M \ll N x^{2\theta-1-3\delta}$. We then have

$$V_4 \ll M^{\frac{1}{3}} T^{\frac{1}{2}} N^{-\frac{1}{6}+\frac{1}{12g}} x^{\frac{2}{3}}$$

$$\ll T^{\frac{1}{2}} N^{\frac{1}{12g}(2g+1)} x^{\frac{2\theta+1}{3} - \delta}$$

$$\ll T^{\frac{2}{3}} \mathcal{L}^A x^{\frac{2\theta}{3} - \delta} \ll x\mathcal{L}^{-A}$$

as required.

The remaining two cases, 4(i) and 4(ii), are analogous to 3(i) and 3(ii) with the roles of M and N reversed. This completes the proof. $\qquad\square$

Clearly, we would like γ to be as large as possible here. To get $\gamma = 4\theta - 2$ (the maximum value possible) we need to choose g as an integer in the range

$$\frac{1}{4(2\theta - 1)} - \frac{1}{2} \leq g \leq \frac{1}{4(2\theta - 1)}.$$

This shows that we can only obtain the maximum permissible value when

$$\theta \in \bigcup_{g=1}^{\infty} \left[\frac{1}{2} + \frac{1}{8g + 4}, \frac{1}{2} + \frac{1}{8g} \right].$$

Later in this book, when we apply Lemma 7.3 and its cognates in other situations, we shall take $\theta = \frac{7}{12}, 0.55, 0.53, \frac{19}{36}$, or 0.525, all of which lie in the above range for θ with $g = 1, 2, 4, 4, 5$, respectively.

We illustrate some of the values obtained for $\gamma_g(\theta)$ in the Table 7.1.

θ	$\frac{3}{4}$	$\frac{2}{3}$	$\frac{7}{12}$	$\frac{11}{20}$	$\frac{21}{40}$
$\gamma_1(\theta)$	$\frac{2}{3}$	$\frac{5}{9}$	$\frac{1}{3}$	$\frac{1}{15}$	$-\frac{2}{15}$
$\gamma_2(\theta)$	$\frac{4}{7}$	$\frac{3}{7}$	$\frac{2}{7}$	$\frac{1}{5}$	$\frac{1}{35}$
$\gamma_3(\theta)$	$\frac{6}{11}$	$\frac{13}{33}$	$\frac{8}{33}$	$\frac{2}{11}$	$\frac{4}{55}$
$\gamma_5(\theta)$	$\frac{10}{19}$	$\frac{7}{19}$	$\frac{4}{19}$	$\frac{14}{95}$	$\frac{1}{10}$

Table 7.1 Values for γ_g

7.3 THE $\frac{7}{12}$ RESULT

We now apply our sieve method to the set $\mathcal{A} = \mathbb{Z} \cap [x - y, x]$ with $y = x^{\theta + \epsilon}, \theta > \frac{1}{2}$. The fact that we have an exponent $+\epsilon$ here enables us to write down inequalities involving θ with no epsilons attached (for example, (7.3.1)); we have already used this technique in Chapters 3 and 5, of course. Our comparison set, as suggested by Lemma 7.2, is $\mathcal{B} = \mathbb{N} \cap I_1$. Our first task, as always, is to ascertain what Type I and Type II information is available. In the previous section we have obtained what will substitute for Type II information in the present context (since genuine Type II information cannot be obtained here). We therefore turn to the Type I information. Consider first

$$\sum_{\substack{mn \in \mathcal{A} \\ m \sim M}} a_m.$$

This sum can be evaluated in an elementary manner when $M < x/T$, for it equals

$$\sum_{m \sim M} a_m \left(\frac{y}{m} + O(1) \right) = \frac{y}{y_1} \sum_{m \sim M} a_m \frac{y_1}{m} + O\left(\sum_{m \sim M} |a_m| \right)$$

$$= \frac{y}{y_1} \sum_{\substack{m \sim M \\ mn \in \mathcal{B}}} a_m + O\left(\sum_{m \sim M} |a_m| \right).$$

On the other hand, if $T > N > T^{1/2}$, we can use the reflection principle as in (5.6.12) with Hölder's inequality and the mean value theorem to obtain

$$\int_0^T \left| \sum_{m \sim M} a_m m^{-s} \sum_{r \sim R} b_r r^{-s} \sum_{n \sim N} n^{-s} \right| dt \ll (T+M)^{\frac{1}{2}} (T+R^2)^{\frac{1}{4}} T^{\frac{1}{4}} \mathcal{L}^B.$$

Using this mean value in tandem with Lemma 7.2 we obtain the following.

Lemma 7.4. *Let* $M \sim x^\alpha, R \sim x^\beta$ *with*

$$\max(2 - 2\theta, 2\alpha) + \max(1 - \theta, 2\beta) \leq 1 + \theta. \tag{7.3.1}$$

Suppose that a_m, c_r *are bounded by divisor functions. Then we have*

$$\sum_{\substack{mnr \in \mathcal{A} \\ m \sim M, r \sim R}} a_m c_r = \frac{y}{y_1} \sum_{\substack{mnr \in \mathcal{B} \\ m \sim M, r \sim R}} a_m c_r + O\left(y\mathcal{L}^{-A}\right)$$

for any $A > 0$.

Remarks. For $\theta = \frac{7}{12}$ this becomes

$$\max\left(\tfrac{5}{6}, 2\alpha\right) + \max\left(\tfrac{5}{12}, 2\beta\right) \leq \tfrac{19}{12}.$$

So, for example, if $\alpha < \frac{1}{2}$, we can take $\beta < \frac{7}{24}$, or if $\alpha \leq \frac{5}{12}$, we can allow $\beta \leq \frac{3}{8}$.

Until further notice we will take $\theta = \frac{7}{12}, z = x^{1/6}$. Our basic building blocks are given by the following lemma.

Lemma 7.5. *Suppose that* $1 \leq R \leq x^{1/2}$ *and the coefficients* c_r *are bounded by a divisor function. Then we have*

$$\sum_{r \sim R} c_r S(\mathcal{A}_r, z) = \frac{y}{y_1} \sum_{r \sim R} c_r S(\mathcal{B}_r, z)(1 + O(\mathcal{L}^{-A})) \tag{7.3.2}$$

for any $A > 0$. *In particular,*

$$S(\mathcal{A}, z) = \frac{y}{y_1} S(\mathcal{B}, z) \left(1 + O\left(\mathcal{L}^{-A}\right)\right).$$

Proof. Let $w = x^{1/u}$ with $u = \log \log x$. By Buchstab's identity

$$\sum_{r \sim R} c_r S(\mathcal{A}_r, z) = \sum_{r \sim R} c_r S(\mathcal{A}_r, w) - \sum_{r \sim R} c_r \sum_{w \leq p < z} S(\mathcal{A}_{rp}, p). \tag{7.3.3}$$

By Theorem 4.3 with $a = \frac{1}{4}$ and u as above, we can use the Type I information from Lemma 7.4 to obtain

$$\sum_{r \sim R} c_r S(\mathcal{A}_r, w) = \frac{y}{y_1} \sum_{r \sim R} c_r S(\mathcal{B}_r, w) \left(1 + O(\mathcal{L}^{-A})\right).$$

To be completely accurate we are really applying Theorem 4.3 twice: once with $\mathcal{A}$ and once for $\mathcal{B}$.

On the other hand, we can treat the final term on the right-hand side of (7.3.3) by our sieve method. We can divide up the range for p in a standard dyadic way:

$p \sim P$. This gives the small prime factor we need for the work in the previous section. Now suppose that

$$x^{\frac{5}{6}} \leq R^2 P \leq x^{\frac{7}{6}}. \tag{7.3.4}$$

This means that we are counting numbers rpn with

$$x^{-\frac{1}{6}} \ll \frac{r}{n} \ll x^{\frac{1}{6}}.$$

Hence we can apply Lemma 7.3 to obtain an asymptotic formula (we have $|\alpha_1 - \alpha_2| < \frac{1}{6} + \frac{1}{2}\epsilon, \alpha_1 + \alpha_2 \geq \frac{5}{6} > \frac{2}{3}$). If (7.3.4) fails, we can apply the method of proof of Theorem 3.1 to take out prime factors one by one and glue them onto the range R until (7.3.4) is satisfied. We can do this, of course, since each prime factor is no more than $x^{1/6}$. The Type I information we need is furnished by our estimates above. This would complete the proof but for a small fly in the ointment that must be removed. When we apply Perron's formula in the application of the sieve method, it interferes with the application of the formula to convert the problem into an estimation of Dirichlet polynomials (Lemma 7.2). To be precise, we end up with Dirichlet polynomials

$$\sum_m a_m m^{-\frac{1}{2}-it+iu}.$$

There are also polynomials involving a $(p + \frac{1}{2})^{iu}$ factor but we can treat these as having a "general unknown coefficient" since we only require *one* Dirichlet polynomial to be prime-factored in our method. The problem comes when $|t - u|$ is very small. We rely on being able to detect a small cancellation in such polynomials, and this requires $|t - u|$ to be greater than something like $\exp(\mathcal{L}^\epsilon)$, say. From Lemma 7.2 we have $|t| > \exp(\mathcal{L}^{1/3})$, and there will be a problem when $|t - u| < \exp(\mathcal{L}^{1/4}) = T_1$, say. Now the type of integral that occurs has the form

$$\int_{-X}^{X} \min\left(1, \frac{1}{|u|}\right) \int_{T_0}^{T} |M(s \pm iu)N(s \pm iu)R(s \pm iu)| \, dt \, du,$$

which is certainly bounded above for any case that could arise by

$$\int_{-X}^{X} \min\left(1, \frac{1}{|u|}\right) \max_{v,w} \int_{T_0}^{T} |M(s + iv)N(s + iw)R(s + iu)| \, dt \, du.$$

We need then only bound

$$\max_{v,w} \int_{T_0}^{T} |M(s + iv)N(s + iw)||R'(s) \, dt \tag{7.3.5}$$

with

$$R'(s) = \int_{-X}^{X} \min\left(1, \frac{1}{|u|}\right) |R(s + iu)| \, du.$$

By considering the cases $|t - u| \leq T_1, |t - u| > T_1$ separately we obtain

$$|R'(s)| \ll R^{\frac{1}{2}}\mathcal{L}^{1-A} + R^{\frac{1}{2}}\mathcal{L}^B T_1 T_0^{-1} \ll R^{\frac{1}{2}}\mathcal{L}^{-A},$$

with our convention on A. Since the mean- and large-value results are unaffected by the presence of an additional n^{iu} factor, we can therefore obtain the same result as Lemma 7.3 for (7.3.5). The fly in the ointment has then been removed, and the proof is complete. We shall not refer again to this technique — whenever Perron's formula is applied elsewhere in this or later chapters, we may have to resort to this method to give a rigorous proof.   □

Now if we write

$$A_r^* = \{mn \in A_r : m \sim M, n \sim N\}$$

with $MN > x^{1/2}$, then we can apply the two-dimensional version of our results to treat

$$\sum_{r \sim R} a_r S(A_r^*, z). \tag{7.3.6}$$

Defining $\mathcal{B}_r^*$ analogously we have the following result.

Lemma 7.6. *Suppose that* $1 \leq R \leq x^{1/2}$, *and the coefficients* c_r *are bounded by a divisor function. Then*

$$\sum_{r \sim R} c_r S(A_r^*, z) = \frac{y}{y_1} \sum_{r \sim R} c_r S(\mathcal{B}_r^*, z) \left(1 + O\left(\mathcal{L}^{-A}\right)\right) \tag{7.3.7}$$

for any $A > 0$.

Proof. The proof goes through as in the case of Lemma 7.5 except for the Type I information. We now have the possibility of two Type I ranges. We use (applying the reflection principle to either or both of the summations over m and n if necessary)

$$\int_0^T \left| \sum_{m \sim M} m^{-s} \sum_{k \sim K} b_k k^{-s} \sum_{n \sim N} n^{-s} \right| dt \ll (T + K^2)^{\frac{1}{2}} T^{\frac{1}{2}} \mathcal{L}^B. \tag{7.3.8}$$

This gives a suitable bound so long as $K \leq x^\theta$. This is adequate for our purposes since the basic principle is still to build up the variable r until

$$x^{\frac{5}{6}} < r^2 p < x^{\frac{7}{6}}.$$

When we prove the general version of this result later in Section 7.6, we shall supply all the details for completeness.   □

Let $X = x^{1/2}$. We begin in our usual way:

$$S(A, X) = S(A, z) - \sum_{z \leq p < X} S(A_p, z) + \sum_{\substack{z \leq q < p < X \\ pq^2 < x}} S(A_{pq}, q)$$

$$= S_1 - S_2 + S_3, \quad \text{say.}$$

We have an asymptotic formula for S_1 and S_2 as above. We now consider various cases for S_3. Some of our cases will overlap: the important point is that there is *at least* one way to treat each part of the sum. We write $U = x^{1/3}$. The significance of this parameter is that we need to count products of three numbers with the smallest being $\leq U$ if we are to apply Lemma 7.3.

(A) $q \leq U, z^5 \leq p^2 q \leq z^7$. This can be dealt with by Lemma 7.3 since, if $pqr \in \mathcal{B}$, then $z^{-1} \ll p/r \ll z$ under these conditions. Indeed, this corresponds to the parallelogram P in Diagram 7.1.

(B) $pq > X, p < z^2$, or $pq > z^4$. In either of these two situations there is a product of two variables exceeding X, but neither exceeds z^2. To see this, note that if $pq > z^4$, then $p > z^2, q < z^2$, and $x/pq < z^2$. We can thus estimate these cases with the two-dimensional result (7.3.6).

(C) $pq > X, r > z^2, p > z^2$. This includes the only remaining case for $p > z^2$ not covered by (A) or (B). We can estimate this as

$$\sum_p S(\mathcal{A}_p^*, z) - \sum_{\substack{pqrs \in \mathcal{A} \\ z \leq r < s}} 1.$$

The first term above is covered by (7.3.7) since $p < X$. We now consider the sum over p, q, r, s. Since $p > z^2$ and $qs < x/(zp)$ we have $qs/p < z$. Also, $p/(qs) < X/z^2 = z$, so the sum over p, q, r, s is covered by our Type II information. This completes the discussion of this case.

(D) $pq < X$. For this case we apply Buchstab's identity once more:

$$\sum_{p,q} S(\mathcal{A}_{pq}, q) = \sum_{p,q} S(\mathcal{A}_{pq}, z) - \sum_{p,q,r} S(\mathcal{A}_{pqr}, r).$$

The first term is covered by the fundamental theorem. The second term counts numbers $pqrs$ where p, q, r are all primes, and s has all its prime factors exceeding r. Now the product of the two largest variables exceeds X, and so we can work in case (C) using (7.3.7) with any other terms introduced being covered by our Type II information.

Huxley's theorem (Theorem 7.1) then follows by combining the above results. We illustrate the regions (B), (C), (D) in Diagram 7.2 with $p = x^{\alpha_1}, q = x^{\alpha_2}$. For the sake of clarity we have not included (A), which overlaps with some of them. Working more carefully (using log powers rather than epsilons and checking how the method breaks down) allows one to recover Heath-Brown's result that one can take $y = x^{7/12}$. For $\theta < \frac{7}{12}$ one obtains

$$S(\mathcal{A}, X) \geq \frac{y}{y_1} S(\mathcal{B}, X) \left(1 - C \left(\frac{7}{12} - \theta\right)^4\right).$$

The worst-case sums that must be discarded at the $\frac{7}{12}$ breakdown point have the form $pqrst \in \mathcal{A}$ with four variables near z and the final one near z^2. There are various other Dirichlet polynomial mean-value results that can be proved as θ reduces and the method becomes increasingly complicated. A thorough discussion of the case $\theta = \frac{11}{20}$ will be given in Chapter 10. The next sections describe the proof of the current "world record" for this problem.

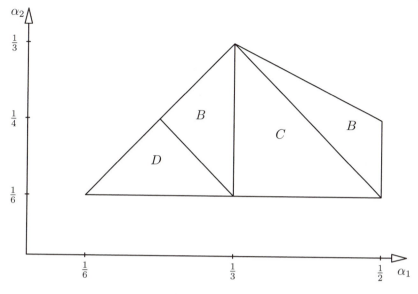

Diagram 7.2

We finish this section with a brief discussion of what as yet unproved results would be useful for the discussion of this problem. The reader has seen the application of the two-dimensional version of the method and presumably has thought "Why not use a three- or higher-dimensional version?" There seems to be no benefit in doing this since we do not have any results having the strength of the sixth-power moment of the Riemann zeta-function. However, if we assumed this, or even the eighth-power moment, this would give extra very useful arithmetical information. It should not take much imagination to envisage that the Lindelöf hypothesis very quickly leads to the value $\theta = \frac{1}{2}$ with an asymptotic formula. To do this we use high-dimensional (essentially up to $[\epsilon^{-1}]$) versions of the method. The very weak Type II sum information available is counterbalanced by the strong Type I_j information, which means that all variables can be assumed either to run over consecutive integers or to be made up of very small prime factors.

7.4 SHORTER INTERVALS

We shall now use our approach to prove the following result of Baker, Harman, and Pintz [12]. In this work the use of Watt's theorem [164] is significant in making an approximation to the conjectured sixth-power moment of the Riemann zeta-function.

Theorem 7.2. *For all $x > x_0$, the interval $[x - x^{0.525}, x]$ contains prime numbers.*

We continue with the notation introduced in previous sections. In particular, we remind the reader that $\mathcal{L} = \log x$, $y_1 = x \exp(-3\mathcal{L}^{1/3})$, $y = x^{\theta+\epsilon}$,

$$\mathcal{A} = [x - y, x] \cap \mathbb{Z}, \quad \mathcal{B} = [x - y_1, x] \cap \mathbb{Z},$$

where ϵ is a sufficiently small positive number. Here, however, we assume that

$$0.524 < \theta \leq 0.535. \tag{7.4.1}$$

We begin by using Buchstab's identity twice to reach the decomposition

$$S(\mathcal{A}, x^{\frac{1}{2}}) = S(\mathcal{A}, x^{\nu(0)}) - \sum_{\nu(0) \leq \alpha_1 < 1/2} S(\mathcal{A}_{p_1}, x^{\nu(\alpha_1)})$$

$$+ \sum_{\substack{\nu(0) \leq \alpha_1 < 1/2 \\ \nu(\alpha_1) \leq \alpha_2 < \min(\alpha_1, (1-\alpha_1)/2)}} S(\mathcal{A}_{p_1 p_2}, p_2). \tag{7.4.2}$$

$$= \sum_1 - \sum_2 + \sum_3, \text{ say.}$$

(Here $p_j = x^{\alpha_j}$.) We give asymptotic formulae for $\sum_1, \sum_2$. The piecewise-linear function $\nu(\dots)$ is at least as large as $2\theta - 1$, which is the value suggested by our work in Lemma 7.5. From this point on, role reversals are employed as described in Section 5.3. In the rest of this chapter we will write $m \asymp M$ to mean $B^{-1}M < m < BM$, where B is a positive absolute constant, that need not have the same value at each occurrence.

As we have already seen (Chapter 3.5), it will generally be beneficial to attempt as many decompositions as possible. There are two reasons for this. First, if there are several variables, there should often be a combination of variables that satisfy one of our criteria for obtaining an asymptotic formula. Second, if there are many variables, the contribution is already quite small. To see this, if $*$ represents $x^\nu < p_n < p_{n-1} < \cdots < p_1 < x^\lambda$, then

$$\sum{}^{*} S(\mathcal{B}_{p_1 \dots p_n}, p_n) = \frac{y_1}{\mathcal{L}}(1 + o(1))\eta,$$

where

$$\eta = \int_{\alpha_1 = \nu}^{\lambda} \int_{\alpha_2 = \nu}^{\alpha_1} \cdots \int_{\alpha_n = \nu}^{\alpha_{n-1}} \omega\left(\frac{1 - \alpha_1 - \cdots - \alpha_n}{\alpha_n}\right) \frac{d\alpha_1 \dots d\alpha_n}{\alpha_1 \dots \alpha_n^2}$$

$$\leq \frac{(\log(\lambda/\nu))^n}{n!\nu} < \left(3 \times 10^{-5}\right) y\mathcal{L}^{-1}$$

for $\theta = 0.525$ when $\nu \geq 0.05, n = 8$, and $\lambda = \frac{1}{10}$. This means the contribution from the terms with $p_1 \leq x^{1/10}$ is automatically very small.

However, when role reversals are used, it may not always be beneficial to perform as many decompositions as possible. The reason for this is that with role reversals, a sum may be replaced by the difference of two sums, each substantially larger than the original one. If not enough combinations of variables lie in "asymptotic formula regions," we have made matters worse. For example, when decomposing in straightforward fashion, we count

$$p_1 \dots p_n m, p|m \Rightarrow p > p_n.$$

When role reversals are used, we may have

$$p_1 \dots p_n k\ell m, p|k \Rightarrow p > p_r, p|\ell \Rightarrow p > p_s, p|m \Rightarrow p > p_n.$$

The first expression gives rise to a term

$$\omega\left(\frac{1-\alpha_1-\cdots-\alpha_n}{\alpha_n}\right)\frac{1}{\alpha_1\ldots\alpha_{n-1}\alpha_n^2},$$

while the second leads to a term

$$\omega\left(\frac{f_1}{\alpha_r}\right)\omega\left(\frac{f_2}{\alpha_3}\right)\omega\left(\frac{f_3}{\alpha_n}\right)\frac{1}{(\alpha_1\ldots\alpha_n)\alpha_r\alpha_s\alpha_n}$$

for certain expressions f_1, f_2, f_3. The corresponding integral can then be larger than the original term under consideration.

7.5 APPLICATION OF WATT'S THEOREM

Let $T = x^{1-\theta-\epsilon/2}$. In this section we seek to extend the work we did in Section 7.2 on Type I information. Henceforth the appearance of A in the exponent of a logarithm will always signify that the result holds for every $A > 0$ with an implied constant that depends on A. For simplicity we shall appeal to the Heath-Brown identity (2.5.2) in the following rather than cascading the applications of our sieve method. We shall thus require Type I information with the variable possibly weighted with a $\log n$ factor. Hence we shall be concerned, as previously, with results of the type (Re $s = \frac{1}{2}$ still)

$$\int_{T_0}^{T} |M(s)N(s)K(s)|\,dt \ll x^{1/2}\mathcal{L}^{-A}, \tag{7.5.1}$$

where $M(s)$ and $N(s)$ are Dirichlet polynomials,

$$M(s) = \sum_{m\sim M} a_m m^{-s}, \quad N(s) = \sum_{n\sim N} b_n n^{-s},$$

and $K(s)$ is a Type I sum in our context (a partial sum for either $\zeta(s)$ or $\zeta'(s)$); that is,

$$K(s) = \sum_{k\sim K} k^{-s} \quad \text{or} \quad \sum_{k\sim K}(\log k)k^{-s}. \tag{7.5.2}$$

If there is a $\log k$ factor, then partial summation may be necessary to remove it in the following; we leave the details for the reader. We assume, as we may, that all Dirichlet polynomials that arise are divisor-bounded. We also write

$$\left\|A_1^{a_1}\ldots A_n^{a_n}\right\|_k = \left(\int_{T_0}^{T}|A_1(s)^{a_1}\ldots A_n(s)^{a_n}|^k\,dt\right)^{\frac{1}{k}}.$$

With this notation we observe that if $K(s)$ is of type (7.5.2) then by the reflection principle or fourth-power moment of the Riemann zeta-function we have

$$\left\|K^4\right\|_1 \ll T^{1+\epsilon}.$$

We would like this result to hold for higher powers of $K(s)$. This would enable us to perfom more iterations of Buchstab's identity, and so the sums thrown away would have large numbers of variables and therefore be "small." The following is the crucial result due to Watt [164] that enables us to perform extra Buchstab decompositions because it enables us to "squeeze" a smallish extra factor in with the $K(s)$ term, thus enabling more flexibility with the remaining factors (since the product of their lengths is now reduced by the size of the factor included with $K(s)$).

Lemma 7.7. *If $K(s)$ is of type (7.5.2), then*

$$\left\| M^2 K^4 \right\|_1 \ll T^{1+\epsilon}(1 + M^2 T^{-\frac{1}{2}}). \tag{7.5.3}$$

The following is a consequence of Heath-Brown's identity that we use to break up certain sums over primes.

Lemma 7.8. *Let*

$$N(s) = \sum_{p_i \sim P_i} (p_1 \dots p_u)^{-s}, \tag{7.5.4}$$

where $u \leq B$, $P_1 \dots P_u \leq x$. Then, for $\operatorname{Re} s = \frac{1}{2}$,

$$|N(s)| \leq g_1(s) + \dots + g_r(s), \ r \leq \mathcal{L}^B, \tag{7.5.5}$$

where each g_i is of the form

$$\mathcal{L}^B \prod_{i=1}^{h} |N_i(s)|, \ h \leq B, \tag{7.5.6}$$

and among the Dirichlet polynomials $N_1, \dots, N_h$ the only polynomials of length $> T^{1/2}$ are of type (7.5.2).

Proof. It clearly suffices to prove (7.5.5) for

$$N(s) = \sum_{n \sim N} \Lambda(n) n^{-s},$$

where Λ is von Mangoldt's function. The result then follows by (2.5.2). □

We now apply Lemma 7.7 in tandem with Lemma 7.8 in the following lemmas that appeared in [12]. We reproduce the proofs from there nearly verbatim.

Lemma 7.9. *Let $MN_1N_2K = x$ and suppose that $K(s)$ is of type (7.5.2). Let $M = x^\alpha, N_j = x^{\beta_j}$ and suppose that*

$$\alpha \leq \theta, \ 2\beta_1 + \beta_2 \leq 1 - \theta - 2\alpha'. \tag{7.5.7}$$

Here and subsequently, $\alpha' = \max(\alpha, 1 - \theta)$. Suppose further that

$$\beta_2 \leq \frac{1 + 3\theta}{4} - \alpha' \tag{7.5.8}$$

and

$$\beta_1 + \frac{3\beta_2}{2} \leq \frac{3 + \theta}{4} - \alpha'. \tag{7.5.9}$$

Then

$$\left\| MN_1N_2K \right\|_1 \ll x^{\frac{1}{2}} \mathcal{L}^{-A}. \tag{7.5.10}$$

Proof. By Hölder's inequality the left-hand side of (7.5.10) is

$$\leq \left\| M \right\|_2 \left\| N_1 N_2^{\frac{1}{2}} \right\|_4 \left\| K N_2^{\frac{1}{2}} \right\|_4$$
$$\ll x^{\frac{\epsilon}{50}} (M + T)^{\frac{1}{2}} (N_1^2 N_2 + T)^{\frac{1}{4}} T^{\frac{1}{4}} (1 + N_2^2 T^{-\frac{1}{2}})^{\frac{1}{4}}$$
$$\ll x^{\gamma}$$

by Lemma 7.7 and the mean value theorem (Lemma 5.2). Here

$$\gamma = \frac{\alpha'}{2} + \frac{1}{4} \max(2\beta_1 + \beta_2, 1 - \theta) + \frac{1 - \theta}{4}$$
$$+ \max \left(0, \frac{\beta_2}{2} - \frac{1 - \theta}{8} \right) - \frac{\epsilon}{25}.$$

The conditions (7.5.7) – (7.5.9) guarantee that $\gamma \leq \frac{1}{2} - \epsilon/25$. □

Lemma 7.10. *The bound (7.5.10) holds if the conditions on β_j in Lemma 7.9 are replaced by*

$$either \quad \beta_1 \leq \frac{1 - \theta}{2} \quad or \; N_1 \; is \; of \; the \; form \; (7.5.2)$$
$$and \quad \beta_2 \leq \frac{1 + 3\theta}{8} - \frac{\alpha'}{2}. \tag{7.5.11}$$

Proof. By Hölder's inequality the left-hand side of (7.5.10) is

$$\leq \left\| M \right\|_2 \left\| N_1 \right\|_4 \left\| K N_2 \right\|_4 \ll x^{\delta},$$

where

$$\delta = \frac{\alpha'}{2} + \frac{1 - \theta}{2} - \frac{\epsilon}{10} + \frac{1}{4} \max \left(0, 4\beta_2 - \frac{1 - \theta}{2} \right).$$

(If $\beta_1 \leq (1 - \theta)/2$, the mean value theorem yields $\left\| N_1 \right\|_4 \ll x^{(1-\theta)/4}$; if N_1 is of type (7.5.2), the same bound follows using the reflection principle if necessary.) Substituting in (7.5.11) gives $\delta < \frac{1}{2} - \epsilon/10$ as required. □

Lemma 7.11. *Make the hypotheses of Lemma 7.8 and let $K(s)$ be of type (7.5.2). Suppose that $M = x^{\alpha}$, $N = x^{\beta}$, $\alpha \leq \theta$, and*

$$\beta \leq \min \left(\frac{3\theta + 1 - 4\alpha'}{2}, \frac{3 + \theta - 4\alpha'}{5} \right). \tag{7.5.12}$$

Then

$$\left\| MNK \right\|_1 \ll x^{\frac{1}{2}} \mathcal{L}^{-A}. \tag{7.5.13}$$

This result also holds if $N(s) \equiv 1$.

Proof. Let

$$a = \min \left(2\theta - 2\alpha', \frac{1 - 3\theta + 2\alpha'}{5} \right).$$

We may suppose that

$$\beta > \frac{1 - \theta}{2}$$

since otherwise the result follows from Lemma 7.10 with $\beta_2 = 0$.

In view of Lemma 7.8, we may suppose that, with the notation $N_j = x^{\delta_j}$, we have

$$N = N_1 \cdots N_t, \; \delta_1 \leq \cdots \leq \delta_t,$$

where any N_j with $\delta_j > (1-\theta)/2$ is of type (7.5.2). We now consider two cases which we will show cover all possible values of α and β.

Case 1. There is a subproduct x^δ of $N_1 \ldots N_t$ that either is of type (7.5.2) or has $\delta \leq (1-\theta)/2$. Moreover,

$$\beta - \delta \leq a.$$

If $\alpha' \geq (13\theta - 1)/12$, then

$$a \leq 2\theta - 2\alpha' \leq \frac{1 + 3\theta}{8} - \frac{\alpha'}{2},$$

while if $\alpha' < (13\theta - 1)/12$, then

$$a \leq \frac{1 - 3\theta + 2\alpha'}{5} \leq \frac{1 + 3\theta}{8} - \frac{\alpha'}{2}.$$

Now (7.5.13) follows on applying Lemma 7.10.

Case 2. There is a subproduct x^δ of $N_1 \ldots N_t$ such that

$$a \leq \delta \leq \beta - a.$$

Let $\beta_2 = \min(\delta, \beta - \delta)$; then $\beta_2 \in (a, \beta/2]$. Let $\beta_1 = \beta - \beta_2$: then

$$\beta_1 + \frac{\beta_2}{2} = \beta - \frac{\beta_2}{2} \leq \beta - \frac{a}{2} \leq \frac{\theta + 1}{2} - \alpha'.$$

Moreover,

$$\beta_2 \leq \frac{1}{2}\left(\frac{3\theta + 1 - 4\alpha'}{2}\right) = \frac{3\theta + 1}{4} - \alpha'$$

and

$$\beta_1 + \frac{3\beta_2}{2} \leq \frac{5\beta}{4} \leq \frac{3 + \theta - 4\alpha'}{4}.$$

Now (7.5.13) follows from Lemma 7.9.

We may now complete the proof of the lemma. If $\delta_t \leq a$, there is evidently a subsum of $\delta_1 + \cdots + \delta_t$ in $[a, 2a]$. Now

$$2a \leq \beta - a$$

since if $\alpha' \geq (13\theta - 1)/12$, then

$$3a \leq 6\theta - 6\alpha' \leq \frac{1 - \theta}{2} < \beta,$$

while if $\alpha' < (13\theta - 1)/12$, then

$$3a \leq \frac{3 - 9\theta + 6\alpha'}{5} \leq \frac{1 - \theta}{2} < \beta.$$

Thus Case 2 holds when $\delta_t \leq a$, and of course Case 2 holds also when $a < \delta_t \leq \beta - a$.

Finally, suppose that $\delta_t > \beta - a$; then we are in Case 1 with $\delta = \delta_t$. This completes the proof. $\qquad \square$

7.6 SIEVE ASYMPTOTIC FORMULAE

Now we use the results from the previous section to establish asymptotic formulae
for sums like

$$\sum_{m \sim M} a_m \sum_{n \sim N} b_n S(\mathcal{A}_{mn}, w) \tag{7.6.1}$$

over larger ranges than possible in Section 7.2. As in the proof of Lemma 7.5, we
take

$$w = \exp\left(\frac{\log x}{\log \log x}\right).$$

Lemma 7.12. *Let* $M(s)$, $N(s)$, $M = x^\alpha$, $N = x^\beta$, *with* $\alpha \le \theta - \epsilon$, *both satisfy
the hypotheses of Lemma 7.8 with*

$$\beta \le \min\left(\frac{3\theta + 1 - 4\alpha'}{2}, \frac{3 + \theta - 4\alpha'}{5}\right) - 2\epsilon. \tag{7.6.2}$$

Then we can obtain an asymptotic formula for (7.6.1).

Proof. The proof is analogous to that of Lemma 7.5. We now must take $a = \epsilon/2$ in
Lemma 4.3, but the main difference is that the Type I/II information is here given
by Lemma 7.11. $\qquad\square$

We now use combinatorial arguments so that Lemma 7.3 (with γ modified to the
value taken in (7.12) with $g = 5$) can be applied. The basic idea is that we want to
have two variables n, p satisfying

$$n^2 p \in [x/z, xz], \qquad w \le p \le x^\gamma,$$

where $z = x^{2\theta - 1}$, while ensuring that the Type I information is valid when $n^2 p <
x/z$. If $n = rp_1 \ldots p_k$, with $(rp_1 \ldots p_{k-1})^2 p_k < x/z$, this translates to conditions
of the form

$$\alpha + \alpha_1 + \cdots + \alpha_{k-1} + \tfrac{1}{2}\alpha_k \le 1 - \theta, \tag{7.6.3}$$

if $r = x^\alpha, p_j = x^{\alpha_j}$. For $h \ge 1$ we write

$$I_h = \left[\tfrac{1}{2} - 2h(\theta - \tfrac{1}{2}), \tfrac{1}{2} - (2h - 2)(\theta - \tfrac{1}{2})\right).$$

We shall find that the argument changes depending on which interval contains the
variable α. We define a function α^* as follows on $\left[0, \tfrac{1}{2}\right]$. If $\alpha \in I_h$ where $h \ge 2$,
then

$$\alpha^* = \max\left(\frac{2h(1 - \theta) - \alpha}{2h - 1}, \frac{2(h - 1)\theta + \alpha}{2h - 1}\right).$$

Note that

$$1 - \theta \le \frac{h - \theta}{2h - 1} \le \alpha^* \le \frac{1}{2}.$$

For $\alpha \in \left[\tfrac{3}{2} - 2\theta, 1 - \theta\right]$, we define

$$\alpha^* = 2(1 - \theta) - \alpha,$$

and for $\alpha \in \left(1 - \theta, \tfrac{1}{2}\right]$ we define $\alpha^* = \alpha$. In all cases,

$$\alpha^* \in \left[1 - \theta, \tfrac{1}{2}\right].$$

Lemma 7.13. *Let* $\alpha \geq 0, \alpha \in I_h = \left[\frac{1}{2} - 2h(\theta - \frac{1}{2}), \frac{1}{2} - (2h - 2)(\theta - \frac{1}{2})\right),$ *where* $h \geq 2$. *Let* $\alpha_1, \ldots, \alpha_{k+1}$ *be positive numbers,* $k \geq 0$, *satisfying* (7.6.3) *and*

$$\alpha_k \leq \cdots \leq \alpha_1 \leq \frac{2(\theta - \alpha)}{2h - 1}. \tag{7.6.4}$$

Then

$$\alpha + \alpha_1 + \cdots + \alpha_k \leq \alpha^*. \tag{7.6.5}$$

Proof. We distinguish two cases.
Case 1: $k \geq h$. Since the α_j are decreasing,

$$\alpha + \left(h - \tfrac{1}{2}\right)\alpha_k \leq 1 - \theta,$$

and so

$$\alpha_k \leq \frac{1 - \theta - \alpha}{h - \frac{1}{2}}.$$

Hence we have

$$\alpha + \alpha_1 + \cdots + \alpha_k \leq 1 - \theta + \frac{\alpha_k}{2} \leq 1 - \theta + \frac{1 - \theta - \alpha}{2h - 1}$$
$$= \frac{2h(1 - \theta) - \alpha}{2h - 1}$$

to complete the proof in this case.
Case 2: $k < h$. From (7.6.4) we have

$$\alpha + \alpha_1 + \cdots + \alpha_k \leq \alpha + (h - 1)\frac{2(\theta - \alpha)}{2h - 1} = \frac{2(h - 1)\theta + \alpha}{2h - 1}, \tag{7.6.6}$$

as desired. $\qquad\qquad\qquad\qquad\qquad\qquad\qquad\qquad\qquad\qquad\qquad\square$

We are now in a position to obtain asymptotic formulae (7.6.8) in quite large ranges. First, we point out what we would obtain from the argument in Lemma 7.5. There is no problem changing the parameters in the proof of that result to establish the formula

$$\sum_{r \sim R} c_r S(\mathcal{A}_r, z) = \frac{y}{y_1} \sum_{r \sim R} c_r S(\mathcal{B}_r, z)(1 + O(\mathcal{L}^{-A}))$$

for $R \leq x^{1/2}$. The next lemma extends this result using the Type I information we have gathered, and the subsequent lemma increases the permissible value for z.

Lemma 7.14. *Let* $\alpha \in [0, \frac{1}{2}],$

$$0 \leq \beta \leq \min\left(\frac{3\theta + 1 - 4\alpha^*}{2}, \frac{3 + \theta - 4\alpha^*}{5}\right) - 2\epsilon.$$

Let $M = x^\alpha, N = x^\beta$, *where* $N(s)$ *satisfies the hypotheses of Lemma 7.8. Then we have*

$$\sum_{m \sim M} \sum_{n \sim N} a_m b_n S(\mathcal{A}_{mn}, z) = \frac{y}{y_1}(1 + o(1)) \sum_{m \sim M} \sum_{n \sim N} a_m b_n S(\mathcal{B}_{mn}, z). \tag{7.6.7}$$

Proof. We begin in the standard way by writing

$$\sum_{m \sim M} \sum_{n \sim N} a_m b_n S(\mathcal{A}_{mn}, z) = \sum_{m \sim M} \sum_{n \sim N} a_m b_n S(\mathcal{A}_{mn}, w)$$

$$- \sum_{m \sim M} \sum_{n \sim N} a_m b_n \sum_{w \le p < z} S(\mathcal{A}_{mnp}, p)$$

$$= S_1 - S_2, \quad \text{say.}$$

From Lemma 7.12 we can obtain an asymptotic formula for S_1. We divide up the p range in a standard dyadic manner, so we can assume that $p \sim P$. If $M^2 P \ge x/z$, then $M^2 P \le xz$, so we can obtain an asymptotic formula for S_2. Otherwise we apply Buchstab's identity again to S_2:

$$S_2 = \sum_{m \sim M} \sum_{n \sim N} a_m b_n \sum_{p \sim P} S(\mathcal{A}_{mnp}, w)$$

$$- \sum_{m \sim M} \sum_{n \sim N} a_m b_n \sum_{p_1 \sim P} \sum_{w \le p_2 < p_1} S(\mathcal{A}_{mnp_1 p_2}, p_2).$$

We can continue until we have $(mp_1 \ldots p_k)^2 p_{k+1} \in [x/z, xz]$. At each stage we have the correct information to obtain a formula for

$$\sum_{m \sim M} \sum_{n \sim N} a_m b_n \sum_{p_1, \ldots, p_k} S(\mathcal{A}_{mnp_1 \ldots p_k}, w)$$

by Lemmas 7.13 and 7.12. As usual, reversing the decomposition produces the right-hand side of (7.6.7). $\qquad \square$

Lemma 7.15. *Given the hypotheses of Lemma 7.14, write*

$$\nu(\alpha) = \min\left(4\theta - 2, 2\theta - 2\alpha, \frac{36\theta - 17}{19}\right) \quad \text{for } \alpha \in I_1,$$

$$\nu(\alpha) = \min\left(\frac{2}{2h-1}(\theta - \alpha), \frac{36\theta - 17}{19}\right) \quad \text{for } \alpha \in I_h, h \ge 2.$$

Then

$$\sum_{m \sim M} \sum_{n \sim N} a_m b_n S(\mathcal{A}_{mn}, x^\nu) = \frac{y}{y_1}(1 + o(1)) \sum_{m \sim M} \sum_{n \sim N} a_m b_n S(\mathcal{B}_{mn}, x^\nu)$$

$$(7.6.8)$$

holds for every $\nu \in [\epsilon, \nu(\alpha)]$.

Remarks. It is clear that $\nu(\alpha) \ge 2\theta - 1$, the value given by Lemma 7.14. The upper bound on β (inherited from Lemma 7.14) never falls below $(3\theta - 1)/2 - 2\epsilon$. For the value $\theta = 0.525$ the values for $\nu(\alpha)$ are as follows:

$$\nu(\alpha) = \min(0.1, 1.05 - 2\alpha) \quad \text{for } \alpha \in I_1,$$

$$\nu(\alpha) = \frac{2(0.525 - \alpha)}{2h - 1} \quad \text{for } \alpha \in I_h, h \ge 2.$$

The values for $\nu(\alpha)$ can be seen in Diagram 7.3.

Proof. The reader will readily verify that the argument of Lemma 7.14 regarding the validity of the Type I information carries over for the proof of this result, so we shall concentrate only on the new working. We have

$$\sum_{m \sim M} \sum_{n \sim N} a_m b_n S(\mathcal{A}_{mn}, x^\nu) = \sum_{m \sim M} \sum_{n \sim N} a_m b_n S(\mathcal{A}_{mn}, z)$$

$$- \sum_{m \sim M} \sum_{n \sim N} a_m b_n \sum_{z \le p < x^\nu} S(\mathcal{A}_{mnp}, p)$$

$$= S_1 - S_2, \quad \text{say.}$$

Lemma 7.14 takes care of S_1. If $\alpha \in I_1$, then using $p \le x^{\nu(\alpha)} \le x^{2\theta - 2\alpha}$, we have $M^2 p \in [x/z, xz]$, and so we can obtain a formula for S_2. For $\alpha \in I_h$ with $h \ge 2$ we proceed by induction. For the inductive step we must show that if $r \sim x^{\alpha_1}$, and we take out another prime factor $p = x^{\alpha_2}$ with $2\theta - 1 \le \alpha_2 \le \nu(\alpha_1)$, then $\alpha_1 + \alpha_2 \in I_{h-1}$ and $\alpha_2 \le \nu(\alpha_1 + \alpha_2)$. Since $\alpha_2 \ge 2\theta - 1$, we only need show that

$$\alpha_1 + \alpha_2 \le \tfrac{1}{2} - (2h - 4)(\theta - \tfrac{1}{2})$$

to obtain $\alpha_1 + \alpha_2 \in I_{h-1}$. We have

$$\alpha_1 + \alpha_2 \le \alpha_1 + \frac{2}{2h - 1}(\theta - \alpha_1) = \frac{2\theta + (2h - 3)\alpha_1}{2h - 1},$$

which is an increasing function of α_1. Putting in the maximum value, $\tfrac{1}{2} - (2h - 2)(\theta - \tfrac{1}{2})$, for α_1 then gives

$$\alpha_1 + \alpha_2 < \tfrac{1}{2} - (2h - 4)(\theta - \tfrac{1}{2})$$

as required.

Now

$$\alpha_2 \le \frac{2}{2h - 1}(\theta - \alpha_1),$$

and so we have

$$(2h - 1)\alpha_2 \le 2(\theta - \alpha_1)$$
$$\Rightarrow \quad (2h - 3)\alpha_2 \le 2(\theta - \alpha_1 - \alpha_2)$$
$$\Rightarrow \quad \alpha_2 \le \nu(\alpha_1 + \alpha_2),$$

as required. This establishes the inductive stage and so completes the proof. $\qquad\square$

Lemma 7.16. *Let $M = x^\alpha$, $N_1 = x^\beta$, and $N_2 = x^\gamma$ and suppose that $\alpha \le \tfrac{1}{2}$ and either*

$(i) \qquad 2\beta + \gamma \le 1 + \theta - 2\alpha^* - 2\epsilon, \qquad \gamma \le \dfrac{1 + 3\theta}{4} - \alpha^* - \epsilon,$

$$2\beta + 3\gamma \le \frac{3 + \theta}{2} - 2\alpha^* - 2\epsilon$$

or

$(ii) \qquad\qquad \beta \le \dfrac{1 - \theta}{2}, \quad \gamma \le \dfrac{1 + 3\theta - 4\alpha^*}{8} - \epsilon.$

Let b_n, the coefficient of $N(s)$, satisfy

$$b_n = \sum_{\substack{n_1 n_2 = n \\ n_1 \sim N_1, \, n_2 \sim N_2}} A_{n_1} B_{n_2}.$$

Then (7.6.8) holds whenever $\nu \le \nu(\alpha)$.

Proof. We follow the proofs of Lemmas 7.14 and 7.15, altering only the discussion where $N(s)$ fails to satisfy the requirements of Lemma 7.12. We then use the fact that we can factorize $N(s)$ as $N_1(s)N_2(s)$. At the point in the proof of Lemma 7.12 where we appealed to Lemma 7.11, we can now use Lemma 7.9 in Case (i) or Lemma 7.10 in Case (ii). The proof quickly follows.   □

7.7 THE TWO-DIMENSIONAL SIEVE REVISITED

Here we improve upon Lemma 7.6. For a given positive integer m we write

$$\mathcal{E}^m = \{r\ell : mr\ell \in \mathcal{A}, r \sim R\}$$

and define $\mathcal{F}^m$ similarly with $\mathcal{A}$ replaced by $\mathcal{B}$. Here we have suppressed the dependence on R for clarity.

Lemma 7.17. *Let $M \asymp x^\alpha, \alpha \le \frac{1}{2}$; let $R \ll x^{1/2-2\epsilon}$. Then, for any divisor-bounded sequence a_m with $a_m \ge 0$, we have*

$$\sum_{m \sim M} a_m S(\mathcal{E}^m, z) = \frac{y}{y_1} \sum_{m \sim M} a_m S(\mathcal{F}^m, z) + O(y\mathcal{L}^{-A}).$$

Proof. We now supply all the details that were suppressed in the proof of Lemma 7.6. Let $J = x^{1/200}$ and write

$$P(w, z) = \prod_{w \le p < z} p.$$

We then have

$$\sum_{m \sim M} a_m S(\mathcal{E}^m, z) = \sum_{m \sim M} a_m \sum_{n \in \mathcal{E}^m} \sum_{d|(n,P(w,z))} \mu(d) \sum_{e|(n,P(w))} \mu(e)$$

$$= \sum\nolimits_1 + \sum\nolimits_2,$$

where $Md \le x^{1/2}$ in $\sum_1$ and $Md > x^{\frac{1}{2}}$ in $\sum_2$. We write $\sum_1 = \sum_3 + \sum_4$, where $e \le J$ in $\sum_3$ and $e > J$ in $\sum_4$. By Lemma 4.1

$$\sum\nolimits_4 \ll \sum_{m \sim M} a_m \sum_{\substack{n \in \mathcal{E}^m \\ }} \sum_{d|(n,P(w,z))} \sum_{\substack{e|(n,P(w)) \\ J \le e \le Jw}} 1 = \sum\nolimits_5, \quad \text{say.}$$

We apply our standard procedure of taking the prime factors of d out one by one in $\sum_2$. Since each prime factor, say p_j, satisfies $p_j < z$, we have

$$M^2(p_1 \ldots p_{k-1})^2 p_k < x/z \implies Mp_1 \ldots p_k < x^{\frac{1}{2}} \implies (Mp_1 \ldots p_k)^2 p_{k+1} < xz.$$

Since $Md \geq x^{1/2}$, we must eventually have $M^2(p_1 \ldots p_{k-1})^2 p_k > x/z$. Our Type II information therefore supplies an asymptotic formula in this case for all the $\ll \log x$ terms that arise with a satisfactory error term.

Now we consider $\sum_3$. We can rewrite this term as

$$\sum_{m \sim M} a_m \sum_{d_1, d_2} \mu(d_1 d_2) \sum_{e_1 e_2} \mu(e_1 e_2) \sum_{\substack{n_1 n_2 d_1 d_2 e_1 e_2 \in \mathcal{E}^m \\ n_1 d_1 e_1 \sim R}} 1.$$

In the above we have suppressed the additional summation conditions:

$$d_1 d_2 | P(w, z), \quad e_1 e_2 | P(w), \quad e_1 e_2 \leq J, \quad d_1 e_1 \leq 2R, \quad Md_1 d_2 \leq x^{\frac{1}{2}}.$$

Now we can remove the condition $n_1 d_1 e_1 \sim R$ by an application of Perron's formula. Since $md_1 d_2 e_1 e_2 \leq x^{1/2} J = x^{0.505}$, we can apply (7.3.8) to obtain the asymptotic formula

$$\sum_3 = \frac{y}{y_1} \sum_{m \sim M} a_m \sum_{d_1, d_2} \mu(d_1 d_2) \sum_{e_1 e_2} \mu(e_1 e_2) \sum_{\substack{n_1 n_2 d_1 d_2 e_1 e_2 \in \mathcal{F}^m \\ n_1 d_1 e_1 \sim R}} 1 + O(yx^{-\eta})$$

$$= \frac{y}{y_1} \sum_6 + O(yx^{-\eta}), \quad \text{say.}$$

Now

$$\sum_6 = \sum_{m \sim M} a_m \sum_{n \in \mathcal{F}^m} \sum_{d | (n, P(w,z))} \mu(d) \sum_{e | (n, P(w))} \mu(e),$$

where $dM \leq x^{1/2}, e \leq J$. An application of Lemma 4.1 makes this equal $\sum_7 + \sum_8$, where $\sum_7$ is the same as $\sum_6$ but without the condition $e \leq J$. The sum $\sum_8$ is

$$\ll \sum_{m \sim M} a_m \sum_{\substack{n \in \mathcal{F}^m \\ dM \leq x^{1/2}}} \sum_{d | (n, P(w,z))} \sum_{\substack{e | (n, P(w)) \\ J \leq e \leq Jw}} 1 = \sum_9, \quad \text{say.}$$

We now observe that by (7.3.8)

$$\sum_5 = \frac{y}{y_1} \sum_9 + O(yx^{-\eta}).$$

It therefore remains to bound $\sum_9$.

We have

$$\sum_9 \leq \sum_{m \sim M} a_m \sum_{\substack{e | P(w) \\ J \leq e \leq Jw}} \sum_{ne \in \mathcal{F}^m} \tau(n)$$

$$\ll \sum_{m \sim M} a_m \frac{y_1}{m} \sum_{\substack{e | P(w) \\ J \leq e \leq Jw}} \frac{\tau(e)}{e} (\log x)^2$$

$$\ll \sum_{m \sim M} a_m S(\mathcal{F}^m, z)(\log x)^3 \exp(2 \log \log w + 4ua - ua \log ua)$$

$$\ll \sum_{m \sim M} a_m S(\mathcal{F}^m, z)\mathcal{L}^{-A}.$$

Here we have applied Lemma 4.3 with $u = \log \log x, a = (200)^{-1}$. This completes the proof. $\qquad \square$

The reader should note the way in the above proof we applied our Type I_2 information and then applied Heath-Brown's fundamental lemma to one sum but took these processes in reverse order for a different sum. For both terms we were left with the same sum to estimate, namely, $\sum_9$ above.

Lemma 7.18. *Let $M \asymp x^\alpha, \alpha \le \frac{1}{2}$; let $R \ll x^{1/2-2\epsilon}$. Then*

$$\sum_{m \sim M} a_m S(\mathcal{E}^m, x^\nu) = \frac{y}{y_1} \sum_{m \sim M} a_m S(\mathcal{F}^m, x^\nu) + O(y\mathcal{L}^{-A})$$

for all $\nu \le \nu(\alpha)$.

Proof. This follows by combining Lemma 7.17 with the method of Lemma 7.15. $\qquad\square$

We now use Lemma 7.18 to deduce that

$$\sum_{p_1 \sim M} \sum_{p_2 \sim R} S(\mathcal{A}_{p_1 p_2}, x^\nu)$$

$$= \frac{y}{y_1} \sum_{p_1 \sim M} \sum_{p_2 \sim R} S(\mathcal{B}_{p_1 p_2}, x^\nu) + O(y\mathcal{L}^{-A}) \tag{7.7.1}$$

for values of R, M not covered by the methods of Section 7.5.

Lemma 7.19. *Suppose that $M = x^\alpha$,*

$$R \le M, \quad M^2 R < x.$$

Then (7.7.1) holds for $\nu = 2\theta - 1$.

Proof. In view of Lemma 7.15 we may suppose that $R \ge x^{(3\theta-1)/2-2\epsilon}$. Lemma 7.18 yields

$$\sum_{p_1 \sim M} S(\mathcal{E}^{p_1}, x^\nu) = \frac{y}{y_1} \sum_{p_1 \sim M} S(\mathcal{F}^{p_1}, x^\nu) + O(y\mathcal{L}^{-A}). \tag{7.7.2}$$

Here

$$\mathcal{E}^{p_1} = \{p_1 r \ell : r \sim R, p_1 r \ell \in \mathcal{A}\}$$

and $\mathcal{F}$ is defined similarly with $\mathcal{B}$ in place of $\mathcal{A}$. Now the left-hand side of (7.7.2) includes the left-hand side of (7.7.1) together with extra terms for which we can apply Lemma 7.15. To be precise, leaving implicit a summation over u with $1 \le u \le (\log R)/(\nu \log x)$, we have

$$\sum_{p_1 \sim M} S(\mathcal{E}^{p_1}, x^\nu) = \sum_{p_1 \sim M} \sum_{\substack{x^\nu \le p_1' \le \cdots \le p_u' \\ p_1' \cdots p_u' \sim R}} S(\mathcal{A}_{p_1 p_1' \cdots p_u'}, x^\nu)$$

$$= \sum_{p_1 \sim M} \sum_{p_1' \sim R} S(\mathcal{A}_{p_1 p_1'}, x^\nu) \tag{7.7.3}$$

$$+ \sum_{\substack{p_1 \sim M \\ x^\nu \le p_1' \le (2R)^{1/2}}} \sum_{\substack{p_1' \le p_2' \le \cdots \le p_u' \\ p_2' \cdots p_u' \sim R/p_1'}} S(\mathcal{A}_{p_1 p_1' p_2' \cdots p_u'}, x^\nu).$$

The second sum in the last expression is

$$S = \frac{y}{y_1} \sum_{\substack{p_1 \sim M \\ x^\nu \leq p_1' \leq (2R)^{1/2}}} \sum_{\substack{p_1' \leq p_2' \leq \cdots \leq p_u' \\ p_2' \cdots p_u' \sim R/p_1'}} S(\mathcal{B}_{p_1 p_1' \ldots p_u'}, x^\nu) + O(y\mathcal{L}^{-A})$$

by an application of Lemma 7.15. For $MR^{1/2} \ll x^{1/2}$,

$$\frac{R}{p_1'} \ll x^{\frac{1}{3}} p_1'^{-1} \ll x^{\frac{1}{3} - (2\theta - 1)} \ll x^{\frac{1}{2}(3\theta - 1) - 2\epsilon}$$

since $\theta > 0.524$. Now

$$\sum_{p_1 \sim M} \sum_{p_1' \sim R} S(\mathcal{A}_{p_1 p_1'}, x^\nu) = \frac{y}{y_1} \left(\sum_{p_1 \sim M} S(\mathcal{F}^{p_1}, x^\nu) - S \right) + O(y\mathcal{L}^{-A})$$

$$= \frac{y}{y_1} \sum_{p_1 \sim M} \sum_{p_1' \sim R} S(\mathcal{B}_{p_1 p_1'}, x^\nu) + O(y\mathcal{L}^{-A}),$$

working as we did to obtain (7.7.3). The formula (7.7.1) then follows. $\qquad\square$

Lemma 7.20. *Let* $M_1 \leq M_2 \leq M_3$, $M_1 M_2 M_3^2 \leq x$, $M_1 \geq x^{2\theta - 1}$. *Then*

$$\sum_{m_1 \sim M_1} \sum_{p_2 \sim M_2} \sum_{m_3 \sim M_3} a_{m_1} b_{m_3} S(\mathcal{A}_{m_1 p_2 m_3}, x^\nu)$$

$$= \frac{y}{y_1} \sum_{m_1 \sim M_1} \sum_{p_2 \sim M_2} \sum_{m_3 \sim M_3} a_{m_1} b_{m_3} S(\mathcal{B}_{m_1 p_2 m_3}, x^\nu) + O(y\mathcal{L}^{-A}).$$

Proof. We have

$$M_1 M_3 \leq (M_1 M_2 M_3^2)^{\frac{1}{2}} \leq x^{\frac{1}{2}}, \quad M_2^3 \ll x M_1^{-1} \ll x^{2 - 2\theta}.$$

If $M_2 \leq x^{(3\theta - 1)/2 - 2\epsilon}$, the result follows from Lemma 7.15. Suppose now that

$$M_2 > x^{\frac{1}{2}(3\theta - 1) - 2\epsilon}.$$

Lemma 7.18 (with $R = M_2$) yields an asymptotic formula for

$$\sum_{m_1 \sim M_1} \sum_{m_3 \sim M_3} a_{m_1} b_{m_3} S(\mathcal{E}^{m_1 m_3}, x^\nu).$$

Working as we did to obtain (7.7.3),

$$\sum_{m_1, m_3} a_{m_1} b_{m_3} S(\mathcal{E}^{m_1 m_3}, x^\nu)$$

$$= \sum_{m_1, m_3} a_{m_1} b_{m_3} \sum_{\substack{x^\nu \leq p_1' \leq \cdots \leq p_u' \\ p_1' \cdots p_u' \sim M_2}} S(\mathcal{A}_{m_1 m_3 p_1' \ldots p_u'}, x^\nu)$$

$$= \sum_{m_1, m_3} a_{m_1} b_{m_3} \sum_{p_1' \sim M_2} S(\mathcal{A}_{m_1 m_3 p_1'}, x^\nu)$$

$$+ \sum_{\substack{m_1, m_3 \\ x^\nu \leq p_1' \leq (2M_2)^{1/2}}} \sum_{\substack{p_1' \leq p_2' \leq \cdots \leq p_u' \\ p_2' \cdots p_u' \sim M_2/p_1'}} S(\mathcal{A}_{m_1 m_3 p_1' \ldots p_u'}, x^\nu).$$

Now we can show that the second sum in the last expression is

$$
\frac{y}{y_1} \sum_{\substack{m_1, m_3 \\ x^\nu \le p_1' \le (2M_2)^{1/2}}} \sum_{\substack{p_1' \le p_2' \le \cdots \le p_u' \\ p_2' \cdots p_u' \sim \frac{M_2}{p_1'}}} S(\mathcal{B}_{m_1 m_3 p_1' \ldots p_u'}, x^\nu) + O(y\mathcal{L}^{-A}).
$$

To do this we must divide it into two subsums defined by

(i) $m_1 m_3 p_1' \le x^{1/2}$,

(ii) $m_1 m_3 p_1' > x^{1/2}$.

If condition (i) holds, then

$$
p_2' \ldots p_u' \ll M_2 x^{-(2\theta-1)} \ll x^{\frac{1}{3}(2-2\theta)-(2\theta-1)} \ll x^{\frac{1}{2}(3\theta-1)-2\epsilon}
$$

since $\theta > 0.524$. We now get the desired result from Lemma 7.15, with the variables combined as $m = m_1 m_3 p_1'$, $n = p_2' \ldots p_u'$.

If condition (ii) holds, then we regroup the variables as follows:

$$
m = m_3 p_2' \ldots p_u' \ll m_3 M_2/p_1' \ll m_3^2 m_1 M_2 x^{-\frac{1}{2}} \ll x^{\frac{1}{2}},
$$

and

$$
n = m_1 p_1' \ll m_1 M_2^{\frac{1}{2}} \ll x M_2^{\frac{5}{2}} \ll x^{1-\frac{5}{4}(3\theta-1)+6\epsilon} \ll x^{\frac{1}{2}(3\theta-1)-2\epsilon}.
$$

Once again, the desired result follows from Lemma 7.15. The proof follows in a similar way to that of Lemma 7.19. $\qquad\square$

7.8 FURTHER ASYMPTOTIC FORMULAE

Let $L_1 \ldots L_4 = x$, $L_j = x^{\alpha_j}$, $\alpha_j \ge \epsilon$. We shall find a region of $(\alpha_1, \ldots, \alpha_4)$ in which

$$
\left\| L_1 \ldots L_4 \right\|_1 \ll x^{\frac{1}{2}} \mathcal{L}^{-A} \tag{7.8.1}
$$

for every $A > 0$. The basic idea is that it is helpful if one of the α_j is near $1 - \theta$ in size and two of the others are near $\frac{1}{2}, \frac{1}{3}$, or $\frac{1}{4}$ of $1 - \theta$, or if a sum of the α_j is near a certain fraction of $1 - \theta$.

Now (7.8.1) permits us to evaluate

$$
\sum_{\substack{p_j \sim L_j \\ 1 \le j \le 3}} S(\mathcal{A}_{p_1 p_2 p_3}, p_3).
$$

This is essentially an application of Lemma 7.2. We have already discussed the removal of the condition $(b_{r+1}, P(p_r)) = 1$ in counting $p_1 \ldots p_r b_{r+1}$ in the last sum.

To avoid stating a result that has so many parameters it is difficult to apply, we are going to prove a less general result than that given in [12]. Effectively we are going back to the method of [11]. Indeed, we first need a lemma that essentially was given in [58].

Lemma 7.21. *Let $L(s)$ be a prime-factored Dirichlet polynomial of length L and suppose that $T \leq L^\gamma$, where $1 \leq \gamma < B$. Then*

$$\left\|L^\beta\right\|_1 \ll L^{\frac{1}{2}\beta}\mathcal{L}^{-A} \tag{7.8.2}$$

for every $A > 0$, provided that

$$\beta \geq 4\gamma - 2h + \epsilon, \tag{7.8.3}$$

where h is an integer satisfying $h \ll 1$,

$$2h - \epsilon < \beta < 6h - \epsilon. \tag{7.8.4}$$

Proof. Denote by I the left-hand side of (7.8.2). Put $M(s) = L(s)^h, M = L^h$. We may suppose that $M(\frac{1}{2} + it) \sim V$ for all $t \in \mathcal{S}$, where $t, u \in \mathcal{S}, t \neq u \Rightarrow |t - u| \geq 1$. Using (7.2.1) we therefore obtain

$$I \ll |\mathcal{S}|V^{\frac{\beta}{h}} = |\mathcal{S}|V^{2+\frac{\beta-2h}{h}}$$
$$\ll \mathcal{L}^B(MV^{\frac{\beta-2h}{h}} + TMV^{-4+\frac{\beta-2h}{h}}). \tag{7.8.5}$$

For the first term in the last expression we have

$$MV^{\frac{\beta-2h}{h}} \leq M(M\mathcal{L}^{-A_3})^{\frac{\beta-2h}{2h}} \ll M^{\frac{\beta}{2h}}\mathcal{L}^{-A} \ll L^{\frac{1}{2}\beta}\mathcal{L}^{-A}$$

from (7.8.5) and the prime-factored hypothesis. For the remaining term, we get the desired bound if

$$V > (TM^{\frac{2h-\beta}{2h}})^{\frac{h}{6h-\beta}}\mathcal{L}^A. \tag{7.8.6}$$

Suppose now (7.8.6) is violated. Then

$$|\mathcal{S}|V^{\frac{\beta}{h}} = V^{\frac{\beta-2h}{h}}|\mathcal{S}|V^2$$
$$\ll (TM^{\frac{2h-\beta}{2h}})^{\frac{\beta-2h}{6h-\beta}}(M + T)\mathcal{L}^A. \tag{7.8.7}$$

We now find that the right-hand side of (7.8.7) is $\ll M^{\frac{\beta}{2h}}\mathcal{L}^{-A}$ provided that $L^{2h+\beta-\epsilon} \geq T^4$, which is a consequence of (7.8.3). This completes the proof of Lemma 7.21. □

Lemma 7.22. *Let $L_j(s), 1 \leq j \leq 4$ be as above. Suppose that $L_j(s)$ is prime-factored for $j \geq 2$ and $\alpha_1 \geq 1 - \theta$. Then* (7.8.1) *holds under any of the following conditions:*

(i) $\alpha_2 \geq (1 - \theta)/2$, $\alpha_3 \geq (1 - \theta)/4$, $\alpha_4 \geq 2(1 - \theta)/7$;

(ii) $\alpha_2 \geq (1 - \theta)/2$, $\alpha_3 \geq (1 - \theta)/3$, $\alpha_4 \geq 2(1 - \theta)/9$;

(iii) $\alpha_2 \geq (1 - \theta)/3$, $2\alpha_3 + \alpha_4 \geq 1 - \theta$, $\alpha_4 \geq 2(1 - \theta)/5$;

(iv) $\alpha_2, \alpha_3 \leq (1 - \theta)/3$, $\alpha_2 + \alpha_3 \geq 4(1 - \theta)/7$, $\alpha_4 + \alpha_2 + \alpha_3 \geq 14(1 - \theta)/13$.

Proof. By the hypothesis $\alpha_1 \geq 1 - \theta$ and the mean value theorem for Dirichlet polynomials we obtain

$$\left\|L_1 \cdots L_4\right\|_1 \leq \left\|L_1\right\|_2 \left\|L_2\right\|_{\beta_1} \left\|L_3\right\|_{\beta_2} \left\|L_4\right\|_{\beta_3}$$
$$\ll \mathcal{L}^B(x^{\alpha_1})^{\frac{1}{2}}\left\|L_2\right\|_{\beta_1} \left\|L_3\right\|_{\beta_2} \left\|L_4\right\|_{\beta_3} \tag{7.8.8}$$

whenever $\beta_1, \beta_2, \beta_3$ are positive and $\sum_j \beta_j^{-1} = \frac{1}{2}$.

We note that when β_{j-1} is an even integer, the inequality

$$\left\| L_j \right\|_{\beta_{j-1}} \ll \mathcal{L}^B L_j^{\frac{1}{2}} \tag{7.8.9}$$

follows from the mean value theorem whenever

$$L_j \gg (T)^{\frac{2}{\beta_1}}. \tag{7.8.10}$$

We now consider the four cases (i)–(iv) one by one.

(i) Take $\beta_1 = 8$, $\beta_2 = 8$, $\beta_3 = 4$. The hypothesis gives (7.8.10) for $j = 2, 3$. We can then apply Lemma 7.21 with $\beta = 8$, $\gamma = \frac{7}{2} - \epsilon$, $h = 3$ in conjunction with (7.8.8) to get the result.

(ii) Take $\beta_1 = 4$, $\beta_2 = 12$, $\beta_3 = 6$. The hypothesis for this case gives (7.8.10) for $j = 2, 3$. We now apply Lemma 7.21 with $\beta = 12$, $\gamma = \frac{9}{2} - \epsilon$, $h = 3$.

(iii) Instead of (7.8.8) we now use

$$\left\| L_1 \cdots L_4 \right\|_1 \le \left\| L_1 \right\|_2 \left\| L_2 \right\|_6 \left\| L_3 L_4^{\frac{1}{2}} \right\|_4 \left\| L_4^{\frac{1}{2}} \right\|_{12}.$$

Now (7.8.9) holds for $j = 2$, from (7.8.10) with $\beta_1 = 6$; similarly

$$\left\| L_3 L_4^{\frac{1}{2}} \right\|_4 \ll (L_3 L_4^{\frac{1}{2}})^{\frac{1}{2}} \mathcal{L}^B$$

since $L_3^2 L_4 \ge T$. We now get the desired inequality because

$$\left\| L_4^{\frac{1}{2}} \right\|_{12} = \left\| L_4 \right\|_6^{\frac{1}{2}} \ll L_4^{\frac{1}{4}} \mathcal{L}^{-A}$$

from Lemma 7.21 with $\beta = 6$, $\gamma = \frac{5}{2} - \epsilon$, $h = 2$.

(iv) Case (a) Write $\tau = 1 - \theta$. Suppose that $\alpha_4 \ge \frac{1}{2}\tau$. If $\min(\alpha_2, \alpha_3) \ge \frac{1}{4}\tau$, we can use (i) since $\max(\alpha_2, \alpha_3) \ge \frac{2}{7}\tau$. If $\min(\alpha_2, \alpha_3) < \frac{1}{4}\tau$, we can suppose without loss of generality that $\alpha_2 \le \alpha_3$. Then the hypotheses

$$\alpha_2, \alpha_3 \le \tfrac{1}{3}\tau, \quad \alpha_2 + \alpha_3 \ge \tfrac{4}{7}\tau,$$

give $\frac{1}{10} < \frac{5}{21}\tau \le \alpha_2 < \frac{1}{4}\tau$ for $\theta < \frac{29}{50}$. We then take $\beta_1 = 10$, $\beta_2 = 4$, $\beta_3 = \frac{20}{3}$. We have (7.8.9) for $j = 2, 3$ and we apply Lemma 7.21 to L_4 with

$$\beta = \frac{20}{3}, \quad \gamma = \frac{\tau}{\frac{4}{7}\tau - \frac{1}{4}\tau} = \frac{28}{9}, \quad h = 3.$$

Case (b) Now $\alpha_4 < \frac{1}{2}\tau$. The hypotheses give $\alpha_4 > \frac{16}{39}\tau$. We may apply Lemma 7.21 to L_4 with

$$\gamma = \frac{\tau}{16\tau/39} = \frac{39}{16}, \quad h = 2, \quad \beta = 4\gamma - 4 + \epsilon.$$

We also apply Lemma 7.21 to L_2 and L_3. In the case of L_2,

$$\gamma = \frac{\tau}{\alpha_2}, \quad h = 3, \quad \beta = \frac{4\tau}{\alpha_2} - 6 + \epsilon,$$

and in the case of L_3,

$$\gamma = \frac{\tau}{\alpha_3}, \quad h = 3,$$

with the following equation giving the value of β:

$$\frac{1}{\beta} + \frac{1}{\frac{4\tau}{\alpha_2} - 6 + \epsilon} + \frac{1}{\frac{4\tau}{\alpha_4} - 4 + \epsilon} = \frac{1}{2}. \tag{7.8.11}$$

This will give the desired result provided that $\beta \geq 4\tau/\alpha_3 - 6 + \epsilon$ in (7.8.11). Thus we must show that

$$S := \frac{1}{\frac{4\tau}{\alpha_4} - 4} + \frac{1}{\frac{4\tau}{\alpha_2} - 6} + \frac{1}{\frac{4\tau}{\alpha_3} - 6} \geq \frac{1}{2} + B\epsilon \tag{7.8.12}$$

where $\frac{16}{39}\tau < \alpha_4 < \frac{1}{2}\tau$, $\frac{5}{21}\tau < \alpha_2$, $\alpha_3 < \frac{1}{3}\tau$, $\alpha_1 + \alpha_2 + \alpha_3 \geq \frac{14}{13}\tau + B\epsilon$. We need only consider a fixed value c of $\alpha_4 + \alpha_2 + \alpha_3$. Now fix α_4. It is easy to show that $S = S(\alpha_2, \alpha_3)$ is least when $\alpha_2 = \alpha_3$. Now

$$S(\alpha, \alpha) = \frac{2\alpha}{4\tau - 6\alpha} + \frac{c - 2\alpha}{4\tau - 4(c - 2\alpha)}$$

has an increasing derivative vanishing at $\alpha = 2c/7$. The minimum value of S is thus $c/(4\tau - 12c/7)$, which is increasing in c and takes the value $\frac{1}{2}$ at $\frac{14}{13}\tau$. The desired result follows readily.   □

7.9 THE FINAL DECOMPOSITION

As usual, we ignore the presence of ϵ for clarity. Let $\theta = 0.525$. We begin with some further notation needed to describe the further decomposition of $\sum_3$ in (7.4.2), following [12] closely. We use the letters G, D, R in a naïve fashion for sets that are, respectively, *good* (an asymptotic formula holds), *decomposable* (we can apply Buchstab's identity again), and *role-reversible* (no other technique works, but role reversals *may* be possible). Write

$$U_n = \{\alpha_n : 0 < \alpha_n < \alpha_{n-1} < \cdots < \alpha_1, 2\alpha_n < 1 - \alpha_1 - \cdots - \alpha_{n-1}\}.$$

Let

$$G = \bigcup_{n=2}^{\infty} G_n,$$

where G_n is the set of α_n such that an asymptotic formula can be obtained for

$$\sum_{\substack{p_1, \ldots, p_n \\ p_j \sim x^{\alpha_j}}} S(\mathcal{A}_{p_1 \ldots p_n}, p_n).$$

Put

$$D_0 = \left\{\alpha_2 : 0 \leq \alpha_1 \leq \tfrac{1}{2}, 0 \leq 2\alpha_2 \leq \min\left(3\theta + 1 - 4\alpha^*, \tfrac{2}{5}(3 + \theta - 4\alpha^*)\right)\right\}$$

with α^* as in Section 7.6,

$$D_1 = \left\{\alpha_3 : 0 \leq \alpha_1 \leq \tfrac{1}{2}, \alpha_3 \leq \tfrac{1}{4}(1 + 3\theta) - \alpha^*, \right.$$
$$\left. \alpha_2 + \alpha_3 \leq 1 + \theta - 2\alpha^*, 2\alpha_2 + 3\alpha_3 \leq \tfrac{1}{2}(3 + \theta) - 2\alpha^*\right\},$$
$$D_2 = \left\{\alpha_3 : 0 \leq \alpha \leq \tfrac{1}{2}, \alpha_2 \leq \tfrac{1}{2}(1 - \theta), \alpha_3 \leq \tfrac{1}{8}(1 + 3\theta) - \tfrac{1}{2}\alpha^*\right\}.$$

We write D^* for the set of α_4 such that $(\alpha_1, \alpha_2, \alpha_3, \alpha_4, \alpha_4)$ can be *partitioned* into D_0 or $D_1 \cup D_2$ in the sense that the five quantities can be grouped into two or three variables in these sets. For example, we might have $(\alpha + 2\alpha_4, \alpha_2 + \alpha_3) \in D_0$ or $(\alpha + \alpha_4, \alpha_2, \alpha_3 + \alpha_4) \in D_1$. Finally, we write, with an obvious extension of this partition terminology,

$$R = \{\alpha_3 : \alpha_3 \notin D_1 \cup D_2, \alpha_3 \text{ cannot be partitioned into } (\eta, \zeta) \in D_0\}.$$

The significance of these sets is that D_0, D_1, D_2 correspond to conditions on variables that allow a further decomposition via Lemma 7.15 or 7.16. On the other hand, in D^* two further decompositions can be made. In regions corresponding to R, no further decompositions are possible without role reversals.

As before, when a sum such as

$$\sum_{p,q} S(\mathcal{A}_{pq}, q)$$

arises, we may be able to give an asymptotic formula for some of the almost-primes counted. We have seen how to make these visible by writing

$$\sum_{p,q} S(\mathcal{A}_{pq}, q) = \sum_{p,q} S\left(\mathcal{A}_{pq}, (x/pq)^{\frac{1}{2}}\right) + \sum_{\substack{p,q \\ q<r<(x/pq)^{1/2}}} S(\mathcal{A}_{pqr}, r).$$

We follow [12] in defining a new function to take into account the possible savings introduced by this technique.

Given $\alpha \in U_n$, write

$$u = \left[\frac{1 - \alpha_1 - \cdots - \alpha_n}{\alpha_n}\right].$$

Then $u \geq 1$ by the definition of U_n, and $S(\mathcal{A}_{p_1 \ldots p_n}, p_n)$ counts numbers with up to u prime factors. Now write

$$w(\alpha, 1) = \frac{1}{\alpha_{n+1}} \quad \text{where} \quad \alpha_{n+1} = 1 - \alpha_1 - \cdots - \alpha_n.$$

We can then define $w(\alpha, k)$ inductively by

$$w(\alpha, k+1) = w(\alpha, k) + \int^* \frac{d\beta_1 \ldots d\beta_k}{\beta_1 \beta_2 \ldots \beta_k(\alpha_{n+1} - \beta_1 - \cdots - \beta_k)},$$

where $*$ denotes the region

$$\alpha_n \leq \beta_1 \leq \beta_2 \leq \cdots \leq \beta_k \leq \frac{\alpha_{n+1} - \beta_1 - \cdots - \beta_{k-1}}{2},$$

with $(\alpha_1, \ldots, \alpha_n, \beta_1, \ldots, \beta_k) \notin G$. We then put

$$w(\alpha) = w(\alpha, u).$$

Clearly

$$w(\alpha) \leq \frac{w(\alpha_{n+1}/\alpha_n)}{\alpha_n}$$

and, for $1 \leq k \leq u$,

$$w(\boldsymbol{\alpha}) \leq w(\boldsymbol{\alpha}, k) + \int^* \omega\left(\frac{\alpha_{n+1} - \beta_1 - \cdots - \beta_k}{\beta_k}\right) \frac{d\beta_1}{\beta_1} \cdots \frac{d\beta_{k-1}}{\beta_{k-1}} \frac{d\beta_k}{\beta_k^2}. \quad (7.9.1)$$

We will use (7.9.1) in some numerical calculations with $k = 2$ or 3. We shall also use the familiar (1.4.16) to substitute for $\omega(u)$ in the integrals.

Until now we have not taken into account role reversals in this notation. We rectify this as follows. Let $\boldsymbol{\alpha}_3 \in U_3$. Put $\boldsymbol{\alpha}_4 = (\alpha_1, \alpha_2, \alpha_3, \alpha_4)$ with $\nu \leq \alpha_4 \leq \frac{1}{2}\alpha_1$. We write

$$u' = \left[\frac{\alpha_1 - \alpha_4}{\alpha_4}\right].$$

We then define new functions $w^*(\boldsymbol{\alpha}_4, j)$, $Y(\boldsymbol{\alpha}_4, j)$ by

$$w^*(\boldsymbol{\alpha}_4, 1) = w(\boldsymbol{\alpha}_3)\frac{1}{\alpha_1 - \alpha_4}, \qquad w^*(\boldsymbol{\alpha}_4, j) = w(\boldsymbol{\alpha}_3)Y(\boldsymbol{\alpha}_4, j),$$

$$w^*(\boldsymbol{\alpha}_4, j+1) = w(\boldsymbol{\alpha}_3)\left(Y(\boldsymbol{\alpha}_4, j) + \int^\dagger \frac{d\gamma_1 \ldots d\gamma_j}{\gamma_1 \ldots \gamma_j(\alpha_1 - \alpha_4 - \gamma_1 - \cdots - \gamma_j)}\right).$$

Now the integration conditions here are rather messy, but the reader has seen similar things before. The last expression indicates a sum of multiple integrals (including those counted by $w(\boldsymbol{\alpha}_3)$) with the integration condition † dependent on which multiple integral from $w(\boldsymbol{\alpha}_3)$ occurs. Let

$$w^*(\boldsymbol{\alpha}_4) = w^*(\boldsymbol{\alpha}_4, u').$$

We note that

$$w^*(\boldsymbol{\alpha}_4) \leq \frac{w(\boldsymbol{\alpha}_3)\omega\left(\frac{\alpha_1 - \alpha_4}{\alpha_4}\right)}{\alpha_4}.$$

For computing the integrals there are various other upper bounds that could be derived using small numbers of integration variables.

We are now ready to split $\alpha_2 \in U_2$ into a number of subregions corresponding to the different techniques that should be applied. Continuing to follow [12] closely, we write

A: $\frac{1}{4} \leq \alpha_1 \leq \frac{2}{5}, \frac{1}{3}(1 - \alpha_1) \leq \alpha_2 \leq \min\left(\alpha_1, \frac{1}{2}(3\theta - 1), 1 - 2\alpha_1\right)$;

B: $\frac{1}{4}(3 - 3\theta) \leq \alpha_1 \leq \frac{1}{2}$,
$\max\left(\frac{1}{2}\alpha_1, 1 - 2\alpha_1\right) \leq \alpha_2 \leq \min\left(\frac{1}{2}(3\theta - 1), \frac{1}{2}(1 - \alpha_1)\right)$;

C: $\nu(0) \leq \alpha_1 \leq \frac{1}{3}, \nu(\alpha_1) \leq \alpha_2 \leq \min(\alpha_1, \frac{1}{3}(1 - \alpha_1))$;

D: $\frac{1}{3} \leq \alpha_1 \leq \frac{1}{2}, \nu(\alpha_1) \leq \alpha_2 \leq \max\left(\frac{1}{3}(1 - \alpha_1), \frac{1}{2}\alpha_1\right)$;

E: $\frac{1}{2}(3\theta - 1) \leq \alpha_1 \leq \frac{1}{4}(3 - 3\theta), \frac{1}{2}(3\theta - 1) \leq \alpha_2 \leq \min(\alpha_1, 1 - 2\alpha_1)$;

F: $\frac{1}{3} \leq \alpha_1 \leq 2 - 3\theta, \max\left(1 - 2\alpha_1, \frac{1}{2}(3\theta - 1)\right) \leq \alpha_2 \leq \frac{1}{2}(1 - \alpha_1)$.

These regions are illustrated in Diagram 7.3.

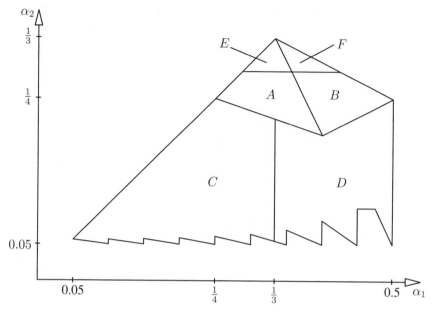

Diagram 7.3

We note that
$$(\alpha_1, \alpha_2) \in A \Leftrightarrow (1 - \alpha_1 - \alpha_2, \alpha_2) \in B.$$
A similar relation also holds between E and F. Since in $A \cup B \cup E \cup F$ only products of three primes are counted, we have
$$\sum_{\alpha_2 \in B} S(\mathcal{A}_{p_1 p_2}, p_2) = \sum_{\alpha_2 \in A} S(\mathcal{A}_{p_1 p_2}, p_2),$$
$$\sum_{\alpha_2 \in F} S(\mathcal{A}_{p_1 p_2}, p_2) = \sum_{\alpha_2 \in E} S(\mathcal{A}_{p_1 p_2}, p_2).$$
Thus, recalling our starting point of (7.4.2), we have
$$\sum\nolimits_3 = 2 \sum_{\alpha_2 \in A} S(\mathcal{A}_{p_1 p_2}, p_3) + 2 \sum_{\alpha_2 \in E} S(\mathcal{A}_{p_1 p_2}, p_2)$$
$$+ \sum_{\alpha_2 \in C} S(\mathcal{A}_{p_1 p_2}, p_2) + \sum_{\alpha_2 \in D} S(\mathcal{A}_{p_1 p_2}, p_2).$$
Our next task is to consider sums over A. Discarding the sum over A leads to a "loss" ≈ 0.1971. The saving that can be made in this region with $\theta = 0.525$ is relatively small. At this point we are "running faster just to stay still," so to speak. We would have had much greater success when $\theta = 0.53$. An application of Buchstab's identity leads to
$$\sum_{\alpha_2 \in A} S(\mathcal{A}_{p_1 p_2}, p_2) = \sum_{\alpha_2 \in A} S(\mathcal{A}_{p_1 p_2}, x^{\nu(\alpha_1)}) - \sum_{\alpha_3 \in A'} S(\mathcal{A}_{p_1 p_2 p_3}, p_3). \quad (7.9.2)$$

Here $A' = \{\alpha_3 \in U_3 : \alpha_2 \in A, \nu(\alpha_1) \leq \alpha_3\}$. We can give an asymptotic formula for the first sum on the right-hand side of (7.9.2). If $\alpha_3 \in D_1 \cup D_2$ or $(\alpha_1 + \alpha_3, \alpha_2) \in D_0$ or $(\alpha_2 + \alpha_3, \alpha_1) \in D_0$, then a further straightforward decomposition of the final sum is possible. In the remaining part of A' we note that $\alpha_1 + \alpha_2 \geq \frac{1}{2}$. Writing h for a number counted by $S(\mathcal{A}_{p_1 p_2 p_3}, p_3)$, we have $h \sim x^{\alpha_4}$ with

$$\alpha_4 + \alpha_3 \leq \tfrac{1}{2}, \quad \alpha_2 \leq \tfrac{1}{2}(3\theta - 1).$$

By reversing roles we then obtain

$$\sum_{p_1, p_2, p_3} S(\mathcal{A}_{p_1 p_2 p_3}, p_3) = \sum_{h, p_2, p_3} S\left(\mathcal{A}_{h p_2 p_3}, (x/h p_2 p_3)^{\frac{1}{2}}\right)$$

$$= \sum_{h, p_2, p_3} S\left(\mathcal{A}_{h p_2 p_3}, x^{\nu(\alpha_4 + \alpha_3)}\right) - \sum_{h, p_2, p_3, q} S(\mathcal{A}_{h p_2 p_3 q}, q).$$

We have suppressed the summation conditions here — the reader will be able to supply them as necessary. Working as in previous problems, we get a loss from region A of

$$\int_{\alpha_2 \in A} \min\left(\frac{1}{\alpha_1 \alpha_2 (1 - \alpha_1 - \alpha_2)}, \frac{1}{\alpha_1 \alpha_2}(I_1 + I_2) + \frac{1}{\alpha_2} I_3\right) d\alpha_1 \, d\alpha_2$$

with

$$I_1 = \int_{\substack{\alpha_3 \in U_3 \setminus R \\ \alpha_4 \notin D^* \cup G}} \frac{w(\alpha_4)}{\alpha_3 \alpha_4} \, d\alpha_3 \, d\alpha_4,$$

$$I_2 = \int_{\substack{\alpha_3 \in U_3 \setminus R \\ \alpha_4 \in D^* \setminus G}} \frac{1}{\alpha_3 \alpha_4} \min\left(w(\alpha_4), \int_{\substack{\alpha_6 \in U_6 \\ \alpha_6 \notin G}} \frac{w(\alpha_6)}{\alpha_5 \alpha_6} \, d\alpha_5 \, d\alpha_6\right) d\alpha_3 \, d\alpha_4,$$

$$I_3 = \int_{\alpha_3 \in R} \frac{1}{\alpha_3} \, \omega\left(\frac{1 - \alpha_1 - \alpha_2 - \alpha_3}{\alpha_3}\right) \int_{\substack{\nu \\ \alpha_4 \notin G}}^{\alpha_1/2} \frac{w^*(\alpha_4)}{\alpha_4} \, d\alpha_4 \, d\alpha_3.$$

Further decompositions may be possible in some subregions, but we have not considered them here. Numerical calculations lead to a loss of size no more than 0.15, and so a loss from regions A and B less than 0.3.

For region E we perform two further decompositions leading to sums of the form

$$\sum_{p_1, p_2} S(\mathcal{A}_{p_1 p_2}, x^\nu), \quad \sum_{h, p_2, p_3} S(\mathcal{A}_{h p_2 p_3}, x^\nu), \quad \sum_{h, p_2, p_3, p_4} S(\mathcal{A}_{h p_2 p_3 p_4}, p_4).$$

$$(7.9.3)$$

Here we may have applied a role reversal to obtain the second sum, so h runs either over primes or over integers coprime to $P(p_3)$. The first sum in (7.9.3) is covered by Lemma 7.19, while the second can be estimated via Lemma 7.20. We discard those parts of the final sum for which no asymptotic formula is available. Calculations yield a loss from this region no more than 0.03 (and so < 0.06 from $E \cup F$). Without using the two-dimensional sieve we would have had to discard all of this region with a loss from $E \cup F$ of ≈ 0.0864, so the saving with $\theta = 0.525$ is again quite small.

We now consider region C. Now it is only necessary to reverse the roles of variables for a small part of the sum

$$\sum_{\substack{(\alpha_1,\alpha_2)\in C \\ \alpha_3 \in U_3}} S(\mathcal{A}_{p_1 p_2 p_3}, p_3).$$

For example, we can perform a further decomposition in a straightforward manner whenever $\alpha_1 \leq 0.2875$ since

$$\alpha_2 + \alpha_3 \leq \tfrac{1}{2}(\alpha_1 + \alpha_2 + 2\alpha_3) \leq \tfrac{1}{2}.$$

This is also possible if $\alpha_1 + \alpha_3 \leq \frac{1}{2}$, since we then have $(\alpha_1 + \alpha_3, \alpha_2) \in D_0$. For the rest of C we have $(1 - \alpha_1 - \alpha_2 - \alpha_3) + \alpha_2 < \frac{1}{2}$, and we can reverse roles to decompose α_1. Working in this manner, the loss from region C is < 0.21. We note that if the whole of C were discarded, we would lose

$$\int_C \omega \left(\frac{1 - \alpha_1 - \alpha_2}{\alpha_2} \right) \frac{d\alpha_1}{\alpha_1} \frac{d\alpha_2}{\alpha_2^2} > 1.$$

The savings in this region are therefore highly significant.

In region D we note that p_1 is often the largest variable. As previously, we can split into different cases according to when straightforward decompositions are possible and when a role reversal becomes essential. After further numerical calculations we find that the loss from region D is no more than 0.34. It can be seen that this is, as with region C, a great saving on the trivial estimate.

Finally, we conclude that for all large x,

$$\pi(x) - \pi(x - x^{0.525}) \geq \frac{x^{0.525}}{\log x}(1 - 0.3 - 0.06 - 0.21 - 0.34)$$

$$= \frac{9}{100} \frac{x^{0.525}}{\log x},$$

and the proof of Theorem 7.2 is complete.   □

7.10 WHERE TO NOW?

As θ decreases below 0.525 in the argument of the previous section, the savings over the trivial bounds from regions A, B, E, and F become negligible, while the contributions from regions C and D rise rapidly. With faster PCs some of the calculations could be performed with greater precision and more work done on six-dimensional integrals. However, too much will still be discarded at the four-dimensional stage to make a great deal of progress. For example, 0.52 might be achievable with an incredibly long and boring argument, carefully going over every case and subcase with a fine-tooth comb. Something like 0.505 looks to be a distant target that requires substantial new arithmetical information. Our problem is the paucity of results that lead to such information. Very little of direct relevance has been proved in this regard, apart from Watt's theorem (and its precursors by Deshouillers and Iwaniec), since Huxley's form of the large-values result. This is the greatest challenge facing researchers in this area. If one could obtain an improved large-values result just for some of the special types of sums that occur in our applications, this would be of great benefit. Of course, so too would be a proof of the sixth-power (or even eighth-power) moment of the Riemann zeta-function.

Exercise

Assume that we have, for any divisor-bounded Dirichlet polynomial $M(s)$, a mean-value result of the form

$$\|M^\lambda\|_1 \ll (TM)^\epsilon \left(T + M^{\frac{1}{2}\lambda}\right), \quad \text{for } 2 \le \lambda \le 4, \tag{7.10.1}$$

where $\epsilon > 0$ is arbitrary. Deduce that, for any $\epsilon > 0$ and $x > x_0(\epsilon)$, there are primes in the interval $[x, x + x^{1/2+\epsilon}]$ (only Heath-Brown's generalized Vaughan identity is needed for this). We note that (7.10.1) is true when $M(s)$ is a Type I sum, and we only need the estimate to hold for the special types of sums generated by one of the sieve methods we have described. See [133, Chapter 7] for a discussion of (7.10.1).

Chapter Eight

The Brun-Titchmarsh Theorem on Average

8.1 INTRODUCTION

We saw in Chapter 2 that, when $(a, q) = 1$, the conjectured asymptotic formula

$$\pi(x; q, a) = \frac{\pi(x)}{\phi(q)}(1 + o(1)) \tag{8.1.1}$$

holds on average over q up to $x^{1/2}(\log x)^{-A}$ for some A. This raises the questions of what can be proved for *every* q and what can be said if $q > x^{1/2}(\log x)^{-A}$. The possible existence of Siegel zeros inhibits progress for the former question, while even knowing the Generalized Riemann Hypothesis does not help with the second. As far as *upper bounds* are concerned we have already stated a result sufficient for our purposes in the present chapter in the exercise in Chapter 4 (the Brun-Titchmarsh inequality). However, we note that there is a very clean result due to Montgomery and Vaughan [134] that uses the linear sieve (in Selberg's formulation) and the large sieve , which we state as follows.

Theorem 8.1. *For every* $x \geq 2$ *and every pair of integers* a, q *with* $(a, q) = 1, 1 \leq q < x$, *we have*

$$\pi(x; q, a) \leq \frac{2x}{\phi(q) \log(x/q)}. \tag{8.1.2}$$

We would like to have (8.1.2) with the upper bound replaced by an equality, the factor 2 reduced to $1 + o(1)$, and $\log(x/q)$ increased to $\log x$, but any of these developments seems far out of reach at present. Various authors have attempted to demonstrate stronger upper bounds on average. Until the work of Rousselet, [152] there were no corresponding lower bounds. The purpose of this chapter is to apply the alternative sieve to this problem, closely following our joint work with Baker [7]. In that paper we used the earlier version of the alternative sieve [59], and we must make some modifications to be consistent with the approach we have used throughout this monograph. We shall also incorporate the small improvement made in [9] and produce two lemmas that were needed in [7] but never appeared. The current problem enables us to show how to apply our techniques to a different situation, as well as presenting the only three-dimensional form of the Alternative Sieve yet to appear. It should be noted that Fouvry [34] was the first to give an application of the alternative sieve to this problem. His work appears to be independent of our own.

Apart from its intrinsic interest, the current problem has attracted much attention since it has many applications. The one we pursue in this chapter is to the greatest prime factor of $p + a$. This example has further applications in cryptography and primality testing that we shall not consider further here. We say that a result holds *for most q up to y* if the number of exceptional q is $O\big(y(\log y)^{-A}\big)$ for any given $A > 0$. The implied constant here will depend on the integer a given in the statement of Theorem 8.2. We note that the Bombieri-Vinogradov theorem demonstrates that (8.1.1) holds for most q up to $x^{1/2}(\log x)^{-B}$ where $B = B(A)$, where here the constant is *independent* of a. The difference in our present work arises from our use of the important work of Bombieri, Friedlander, and Iwaniec [16, 17, 18]. We also write, as in Chapter 6, $P(n)$ for the greatest prime factor of an integer n. Our main results are then as follows.

Theorem 8.2. *There are functions* $C_1(\theta), C_2(\theta)$ *where* $0.5 \leq \theta \leq 0.6$ *such that, given* $a \in \mathbb{Z} \setminus 0$*, for most* $q \leq x^{\theta}$*, we have*

$$\frac{C_1(\theta)x}{\phi(q)\log x} < \pi(x; q, a) < \frac{C_2(\theta)x}{\phi(q)\log x}. \tag{8.1.3}$$

These functions satisfy, for any $\epsilon > 0$,

   (i) $C_2(\theta)$ *monotonic increasing,* $C_1(\theta)$ *monotonic decreasing,*

   (ii) $C_2(0.5) = 1 + \epsilon, C_2(0.51) \leq 1.015, C_2(0.533) < 2$,

   (iii) $C_1(0.5) = 1 - \epsilon, C_1(0.52) > 0.16$,

   (iv) *We have*

$$\int_{1/2}^{3/5} C_2(\theta)\, d\theta < 0.239. \tag{8.1.4}$$

Theorem 8.3. *Let* $a \in \mathbb{Z} \setminus 0$*. Then, for infinitely many primes* p*, we have*

$$P(p + a) > p^{0.677}. \tag{8.1.5}$$

The first author to obtain an exponent in excess of $\frac{1}{2}$ for this problem was Goldfeld [46]. Various authors have since improved this result (see [7] for the details); we mention just two results in particular. Motohashi [137] obtained an exponent $0.6105\ldots$ as a result of combining Theorem 8.1 and the Bombieri-Vinogradov theorem with Chebychev's method (given below, adapted from Chapter Five). The author has shown [66] that this result holds with *effective* constants; that is, one can avoid any appeal to Siegel's theorem. After several authors improved this exponent over the next 15 years, eventually Fouvry [34] obtained 0.6687 (actually 0.6683 after the correction of an error), which result was of interest since exceeding the exponent $\frac{2}{3}$ enabled Adleman and Heath-Brown [1] to conclude that the first case of Fermat's Last Theorem was true for infinitely many exponents. This last result is now only of historical interest in view of Wiles' celebrated theorem. Mikawa [129] has improved the lower bound in (8.1.3) so that $C_1\left(\frac{17}{32}\right) > 0$, but no improvement has yet been made in the upper bound.

The start of the proof of Theorem 8.3 is similar to the framework employed in Chapter 6. We reduce the problem to an application of an upper-bound sieve as follows. We have

$$\sum_{p \leq x} \log(p+a) = \sum_{p \leq x} \sum_{n|(p+a)} \Lambda(n)$$

$$= \sum_{p_1 \leq x} \sum_{p_2|(p_1+a)} \log p_2 \ + \ o(x)$$

$$= \sum_{p \leq x+a} (\log p)\pi(x; p, -a) \ + \ o(x).$$

Since we have

$$\sum_{p \leq x} \log(p+a) \sim x$$

by the PNT, for all large x we have

$$\sum_{p \leq x+a} (\log p)\pi(x; p, -a) > (1 - \epsilon)x. \tag{8.1.6}$$

From the Bombieri-Vinogradov theorem and Theorem 8.1 we have

$$\sum_{p \leq x^{1/2}} (\log p)\pi(x; p, -a) < \left(\frac{1}{2} + \epsilon\right) x$$

for all sufficiently large x. We can thus establish that

$$\sum_{x^{0.677} \leq p \leq x+a} (\log p)\pi(x; p, -a) > \epsilon x \tag{8.1.7}$$

(which proves Theorem 8.3) once we have shown that

$$\sum_{x^{1/2} < p \leq x^{0.677}} (\log p)\pi(x; p, -a) < \left(\frac{1}{2} - 3\epsilon\right) x. \tag{8.1.8}$$

As in Chapter 6, the key to success is to use the alternative sieve over part of the range — up to 0.56 to be precise — and the Rosser-Iwaniec upper-bound sieve over the remaining range. In the next section we shall assemble the arithmetic information required for the Rosser-Iwaniec sieve and the one-dimensional form of the alternative sieve. In later sections we shall obtain further arithmetical information relevant to two- and three-dimensional applications of our approach.

8.2 THE ARITHMETICAL INFORMATION

As usual, our first task is to identify the relevant Type I and Type II information available. We classify the sums of interest as follows:

$$\text{Type I}_j: \qquad \sum_q \gamma_q \sum_m a_m \sum_{mn_1 \ldots n_j \equiv a \,(\mathrm{mod}\, q)} 1;$$

$$\text{Type II}_j: \qquad \sum_q \gamma_q \sum_{m_1 \ldots m_j \equiv a \,(\mathrm{mod}\, q)} a(1, m_1) \ldots a(j, m_j);$$

$$\text{Type I/II}: \qquad \sum_q \gamma_q \sum_{m,n} a_m b_n \sum_{\ell mn \equiv a \,(\mathrm{mod}\, q)} 1.$$

Here we assume that all variables are constrained in the usual dyadic way and that all the coefficients are *divisor-bounded* in the sense that they are no more than a power of a divisor function. The results we need follow from deep work by Bombieri, Friedlander, and Iwaniec [16, 17, 18] using mean values of Kloosterman-type sums. It would take us too far from our main theme to say anything here about the important ideas introduced by these authors.

At this point we must produce some more notation to describe the conditions that must be placed on the coefficients occurring in our sums. We write

$$\mathcal{A}^q = \{n \equiv a \,(\mathrm{mod}\, q) : n \sim x\}, \qquad \mathcal{B}^q = \{(n, q) = 1 : n \sim x\}.$$

Our expectation for Type II$_j$ information, for example, then becomes

$$\sum_q \gamma_q \sum_{m_1 \dots m_j \in \mathcal{A}^q} a(1, m_1) \dots a(j, m_j)$$
$$= \frac{1}{\phi(q)} \sum_q \gamma_q \sum_{m_1 \dots m_j \in \mathcal{B}^q} a(1, m_1) \dots a(j, m_j) + E, \tag{8.2.1}$$

where E is a smaller order error (usually by an arbitrary power of $\log x$). In summation conditions we henceforth abbreviate $(\mathrm{mod}\, q)$ simply to (q). The conditions we may want the coefficients to satisfy are as follows (the summation condition $\ell \sim L$ is assumed throughout):

(A1) *For any $d \geq 1, k \geq 1, b \neq 0, (k, b) = 1$, we have*

$$\sum_{\substack{\ell \equiv b(k) \\ (\ell, d) = 1}} \lambda_\ell = \frac{1}{\phi(k)} \sum_{(\ell, dk) = 1} \lambda_\ell + O_A \left(\|\lambda\| \tau^B(d) L^{\frac{1}{2}} (\log L)^{-A} \right) \tag{8.2.2}$$

for any $A > 1$, where

$$\|\lambda\| = \left(\sum_\ell |\lambda_\ell|^2 \right)^{\frac{1}{2}}.$$

(A2) $\lambda_\ell = 0$ *whenever ℓ has a prime factor less than* $\exp\left((\log x)(\log \log x)^{-2} \right)$.

(A3) *We have*

$$L^{1-\epsilon} \sum_\ell |\lambda_\ell|^4 \ll \left(\sum_\ell |\lambda_\ell|^2 \right)^2.$$

We note that (A1) is satisfied if $\lambda_\ell \equiv 1$ or, by the Siegel-Walfisz theorem, if

$$\lambda_\ell = \sum_{\substack{p_1 \dots p_t = \ell \\ p_j \sim D_j}} 1,$$

which will usually be the case here. Condition (A2) can be met by the technique of sieving out small primes separately that we employed in the previous chapter. Condition (A3) will be met for any divisor-bounded sequence that takes values at least 1 in size with sufficient frequency (again this happens for the types of sequences generated by our sieve method). In all the following lemmas we assume that variables m_j are supported on intervals $[M_j, 2M_j)$ and γ_q is supported on $[Q, 2Q)$. As in earlier chapters, we write $\mathcal{L} = \log x$. We write $\eta = \epsilon^2$ and put

$$\mu = (\log \log x)^{-1}, \qquad w = x^{\mu}.$$

We remark at the outset that Lemma 8.1 is only nontrivial up to $\frac{11}{20}$, Lemma 8.2 up to $\frac{3}{5}$, and Lemma 8.3 up to $\frac{4}{7}$.

Lemma 8.1. *Let $M_1 M_2 = x$, $\min(M_1, M_2) > x^{\eta}$ and suppose that $a(2, m)$ satisfies (A1), (A2), and (A3). Then (8.2.1) holds provided that*

$$x^{\eta-1} Q^2 < M_2 < x^{\frac{5}{6}-\eta} Q^{\frac{4}{3}}. \tag{8.2.3}$$

Proof. This follows from [16, Theorem 3]. $\square$

Lemma 8.2. *Let $A > 0$, $M_1 M_2 M_3 = x$, $\min_j M_j > x^{\eta}$ and suppose that $a(j, m_j)$ satisfy (A2) ($j = 1, 2, 3$) while $a(2, m_2)$ also satisfies (A1). Then there is a number $B = B(A)$ such that if*

$$Q\mathcal{L}^B < M_1 M_2, \tag{8.2.4}$$

$$M_1^2 M_2^3 < Q x^{1-\eta}, \tag{8.2.5}$$

$$M_1^4 M_2^2 (M_1 + M_2) < x^{2-\eta} \tag{8.2.6}$$

then (8.2.1) holds.

Proof. See [17, Theorem 3]. $\square$

Lemma 8.3. *Lemma 8.2 holds with (8.2.5) and (8.2.6) replaced by*

$$M_1 M_2^2 Q^2 < x^{2-3\eta} \tag{8.2.7}$$

and

$$M_1^5 M_2^2 < x^{2-3\eta}. \tag{8.2.8}$$

Proof. This is a slight variant of Theorem 4 in [17]. The modifications necessary are explained in [7, Lemma 5]. $\square$

Lemma 8.4. *Let $A > 0$ be given. Suppose that $M_1 M_2 M_3 = x$ and*

$$a(2, m_2) = \begin{cases} 1 & \text{if } M_2 \le m_2 < L, \\ 0 & \text{otherwise,} \end{cases} \tag{8.2.9}$$

where $L \in [M_2, 2M_2]$. Then there is a $B = B(A)$ such that (8.2.1) holds provided that

$$Q\mathcal{L}^B < M_1 M_2, \tag{8.2.10}$$

$$M_3 M_1^4 Q < x^{2-\eta}, \tag{8.2.11}$$

and

$$M_3 M_1^2 Q^2 < x^{2-\eta}. \tag{8.2.12}$$

Proof. This is [17, Theorem 5]. ◻

There is a different Type I/II bound that is much more powerful for larger θ. It is given as Lemme 10 in [34] as a consequence of Theorem 1 in [16], although one needs to swap the roles of the variables to put it in the form given here. By swapping roles we mean that

$$mn\ell \equiv a \,(\mathrm{mod}\,q),$$

where (8.2.9) holds with $m_2 = \ell$, is detected by

$$qs \equiv -a \,(\mathrm{mod}\,mn).$$

Now s is the "unrestricted" variable. The result is then as follows.

Lemma 8.5. *In Lemma 8.4 the conditions* (8.2.10)–(8.2.12) *can be replaced by*

$$M_1 < x^{1-\theta-\eta},$$

$$M_3 < x^\eta \min\left(Q^2 x^{-1}, \left(\frac{xQ}{M_1}\right)^{\frac{1}{5}}, \left(\frac{Q^2}{M_1}\right)^{\frac{1}{4}}\right). \tag{8.2.13}$$

Remark. For $\theta \in [3/5, 5/7]$, which is the range where the above is applied, the condition on M_3 reduces simply to

$$M_3 < x^\epsilon \left(\frac{Q^2}{M_1}\right)^{\frac{1}{4}}. \tag{8.2.14}$$

As previously, we shall associate prime variables p_j with variables α_j by $p_j = x^{\alpha_j}$, write $\boldsymbol{\alpha}_j = (\alpha_1, \ldots, \alpha_j)$, and thus turn conditions $(p_1, \ldots, p_j) \in \mathcal{D}$ into $\boldsymbol{\alpha}_j \in \mathcal{E}$. We write

$$T_j = \{\boldsymbol{\alpha}_j : \alpha_k \geq 0, \alpha_1 + cldots + \alpha_j \leq 1\}$$

and

$$A_j = A_j(\beta_0, \beta_1) = \{\boldsymbol{\alpha}_j \in T_j : \beta_0 \leq \alpha_j < \cdots < \alpha_1 < \beta_1\}.$$

We now use a modified version of terminology introduced at the end of Chapter 7. Given any set $\mathcal{H} \subset T_j$, we say that $\mathcal{H}$ *partitions* into $\mathcal{D} \subset T_2$ if, for every $\boldsymbol{\alpha}_j \in \mathcal{H}$, there exist $\mathfrak{I}, \mathfrak{J}$ with $\mathfrak{I} \cup \mathfrak{J} \subseteq \{1, \ldots, j\}, \mathfrak{I} \cap \mathfrak{J} = \emptyset$, and

$$\left(\sum_{i\in\mathfrak{I}} \alpha_i, \sum_{j\in\mathfrak{J}} \alpha_j\right) \in \mathcal{D}.$$

If $\mathfrak{I} \cup \mathfrak{J} = \{1, \ldots, j\}$ for every $\boldsymbol{\alpha} \in \mathcal{H}$, then we say that $\mathcal{H}$ partitions *exactly* into $\mathcal{D}$. If $\boldsymbol{\alpha}_j \in T_j$ we say that $\boldsymbol{\alpha}_j$ partitions into $\mathcal{D}$ if the set $\{\boldsymbol{\alpha}_j\}$ partitions into $\mathcal{D}$.

We define sets $\mathcal{G}, \mathcal{G}_1, \mathcal{G}_2, \mathcal{G}_3 \subset T_2$ and $G_j \subset T_j$ as follows. First, $(s, t) \in \mathcal{G}_1$ implies

$$2\theta - 1 + \epsilon \leq s \leq \tfrac{1}{6}(5 - 8\theta) - \epsilon. \tag{8.2.15}$$

Then $(s, t) \in \mathcal{G}_2$ if the following four inequalities are satisfied:

$$s + t \geq \theta + \epsilon \tag{8.2.16}$$

$$2s + 3t \leq 1 + \theta - \epsilon \tag{8.2.17}$$

$$5s + 2t \leq 2 - \epsilon \tag{8.2.18}$$

$$4s + 3t \leq 2 - \epsilon. \tag{8.2.19}$$

We also write $(s, t) \in \mathcal{G}_3$ to indicate that (s, t) satisfies (8.2.16), (8.2.18), and

$$s + 2t \leq 2 - 2\theta - \epsilon. \tag{8.2.20}$$

Finally, let $\mathcal{G}$ be the set of (s, t) such that either (s, t) or $(s, 1 - s - t)$ belongs to $\mathcal{G}_1 \cup \mathcal{G}_2 \cup \mathcal{G}_3$ and write $G_j = \{\alpha_j \in T_j : \alpha_j \text{ partitions into } \mathcal{G}\}$. We then have the following result.

Lemma 8.6. *Let $\mathcal{C}$ be a polyhedral subset of G_j and suppose that $\min \alpha_i \geq \mu$ when $\alpha_j \in \mathcal{C}$. Then*

$$\sum_{\alpha_j \in \mathcal{C}} S(\mathcal{A}^q_{p_1 \ldots p_j}, p_j) = \frac{(1 + O(\mathcal{L}^{-A}))}{\phi(q)} \sum_{\alpha_j \in \mathcal{C}} S(\mathcal{B}^q_{p_1 \ldots p_j}, p_j) \tag{8.2.21}$$

for most $q \sim x^\theta$.

Proof. The reader will have noticed that for $j = 1, 2, 3$, the conditions for $\mathcal{G}_j$ correspond to Lemma 8.j. We can therefore apply our usual techniques of splitting up the variables into dyadic ranges and use Perron's formula to complete the proof. More details are given in [7], but the reader has seen these arguments several times by now! We have $\min_i \alpha_i \geq \mu$ here, whereas in [7] we had $\min_i \alpha_i \geq \eta$, but this causes no real difficulties. $\qquad \square$

Definitions. We first define certain functions of θ that occur frequently. We write

$$\kappa = \begin{cases} (5 - 8\theta)/6 - \epsilon & (\theta \leq \frac{17}{32} - \epsilon), \\ (5 - 8\theta)/12 - 3\epsilon & (\frac{17}{32} - \epsilon < \theta \leq \frac{7}{13} - \epsilon), \\ (3 - 5\theta)/7 - 2\epsilon & (\frac{7}{13} - \epsilon < \theta \leq \frac{4}{7} - \epsilon), \end{cases}$$

and

$$\tau = \begin{cases} 3(1 - \theta)/5 - \epsilon & (\theta \leq \frac{11}{21}), \\ 2/7 - \epsilon & (\frac{11}{21} < \theta \leq \frac{6}{11} - \epsilon), \\ (5 - 6\theta)/7 - \epsilon & (\theta > \frac{6}{11} - \epsilon). \end{cases}$$

We also write

$$\kappa' = \begin{cases} (11 - 20\theta)/6 - 2\epsilon & \text{if } \frac{7}{13} - \epsilon < \theta \leq \frac{11}{20} - \epsilon, \\ \kappa & \text{otherwise,} \end{cases}$$

and

$$\tau' = \begin{cases} (5 - 6\theta)/7 & \text{if } \frac{7}{13} - \epsilon < \theta \leq \frac{11}{20} - \epsilon, \\ \tau & \text{otherwise.} \end{cases}$$

For $\frac{1}{2} \leq \theta \leq 0.56$ it is convenient to define the following subsets of T_2; these are illustrated in Diagrams 8.1 and 8.2 for $\theta = 0.5$ and $\theta = 0.54$. Note that for $\theta < \frac{29}{56} - \epsilon$, $\mathcal{G}_2$ lies within $\mathcal{G}_3$. The region $\mathcal{G}_1$ is empty for $\theta > \frac{11}{20}$, and $\mathcal{G}_3$ is empty for $\theta > \frac{4}{7}$. The region $\mathcal{G}_2$ is small, but nonempty, for $\theta \in [\frac{4}{7}, \frac{3}{5})$.

$A = \{\alpha_2 : \kappa \leq \alpha_2 < \alpha_1, \alpha_1 + \alpha_2 < \theta + \epsilon\},$

$B = \{\alpha_2 : \kappa \leq \alpha_2 < \alpha_1, 5\alpha_2 + 2\alpha_1 \leq 2 - \epsilon, \alpha_1 + \alpha_2 \geq \theta + \epsilon;$
$\qquad \alpha_2 \text{ satisfies either } \{(8.2.17) \text{ and } (8.2.19)\} \text{ or } (8.2.20)\},$

$B' = \{\alpha_2 : \alpha_2 > \alpha_1, \alpha_1 + \alpha_2 \geq \theta + \epsilon, 5\alpha_2 + 2\alpha_1 \leq 2 - \epsilon\},$

$E = \{\alpha_2 : \alpha_1 \leq 1 - \theta - \epsilon, \alpha_1 + 2\alpha_2 < 1, 3\alpha_2 \leq 2\alpha_1 - \epsilon;$
$\qquad (\alpha_2, 1 - \alpha_1 - \alpha_2) \text{ satisfies either } \{(8.2.17) \text{ and } (8.2.19)\} \text{ or } (8.2.20)\},$

$E' = \{\alpha_2 : \alpha_1 \leq 1 - \theta - \epsilon, \alpha_1 + 2\alpha_2 \geq 1, 3\alpha_2 \leq 2\alpha_1 - \epsilon\},$

$C = \{\alpha_2 : \kappa \leq \alpha_2 < \alpha_1, \alpha_1 + 4\alpha_2 \leq 3 - 3\theta - \epsilon, \alpha_2 \notin A \cup B \cup E\},$

$D = \{\alpha_2 : \alpha_2 < \alpha_1, \alpha_1 + 2\alpha_2 < 1, \alpha_2 \notin A \cup B \cup C \cup E\}.$

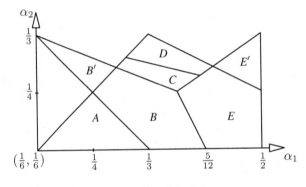

Diagram 8.1: $\theta = 0.5$

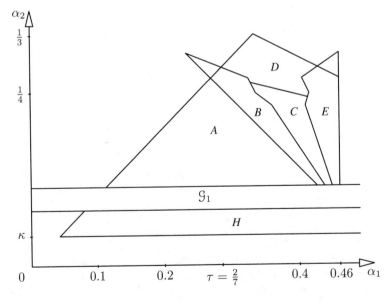

Diagram 8.2: $\theta = 0.54$

The scales for these diagrams are the same to give a faithful impression of how increasing θ has reduced the asymptotic formula regions in size. The regions $B \cup B'$ and $E \cup E'$ are asymptotic formula regions from Lemma 8.6. The region $\mathcal{G}_1$ has not been included in Diagram 8.1; it is just $\alpha_2 < \frac{1}{6}$, of course. The region H will be defined later in this chapter. For clarity B' and E' are unmarked on Diagram 8.2.

8.3 THE ALTERNATIVE SIEVE APPLIED

We shall only consider the upper bound for the present. We also suppress the superscript q on $\mathcal{A}$ and $\mathcal{B}$ for clarity. Write $V_1 = [\kappa, \frac{1}{2}]$ and let $R_1 \subseteq V_1$. Put

$$R_2 = \{\boldsymbol{\alpha}_2 \in T : \alpha_1 \in R_1, \kappa \le \alpha_2 < \min(\alpha_1, (1 - \alpha_1)/2)\},$$
$$V_3 = \{\boldsymbol{\alpha}_3 : (\alpha_1, \alpha_2) \in R_2, \kappa \le \alpha_3 < \min(\alpha_2, (1 - \alpha_1 - \alpha_2)/2)\}.$$

Finally, suppose $R_3 \subseteq V_3$. Then, by Buchstab's identity (and discarding sums corresponding to $\boldsymbol{\alpha}_1 \notin R_1, \boldsymbol{\alpha}_3 \notin R_3$), we have

$$S(\mathcal{A}, (2x)^{\frac{1}{2}}) \le S(\mathcal{A}, z) - \sum_{\alpha_1 \in R_1} S(\mathcal{A}_{p_1}, z) + \sum_{\alpha_2 \in R_2} S(\mathcal{A}_{p_1 p_2}, z)$$
$$- \sum_{\alpha_3 \in R_3} S(\mathcal{A}_{p_1 p_2 p_3}, p_3)$$
$$= S_0 - S_1 + S_2 - S_3, \quad \text{say},$$

where $z = x^\kappa$. As usual, we choose R_1, R_3 so that we can give an asymptotic formula for $S_j, 1 \le j \le 3$. Working as in Chapter 6, for example, we then have (for most q)

$$S(\mathcal{A}, (2x)^{\frac{1}{2}}) \le \frac{x}{\phi(q) \log x}(1 + I_1 + I_3 + O(\epsilon)),$$

with

$$I_1 = \int_{V_1 \setminus R_1} \frac{\omega\left(\frac{1 - \alpha_1}{\alpha_1}\right)}{\alpha_1^2} \, d\alpha_1,$$

$$I_3 = \int_{V_3 \setminus R_3} \frac{\omega\left(\frac{1 - \alpha_1 - \alpha_2 - \alpha_3}{\alpha_3}\right)}{\alpha_1 \alpha_2 \alpha_3^2} \, d\alpha_1 \, d\alpha_2 \, d\alpha_3.$$

We shall be able to use a one-dimensional approach for $\alpha_1 < 1 - \theta$. Indeed, this method works for all the regions we need to consider when $\kappa \le \alpha_1 \le \tau$ or $\frac{3}{7} \le \alpha_1 \le 1 - \theta$. In later sections we shall apply two- and three-dimensional forms of the method to treat $1 - \theta \le \alpha_1 \le \frac{1}{2}$ and certain parts of S_2 (we shall find that the region D requires a two-dimensional sieve). We will find that a similar phenomenon occurs as we encountered in Chapter 6. For θ close to $\frac{1}{2}$ we can get a good approximation to $S(\mathcal{A}, (2x)^{1/2})$, but as θ increases we will be effectively bounding a sum $S(\mathcal{A}, x^\tau)$ but "deducting" the sum

$$\sum_{3/7 < \alpha_1 < 1 - \theta} S(\mathcal{A}_{p_1}, p_1).$$

We write S for the subset of $\alpha_2 \in T$ with

$$\alpha_1 \le 1 - \theta - \epsilon, \quad \alpha_1 + 2\alpha_2 \le 2 - 2\theta - \epsilon, \quad \alpha_1 + 4\alpha_2 \le 2 - \theta - \epsilon. \quad (8.3.1)$$

For now $A_j = A_j(\mu, \tau)$, and we let S_j be the subset of $\mathbb{R}^j$, which partitions exactly into S. The significance of this is that $\alpha_j \in S_j$ implies that the variables $p_1, \dots, p_j$ can be grouped into two variables m, k where $m \sim M = x^s, k \sim K = x^t$, satisfying the conditions of Lemma 8.4. In other words this is what is needed for Type I/II information.

We now prove some elementary results that give conditions for points to belong to G_j or S_j.

Lemma 8.7. *Suppose* $\theta \le \frac{11}{20} - \epsilon$. *Let* $\alpha_k \in A_k$ *with*

$$\alpha_{j+1} + \cdots + \alpha_{k-1} < 2\theta - 1 + \epsilon \le \alpha_{j+1} + \cdots + \alpha_k \quad (8.3.2)$$

and

$$\alpha_{j+1} < \kappa' \quad (8.3.3)$$

for some j satisfying $0 \le j \le k - 1$. Then $\alpha_k \in G_k$.

Remark. We note that since $\kappa = \kappa'$ for $\theta < \frac{7}{13} - \epsilon$, the above result holds with (8.3.3) replaced by

$$\alpha_{j+1} < \kappa, \quad \theta < \frac{7}{13} - \epsilon. \quad (8.3.4)$$

Proof. First suppose that $\alpha_k < (11 - 20\theta)/6 - 2\epsilon$. It then follows that

$$2\theta - 1 + \epsilon \le \alpha_{j+1} + \cdots + \alpha_k < \tfrac{1}{6}(5 - 8\theta) - \epsilon.$$

Thus (8.2.15) is verified, and so $\alpha_k \in G_k$.

Now suppose that $\alpha_k \ge (11 - 20\theta)/6 - 2\epsilon, \theta \le \frac{17}{32} - \epsilon$. From (8.3.3) we therefore deduce that

$$2\theta - 1 + \epsilon \le \alpha_k \le \tfrac{1}{6}(5 - 8\theta) - \epsilon.$$

Again (8.2.15) is verified, so that $\alpha_k \in G_k$.

Finally, we take the case $\alpha_k \ge (11 - 20\theta)/6 - 2\epsilon, \frac{17}{32} - \epsilon < \theta \le \frac{7}{13} - \epsilon$. We now have

$$2\alpha_k \ge \frac{11 - 20\theta}{3} - 4\epsilon > 2\theta - 1 + \epsilon.$$

By our definition of κ in this region, $\alpha_{j+1} < (5 - 8\theta)/12 - \epsilon < 2\theta - 1 + \epsilon$. It then follows from (8.3.2) that $k \ge j + 2$. We then have

$$2\theta - 1 + \epsilon \le 2\alpha_k \le \alpha_{j+1} + \alpha_{j+2} < 2\kappa = \tfrac{1}{6}(5 - 8\theta) - 2\epsilon.$$

As before, this gives $\alpha_k \in G_k$, which completes the proof. □

Lemma 8.8. *Let* $\theta \in (\frac{7}{13} - \epsilon, \frac{4}{7} - \epsilon]$. *Let* $\alpha_k \in A_k$, *where $k \ge 1$, and suppose that*

$$\alpha_1 + \cdots + \alpha_k < \theta + \epsilon. \quad (8.3.5)$$

Then $\alpha_k \in S_k$.

Proof. If $\alpha_1 + \cdots + \alpha_k < 1 - \theta - \epsilon$, the result is immediate with $t = 0, s = \alpha_1 + \cdots + \alpha_k$. Otherwise we have

$$\alpha_1 + \cdots + \alpha_k \geq 1 - \theta - \epsilon > 2\theta - 1 + 2\epsilon. \qquad (8.3.6)$$

We choose i to be the smallest integer such that $\alpha_1 + \cdots + \alpha_i > 2\theta - 1 + 2\epsilon$. Thus

$$\alpha_1 + \cdots + \alpha_{i-1} < 2\theta - 1 + 2\epsilon \leq \alpha_1 + \cdots + \alpha_i. \qquad (8.3.7)$$

Write $v = \alpha_1 + \cdots + \alpha_i, u = \alpha_1 + \ldots + \alpha_k - v$.

If $\alpha_1 > 2\theta_1 - 1 + \epsilon$, then $v = \alpha_1$. Otherwise, $v \leq 4\theta - 2 + 2\epsilon$. Now

$$u < \theta + \epsilon - v \leq 1 - \theta - \epsilon$$

from (8.3.5) and (8.3.7). Also,

$$u + 2v < \theta + \epsilon + v \leq \max(5\theta - 2 + 3\epsilon, \theta + \epsilon + \alpha_1) \leq 2 - 2\theta - \epsilon$$

since $\alpha_1 \leq \tau$. In the same way we have

$$u + 4v < \theta + \epsilon + 3v \leq \max(13\theta - 6 + 7\epsilon, \theta + \epsilon + 3\alpha_1) \leq 2 - \theta - \epsilon.$$

This completes the proof of Lemma 8.8. $\square$

Now put

$$T^* = \{(s,t) : 0 \leq s \leq \tfrac{1}{7}(8\theta - 2), 0 \leq t \leq \tfrac{1}{7}(5 - 6\theta)\},$$

$$U'_j = U'_j(\theta) = \{\boldsymbol{\alpha}_j \in A_j : (\alpha_1, \ldots, \alpha_j, 2\theta - 1 + \epsilon) \in S_{j+1}\}.$$

If $\theta < \tfrac{7}{13} - \epsilon$, we put $U_j = U_j(\theta) = U'_j(\theta)$. Otherwise we put

$$U_j = U_j(\theta) = \{\boldsymbol{\alpha}_j \in A_j : \boldsymbol{\alpha}_j \text{ partitions exactly into } T^*\}.$$

Lemma 8.9. *Let $\boldsymbol{\alpha}_k \in A_k$, $\theta > \tfrac{7}{13} - \epsilon$, and suppose that for some $j, 0 \leq j \leq k-1$, we have $\boldsymbol{\alpha}_j \in U_j$ and*

$$\alpha_{j+1} < \kappa. \qquad (8.3.8)$$

Suppose, in addition, that

$$\alpha_1 + \cdots + \alpha_k \geq \theta + \epsilon > \alpha_1 + \cdots + \alpha_{k-1}. \qquad (8.3.9)$$

Then $\boldsymbol{\alpha}_k \in G_k$.

Proof. From the definition of U_j we can combine $\alpha_1, \ldots, \alpha_j$ into two variables s, t with $(s,t) \in T^*$. Recalling that $\kappa = (3 - 5\theta)/7 - 2\epsilon$ for $\theta > \tfrac{7}{13} - \epsilon$ we have

$$s \leq 1 - \theta - 3\kappa - 6\epsilon.$$

Thus

$$t + \alpha_{j+1} + \cdots + \alpha_k \geq \theta + \epsilon - s \geq 2\theta - 1 + 3\kappa + 4\epsilon. \qquad (8.3.10)$$

Using this, the restriction $t \leq (5 - 6\theta)/7 \leq 2 - 3\theta - 3\kappa - 4\epsilon$ and the fact that $\alpha_t \ (j + 1 \leq t \leq \ell)$ are all bounded above by κ from (8.3.8), we conclude that, for some $\ell \ (j \leq \ell \leq k)$, we must have

$$\beta_\ell := t + \sum_{j+1 \leq t \leq \ell} \alpha_t \in [2\theta - 1 + 3\kappa + 4\epsilon, 2 - 3\theta - 3\kappa - 4\epsilon]. \qquad (8.3.11)$$

We now show that

$$(\beta_\ell, \alpha_1 + \cdots + \alpha_k - \beta_\ell) \in \mathcal{G}_2. \tag{8.3.12}$$

From (8.3.9) we obtain (8.2.16). Combining (8.3.8) and (8.3.9), we get

$$\alpha_1 + \cdots + \alpha_k \le \theta + \kappa + \epsilon. \tag{8.3.13}$$

Using this with (8.3.11) we derive (8.2.17) since

$$2\beta_\ell + 3(\alpha_1 + \cdots + \alpha_k - \beta_\ell) \le 3(\theta + \kappa + \epsilon) - (2\theta - 1 + 3\kappa + 4\epsilon)$$
$$= 1 + \theta - \epsilon.$$

Finally, to establish (8.2.18) and (8.2.19) we note that

$$5\beta_\ell + 2(\alpha_1 + \cdots + \alpha_k - \beta_\ell) \le 4\beta_\ell + 3(\alpha_1 + \cdots + \alpha_k - \beta_\ell)$$
$$\le 2 - 3\theta - 3\kappa - 4\epsilon + 3(\theta + \kappa + \epsilon)$$
$$= 2 - \epsilon.$$

This establishes (8.3.12), and so $\alpha_k \in G_k$. $\qquad\square$

At last we can now establish our basic asymptotic formulae. The following result follows from Theorem 5.2 with Lemmas 8.1 and 8.4. We shall not have cause to use it further in this section, but it will play a vital role in the next.

Lemma 8.10. *Suppose that $\theta < \frac{11}{20} - \epsilon$ and $R < x^{2-2\epsilon}Q^{-3}$. Let c_r be a divisor-bounded sequence. Put*

$$z = x^{\frac{1}{6}(11-20\theta)-\epsilon}.$$

Then, for most q,

$$\sum_{r \sim R} c_r S(\mathcal{A}_r, z) = \frac{1}{\phi(q)} \sum_{r \sim R} c_r S(\mathcal{B}_r, z) \left(1 + O(\mathcal{L}^{-A})\right). \tag{8.3.14}$$

When $\alpha_1 < \tau$, we can always do at least as well as the above result by the following lemma that uses more of the results we have derived on asymptotic formulae.

Lemma 8.11. *Let H be a polyhedral subset of U_j where $\frac{1}{2} \le \theta \le \frac{7}{13} - \epsilon$. Then we have, for most q,*

$$\sum_{\alpha_j \in H} S(\mathcal{A}_{p_1 \ldots p_j}, x^\kappa) = \frac{1}{\phi(q)} \sum_{\alpha_j \in H} S(\mathcal{B}_{p_1 \ldots p_j}, x^\kappa) \left(1 + O(\mathcal{L}^{-A})\right). \tag{8.3.15}$$

We can also obtain an asymptotic formula for $S(\mathcal{A}, x^\kappa)$. Moreover, (8.3.15) remains valid for $\frac{7}{13} - \epsilon < \theta < \frac{11}{20} - \epsilon$ if we replace U_j by U_j' and κ by κ'.

Proof. We write $r_t = p_1 \ldots p_t$, $z = x^\kappa$. We suppose first that $\theta \le \frac{7}{13} - \epsilon$. We also note that the evaluation of $S(\mathcal{A}, z)$ will follow in a similar manner to our derivation of (8.3.15). Working as in previous chapters we put

$$\rho(n) = \begin{cases} 1 & \text{if } (n, P(w)) = 1, \\ 0 & \text{otherwise.} \end{cases}$$

Here we revert to the notation

$$P(w) = \prod_{p<w} p$$

and use, as in Chapter 7,

$$P(w,z) = \prod_{w\le p<z} p = \frac{P(z)}{P(w)}.$$

Now

$$S(\mathcal{A}_{r_j}, z) = \sum_{\substack{d\mid P(w,z) \\ dnr_j\in\mathcal{A}}} \mu(d)\rho(n)$$

$$= \frac{1}{\phi(q)} \sum_{\substack{d\mid P(w,z) \\ dnr_j\in\mathcal{B}}} \mu(d)\rho(n) + \sum_{d\mid P(w,z)} \mu(d)\psi(dr_j)$$

$$= \frac{1}{\phi(q)} S(\mathcal{B}_{r_j}, z) + S(r_j), \quad \text{say.}$$

Here

$$\psi(m) = \sum_{mn\in\mathcal{A}} \rho(n) - \frac{1}{\phi(q)} \sum_{mn\in\mathcal{B}} \rho(n).$$

We may then handle

$$\sum_{\alpha_j\in H} S(r_j)$$

in our usual manner. We take out the prime factors of d starting with the largest. Let these be $p_{j+1}, \ldots$ So long as

$$\alpha_{j+1} + \cdots + \alpha_k < 2\theta - 1 + \epsilon, \tag{8.3.16}$$

we have $\alpha \in S_k$ (by the definition of U_j'). We can therefore apply Lemma 8.4 in tandem with the fundamental lemma from Chapter 4 when necessary. If (8.3.16) fails, then we note that the conditions of Lemma 8.7 are met. Hence we have an asymptotic formula for this part of the sum. We therefore have our standard procedure: We can keep taking out prime factors until at last an asymptotic formula is obtained for every part of the sum. We thus arrive at

$$\sum_{\alpha_j\in H} S(r_j) \ll \mathcal{L}^{-A} \frac{1}{\phi(q)} \sum_{\alpha_j\in H} S(\mathcal{B}_{r_j}, z)$$

as required to complete the proof when $\theta < \frac{7}{13} - \epsilon$.

For $\frac{7}{13} - \epsilon \le \theta < \frac{11}{20} - \epsilon$ the proof is identical except that we use Lemma 8.1 in place of Lemma 8.7. We thus need

$$\alpha_{j+1} + \cdots + \alpha_k \in [2\theta - 1 + \epsilon, 5/6 - \epsilon - 4\theta/3].$$

In this case the result is very much like Lemma 8.10 except that we have used the Type I/II information more efficiently. $\qquad\square$

Lemma 8.12. *Let $\frac{7}{13} - \epsilon < \theta \le \frac{4}{7} - \epsilon$. We can obtain an asymptotic formula for $S(\mathcal{A}, x^\kappa)$ and derive (8.3.15) provided that H is a polyhedral subset of U_j.*

Remark. We have increased κ' to κ compared to Lemma 8.11, but at the expense of deleting $U_j' \setminus U_j$.

Proof. The proof is similar to the previous result, except that in place of (8.3.16) as the deciding inequality at each stage we use

$$\alpha_1 + \cdots + \alpha_k < \theta + \epsilon. \tag{8.3.17}$$

This condition with Lemma 8.8 enables us to have the required Type I/II information. When (8.3.17) fails, we can use Lemmas 8.9 and 8.6 to obtain an asymptotic formula. The reader should now have no difficulty filling in all the details to complete the proof if so required. $\qquad\square$

So far our results have been useful for $\alpha_2 \in A \cup C$. We now obtain some results that are useful when one variable (or product of variables) lies between $x^{3/7+\epsilon}$ and $x^{1-\theta-\epsilon}$. We write

$$T^{**} = \{(s,t) : \tfrac{3}{7} + \epsilon \le s \le 1 - \theta - \epsilon, 0 \le t \le \tfrac{1}{2}(1-s)\},$$

$$U_j^* = \{\alpha_j \in \mathbb{R}^j : \alpha_j \text{ partitions exactly into } T^{**}\},$$

$$\gamma(\alpha) = \tfrac{2}{3}\alpha - \min(2 - \theta - 3\alpha, 2\theta - 2\alpha) - 2\epsilon.$$

The reader will see that in the following $\gamma(\alpha)$ takes the place of κ and U_j^* replaces U_j or U_j'.

Lemma 8.13. *Let $\frac{1}{2} \le \theta \le \frac{4}{7} - \epsilon$. Let H be a polyhedral subset of U_j^*. Suppose that $\alpha_1, \ldots, \alpha_j$ partitions into $(s,t) \in T^{**}$ (there must be at least one such partition from the definition of U_j^*). Then, for any $\lambda \in [\eta, \gamma(s)]$ we have*

$$\sum_{\alpha_j \in H} S(\mathcal{A}_{p_1 \ldots p_j}, x^\lambda) = \frac{1}{\phi(q)} \sum_{\alpha_j \in H} S(\mathcal{B}_{p_1 \ldots p_j}, x^\lambda)\left(1 + O\left(\mathcal{L}^{-A}\right)\right). \tag{8.3.18}$$

Proof. We may again follow the proof of Lemma 8.11. This time the deciding inequality in the decomposition of d is

$$t + \alpha_{j+1} + \cdots + \alpha_k < \min(2 - \theta - 3s, 2\theta - 2s) + \epsilon. \tag{8.3.19}$$

The reader may verify that when (8.3.19) is satisfied, then Lemma 8.4 can be used as before with $M_3 = x^s$, $M_1 = x^u$ with $u = t + \alpha_j + \cdots + \alpha_k$ (note that $s \ge \frac{3}{7}$ is vital to make the inequalities work here). On the other hand, when (8.3.19) fails, $(t', 1 - s - t') \in \mathcal{G}$, where

$$t' = t + \alpha_{j+1} + \cdots + \alpha_k.$$

Hence Lemma 8.6 supplies the necessary asymptotic formula. Full details are given in [7, Lemma 17]. $\qquad\square$

We now use Lemma 8.13 to establish the following result, which is very useful when performing the calculations needed later. This result demonstrates that it is often possible to obtain an asymptotic formula if some of the variables can be grouped together to form a product lying between $x^{3/7+\epsilon}$ and $x^{1-\theta-\epsilon}$.

Lemma 8.14. *Let $\frac{1}{2} \le \theta \le \frac{4}{7} - \epsilon$. Let H be a polyhedral subset of U_j^* and let $\xi(\boldsymbol{\alpha}_j)$ be a continuous function on H with $\eta \le \xi(\boldsymbol{\alpha}_j) \le \frac{1}{2}$. Then we have*

$$\sum_{\boldsymbol{\alpha}_j \in H} S(\mathcal{A}_{p_1 \ldots p_j}, x^{\xi(\boldsymbol{\alpha}_j)}) = \frac{1}{\phi(q)} \sum_{\boldsymbol{\alpha}_j \in H} S(\mathcal{B}_{p_1 \ldots p_j}, x^{\xi(\boldsymbol{\alpha}_j)}) \left(1 + O(\mathcal{L}^{-A})\right).$$

(8.3.20)

Corollary. *For $\frac{1}{2} \le \theta \le \frac{4}{7} - \epsilon$ we can give an asymptotic formula for*

$$\sum_{x^{3/7-\epsilon} < p < x^{1-\theta-\epsilon}} S(\mathcal{A}_p, p).$$

Proof. Let $s = s(\boldsymbol{\alpha}_j)$ be as in the statement of Lemma 8.13. We can then assume, in view of that result, that $\xi(\boldsymbol{\alpha}_j) > \gamma(\boldsymbol{\alpha}_j)$. The sum is also empty when $\xi(\boldsymbol{\alpha}_j) > 1 - \alpha_1 - \cdots - \alpha_j$. To simplify the proof we shall henceforth assume that $\xi(\boldsymbol{\alpha}_j)$ is fixed, say ξ, with $\gamma(\boldsymbol{\alpha}_j) < \xi \le 1 - \alpha_1 - \cdots - \alpha_j$. We can also assume that $\boldsymbol{\alpha} \notin G_j$, for otherwise there is nothing left to prove.

We combine s, t into two variables as in Lemma 8.13. An application of Buchstab's identity to (8.3.20) yields

$$\sum_{\boldsymbol{\alpha}_j \in H} S(\mathcal{A}_{p_1 \ldots p_j}, x^{\gamma(s)}) - \sum_{\substack{\boldsymbol{\alpha}_j \in H \\ \gamma(s) < \alpha_{j+1} \le \xi}} S(\mathcal{A}_{p_1 \ldots p_{j+1}}, p_{j+1}). \qquad (8.3.21)$$

The first sum in (8.3.21) can be estimated by Lemma 8.13. We can assume in the second sum that

$$\alpha_1 + \cdots + \alpha_j + 2\alpha_{j+1} < 1 \qquad (8.3.22)$$

since we cannot have $p_1 \ldots p_j p_{j+1}^2 > x$.

We now test to see if (s, α_{j+1}) belongs to E (recall that E is a subset of $\mathcal{G}$). We note that $s \le 1 - \theta - \epsilon$. By (8.3.22) $s + 2\alpha_{j+1} < 1$. Combining this with $s \ge \frac{3}{7} + \epsilon$ then yields $3\alpha_{j+1} \le 2s - \epsilon$. Hence $(s, \alpha_{j+1}) \in E$ unless $(\alpha_{j+1}, 1 - s - \alpha_{j+1})$ satisfies neither $\{(8.2.17), (8.2.19)\}$ nor (8.2.20). A little algebraic manipulation reduces this to

$$\alpha_{j+1} < \min(2 - \theta - 3s, 2\theta - 2s) + \epsilon = \rho(\theta, s), \text{ say.}$$

We note that t must also satisfy this inequality, as we have assumed $(s, t) \notin \mathcal{G}$. Now let the remaining sum be S_1 after removing that part with $\alpha_{j+1} \ge \rho(\theta, s)$, We write

$$S_1 = S_2 + S_3 + S_4,$$

where $t' = \alpha_{j+1} + t$ satisfies $\rho(\theta, s) \le t' \le \frac{2}{3}s - \epsilon$ in S_2, $t' < \rho(\theta, s)$ in S_3, and $t' > \frac{2}{3}s - \epsilon$ in S_4. The proof for S_2 goes through as the previous lemma: we can show that $(t', 1 - s - t') \in \mathcal{G}$.

For S_4 we note that $s > \frac{3}{7} + \epsilon$ gives $2s - \epsilon \ge (1 - s)/2$, which then implies $s + 2t' \ge 1$ in this case. We now show that $(t', 1 - s - t') \in \mathcal{G}_3$. Since $s \le 1 - \theta - \epsilon$, we have $t' + (1 - s - t') \ge \theta + \epsilon$. Then

$$(1 - s - t') + 2t' = 1 + t' - s$$
$$\le 1 + 2(2 - \theta - 3s) - s = 5 - 2\theta - 7s$$
$$< 2 - 2\theta - \epsilon$$

since $s \geq \frac{3}{7} + \epsilon$. Finally,

$$5(1 - s - t') + 2t' = 5 - \tfrac{7}{2}s - \tfrac{3}{2}(s + 2t')$$
$$\leq \tfrac{7}{2}(1 - s) < 2 - \epsilon,$$

again since $s \geq \frac{3}{7} + \epsilon$. Thus $(t', 1 - s - t') \in \mathcal{G}_3$ as claimed, and we can obtain an asymptotic formula for S_4.

Finally, in S_3 we apply Buchstab's identity once more to obtain

$$\sum_{\substack{\alpha_j \in H \\ \gamma(s) < \alpha_{j+1} \leq \xi \\ t' < \rho(\theta, s)}} S(\mathcal{A}_{p_1 \ldots p_{j+1}}, x^{\gamma(s)}) - \sum_{\substack{\alpha_j \in H \\ \gamma(s) < \alpha_{j+2} < \alpha_{j+1}}} S(\mathcal{A}_{p_1 \ldots p_{j+2}}, p_{j+2}).$$

The first sum above can be estimated by Lemma 8.13, while the second can be split into two parts $S_5 + S_6$ depending on whether or not $(s, \alpha_{j+1} + \alpha_{j+2}) \in E$. If $(s, \alpha_{j+1} + \alpha_{j+2}) \notin E$, then working as for S_4 we can show that $(t', 1 - s - t') \in \mathcal{G}_3$ with $t' = \alpha_{j+1} + \alpha_{j+2}$. We thus obtain an asymptotic formula for each sum, which completes the proof of this lemma. □

8.4 THE ALTERNATIVE SIEVE FOR $\tau \leq \alpha_1 \leq \frac{3}{7}, \theta \leq \frac{11}{21}$

After reading an earlier draft of this book, Kaisa Matomäki pointed out that there was a gap in the proof that had lain undetected in the original paper [7] for 10 years. The results of this section correct this oversight.

Lemma 8.15. *Suppose that $\theta \leq \frac{17}{32} - \epsilon, R < x^{2-\epsilon}Q^{-3}$. Let c_r be a divisor-bounded sequence. Then, for most q,*

$$\sum_{r \sim R} c_r S(\mathcal{A}_r, x^\kappa) = \frac{1}{\phi(q)} \sum_{r \sim R} c_r S(\mathcal{B}_r, x^\kappa) \left(1 + O(\mathcal{L}^{-A})\right). \qquad (8.4.1)$$

Remark. We have $x^{2-\epsilon}Q^{-3} > x^{3/7}$ for $\theta < \frac{11}{21} - \epsilon$.

Proof. We have, with $z = x^{(11-20\theta)/6 - \epsilon}$,

$$\sum_{r \sim R} c_r S(\mathcal{A}_r, x^\kappa) = \sum_{r \sim R} c_r S(\mathcal{A}_r, z) - \sum_{r \sim R} \sum_{z \leq p < x^\kappa} S(\mathcal{A}_{rp}, p)$$
$$= S_1 - S_2, \quad \text{say.}$$

We can evaluate S_1 via Lemma 8.10. Also, since $\theta \leq \frac{17}{32} - \epsilon$, we have

$$\frac{11 - 20\theta}{6} > 2\theta - 1.$$

Hence S_2 can be evaluated using Lemma 8.1. □

Lemma 8.16. *Suppose that $\theta \leq \frac{11}{21} - \epsilon$. Then, for most q,*

$$\sum_{\alpha_2 \in A \cup C} S(\mathcal{A}_{p_1 p_2}, x^\kappa) = \frac{1}{\phi(q)} \sum_{\alpha_2 \in A \cup C} S(\mathcal{B}_{p_1 p_2}, x^\kappa) \left(1 + O(\mathcal{L}^{-A})\right).$$

Proof. The case $\alpha \leq \tau$ was covered by the work in the previous section. We can therefore assume henceforth that

$$\alpha_1 > \tau = \frac{3(1-\theta)}{5} - \epsilon.$$

Suppose first that $\alpha_2 \in A$. Then, by our usual procedure, we need only show that we can evaluate

$$\sum_{\alpha_2 \in A} \sum_{p,\ldots,p_r} S(\mathcal{A}_{p_1\ldots p_r}, w) \tag{8.4.2}$$

when $\alpha_3 + \cdots + \alpha_r \leq 2\theta - 1 + \eta$ (also including the "empty" case when p_j does not appear for $j \geq 3$). This suffices since

$$2(2\theta - 1 + \eta) < \frac{5 - 8\theta}{6}$$

certainly holds for $\theta \leq \frac{11}{21}$. We can treat (8.4.2) with the fundamental lemma from Chapter 4 using Lemma 8.4 so long as $\alpha_2 \in A, \alpha_3 \leq 2\theta - 1 + \eta$ implies that we can partition $(\alpha_1, \alpha_2, \alpha_3)$ into s, t with (compare the hypotheses of Lemma 8.4 and the definition of S in (8.3.1))

$$t \leq 1 - \theta - \eta, \quad t + 4s + \theta \leq 2 - \eta, \quad t + 2s + 2\theta \leq 2 - \eta.$$

We let $t = \alpha_1 + \alpha_3, s = \alpha_2$. Then

$$t \leq \tfrac{3}{7} + 2\theta - 1 + \eta \leq 1 - \theta - \epsilon$$

if $\theta \leq \frac{11}{21} - \epsilon$. Also

$$\begin{aligned} t + 4s + \theta &\leq \alpha_1 + 4\alpha - 2 + 3\theta + \eta - 1 \\ &= 4(\alpha_1 + \alpha_2) - 3\alpha_1 + 3\theta + \eta - 1 \\ &\leq 4\theta - \frac{9(1-\theta)}{5} + 3\theta + 4\epsilon - 1 \\ &= 2 + \frac{44\theta - 24}{5} + 4\epsilon < 2 - \epsilon \end{aligned}$$

if $\theta \leq \frac{6}{11} - \epsilon$, which is guaranteed by $\theta < \frac{11}{21}$. To finish our consideration of $\alpha_2 \in A$ we note that

$$\begin{aligned} t + 2s + 2\theta &\leq \alpha_1 + 2\alpha_2 + 4\theta - 1 + \eta \\ &= 2(\alpha_1 + \alpha_2) - \alpha_1 + 4\theta - 1 + \eta \\ &\leq 2\theta - \frac{3(1-\theta)}{5} + 4\theta - 1 + 2\epsilon \\ &= 2 + 2\epsilon + \frac{33\theta - 18}{5} < 2 - \epsilon \end{aligned}$$

since $\theta \leq \frac{11}{21} - \epsilon$.

Now suppose that $\alpha_2 \in C$. We start off as in the previous case. However, we note that if $\alpha_1 + \alpha_3 > \frac{3}{7}$, then $\alpha_1 + \alpha_3 < 1 - \theta$ and so we could apply Lemma 8.14. We can therefore assume henceforth that $\alpha_1 + \alpha_3 \leq \frac{3}{7}$. The condition $t < 1 - \theta$

holds as for the case $\alpha_2 \in A$. To verify the other two conditions we recall that for $\alpha_2 \in C$ we have $\alpha_1 + 4\alpha_2 \leq 3 - 3\theta - \epsilon$. We thus have

$$t = 4s + \theta \leq \alpha_1 + 4\alpha_2 + 3\theta_1 + \eta \leq 2 - \eta.$$

Finally,

$$\begin{aligned}
t + 2s + 2\theta &= \tfrac{1}{2}(\alpha_1 + \alpha_3) + \tfrac{1}{2}(\alpha_1 + 4\alpha_2) + \tfrac{1}{2}\alpha_3 + 2\theta \\
&\leq \tfrac{3}{14} + \tfrac{1}{2}(2\theta - 1 + \eta) + \tfrac{1}{2}(3 - 3\theta - \epsilon) + 2\theta \\
&< 2 + \frac{21\theta - 11}{14} < 2 - \epsilon
\end{aligned}$$

since $\theta \leq \tfrac{11}{21} - \epsilon$. This completes the proof. $\qquad\square$

8.5 THE ALTERNATIVE SIEVE IN TWO DIMENSIONS

We now consider the region $1 - \theta - \epsilon < \alpha_1 \leq \tfrac{1}{2}$. We need to deal with this region, for otherwise it would be impossible to give the lower bound on average and the upper bound would be increased. In the following we suppose that $\tfrac{1}{2} < \theta < \tfrac{17}{32} = 0.53125$. We write

$$\mathcal{C} = \{mn : x^{1-\theta-\epsilon} \leq m \leq x^{\frac{1}{2}}, mn \sim x, (mn, q) = 1\}$$

and put $\mathcal{F} = \mathcal{C} \cap A$. The next lemma gives the basic formula for the alternative sieve in two dimensions applied to the current problem. It can be proved in a more general form, but this would not lead to any improvement in our result.

Lemma 8.17. *For most $q \sim x^\theta$ we have*

$$S(\mathcal{F}, x^\kappa) = \frac{1}{\phi(q)} S(\mathcal{C}, x^\kappa) \left(1 + O(\mathcal{L}^{-A})\right). \tag{8.5.1}$$

Proof. We need to combine the sieve framework of Lemma 7.17 with some Type I_2 information appropriate for the current problem. The main new work in this lemma is to establish an asymptotic formula for $|\mathcal{F}_d|$ when $d < x^{1-4\eta}q^{-3/2}$.

Suppose that $(d, q) = 1$ with $\mu(d) \neq 0$. Then

$$|\mathcal{F}_d| = \sum_{e|d} \sum_{\substack{(n,d/e)=1 \\ emn \in \mathcal{F} \\ m \equiv 0(d/e) \\ x^{1-\theta-\epsilon} \leq en \leq x^{1/2}}} 1 = \sum_{e|d} \sum_{f|d/e} \mu(f) \sum_{\substack{nfdm \equiv a(q) \\ x^{1-\theta-\epsilon} \leq nfe \leq x^{1/2} \\ mndf \sim x}} 1.$$

To evaluate this last expression we first apply the following subdivision argument. We restrict nfe and md/e to the intervals $\mathcal{I} = [u, u + I)$ and $\mathcal{J} = [v, v + J)$, respectively, with $uv \sim x$ and $I = u\mathcal{L}^{-A}, J = v\mathcal{L}^{-A}$, to produce a subsum F_d, say. So

$$F_d = \sum_{f|d} \mu(f) \sum_{e|d/f} \left(\frac{\phi(q)IJ}{q^2 df} + S(d, e, f), \right)$$

where

$$S(d, e, f) = \sum{}^{*} 1 - \frac{\phi(q)IJ}{q^2 df},$$

and we have written $*$ for the summation conditions

$$nfdm \equiv a(q), \; nfe \in \mathfrak{I}, \; md/e \in \mathfrak{J}.$$

By a simple calculation we have

$$\sum_{f|d} \mu(f) \sum_{e|d/f} \frac{\phi(q)}{q^2 df} = \frac{\phi(q)}{q^2 d} \rho(d)$$

with

$$\rho(d) = \tau(d) \prod_{p|d} \left(1 - \frac{1}{2p}\right).$$

Note, for future reference, that for a prime p we have $\rho(p) = 2 - 1/p$. It is elementary that

$$\sum_{\substack{f|d \\ f>w}} \sum{}^{*} 1 < \frac{x^{1+\eta}}{dwq}.$$

We thus have

$$F_d = \frac{\phi(q)\rho(d)IJ}{q^2 d} + O\left(\frac{x^{1-\eta}}{qd}\right) + O(S_d),$$

where

$$S_d \ll \max_{\substack{f<x^{2\eta} \\ e|d/f}} x^{\eta} |S(d,e,f)|.$$

We recall the well-known estimate for an incomplete Kloosterman sum of a special type (see [93, p. 36]):

$$\sum_{\substack{A \leq n \leq B \\ (n,q)=1}} e\left(\frac{b\bar{n}}{q}\right) \ll q^{\frac{1}{2}+\eta}(b,q)^{\frac{1}{2}}. \tag{8.5.2}$$

Here, and in the following, $\bar{n}$ indicates inverse (mod q). Using the elementary formula

$$\frac{1}{q} \sum_{r=0}^{q-1} e\left(\frac{r(a-n)}{q}\right) = \begin{cases} 1 & \text{if } n \equiv a(q), \\ 0 & \text{otherwise}, \end{cases}$$

we then obtain

$$\sum{}^{*} 1 = \frac{1}{q} \sum_{r=0}^{q-1} \sum_{md/e \in \mathfrak{J}} e\left(\frac{-rm}{q}\right) \sum_{\substack{(n,q)=1 \\ nfe \in \mathfrak{I}}} e\left(\frac{ar\overline{df}n}{q}\right)$$

$$= S_0 + S_1,$$

where S_0 corresponds to $r = 0$, and S_1 to $r > 0$. Since

$$\sum_{\substack{(n,q)=1 \\ U \leq n < V}} 1 = \frac{\phi(q)}{q}(V - U) + O(q^{\eta}),$$

we have

$$S_0 = \frac{\phi(q)}{q^2 df} IJ + O\left(q^{\eta-1} + \frac{I}{feq} + \frac{Jeq^{\eta-1}}{d}\right).$$

Also, if $|r| \leq q/2$,

$$\sum_{md/e \in \mathcal{J}} e\left(\frac{-rm}{q}\right) \leq \min(J, q/|r|).$$

Thus, using (8.5.2), we have

$$S_1 \leq \sum_{1 \leq |r| \leq q} \frac{1}{|r|} (r,q)^{\frac{1}{2}} q^{\frac{1}{2}+\eta} \ll q^{\frac{1}{2}+2\eta}.$$

Thus, supposing that $d < x^{1-4\eta} q^{-3/2}$ and $\max(I,J) \leq q^{3/2}$, we have

$$F_d = \frac{\phi(q)\rho(d)}{q^2 d} IJ \left(1 + O(\mathcal{L}^{-A})\right).$$

It is then easy to deduce that

$$|\mathcal{F}_d| = \frac{\phi(q)\rho(d)}{q^2 d} \sum_{\substack{mn \sim x \\ x^{1-\theta-\epsilon} \leq m \leq x^{1/2}}} \left(1 + O(\mathcal{L}^{-A})\right)$$

$$= \frac{1}{\phi(q)} |\mathcal{C}_d| \left(1 + O(\mathcal{L}^{-A})\right).$$

This estimate, along with the method used in Lemma 7.17, then completes the proof. Of course we need to inject the Type II information as in the proof of Lemma 8.11. The condition $Md \leq x^{1/2}$ of Lemma 7.17 now must be replaced by $d < x^{1-4\eta} q^{-3/2}$. We should also remark that

$$\sum_{d < x^{1-4\eta} q^{-3/2}} \tau(d)|\mathcal{C}_d| \ll S(\mathcal{C}, \kappa)\mathcal{L}^B,$$

and so the $O(\mathcal{L}^{-A})$ error term in the estimation of $|\mathcal{F}_d|$ causes no problems. $\square$

Lemma 8.18. *For most $q \sim x^\theta$ with $\frac{1}{2} \leq \theta \leq \frac{17}{32} - \epsilon$, we have*

$$\sum_{x^{1-\theta-\epsilon} \leq p \leq x^{1/2}} S(A_p, p) - \sum_{(*)} S(A_{ptr}, r)$$

$$= \frac{1}{\phi(q)} \sum_{x^{1-\theta-\epsilon} \leq p \leq x^{1/2}} S(B_p, p) - \frac{1}{\phi(q)} \sum_{*} S(B_{ptr}, r) + O\left(\frac{x\mathcal{L}^{-A}}{\phi(q)}\right), \qquad (8.5.3)$$

where $(*)$ *indicates the summation conditions*

$$x^\kappa < r < p < x^{\frac{1}{2}(\epsilon+\theta)}, \quad s|t \Rightarrow s \geq p, \quad x^{1-\theta-\epsilon} \leq pt \leq x^{\theta+\epsilon}.$$

Proof. We have

$$\sum_{x^{1-\theta-\epsilon} \leq p \leq x^{1/2}} S(A_p, p) = S(\mathcal{F}, x^{\frac{1}{2}(\theta+\epsilon)})$$

$$= S(\mathcal{F}, x^\kappa) - \sum_{\kappa \leq p \leq x^{(\theta+\epsilon)/2}} S(\mathcal{F}_p, p). \qquad (8.5.4)$$

Lemma 8.17 covers the first sum on the right of (8.5.4), while the second sum is

$$\sum_{\substack{x^{\kappa}\leq p<t \\ s|t\Rightarrow s\geq p \\ x^{1-\theta-\epsilon}\leq pt\leq x^{\theta+\epsilon}}} S(\mathcal{A}_{pt},p). \tag{8.5.5}$$

An application of Buchstab's identity reduces this to

$$\sum_{\substack{x^{\kappa}\leq p<t \\ s|t\Rightarrow s\geq p \\ x^{1-\theta-\epsilon}\leq pt\leq x^{\theta+\epsilon}}} S(\mathcal{A}_{pt},x^{\kappa}) - \sum_{(*)} S(\mathcal{A}_{ptr},r). \tag{8.5.6}$$

After some checking of the algebra we can give an asymptotic formula for the first sum in (8.5.6) by Lemma 8.11. The formula (8.5.3) then follows. $\qquad\square$

Using Lemma 8.6 or 8.14 we can give an asymptotic formula for part of the second sum on the left side of (8.5.3). Recalling that the sum we are considering has a negative sign attached to it, this gives

$$-\sum_{x^{1-\theta-\epsilon}\leq p\leq x^{1/2}} S(\mathcal{A}_p,p) \leq -\frac{1}{\phi(q)}\sum_{x^{1-\theta-\epsilon}\leq p\leq x^{1/2}} S(\mathcal{B}_p,p)(1+O(\mathcal{L}^{-A}))$$

$$+\frac{x}{\phi(q)\log x}\int_{\mathcal{D}} \omega\left(\frac{\alpha_2}{\alpha_1}\right)\omega\left(\frac{1-\alpha_1-\alpha_2-\alpha_3}{\alpha_3}\right)\frac{d\alpha_1\,d\alpha_2\,d\alpha_3}{\alpha_1^2\alpha_3^2}, \tag{8.5.7}$$

where

$$\mathcal{D} = \{\boldsymbol{\alpha}_3 \notin G_3 : \kappa \leq \alpha_1 \leq \tfrac{1}{2}\theta, 1-\theta-\alpha_1 \leq \alpha_2 \leq \theta-\alpha_1,$$
$$\kappa \leq \alpha_3 \leq \min(\alpha_1,(1-\alpha_1-\alpha_2)/2\}.$$

If we discard the left-hand side of (8.5.7), we would suffer a loss $\log(\theta/(1-\theta))$, which is > 0.08 at $\theta = 0.52$, for example. On the other hand, the integral over $\mathcal{D}$ is < 0.025 at this point. There is no difficulty showing, from the volume of $\mathcal{D}$, that the integral has size $O((\theta-\tfrac{1}{2}+\epsilon)^2)$.

We now give the corresponding upper-bound for $S(\mathcal{A},p)$ that we will need in proving the lower-bound part of our theorem. To obtain the following result, simply discard all of the sum in (8.5.5).

Lemma 8.19. *For most $q \sim x^{\theta}$ we have*

$$\sum_{x^{1-\theta-\epsilon}\leq p\leq x^{1/2}} S(\mathcal{A}_p,p) \leq \frac{1}{\phi(q)}\sum_{x^{1-\theta-\epsilon}\leq p\leq x^{1/2}} S(\mathcal{B}_p,p)(1+O(\mathcal{L}^{-A}))$$

$$+\frac{x}{\phi(q)\log x}\int_{\kappa}^{\theta/2}\int_{1-\theta-\alpha_1}^{\theta-\alpha_1} \omega\left(\frac{\alpha_2}{\alpha_1}\right)\omega\left(\frac{1-\alpha_1-\alpha_2}{\alpha_1}\right)\frac{d\alpha_2\,d\alpha_1}{\alpha_1^3}. \tag{8.5.8}$$

Remarks. From the α_2 range length we see that the integral in (8.5.8) is a continuous function of θ of size $O(\theta-\tfrac{1}{2}+\epsilon)$. We have $\boldsymbol{\alpha}_2 \notin G_2$ in the integral. However, a further saving is possible by the technique we have used in previous chapters — making visible the almost-primes counted.

8.6 THE ALTERNATIVE SIEVE IN THREE DIMENSIONS

We now consider the region D in Diagram 8.1. Since $3\alpha_2 \geq 2\alpha_1$, we are not in the asymptotic formula region E. Also, in this region $\alpha_1 + 4\alpha_2 \geq 3 - 3\theta$, and this prevents us from applying Buchstab's identity by any method given hitherto. Our goal is thus to evaluate

$$\sum_{\alpha_2 \in D} S(\mathcal{A}_{p_1 p_2}, x^\kappa). \tag{8.6.1}$$

To this end we let R be a polygonal subset of

$$\{(s,t) : \tfrac{1}{4} \leq t \leq s \leq \min(\tfrac{3}{7}, 4 - 7\theta - 3\epsilon), 7\theta - 3 + 3\epsilon \leq s + t\}. \tag{8.6.2}$$

Write

$$\mathcal{C}' = \{\ell m n : (\log \ell, \log m) \in (\log x)R, \ell m n \sim x, (\ell m n, q) = 1\},$$

and $\mathcal{F}' = \mathcal{C}' \cap \mathcal{A}$.

In order to apply a three-dimensional form of our method we must study $|\mathcal{F}'_d|$. We do this via the following lemma, which is deduced from work of Heath-Brown [79].

Lemma 8.20. *Suppose $\mathcal{I}, \mathcal{J}, \mathcal{K}$ are subintervals of $[1, y)$ of length I, J, K, respectively, such that $\ell \in \mathcal{I}, m \in \mathcal{J}, n \in \mathcal{K} \Rightarrow \ell m n \leq y$. Then for $(b, q) = 1$ we have*

$$\sum_{\substack{\ell m n \equiv b(q) \\ \ell \in \mathcal{I}, m \in \mathcal{J}, n \in \mathcal{K}}} 1 = \frac{1}{\phi(q)} \sum_{\substack{(\ell m n, q)=1 \\ \ell \in \mathcal{I}, m \in \mathcal{J}, n \in \mathcal{K}}} 1 + O(V), \tag{8.6.3}$$

where

$$V = y^\eta \left(q^{\frac{5}{6}} + q^{\frac{1}{4}} (\min(IJ, KJ, IK))^{\frac{1}{2}} + q^{\frac{1}{2}} (\max(K, I, J))^{\frac{1}{2}} \right).$$

Proof. By [79, Lemma 5] we have the left-hand side of (8.6.3) equal to

$$N_1 + O(V)$$

for some N_1 independent of b. This last point is crucial, for it allows us to sum over all $b, 1 \leq b \leq q - 1, (b, q) = 1$ to obtain

$$\sum_{\substack{(\ell m n, q)=1 \\ \ell \in \mathcal{I}, m \in \mathcal{J}, n \in \mathcal{K}}} 1 = \phi(q)N_1 + O(\phi(q)V).$$

The result then follows immediately. $\qquad \square$

Lemma 8.21. *We have, for $d < x^{2\theta - 1 + \epsilon}$ and for most $q \sim x^\theta$ with $\theta < \frac{16}{31} - \epsilon$,*

$$|\mathcal{F}'_d| = \frac{|\mathcal{C}'_d|}{\phi(q)} (1 + O(\mathcal{L}^{-A})). \tag{8.6.4}$$

Proof. Working as in the previous section, we can remove the condition $\ell mn \sim x$ by constraining the variables to ranges of the form $[V, V(1 + O(\mathcal{L}^{-A})))$. Let F_d, C_d be the subsets of $\mathcal{F}_d', \mathcal{C}_d'$, respectively, subject to this restriction, with $\ell \in \mathcal{I}, m \in \mathcal{J}, n \in \mathcal{K}$. Working as before (except that it is more complicated having three variables) we have

$$|F_d| = \sum_{\substack{efg=d}} \sum_{\substack{(n,ef)=1 \\ ng\in\mathcal{K}}} \sum_{\substack{(m,e)=1 \\ mf\in\mathcal{J}}} \sum_{\substack{\ell e\in\mathcal{I} \\ \ell mnd\equiv a(q)}} 1$$

$$= \sum_{\substack{efg=d}} \sum_{\substack{k|e \\ j|fe}} \mu(k)\mu(j) \sum_{\substack{\ell mn\equiv a(q) \\ e\ell\in\mathcal{I}, mkf\in\mathcal{J} \\ njg\in\mathcal{K}}} 1$$

$$= \frac{1}{\phi(q)} \sum_{\substack{efg=d}} \sum_{\substack{k|e \\ j|fe}} \mu(k)\mu(j) \sum_{\substack{(\ell mn,q)=1 \\ e\ell\in\mathcal{I}, mkf\in\mathcal{J} \\ njg\in\mathcal{K}}} 1 + E$$

$$= \frac{1}{\phi(q)}|C_d'| + E.$$

Here, by Lemma 8.20, we have

$$E \ll \sum_{\substack{efg=d}} \sum_{\substack{k|e \\ j|fe}} x^\eta \left(q^{\frac{5}{6}} + q^{\frac{1}{4}} \left(\min\left(\frac{IJ}{ekf}, \frac{KJ}{kfjg}, \frac{IK}{ejg} \right) \right)^{\frac{1}{2}} \right.$$

$$\left. + q^{\frac{1}{2}} \left(\max\left(\frac{K}{jg}, \frac{I}{e}, \frac{J}{kf} \right) \right)^{\frac{1}{2}} \right)$$

$$\ll x^{2\eta} \left(q^{\frac{5}{6}} + q^{\frac{1}{4}} \left(\frac{x}{d} \right)^{\frac{1}{3}} + q^{\frac{1}{2}} (\max(I, J, K))^{\frac{1}{2}} \right)$$

$$= x^{2\eta}(E_1 + E_2 + E_3), \quad \text{say.}$$

Now, if $d \le x^{2\theta-1+\epsilon}$, then $x^{2\eta}E_1$ is a suitable error for $\theta < \frac{12}{23}$. Since $\frac{12}{23} > \frac{16}{31}$, this term is alright. To obtain

$$x^{2\eta}q^{\frac{1}{4}} \left(\frac{x}{d} \right)^{\frac{1}{3}} < \frac{x^{1-\epsilon}}{dq}$$

we need, if $d \le x^{2\theta-1+\epsilon}$, after some simple manipulation,

$$31\theta < 16 - 12\epsilon - 24\eta.$$

Hence this leads to a suitable error term. Finally, to tackle $x^{2\eta}E_3$, we note that if $(s,t) \in R$, then $\max(s, t, 1 - s - t) \le 4 - 7\theta - 3\epsilon$. Thus

$$x^{2\eta}q^{\frac{1}{2}} \max(I, J, K)^{\frac{1}{2}} \le x^{2\eta-\frac{3}{2}\epsilon+2-3\theta} \le \frac{x^{1-\frac{1}{2}\epsilon+2\eta}}{dq},$$

which again is a suitable error to complete the proof. $\qquad\square$

The reader should already have seen in their mind's eye that the above result gives an asymptotic formula by our sieve method. The details are very much like those for the two-dimensional sieve. The function $\rho(d)$ now becomes

$$\sum_{efg=d} \sum_{\substack{k|e \\ j|fe}} \frac{\mu(k)\mu(j)}{kj} = \prod_{p|d}\left(3 - \frac{3}{p} + \frac{1}{p^2}\right).$$

We thus obtain the following result.

Lemma 8.22. *For most $q \sim x^\theta$ with $\theta < \frac{16}{31} - \epsilon$ we have*

$$S(\mathcal{F}', x^\kappa) = \frac{1}{\phi(q)} S(\mathcal{C}', x^\kappa)\left(1 + O(\mathcal{L}^{-A})\right). \tag{8.6.5}$$

This result leads to an upper bound for (8.6.1), as we now show. We tacitly assume that $\theta < \frac{16}{31} - \epsilon$ in the following.

Lemma 8.23. *For most $q \sim x^\theta$ we have*

$$\sum_{\alpha_2 \in R} S(\mathcal{A}_{p_1 p_2}, x^\kappa) \le \frac{1}{\phi(q)}\sum_{\alpha_2 \in R} S(\mathcal{B}_{p_1 p_2}, x^\kappa)\left(1 + O(\mathcal{L}^{-A})\right)$$

$$+ \frac{x}{\kappa\phi(q)\log x}(I_1 + I_2), \tag{8.6.6}$$

where

$$I_1 = \int\limits_{\alpha_3 \in \mathcal{D}} \omega\left(\frac{1 - \alpha_1 - \alpha_2 - \alpha_3}{\kappa}\right) d\mu_3,$$

$$I_2 = \int\limits_{\substack{\alpha_4 \notin G_4 \\ \alpha_j \ge \kappa \\ \alpha_1 + \alpha_2, \alpha_3 + \alpha_4 \in R}} \omega\left(\frac{1 - \alpha_1 - \alpha_2 - \alpha_3 - \alpha_4}{\kappa}\right) d\mu_4,$$

and

$$\mathcal{D} = \{\alpha_3 \notin G_3 : (\alpha_1, \alpha_2 + \alpha_3) \in R \text{ with } \alpha_2 \ge \alpha_3$$
$$\text{or } (\alpha_1 + \alpha_2, \alpha_3) \in R \text{ with } \alpha_1 \ge \alpha_2, \alpha_j \ge \kappa\}.$$

Proof. Clearly

$$\sum_{\alpha_2 \in R} S(\mathcal{A}_{p_1 p_2}, x^\kappa) = S(\mathcal{F}', x^\kappa) - \sum_{\substack{\ell m n \in \mathcal{F} \\ p|\ell m n \Rightarrow p \ge x^\kappa \\ \Omega(\ell m) \ge 3}} 1, \tag{8.6.7}$$

where $\Omega(n)$ denotes the total number of prime factors of n counted with multiplicity. Since κ is a decreasing function of θ and at $\theta = \frac{29}{56}$ it equals $\frac{1}{7} - \epsilon$, we have $\kappa > \frac{1}{7}$ for $\theta < \frac{16}{31} - \epsilon$. Now $m \le \ell \le x^{3/7}$, so we have $\Omega(\ell) \le 2, \Omega(m) \le 2$. By Lemma 8.22 we can give an asymptotic formula for the first sum on the right-hand side of (8.6.7). It follows that we can give an upper bound for the left-hand side of (8.6.6), which is the correct term (the first sum on the right-hand side of (8.6.6)) plus the terms coming from the second sum on the right-hand side of (8.6.7) with $\Omega(\ell m) = 3$ or $= 4$ that do not lie in our asymptotic formula regions. The reader will quickly verify that these lead to I_1 and I_2, respectively. $\qquad \square$

We now define the region R. It must lie in the set (8.6.2), and we may assume that $\alpha_2 \notin S$, for otherwise we could use the method of Section 8.3. We put

$$R = \{\alpha_2 : \alpha_2 \leq \alpha_1, \alpha_1 + 2\alpha_2 \leq 1, \alpha_1 + 4\alpha_2 \geq 3 - 3\theta - \epsilon, 3\alpha_2 \geq 2\alpha_1,$$

$$\max(\tau, \tfrac{1}{3}(31\theta - 15) + 5\epsilon) \leq \alpha_1 \leq \min(\tfrac{3}{7} + \epsilon, 4 - 7\theta - 3\epsilon)\}.$$

For $\alpha_2 \in R$ we have

$$\alpha_1 + \alpha_2 = \tfrac{3}{4}\alpha_1 + \tfrac{1}{4}(\alpha_1 + 4\alpha_2)$$

$$\geq \tfrac{1}{4}(31\theta - 15) + \tfrac{3}{4}(1 - \theta) + 3\epsilon = 7\theta - 3 + 3\epsilon.$$

For $\theta < \frac{25}{49}$ the range for α_1 is $\tau \leq \alpha_1 \leq \frac{3}{7}$.

By Buchstab's identity with (8.6.6), we have

$$\sum_{\alpha_2 \in R} S(\mathcal{A}_{p_1 p_2}, p_2)$$

$$\leq \frac{1}{\phi(q)} \sum_{\alpha_2 \in R} S(\mathcal{B}_{p_1 p_2}, p_2) + \frac{x}{\phi(q) \log x} \left(\frac{1}{\kappa}(I_1 + I_2) + I_3 \right), \qquad (8.6.8)$$

with I_1, I_2 as in the statement of Lemma 8.23 and

$$I_3 = \int\limits_{\alpha_3 \in \mathcal{E}} \omega \left(\frac{1 - \alpha_1 - \alpha_2 - \alpha_3}{\alpha_3} \right) \frac{d\mu_3}{\alpha_3},$$

where

$$\mathcal{E} = \{\alpha_3 \notin G_3 : \alpha_2 \in R, \kappa \leq \alpha_3 \leq \min(\alpha_2, (1 - \alpha_1 - \alpha_2)/2)\}.$$

The three integrals I_1, I_2, I_3 are all continuous functions of θ and, using the elementary argument we have seen before, each is $O((\theta - \frac{1}{2} + \epsilon)^2)$.

Now let R' be the subset of R, which gives no contribution to any I_j. That is,

$$R' = R \cap R_1 \cap R_2 \cap R_3 \cap R_4,$$

with

$$R_1 = \{\alpha_2 : \beta + \gamma = \alpha_2 \text{ with } \kappa \leq \beta \leq \gamma \Rightarrow (\alpha_1, \beta, \gamma) \in G_3\},$$

$$R_2 = \{\alpha_2 : \beta + \gamma = \alpha_1 \text{ with } \kappa \leq \beta \leq \gamma \Rightarrow (\alpha_2, \beta, \gamma) \in G_3\},$$

$$R_3 = \{\alpha_2 : \beta + \gamma = \alpha_1, \lambda + \mu = \alpha_2$$

$$\text{with } \kappa \leq \beta \leq \gamma, \kappa \leq \lambda \leq \mu \Rightarrow (\beta, \gamma, \lambda, \mu) \in G_4\},$$

$$R_4 = \{\alpha_2 : \kappa \leq \alpha_3 \leq \min(\alpha_2, \tfrac{1}{2}(1 - \alpha_1 - \alpha_2)) \Rightarrow \alpha_3 \in G_3\}.$$

Then, working as above, except that we now consider lower bounds, we obtain

$$\sum_{\alpha_2 \in R} S(\mathcal{A}_{p_1 p_2}, p_2) \geq \frac{1}{\phi(q)} \sum_{\alpha_2 \in R} S(\mathcal{B}_{p_1 p_2}, p_2) - \frac{x}{\phi(q) \log x} I_4,$$

where

$$I_4 = \int\limits_{R \setminus R'} \omega \left(\frac{1 - \alpha_1 - \alpha_2}{\alpha_2} \right) \frac{d\mu_2}{\alpha_2}.$$

As expected, I_4 is a continuous function of θ with $I_4 = O(\theta - \frac{1}{2} + \epsilon)$. The reader has seen this phenomenon several times already. More details for this integral are given in [7, p. 85].

8.7 AN UPPER BOUND FOR LARGE θ

In this section we shall consider $\theta \geq 0.56$, a region where we cannot apply the alternative sieve owing to the dearth of Type II information. We shall therefore fall back on the Rosser-Iwaniec sieve with a flexible error term as in Chapter 6. The results here essentially go back to Fouvry [34] with the corrections he supplied to the author and R. C. Baker that were published in [7].

Theorem 8.4. *Let* $\theta + \epsilon \in \left[\frac{17}{32}, \frac{5}{7}\right]$. *For most* $q \sim x^\theta$ *we have*

$$\pi(x; q, a) \leq \frac{x(C(\theta) + O(\epsilon))}{\phi(q) \log x}, \tag{8.7.1}$$

where

$$C(\theta) = \begin{cases} f_1(\theta) & \theta + \epsilon \in \left[\frac{17}{32}, \frac{4}{7}\right], \\ f_2(\theta) & \theta + \epsilon \in \left[\frac{4}{7}, \frac{3}{5}\right], \\ f_3(\theta) & \theta + \epsilon \in \left[\frac{3}{5}, \frac{5}{7}\right]. \end{cases} \tag{8.7.2}$$

Here

$$f_2(\theta) = \frac{14}{12 - 13\theta}, \quad f_1(\theta) = f_2(\theta) - \log\left(\frac{4(1-\theta)}{3\theta}\right), \quad f_3(\theta) = \frac{8}{3 - \theta}.$$

Proof. We consider first the case $\theta + \epsilon \in \left[\frac{3}{5}, \frac{5}{7}\right]$. By Lemma 8.5 we can give a satisfactory estimate for the remainder term in the Rosser-Iwaniec sieve of the form

$$\sum_{m \leq M, n \leq N} |r_{mn}|, \tag{8.7.3}$$

where $M = x^{1-\theta-\epsilon}, N = x^{(3\theta-1)/4-5\epsilon/4}$. We can thus take $D = x^{\frac{1}{4}(3-\theta)-2\epsilon}$ in Theorem 4.2 (or, to be precise, the analogue of this theorem as described at the end of Chapter 4). The required bound quickly follows.

Now consider the case $\theta + \epsilon \in \left[\frac{4}{7}, \frac{3}{5}\right]$. Write $\chi(\theta) = (12 - 13\theta)/7$. We must make more use of the special nature of the remainder term in the Rosser-Iwaniec sieve as given in [103] and applied in [7, 34]. One can replace the variables mn in (8.7.3) by

$$d = np_1 \dots p_t, \quad D_i \leq p_i < D_i^{1+\eta^9}, \quad p_i < z, \quad n \leq x^\eta,$$

and $|r_{mn}|$ is replaced by $c_n r_d$, with $|c_n| \leq 1$. It is important here (see our construction of the Rosser-Iwaniec sieve in Chapter 4), that the D_i satisfy

$$D_1 \dots D_{t-1} D_t^2 < D \quad (1 \leq t \leq \ell). \tag{8.7.4}$$

Now let $D_1 \dots D_t = x^u$. Using (8.7.4) we can partition $D_1, \dots, D_t$ into two products M_1, M_3 satisfying

$$M_3 \leq x^{1-\theta-3\eta}, \quad M_1 \leq x^{\theta+\chi(\theta)-1-97\eta}, \quad M_1 M_3 = x^u.$$

If $u \leq (9 - 8\theta)/7$, it is straightforward to verify the hypotheses of Lemma 8.4 (see [7, pp. 56–57]).

Suppose, on the other hand, that $u > (9 - 8\theta)/7$. Write

$$U = x^{-1+u+\theta+4\eta}, \quad V = x^\lambda,$$

where

$$\lambda = \min\left(-1 + \frac{3u}{2} + \frac{\theta}{2} - 2\eta, \frac{2u}{5} - \frac{4\eta}{5}, -\frac{1}{4} + \frac{3u}{4} - \eta\right).$$

The reader will be able to verify that $1 \leq U \leq V \leq D$. In addition (see [7, p. 57]) we have $VU^{-1} > D(D_1 \ldots D_t)^{-1}$. It follows from (8.7.4) that there is some subproduct of the D_i, say $x^w = M_1$, that lies in $[U, V]$. Let $M_3 = x^u/M_1$. It then follows that the hypotheses of Lemma 8.2 hold, and the proof for this case is complete.

Finally, suppose that $\theta + \epsilon \in \left[\frac{17}{32}, \frac{4}{7}\right]$. The application of the Rosser-Iwaniec sieve goes through as before, but now Lemma 8.14 and its corollary enable us to count some of the almost-primes discarded by the sieve application. To be precise, we can deduct those products $p_1 p_2$ with $p_1 \in [x^{\theta+\epsilon}, x^{4/7-\epsilon}]$. Using our standard procedure for converting sums over primes into integrals then gives the desired result. □

Proof. (Proof of Theorem 8.3) Assuming the result of Theorem 8.2, we are left to evaluate from Theorem 8.4 the integrals

$$\int_{0.56}^{0.6} C(\theta)\, d\theta \quad \text{and} \quad \int_{0.6}^{0.677} C(\theta)\, d\theta.$$

Both integrals can be expressed in closed form (for example, the second is simply $8\log(2.4/2.323)$). A straightforward calculation thus yields upper bounds for these integrals, which are, respectively, < 0.12544 and < 0.26088. The second value with (8.1.4) gives (8.1.8) and so completes the proof of Theorem 8.3. □

8.8 COMPLETION OF PROOF

We begin with the upper bound in (8.1.3) by splitting the range for $\theta \in [0.5, 0.56]$ into subranges as follows, closely following [7].

(i) $\frac{1}{2} \leq \theta \leq \frac{25}{49} - \epsilon$. In this first range we use Buchstab's identity to write

$$S(\mathcal{A}, (2x)^{\frac{1}{2}}) = S(\mathcal{A}, x^\kappa) - \sum_{\kappa \leq \alpha_1 \leq 3/7+\epsilon} S(\mathcal{A}_{p_1}, p_1)$$

$$- \sum_{3/7+\epsilon \leq \alpha_1 \leq 1-\theta-\epsilon} S(\mathcal{A}_{p_1}, p_1) - \sum_{1-\theta-\epsilon \leq \alpha_1 \leq 1/2} S(\mathcal{A}_{p_1}, p_1)$$

$$= S_{1,0} - S_{1,1} - S_{1,2} - S_{1,3}, \quad \text{say.}$$

$$(8.8.1)$$

We can obtain formulae for $S_{1,0}$ and $S_{1,2}$ by Lemmas 8.11 and 8.14, respectively. We can get an upper bound for $-S_{1,3}$ by (8.5.7). We can apply Buchstab's identity again to $S_{1,1}$ to produce

$$S_{1,1} = \sum_{\kappa \leq \alpha_1 \leq 3/7+\epsilon} S(\mathcal{A}_{p_1}, x^\kappa) - \sum_{\substack{\alpha_2 \in A \cup C \\ \alpha_1 \leq 3/7+\epsilon}} S(\mathcal{A}_{p_1 p_2}, p_2)$$

$$- \sum_{\substack{\alpha_2 \in B \cup E \\ \alpha_1 \leq 3/7+\epsilon}} S(\mathcal{A}_{p_1 p_2}, p_2) - \sum_{\alpha_2 \in D} S(\mathcal{A}_{p_1 p_2}, p_2) \qquad (8.8.2)$$

$$= S_{2,0} - S_{2,1} - S_{2,2} - S_{2,3}, \quad \text{say}.$$

We can give asymptotic formulae for $S_{2,0}$ (by Lemma 8.15) and $S_{2,2}$ (since $B \cup E$ is an asymptotic formula region). The term $S_{2,3}$ can be dealt with by (8.6.8). In view of Lemma 8.16 we can apply Buchstab's identity again to $S_{2,1}$ to obtain

$$S_{2,1} = \sum_{\alpha \in A \cup C} S(\mathcal{A}_{p_1 p_2}, x^\kappa) - \sum_{\alpha_3 \in \Theta_1} S(\mathcal{A}_{p_1 p_2 p_3}, p_3) - \sum_{\alpha_3 \in \Theta_2} S(\mathcal{A}_{p_1 p_2 p_3}, p_3)$$

$$= S_{3,0} - S_{3,1} - S_{3,2}, \quad \text{say}.$$

Here Θ_2 is the region where an asymptotic formula can be obtained by Lemma 8.6 or 8.14, and Θ_1 corresponds to the remaining ranges. We can thus evaluate $S_{3,0}$ and $S_{3,2}$. We now write

$$S_{3,1} = S_{3,4} + S_{3,5}.$$

Here $S_{3,4}$ is the part of $S_{3,1}$ for which it is impossible to perform further decompositions, or if such a decomposition is possible, one is still left with terms of the wrong sign that cannot be evaluated. Our only hope to salvage anything from this sum is our technique of making the almost-primes visible by applying Buchstab's identity in reverse. Thus $S_{3,4}$ splits into sums over sets of integers with four, five or at least six prime factors. For some of these an asymptotic formula may be derived using Lemma 8.6 or 8.17.

By its definition and Buchstab's identity, $S_{3,5} = S_{4,0} - S_{4,1}$ with $S_{4,1} = S_{5,0} - S_{5,1}$, where asymptotic formulae can be obtained for $S_{4,0}$ and $S_{5,0}$. For most of $S_{5,1}$ an asymptotic formula can be given, and the rest we discard. We thus arrive at the bound

$$S(\mathcal{A}, (2x)^{\frac{1}{2}}) \leq \frac{x}{\phi(q) \log x} \left(1 + O(\mathcal{L}^{-A}) + E_{1,3} + E_{2,3} + E_{3,4} + E_{5,1}\right),$$

where $E_{j,k}$ denotes the loss from discarding all or part of $S_{j,k}$. Each of these losses is a continuous function of θ with $E_{j,k} = O(\theta - \frac{1}{2} + \epsilon)$.

The problem has now been reduced to numerical integration, as we have seen in previous chapters. In [7] we gave a detailed analysis to justify the extremely good value for $C_2(0.51)$. It was shown there that, for $0.5 \leq \theta \leq 0.51$, we have $C_2(\theta) \leq 1 + 150(\theta - \frac{1}{2} + \epsilon)^2$.

(ii) $\frac{25}{49} - \epsilon < \theta \leq \frac{21}{41}$. In this interval the upper bound on α_1 in R is $4 - 7\theta + O(\epsilon) = g(\theta, \epsilon)$, say. This means that in (8.8.1) we must introduce another sum

$$\sum_{g(\theta,\epsilon) \leq \alpha_1 \leq 3/7+\epsilon} S(\mathcal{A}_{p_1}, p_1).$$

The additional error introduced here would be

$$\int_{4-7\theta}^{3/7} \frac{d\alpha}{\alpha(1-\alpha)} + O(\epsilon) = \log\left(\frac{3}{4} \cdot \frac{7\theta-3}{4-7\theta}\right) + O(\epsilon) \qquad (8.8.3)$$

if we had to discard this new sum, as was done in [7]. The introduction of this and other similar terms as θ increases from this point causes the sharp rise in the value for $C_2(\theta)$ from this point on. However, some parts of this term can be reclaimed, as was demonstrated in [9]. We note that the contribution from integrals of dimension 3 and higher remains small until θ exceeds 0.53.

We now briefly describe the improvement made in [9] for this and the following regions. Let $D(\theta)$ be the set of α_1 in $\left[\kappa, \frac{3}{7}\right]$ for which $S(\mathcal{A}_{p_1}, p_1)$ would otherwise be discarded. Write

$$D(\theta) = \begin{cases} \left[4 - 7\theta, \frac{3}{7}\right] & \text{for } \frac{25}{49} \le \theta \le \frac{21}{41}, \\ \left[(3-3\theta)/5, (31\theta-15)/3\right] \cup \left[4-7\theta, \frac{3}{7}\right] & \text{for } \frac{21}{41} < \theta \le \frac{16}{31}, \\ \left[(3-3\theta)/5, \frac{3}{7}\right] & \text{for } \frac{16}{31} < \theta \le \frac{11}{21}, \\ \left[\frac{2}{7}, \frac{3}{7}\right] & \text{for } \frac{11}{21} \le \theta \le \frac{92}{175}. \end{cases}$$

We can treat the sums corresponding to the intersection of $D(\theta)$ with

$$I(\theta) = \begin{cases} \left[(19\theta-7)/7, \frac{3}{7}\right] & \text{for } \frac{25}{49} \le \theta \le \frac{14}{27}, \\ \left[(50\theta-19)/17, \frac{3}{7}\right] & \text{for } \frac{14}{27} < \theta \le \frac{92}{175}. \end{cases}$$

The value $\frac{92}{175}$ arises as the point at which $I(\theta)$ vanishes. We have

$$-\sum_{\alpha_1 \in D(\theta) \cap I(\theta)} S(\mathcal{A}_{p_1}, p_1)$$

$$= -\sum_{\alpha_1 \in D(\theta) \cap I(\theta)} S(\mathcal{A}_{p_1}, x^\kappa) + \sum_{\substack{\alpha_1 \in D(\theta) \cap I(\theta) \\ \alpha_2 \in C \cup D}} S(\mathcal{A}_{p_1 p_2}, x^\kappa)$$

$$+ \sum_{\substack{\alpha_1 \in D(\theta) \cap I(\theta) \\ \alpha_2 \in B \cup E}} S(\mathcal{A}_{p_1 p_2}, p_2) - \sum_{\substack{\alpha_1 \in D(\theta) \cap I(\theta) \\ \kappa \le \alpha_3 < \alpha_2 \\ \alpha_2 \in C \cup D}} S(\mathcal{A}_{p_1 p_2 p_3}, p_3).$$

$$= -T_1 + T_2 + T_3 - T_4, \text{ say.}$$

We can give asymptotic formulae for T_1 by Lemma 8.11, and for T_3 and that part of T_4 for which $\alpha_3 \in G_3$, by Lemma 8.6. We apply Lemma 8.11 to the part of T_2 with $\alpha_2 \in C$. We then apply a variant of our three-dimensional sieve procedure to the part of T_2 with $\alpha_2 \in D$: the details are given in [9, Lemma 25]. Finally, we discard the portion of T_4 with $\alpha_3 \notin G_3$. We thus replace the loss (8.8.3) with

$$\int_{\mathcal{S}} \frac{d\alpha}{\alpha(1-\alpha)} + O(\epsilon),$$

where $\mathcal{S} = \emptyset$ when $\theta < \frac{35}{68} = 0.5147\ldots$ and otherwise

$$\mathcal{S} = \left[4 - 7\theta, \max\left(\tfrac{1}{7}(19\theta-7), \tfrac{1}{17}(50\theta-19)\right)\right].$$

There are in addition some higher-dimensional integrals whose size is very small (they lead to a contribution $< 4.10^{-5}$ overall).

(iii) $\frac{21}{41} < \theta \le \frac{16}{31}$. For θ in this interval, the lower bound on $\alpha_1 \in R$ is now $\frac{1}{3}(31\theta - 15) + O(\epsilon)$. This introduces a further sum into (8.8.1) with

$$\tau \le \alpha_1 \le \tfrac{1}{3}(31\theta - 15) + O(\epsilon).$$

Since $\frac{1}{3}(31\theta - 15) - \tau = \frac{4}{15}(41\theta - 21)$, the additional error introduced here is $O(\theta - \frac{21}{41} + \epsilon)$.

(iv) $\frac{16}{31} < \theta < \frac{29}{56}$. An exception now occurs to our continuity of constant principle that we would expect as θ varies. To be precise, there is a discontinuity at $\frac{16}{31}$ since the middle term of V in Lemma 8.20 becomes too large for α_1 across a range of values at this point. Hence the whole sum

$$\sum_{\tau < \alpha_1 < 3/7 + \epsilon} S(\mathcal{A}_p, p)$$

must be discarded because we are unable to deal with the region D.

(v) $\frac{29}{56} < \theta \le \frac{11}{21}$. In this region it is possible for products of seven primes to be counted in our sums. With more effort a small saving could be made by analysing these sums more carefully.

(vi) $\frac{11}{21} < \theta \le \frac{17}{32}$. The choice of τ for $\theta < \frac{11}{21}$ was determined by the simultaneous solution to $\alpha_1 = \alpha_2$, $\alpha_1 + 4\alpha_2 = 3 - 3\theta$. However, for $\theta > \frac{11}{21}$ we can evaluate

$$\sum_{\substack{3(1-\theta)/5 \le \alpha_1 \le 2/7 \\ \theta - \alpha_1 + \epsilon \le \alpha_2 < \alpha_1}} S(\mathcal{A}_{p_1 p_2}, p_2)$$

via Lemma 8.6. On the other hand, if $\alpha_2 < \theta - \alpha_1 + \epsilon$, we have $\alpha_1 + 4\alpha_2 \le 3 - 3\theta - \epsilon, \alpha_1 + 2\alpha_2 \le 3 - 4\theta - \epsilon$, so $\alpha_2 \in U_2$.

There is now no contribution from region C, and the contribution from one-dimensional integrals is fixed (at ≈ 0.676). Only $E_{1,2}$ and $E'_{3,4}$ are increasing functions of θ in this region. Here we have written $E'_{3,4}$ for the contribution to $E_{3,4}$ and $E_{5,1}$ from $\alpha_2 \in A$. We find that in this interval we have

$$C_2(\theta) \le 1.725 + 13.125(\theta - 0.523).$$

(vii) $\frac{17}{32} < \theta \le \frac{7}{13} - \epsilon$. At $\frac{17}{32}$ the value of κ is halved. This does not introduce a discontinuity into the method since the sums introduced are $O\left(\theta - \frac{17}{32} + \epsilon\right)$. These new sums come from regions like H in Diagram 8.2. We are now able to make explicit use of the asymptotic formula region given by $2\theta - 1 + \epsilon \le s \le \frac{1}{6}(5 - 8\theta) - \epsilon$. However, the region $\alpha_1 \ge 1 - \theta$ must be discarded. Our opening move is thus

$$S(\mathcal{A}, (2x)^{\frac{1}{2}}) \le S(\mathcal{A}, x^\gamma) - \sum_{x^{3/7+\epsilon} \le p \le x^{1-\theta-\epsilon}} S(\mathcal{A}_p, p). \qquad (8.8.4)$$

Here we are giving an upper bound for the first sum on the right of (8.8.4) and an asymptotic formula for the second.

Since κ has halved in value, we may now be counting products of more than 16 primes. Evidently there must be many different ways that variables can be combined to lie in "good" ranges, and it may be possible to apply Buchstab's identity several more times. With this number of variables it may not be practical to perform the calculations in the most efficient manner. However, we can use the approximations leading to (3.5.1) to show that the integrals involving many variables are already very small, and so we lose very little by discarding them all.

(viii) $\frac{7}{13} - \epsilon < \theta \leq \frac{6}{11} - \epsilon$. At $\frac{7}{13}$ the value of κ reduces again. Indeed, the value reduces to κ' if we wish to continue carrying out decompositions in the same way. We begin by writing

$$S(\mathcal{A}, (2x)^{\frac{1}{2}}) \leq S_1 - S_2 - S_3 - S_4, \tag{8.8.5}$$

where

$$S_1 = S(\mathcal{A}, x^\kappa), \qquad S_2 = \sum_{\kappa \leq \alpha_1 \leq \tau'} S(\mathcal{A}_{p_1}, p_1),$$

$$S_3 = \sum_{\tau' < \alpha_1 \leq \tau} S(\mathcal{A}_{p_1}, p_1), \qquad S_4 = \sum_{3/7 + \epsilon \leq \alpha_1 \leq 1 - \theta - \epsilon} S(\mathcal{A}_{p_1}, p_1).$$

For S_2 we can decompose further:

$$S_2 = \sum S(\mathcal{A}_{p_1}, x^\kappa) - \sum S(\mathcal{A}_{p_1 p_2}, p_2),$$

and we can do the same for S_3 with κ replaced by κ'. The sum S_4 can be evaluated as in previous regions.

We note that the above approach works because the line $\alpha_1 + 4\alpha_2 = 3 - 3\theta - \epsilon$ lies in or above the region B. This fails for $\theta > \frac{6}{11} - \epsilon$.

(ix) $\frac{6}{11} - \epsilon < \theta < 0.56$. For θ near to $\frac{6}{11} - \epsilon$ we need only make a minor modification to (8.8.5). If the sum with

$$\tau'' = \frac{3 - 3\theta}{5} - \epsilon \leq \alpha_1 \leq \frac{7\theta - 3}{3} = \tau'''$$

is discarded, it is still possible to perform further decompositions with $\tau' \leq \alpha_1 \leq \tau''$ and $\tau''' \leq \alpha_1 \leq \frac{2}{7}$. One runs into serious practical difficulties quickly since κ is decreasing to zero at $\theta = 0.55$. It is best therefore to begin with

$$S(\mathcal{A}, (2x)^{\frac{1}{2}}) \leq S(\mathcal{A}, x^\tau) - \sum_{3/7 + \epsilon \leq \alpha_1 \leq 1 - \theta - \epsilon} S(\mathcal{A}_{p_1}, p_1)$$

for $\theta > 0.546$.

The values obtained for $C_2(\theta)$ are given in Table 8.1 at 0.005 intervals. Working with shorter intervals we calculated

$$\int_{0.5}^{0.56} C_2(\theta) \, d\theta < 0.1131.$$

This, together with our calculations at the end of Section 8.6, gives (8.1.4).

Lower Bounds. We now suppose that $\frac{1}{2} < \theta < \frac{25}{49} - \epsilon$. We begin with (8.8.1) but now must use (8.5.8) to supply a lower bound for $-S_{1,3}$. We can apply Buchstab's identity again to $S_{1,1}$ to obtain (8.8.2) as before. Running with our usual

θ	0.505	0.51	0.515	0.52	0.525	0.53
$C_2(\theta)$	1.004	1.015	1.223	1.632	1.75	1.82

θ	0.535	0.54	0.545	0.55	0.555	0.56
$C_2(\theta)$	2.09	2.25	2.47	2.66	2.76	2.88

Table 8.1 Values of $C_2(\theta)$

philosophy, we can produce a lower bound for $\pi(x; q, a)$ for most q when

$$\int_\kappa^{\theta/2} \int_{1-\theta-\alpha_1}^{\theta-\alpha_1} \omega\left(\frac{\alpha_2}{\alpha_1}\right) \omega\left(\frac{1-\alpha_1-\alpha_2}{\alpha_2}\right) \frac{d\alpha_1\,d\alpha_2}{\alpha_1^3}$$
$$+ \int_{\alpha_2 \in A\cup C\cup D} \omega\left(\frac{1-\alpha_1-\alpha_2}{\alpha_2}\right) \frac{d\mu_2}{\alpha_2} < 1.$$

At $\theta = 0.5$ the first integral is zero, of course, while the second is less than 0.61. However, one can do much better by considering the regions A, C, D in turn. We leave it as an exercise for the reader to show that, for F as either A or C, we have an estimate for

$$\sum_{\alpha_2 \in F, \alpha \le 3/7+\epsilon} S(\mathcal{A}_{p_1 p_2}, p_2)$$

with error $O(\delta)$ when $\theta = \frac{1}{2} + \delta$. More details are given in [7]. We have already established the required bound corresponding to the region D in Section 8.5 using the three-dimensional sieve. Numerical integration finishes the proof, as in the upper-bound case, with $C_1(0.51) > 0.6, C_1(0.515) > 0.39$, and $C_1(0.52) > 0.16$. This completes the proof of Theorem 8.2.

$\square$

Chapter Nine

Primes in Almost All Intervals

9.1 INTRODUCTION

We now consider what happens if, instead of asking for primes in *all* intervals $[x - y, x]$, we only demand that there are primes in *most* intervals. Our notion of *most* here is similar to that deployed in the previous chapter. We will say that a property holds for almost all intervals $[x - y, x]$ if it holds for all *real* $x \leq X$ except on an exceptional set of measure $o(X)$. Equivalently, it will hold for all $x \in \mathbb{N}, x \leq X$ with $o(X)$ exceptions. The methods usually make the exceptional set of measure $O(X(\log X)^{-A})$ for any $A > 0$. We show first how the exponent for this problem is related to the exponent for all intervals when using zero-density methods. Then we shall adapt the work of Chapter 7 to this problem and discover that the two exponents no longer appear to satisfy the same relation.

To prove "almost all" results we need to integrate the square of the error in a formula for the number of primes in an interval. The following lemma is then useful.

Lemma 9.1. *Let $D > C \geq 2, B > A \geq 1$ and suppose that $g(t)$ is a continuous function on $[C, D]$. Then*

$$\int_A^B \left| \int_C^D g(t) y^{it} \, dt \right|^2 dy \ll B \log D \int_C^D |g(t)|^2 \, dt. \qquad (9.1.1)$$

Proof. Write I for the left-hand side of (9.1.1). Then

$$I = \int_A^B \int_C^D \int_C^D g(t_1) y^{iyt_1} \overline{g}(t_2) y^{-iy_2} \, dt_1 \, dt_2 \, dy$$

$$= \int_C^D \int_C^D g(t_1) \overline{g}(t_2) \left[\frac{y^{i(t_1 - t_2) + 1}}{i(t_1 - t_2) + 1} \right]_A^B \, dt_1 \, dt_2.$$

Since

$$|g(t_1) g(t_2)| \leq \tfrac{1}{2} \left(|g(t_1)|^2 + |g(t_2)|^2 \right),$$

we have

$$I \leq \int_C^D \int_C^D |g(t_1)|^2 \frac{2B}{\sqrt{1 + (t_1 - t_2)^2}} \, dt_1 \, dt_2.$$

Since

$$\int_C^D \frac{1}{\sqrt{1 + (t_1 - t_2)^2}} \, dt_2 \ll \log D \qquad (9.1.2)$$

for any value of $t_1 \in [C, D]$, this completes the proof. $\qquad \square$

For the zero-density method we need the following analogue of the above result.

Lemma 9.2. *Let $B > A \geq 1, T \geq 2$. For $0 \leq \sigma_1 \leq \sigma_2$ write*

$$\sum_{\sigma_1,\sigma_2,T} = \sum_{\substack{\rho \\ |\gamma| \leq T \\ \sigma_1 \leq \beta \leq \sigma_2}} ,$$

where $\rho = \beta + i\gamma$ denotes a zero of the Riemann zeta-function. Then, for any function $h(\rho)$ with $|h(\rho) \leq H$, we have

$$\int_A^B \left| \sum_{\sigma_1,\sigma_2,T} h(\rho)x^\rho \right|^2 dx \ll (\log B)^2 H^2 \sum_{\sigma_1,\sigma_2,T} B^{2\beta+1}. \tag{9.1.3}$$

Proof. The only substantial change from the previous lemma is that (9.1.2) is to be replaced by

$$\sum_{\sigma_1,\sigma_2,T} \frac{1}{\sqrt{1 + (\gamma_1 - \gamma_2)^2}} \ll (\log T)^2.$$

This reflects the well-known formula ([27, p. 98])

$$\sum_{0,1,T} 1 \sim \frac{T}{2\pi} \log\left(\frac{T}{2\pi}\right).$$

□

Working as in Chapter 7 we have

$$\sum_{x-y \leq n \leq x} \Lambda(n) = y + E_1 + E_2,$$

where

$$E_1 \ll \frac{y(\log x)^2}{T}, \qquad E_2 = \sum_{\substack{\rho \\ |\gamma| \leq T}} \frac{x^\rho - (x-y)^\rho}{\rho}.$$

We write $E_2(\sigma)$ for that part of E_2 with $\sigma \leq \beta \leq \sigma + 1/\log X$. If we take $T = y(\log X)^4$ and show that for any $\sigma \in (0,1)$ we have

$$\int_X^{2X} |E_2(\sigma)|^2 dx \ll Xy^2(\log X)^{-A}, \tag{9.1.4}$$

then

$$\sum_{x-y \leq n \leq x} \Lambda(n) > \frac{y}{2},$$

except for a set of $x \in [X, 2X]$ with measure $O(X(\log X)^{1-A})$. Summing X over powers of 2 will then give a result of the desired form. We note that the prime powers counted by $\Lambda(n)$ give a smaller order contribution. Indeed, for $y < x^{1/3}$ say, almost all intervals $[x - y, x]$ contain no prime powers.

Now let $\delta = (2X)^{\theta-1}$, $y = \delta x$, so we are considering primes in intervals no bigger than $[x - x^\theta, x]$. We have

$$\frac{x^\rho - (x-y)^\rho}{\rho} = x^\rho \frac{1 - (1-\delta)^\rho}{\rho}$$

and

$$\frac{1 - (1-\delta)^\rho}{\rho} = \int_{1-\delta}^{1} u^{\rho-1} du,$$

so that (since $u \geq \frac{1}{2}, \mathrm{Re}\,(\rho - 1) > -1$)

$$\left| \frac{1 - (1-\delta)^\rho}{\rho} \right| \leq 2\delta. \tag{9.1.5}$$

By Lemma 9.2 we then have

$$\int_X^{2X} |E_2(\sigma)|^2 \, dx \ll (\log X)^2 \delta^2 \sum_{\sigma, \sigma + 1/\log X, T} X^{2\beta+1}$$
$$\ll (\log X)^2 X^{2\sigma+1} N(\sigma, T) \delta^2.$$

We have $N(\sigma, T) = 0$ for $\sigma > 1 - (\log X)^{-2/3}$, and otherwise

$$N(\sigma, T) \ll T^{A(1-\sigma)},$$

where $A \leq \frac{12}{5} + \epsilon$. Comparing this with (9.1.4) shows that we require

$$2\sigma + 1 + A(1 - \sigma)(1 - \theta) - 2 < 1.$$

Rearranging this becomes $(1 - \sigma)A(1 - \theta) < 2 - 2\sigma$; that is, $A(1 - \theta) < 2$. Hence the result follows if $\theta > \frac{1}{6}$. We have thus proved the following result.

Theorem 9.1. *Given $\epsilon > 0$, almost all intervals of the form $[x - x^{1/6+\epsilon}, x]$ contain primes.*

A slight modification of the above working gives the expected asymptotic formula for the number of primes in such intervals. Let $\theta_1 + \epsilon$ be the exponent obtained for primes in short intervals by the zero-density method (so $\theta_1 = \frac{7}{12}$ is the best currently known), and $\theta + \epsilon$ the corresponding exponent for primes in almost all intervals (so we have $\theta = \frac{1}{6}$ by the above). Then we have the simple relation $\theta = 2\theta_1 - 1$ between them. This relation appeared for the first time in the work of Selberg [154]. In the following sections we apply our sieve method to the problem, and it is not clear that this relationship should continue to hold. We leave it as an exercise for the reader to show that Theorem 9.1 can be established by the alternative sieve with an asymptotic formula.

9.2 THE ARITHMETICAL INFORMATION

Let $\mathcal{B} = [x - y_1, x] \cap \mathbb{N}$, $\mathcal{A} = [x - y, x] \cap \mathbb{N}$ where $X \leq x \leq 2X$, $\delta = (2X)^{\theta-1+\epsilon}$, $y = \delta x$, $y_1 = x \exp(-3(\log X)^{1/3})$. Here $\frac{1}{25} < \theta < \frac{1}{4}$.

We may combine Lemma 9.1 with the method of proof of Lemma 7.2 to produce the following result. We adopt the notation of Chapter 7 again:

$$T_0 = \exp(\mathcal{L}^{\frac{1}{3}}), \quad \mathcal{L} = \log x,$$

but now $T = \delta^{-1} X^\epsilon$.

Lemma 9.3. *Let*

$$F(s) = \sum_{k \sim x} c_k k^{-s}$$

be a divisor-bounded Dirichlet polynomial. Suppose that

$$\int_{T_0}^{T} \left| F\left(\frac{1}{2} + it\right) \right|^2 dt \ll X\mathcal{L}^{-A} \quad (all \ A > 0). \tag{9.2.1}$$

Then

$$\sum_{k \in A} c_k = \frac{y}{y_1} \sum_{k \in B} c_k + O(Y\mathcal{L}^{-A}) \quad (all \ A > 0) \tag{9.2.2}$$

for all $x \in [X, 2X]$ except on a set with measure at most $O(X\mathcal{L}^{-A})$.

Henceforth in this chapter whenever we say "asymptotic formula" we will mean "an asymptotic formula for almost all $x \in [X, 2X]$," with the understanding that the exceptional set can be made to have measure $O(X\mathcal{L}^{-A})$ for any $A > 0$. In view of the square in (9.2.1) it is possible for this problem to obtain an asymptotic formula for

$$\sum_{mn \in A} a_m b_n,$$

provided that b_n is prime-factored and $x^{1-\epsilon} > M > x^{1-\theta}$, just by using the mean value theorem (Lemma 5.2) and (7.2.3). This quickly leads to an asymptotic formula for

$$S(A, X^\theta).$$

However, we can do much better by using triple sums and employing Lemma 7.3.

Lemma 9.4. *Suppose that all the Dirichlet polynomials*

$$M(s) = \sum_{m \sim M} a_m m^{-s}, \qquad N(s) = \sum_{n \sim N} b_n n^{-s}, \qquad R(s) = \sum_{r \sim R} c_r r^{-s}$$

are divisor-bounded. Suppose that $RMN = x$, $M = x^{\beta_1}$, $N = x^{\beta_2}$, $0 \le \beta_2 \le \beta_1 \le \frac{1}{2}$. Write

$$\gamma'(\theta) = \gamma\left(\frac{1}{2}(1 + \theta)\right),$$

where $\gamma(x)$ is the function given by (7.2.13). If either

$$1 - (\log x)^{-\frac{1}{2}} \ge \beta_1 + \beta_2 \ge 1 - 2\theta, \tag{9.2.3}$$

with $R(s)$ satisfying (7.2.3), or (β_1, β_2) lies in the triangle R defined by

$$R: \qquad \beta_2 > 2\theta, \quad 2\beta_2 - 2\theta + \epsilon \le \beta_1 \le 2\gamma'(\theta), \tag{9.2.4}$$

with $M(s)$ satisfying (7.2.3), or (β_1, β_2) lies in either of the parallelograms P_1, P_2 defined by

$$P_1: \qquad |4\beta_1 + 3\beta_2 - 2| \le 2\theta - \epsilon, \quad 2\theta \le \beta_2 \le 2\gamma'(\theta),$$
$$P_2: \qquad |4\beta_1 + \beta_2 - 2| \le 2\theta - \epsilon, \quad 2\theta \le \beta_2 \le 2\gamma'(\theta), \tag{9.2.5}$$

with $N(s)$ satisfying (7.2.3), then (9.2.1) holds with $F(s) = M(s)N(s)R(s)$.

Proof. We apply Lemma 7.3 with x there replaced by $x_1 = x^2$. We have $T = x_1^{(1-\theta)/2}$, so we are applying the lemma with θ replaced by $(1+\theta)/2$. The different cases for the current lemma arise from different combinations of $M(s), M(s), N(s), N(s), R(s), R(s)$. First consider $M(s)N(s), M(s)N(s), R(s)^2$. We then have in the notation of Lemma 7.3:

$$\alpha_1 = \alpha_2 = \tfrac{1}{2}(\beta_1 + \beta_2).$$

We thus appear to need $\alpha_1 + \alpha_1 \geq 1 - \gamma((1+\theta)/2)$ to obtain (9.2.1). However, since we have a factor $R(s)^2$, our choice for g in Case 2(ii) in the proof of Lemma 7.3 now also includes values that are half an odd integer. We can thus assume that $\gamma((1+\theta)/2)$ always equals $4(1+\theta/2)/2 - 2 = 2\theta$. This leads to (9.2.3).

For the next case we use the combination $N(s)^2 R(s), M(s)R(s), M(s)$. We now have

$$\alpha_1 = \tfrac{1}{2}(2\beta_2 + 1 - \beta_1 - \beta_2) = \tfrac{1}{2}(1 + \beta_2 - \beta_1),$$
$$\alpha_2 = \tfrac{1}{2}(1 - \beta_2).$$

To fulfil the conditions of Lemma 7.3 we then need

$$|\alpha_1 - \alpha_2| < \theta \quad \text{and} \quad \alpha_1 + \alpha_2 > 1 - \gamma'(\theta),$$

that is,

$$-2\theta < 2\beta_2 - \beta_1 < 2\theta, \quad \text{and} \quad \beta_1 < 2\gamma'(\theta).$$

This includes region R; the remaining region has $\beta_2 < 2\theta$ and so is already covered.

In the next case, to get P_1 we use $M(s)^2 N(s), R(s)^2$, and $N(s)$. Thus

$$\alpha_1 = \tfrac{1}{2}(2\beta_1 + \beta_2), \quad \alpha_2 = 1 - \beta_1 - \beta_2.$$

Our conditions now reduce to

$$|4\beta_1 + 3\beta_2 - 2| \leq 2\theta - \epsilon \quad \text{and} \quad \beta_2 \leq 2\gamma'(\theta),$$

which is the parallelogram stated (again we can suppose that $\beta_2 \geq 2\theta$).

We get P_2 by replacing β_1 by $1 - \beta_1 - \beta_2$ in P_1.

□

We have illustrated the regions R, P_1, P_2 for $\theta = \tfrac{1}{18}$ in Diagram 9.1. The regions outlined by dashed lines are further asymptotic formulae regions that will be established in the next section. For this value of θ we have $2\gamma'(\theta) = \tfrac{2}{9}$.

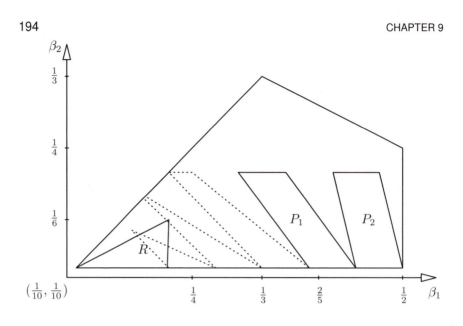

Diagram 9.1

Clearly the last lemma should lead to an asymptotic formula for $S(\mathcal{A}, x^{2\theta})$. At this point the problem so far is looking easier than the question of primes in all short intervals. However, when we turn to the Type I information available, we run into some difficulties. The squaring of the Dirichlet polynomials that worked to our advantage above now means that we obtain the fourth power of the polynomial just by applying Cauchy's inequality. Put another way, the reflection principle that meant we could reduce N to T/N is now inefficient since N will have to be close to $x^{1/2}$ in size for this to be useful (whereas N only had to be around $X^{1/4}$ in Chapter 7). The one slight compensation is that we can use Watt's theorem with larger factors because T is larger.

Lemma 9.5. *Suppose that both the Dirichlet polynomials*

$$M(s) = \sum_{m \sim M} a_m m^{-s}, \qquad N(s) = \sum_{n \sim N} b_n n^{-s}$$

are divisor-bounded and that

$$R(s) = \sum_{r \sim R} r^{-s}.$$

Suppose that $RMN = x$, $M = x^{\alpha_1}$, $N = x^{\alpha_2}$, $0 \leq \alpha_2 \leq \alpha_1 \leq \frac{1}{2}$. Suppose in addition that

$$2\alpha_1 + \alpha_2 \leq 1 + \theta, \tag{9.2.6}$$

$$4\alpha_2 \leq 1 + 3\theta, \tag{9.2.7}$$

$$4\alpha_1 + 6\alpha_2 \leq 3 + \theta. \tag{9.2.8}$$

Then (9.2.1) holds with $F(s) = M(s)N(s)R(s)$.

Remarks. If $\alpha_2 \leq (1-\theta)/4$, then (9.2.8) automatically follows from (9.2.6).

Proof. By Cauchy's inequality

$$\int_{T_0}^{T} |M(s)N(s)R(s)|^2 \, dt$$

$$\leq \left(\int_{T_0}^{T} |M(s)^2 N(s)|^2 \, dt \right)^{\frac{1}{2}} \left(\int_{T_0}^{T} |R(s)^2 N(s)|^2 \, dt \right)^{\frac{1}{2}}.$$

By the mean value theorem for Dirichlet Polynomials and Lemma 7.8 this is

$$\ll \left(M^2 N + T \right)^{\frac{1}{2}} T^{\frac{1}{2}} \left(1 + N^2 T^{-\frac{1}{2}} \right)^{\frac{1}{2}} \mathcal{L}^B.$$

The exponent of x in the above is half of

$$\max(2\alpha_1 + \alpha_2, 1 - \theta) + (1 - \theta) + \max(0, 2\alpha_1 - \tfrac{1}{2}(1 - \theta)) < 2,$$

using (9.2.6)–(9.2.8). Thus (9.2.1) holds as required.   □

9.3 THE ALTERNATIVE SIEVE APPLIED

We can now prove our Fundamental Theorem for the present situation, although the proof is considerably simplified by the Type II information available. From now on $\theta = \frac{1}{18}$ and $z = x^{1/9}$.

Lemma 9.6. *Suppose that $M = x^{\alpha_1}$ and $N = x^{\alpha_2}$ and that the sequences a_m, b_n are divisor-bounded. Suppose that α_1, α_2 satisfy (9.2.6)–(9.2.8). Then*

$$\sum_{m \sim M, n \sim N} a_m b_n S(\mathcal{A}_{mn}, z) = \frac{y}{y_1} \sum_{m \sim M, n \sim N} a_m b_n S(\mathcal{B}_{mn}, z) + O\left(y \mathcal{L}^{-A} \right).$$

$$(9.3.1)$$

Proof. Let

$$w = \exp\left((\log x)^{\frac{1}{2}} \right).$$

Then

$$\sum_{m \sim M, n \sim N} a_m b_n S(\mathcal{A}_{mn}, z) = \sum_{m \sim M, n \sim N} a_m b_n S(\mathcal{A}_{mn}, w)$$

$$- \sum_{m \sim M, n \sim N} a_m b_n \sum_{w \leq p < z} S(\mathcal{A}_{mnp}, p).$$

We can obtain an asymptotic formula for the first sum on the right-hand side above using Lemma 9.5 with Theorem 4.3 (take $a = \epsilon/4$, say). The second sum can be estimated as a Type II sum using Lemma 9.4, condition (9.2.3). The formula (9.3.1) then follows.   □

We can combine Lemma 9.6 with the asymptotic formula regions P_1, P_2 in Diagram 9.1 to prove the following.

Lemma 9.7. *Suppose that $M = x^{\alpha_1}$ and $N = \mathbf{x}^{\alpha_2}$ and that the sequences a_m, b_n are divisor-bounded. In addition, b_n satisfies the prime-factored condition (7.2.3). Suppose that α_1, α_2 lie in either of the two quadrilaterals*

$$Q_1 : \tfrac{1}{9} \leq \alpha_2 \leq \alpha_1, \quad 4\alpha_1 + 7\alpha_2 \leq \tfrac{19}{9}, \quad \alpha_1 + \alpha_2 \geq \tfrac{7}{18}; \tag{9.3.2}$$

$$Q_2 : \tfrac{1}{9} \leq \alpha_2 \leq \min\left(\alpha_1, \tfrac{2}{9}\right), \quad 4\alpha_1 + 5\alpha_2 \leq \tfrac{19}{9}, \quad \alpha_1 + \alpha_2 \geq \tfrac{4}{9}; \tag{9.3.3}$$

or the triangle

$$Q_3 : \tfrac{1}{9} \leq \alpha_2, \quad 4\alpha_1 + 9\alpha_2 \leq \tfrac{19}{9}, \quad \alpha_1 + \alpha_2 \geq \tfrac{1}{3}. \tag{9.3.4}$$

Then

$$\sum_{m \sim M, n \sim N} a_m b_n S(\mathcal{A}_{mn}, n) = \frac{y}{y_1} a_m b_n \sum_{m \sim M, n \sim N} S(\mathcal{A}_{mn}, n) + O\left(y\mathcal{L}^{-A}\right). \tag{9.3.5}$$

Remark. The three regions Q_1, Q_2, Q_3 are the areas enclosed by dashed lines in Diagram 9.1.

Proof. We have

$$\sum_{m \sim M, n \sim N} S(\mathcal{A}_{mn}, n) = \sum_{m \sim M, n \sim N} S(\mathcal{A}_{mn}, z) - \sum_{m \sim M, n \sim N} \sum_{z \leq p < n} S(\mathcal{A}_{mnp}, p). \tag{9.3.6}$$

The first sum on the right-hand side of (9.3.6) is handled by Lemma 9.6. Now write $mp = x^{\alpha_3}$. If $(\alpha_1, \alpha_2) \in Q_j, j = 1, 2$, it is a simple matter to show that $(\alpha_3, \alpha_2) \in P_j$. If $(\alpha_1, \alpha_2) \in Q_3$, we can similarly show that $(\alpha_3, \alpha_2) \in Q_2$. We thus can give an asymptotic formula for this sum too, and this completes the proof.   □

When applying our standard Buchstab decomposition, we would like to apply Buchstab's identity twice more to sums of the form

$$\sum_{\substack{p \sim M \\ q \sim N}} S(\mathcal{A}_{pq}, q), \tag{9.3.7}$$

where $M = x^{\alpha_1}, N = x^{\alpha_2}$, whenever (α_1, α_2) does not belong to any of the asymptotic formulae regions R, P_j, Q_k. This is where our problems with the Type I information come to the fore. Nevertheless we can establish the following two lemmas.

Lemma 9.8. *We can apply Buchstab's identity twice to (9.3.7) provided that either*

$$\alpha_2 \leq \tfrac{19}{90} \quad \text{and} \quad 2\alpha_1 + 3\alpha_2 \leq \tfrac{19}{18} \tag{9.3.8}$$

or

$$\alpha_1 \leq \tfrac{7}{24}, \quad \alpha_1 + 4\alpha_2 \leq \tfrac{19}{18}, \quad 3\alpha_1 + 4\alpha_2 \leq \tfrac{55}{36}. \tag{9.3.9}$$

Remark. The union of the regions given by (9.3.8) and (9.3.9) is labelled D in Diagram 9.2.

Proof. We need to be able to estimate

$$\sum_{\substack{p\sim M \\ q\sim N}} \sum_{z\leq r<q} S(\mathcal{A}_{pqr}, z)$$

in order to apply Buchstab's identity twice. We thus must ensure that we can group the variables p, q, r in such a way that (9.2.6)–(9.2.8) are satisfied. When (9.3.8) holds, we let $m = pq, n = r$ in Lemma 9.6, while we put $n = p, m = qr$ if (9.3.9) is valid. $\qquad\square$

Lemma 9.9. *We can apply Buchstab's identity twice to (9.3.7) by reversing the roles of the variables r and p (where r is the implicit variable given by $pqr \in \mathcal{A}$) provided that $(\alpha_1, \alpha_2) \in D^*$, where D^* is the pentagon with vertices given by*

$$\left(\tfrac{5}{12}, \tfrac{1}{9}\right), \quad \left(\tfrac{17}{48}, \tfrac{17}{72}\right), \quad \left(\tfrac{53}{144}, \tfrac{19}{72}\right), \quad \left(\tfrac{59}{144}, \tfrac{17}{72}\right), \quad \left(\tfrac{17}{36}, \tfrac{1}{9}\right).$$

Remark. The region D^* is shown in Diagram 9.2.

Proof. To apply Buchstab's identity twice with a role reversal we must verify (9.2.6)–(9.2.8) for α_1, α_2, and with α_1 replaced by $1 - \alpha_1 - \alpha_2$. For $\frac{1}{9} \leq \alpha_2 \leq \frac{17}{72}$ this simply reduces to $\frac{17}{18} \leq 2\alpha_1 + \alpha_2 \leq \frac{19}{18}$. For larger α we must check the following five inequalities:

$$2\alpha_1 + \alpha_2 \leq \tfrac{19}{18}, \quad \alpha_2 \leq \tfrac{7}{24}, \quad 2\alpha_1 + 3\alpha_2 \leq \tfrac{55}{36},$$
$$2(1 - \alpha_1 - \alpha_2) + \alpha_2 \leq \tfrac{19}{18}, \quad 2(1 - \alpha_1 - \alpha_2) + 3\alpha_2 \leq \tfrac{55}{36}.$$

It is then a simple matter to check where certain lines intersect to find the region D^* where all the inequalities are satisfied. $\qquad\square$

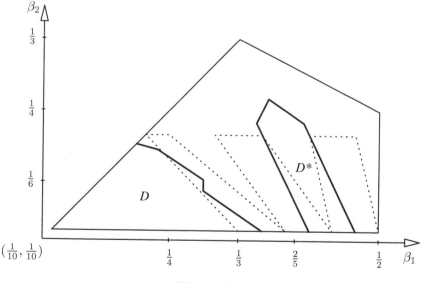

Diagram 9.2

In Diagram 9.2 we have indicated P_1, P_2, Q_2 by dashed lines; the regions Q_2, Q_3, and R lie entirely within D. The reader will notice that a large part of regions D

and D^* intersects with the asymptotic formulae regions P_j, Q_k. This is unfortunate since it leaves large regions of (α_1, α_2) (almost the whole of the region $\alpha_2 > \frac{2}{9}$) where we are unable to handle the sum (9.3.7).

9.4 THE FINAL DECOMPOSITION

We have now assembled all the information we require. Since we are using X as the size of x, we write $\xi = (2X)^{1/2}$. We have

$$
S(\mathcal{A}, \xi) = S(\mathcal{A}, z) - \sum_{z \le p < \xi} S(\mathcal{A}_p, z) + \sum_{\substack{z < p < \xi \\ z \le q < \min(p, (x/p)^{1/2})}} S(\mathcal{A}_{pq}, q)
$$

$$
= S_1 - S_2 + S_3, \quad \text{say.}
$$

Now we can give asymptotic formulae for S_1 and S_2 by Lemma 9.6. We can also give formulae for those parts of S_3 corresponding to $P_1, P_2, Q_1, Q_2, Q_3, R$.

For that part of the sum corresponding to D^* included in neither P_1 nor P_2, we decompose twice more, giving a four-dimensional integral to estimate

$$
\int_W \frac{1}{\alpha_2 \alpha_3^2 \alpha_4^2} \omega \left(\frac{1 - \alpha_1 - \alpha_2 - \alpha_3}{\alpha_3} \right) \omega \left(\frac{\alpha_1 - \alpha_4}{\alpha_4} \right) d\alpha_4 \ldots d\alpha_1,
$$

where W is the four-dimensional region bounded by the inequalities

$$
\frac{17}{48} \le \alpha_1 \le \frac{4}{9}, \quad \alpha_2 > \alpha_3 > \max \left(\frac{17}{18} - 2\alpha_1, \frac{19 - 36\alpha_1}{27} \right),
$$

$$
\alpha_2 \le \min \left(2\alpha_1 - \frac{17}{36}, \frac{55 - 72\alpha_1}{108}, \frac{19}{18} - 2\alpha_1, \frac{17}{9} - 4\alpha_1 \right),
$$

$$
\alpha_3 \le \min \left(\alpha_2, \frac{1 - \alpha_1 - \alpha_2}{2} \right), \quad \frac{1}{9} \le \alpha_4 \le \frac{\alpha_2}{2}.
$$

It is further assumed that no combination of the variables from W lies in $P_1 \cup P_2$. A computer calculation gives the estimation for this integral < 0.028.

The calculation of the four-dimensional integral corresponding to

$$
(\alpha_1, \alpha_2) \in D, \quad (\alpha_1, \alpha_2) \notin Q_1 \cup Q_2 \cup Q_3 \cup R
$$

is relatively straightforward, giving a contribution < 0.004. The main errors arising from sums we must discard come from the remaining parts of S_3 where we can neither decompose twice more nor give an asymptotic formula. We split this region into four areas. In the first, $\alpha_2 \ge \frac{2}{9}$ and we obtained a bound < 0.479. The reader should note that since all the variables have all their prime factors $> x^{2/9}$ in this region, we cannot make any savings on this term. In the second, we had $\frac{1}{4} \le \alpha \le \frac{17}{18}$ (the region between Q_2 and P_1 with $\alpha_2 \le \frac{2}{9}$). This leads to an integral that is < 0.147. For this region we can apply Buchstab's identity in reverse to make visible the almost-primes counted and obtain a small saving (the estimate would have been 0.170 otherwise). In the third, we had $\frac{17}{36} \le \alpha_1 \le \frac{1}{2}$ (the region to the right of P_2). This leads to an integral < 0.059. Finally, there is a very small region bounded by $\alpha_1 = \alpha_2$ and the regions D and Q_2 that leads to an error < 0.01. We thus have shown that the sum of the integrals discarded is < 0.727, and so, using the continuity of the method in θ, we obtain the following result.

Theorem 9.2. *For all $n \leq N$ with at most $O\left(N(\log N)^{-A}\right)$ exceptions, the interval $[n - n^{1/18}, n]$ contains*

$$> (0.273)\frac{n^{\frac{1}{18}}}{\log n} \tag{9.4.1}$$

primes.

A result with this exponent (although a smaller constant) was first obtained by Wong [167]. The exponent was improved to $\frac{1}{20}$ by Jia [111], although the details and calculation of the integrals are quite formidable!

9.5 AN UPPER-BOUND RESULT

We finish this chapter with a look at the corresponding upper bound; we shall need this result for our work in the next chapter. We begin by writing

$$S(\mathcal{A}, \xi) = S(\mathcal{A}, z) - \sum_{z \leq p < \xi} S(\mathcal{A}_p, p).$$

The first sum we have already dealt with on the right-hand side above. Our chief difficulty is that we cannot give an asymptotic formula for any of the final terms above. We thus need to apply Buchstab's identity twice, which will lead to some savings from asymptotic formula regions. To do this we need $p \leq x^{7/24}$. We then have, with $p = x^\alpha$,

$$2\alpha + \alpha \leq 1 + \theta, \quad 4\alpha \leq 1 + 3\theta, \quad 4\alpha + 6\alpha \leq 3 + \theta,$$

and so (9.2.6)–(9.2.8) will always hold after one Buchstab decomposition. We thus obtain

$$S(\mathcal{A}, \xi) \leq S(\mathcal{A}, x^{\frac{7}{24}}) + E\frac{y}{y_1}S(\mathcal{B}, \xi),$$

where

$$E = \int_V \frac{1}{\alpha_1 \alpha_2 \alpha_3^2}\omega\left(\frac{1 - \alpha_1 - \alpha_2 - \alpha_3}{\alpha_3}\right) d\alpha_3\, d\alpha_2\, d\alpha_1.$$

Here the region V is given by:

$$\tfrac{1}{9} \leq \alpha_3 \leq \alpha_2 \leq \alpha_1 \leq \tfrac{7}{24}, \quad \alpha_1 + \alpha_2 + 2\alpha_3 \leq 1,$$

and neither (α_1, α_2) nor (α_1, α_3) belongs to $R \cup Q_1 \cup Q_2 \cup Q_3$. A computer calculation gives

$$\frac{\omega\left(\frac{24}{7}\right)}{\frac{7}{24}} < 1.936$$

and

$$E < 0.05.$$

We thus obtain the following theorem.

Theorem 9.3. *For all $n \leq N$ with at most $O\left(N(\log N)^{-A}\right)$ exceptions, the interval $[n - n^{1/18}, n]$ contains*

$$\leq 2\frac{n^{\frac{1}{18}}}{\log n} \tag{9.5.1}$$

primes.

9.6 OTHER MEASURES OF GAPS BETWEEN PRIMES

The results in Chapter 7 dealt with gaps between primes *never* being too large. In this chapter we have demonstrated that *usually* the gaps cannot be large. There are two other measures to quantify how often there can be large gaps between primes. Writing p_n for the nth prime and $d_n = p_n - p_{n-1}$ (with $p_0 = 1$), these are

$$\sum_{\substack{p_n < x \\ d_n > x^{1/2}}} d_n, \tag{9.6.1}$$

and

$$\sum_{p_n < x} d_n^2. \tag{9.6.2}$$

Wolke [166] was the first to obtain a nontrivial estimate for (9.6.1), obtaining an upper bound $\ll x^{29/30+\epsilon}$. The exponent was improved to $\frac{3}{4} + \epsilon$ by Heath-Brown [72] and further improved, using Heath-Brown's sieve identity method, by Peck [139] to $\frac{25}{36} + \epsilon$. Recently Matomäki [124] reduced the exponent still further to $\frac{2}{3}$ using the alternative sieve. Heath-Brown obtained the bound $\ll x^{23/18+\epsilon}$ for (9.6.2) [72], which was improved in [140] to $\frac{5}{4} + \epsilon$. It is expected that this exponent could be further reduced using the alternative sieve and variants of the mean value results we have employed. For the latest work on *small* gaps between primes we refer the reader to the ground-breaking work of Goldston, Motohashi, Pintz, and Yildirim [48].

Chapter Ten

Combination with the Vector Sieve

10.1 INTRODUCTION

In this chapter we are going to introduce two major developments of the method that were incorporated in the paper [11]. The first is the combination of the alternative sieve with the *vector sieve*. The second is the yoking together of the method with the Hardy-Littlewood circle method (see [161] for a full introduction to this subject). Along the way we must revisit work we did in Chapters 7 and 9 on primes in short intervals, although now we need the primes to be constrained to residue classes to a small modulus.

First, before we introduce the problem where the techniques will be exemplified, we explain the vector sieve. Suppose that we write $\rho(n)$ for the characteristic function of the set of primes. Our work to date has consisted of the construction of upper- and lower-bound approximations to ρ, say ρ^- and ρ^+, with

$$\rho^-(n) \leq \rho(n) \leq \rho^+(n)$$

for all n. Now suppose that we are interested in finding pairs of primes p_1, p_2 with $p_j \in \mathcal{B}_j$ satisfying $(p_1, p_2) \in \mathcal{C}$ where $\mathcal{C}$ is some set of ordered pairs. The example we introduce will be $\{(m, n) : m + n = 2N\}$, but for our discussion at present any set would do. We want to estimate

$$\sum_{\substack{n_j \in \mathcal{B}_j \\ (n_1, n_2) \in \mathcal{C}}} 1.$$

Suppose that

$$a_0(k) \leq \rho(k) \leq a_1(k), \ k \in \mathcal{B}_1; \tag{10.1.1}$$

$$b_0(m) \leq \rho(m) \leq b_1(m), \ m \in \mathcal{B}_2. \tag{10.1.2}$$

Then we have the obvious upper bound

$$\sum_{\substack{n_j \in \mathcal{B}_j \\ (n_1, n_2) \in \mathcal{C}}} \rho(n_1)\rho(n_2) \leq \sum_{\substack{n_j \in \mathcal{B}_j \\ (n_1, n_2) \in \mathcal{C}}} a(n_1)b(n_2). \tag{10.1.3}$$

Since the lower-bound functions we construct can be negative, we cannot simply multiply together $a_0(k)$ and $b_0(m)$ to get a lower bound. However, the following inequality from Brüdern and Fouvry [19] (compare the inequality given earlier by Iwaniec [101]), provides us with the lower bound.

Lemma 10.1. *With the above notation we have*

$$a_0(k)b_1(m) + a_1(k)b_0(m) - a_1(k)b_1(m) \le \rho(k)\rho(m). \qquad (10.1.4)$$

Proof. For clarity we omit the dependence on k and m in the following. We first note that $(a_1 - a_0)(b_1 - b_0)$ is always non-negative. Thus

$$a_0b_1 + a_1b_0 - a_1b_1 = a_0b_0 - (a_1 - a_0)(b_1 - b_0) \le a_0b_0.$$

We thus have (10.1.4) unless both a_0 and b_0 are negative. In that case we note that $(a_1 - a_0)(b_1 - b_0) \ge a_0b_0$, and so

$$a_0b_0 - (a_1 - a_0)(b_1 - b_0) \le 0.$$

Since $\rho(k)\rho(m) \ge 0$, this completes the proof. $\qquad\qquad\square$

From (10.1.4) we obtain

$$\sum_{\substack{n_j \in \mathcal{B}_j \\ (n_1,n_2) \in \mathcal{C}}} \rho(n_1)\rho(n_2) \ge S_{0,1} + S_{1,0} - S_{1,1} \qquad (10.1.5)$$

where

$$S_{i,j} = \sum_{\substack{n_j \in \mathcal{B}_j \\ (n_1,n_2) \in \mathcal{C}}} a_i(n_1)b_j(n_2).$$

We now have to construct our upper- and lower-bound functions so that $S_{0,1} + S_{1,0} - S_{1,1}$ is positive, whereas previously we only had to ensure that sums like

$$\sum_{n \in \mathcal{A}} a_0(n)$$

were positive. The additional restriction this brings means that we will not be able to use results where such sums are barely positive; we need more room to manoeuvre in the present context!

10.2 GOLDBACH NUMBERS IN SHORT INTERVALS

Define a *Goldbach number* to be an even number that can be written as the sum of two primes. Ramachandra [146] proved that almost all even numbers in an interval of the form $[x, x + x^\theta]$ are Goldbach numbers provided that

$$\theta > \tfrac{3}{5}. \qquad (10.2.1)$$

Here and below "almost all" means 'with less than $x^\theta(\log x)^{-A}$ exceptions, provided that $A > 0$ and $x > C_1(A, \theta)$."

In the 1990s this problem attracted a great deal of attention [141, 128, 108, 109, 121, 110], in chronological order. In successive steps the condition (1.1) was weakened to $\theta > \frac{7}{108}$ (Hua, [110]). The paper [11] by the author with Baker and Pintz, on which the material here is based, gave an exponent $\frac{11}{160}$. Although not as good as Hua's exponent, this paper introduced the important new idea of sieving both variables, opening the door for further improvements. We shall push the method close to what is possible with our existing knowledge of mean and large values of Dirichlet polynomials and so establish a stronger result than has previously appeared.

Theorem 10.1. *For* $\theta > \frac{11}{180} = 0.06111\ldots$ *almost all even numbers in* $[x, x + x^\theta]$
are Goldbach numbers.

Remark. Note that $\frac{7}{108} = 0.06481\ldots$.

Our aim, like that of previous workers on this question, is to show that, for almost
all even integers $2n$ in $K = [x, x + x^{\theta_1 \theta_2}]$,

$$S(n) := \sum_{\substack{k+m=2n \\ k \in \mathcal{B}_1, \, m \in \mathcal{B}_2}} \rho(k)\rho(m) > 0, \qquad (10.2.2)$$

where ρ is the characteristic function of the prime numbers, and

$$\mathcal{B}_1 = (x - 2Y, x] \cap \mathbb{N} \quad \text{with} \quad Y = x^{\theta_1}, \qquad \mathcal{B}_2 = (Y, 2Y] \cap \mathbb{N}.$$

Thus, for example, Perelli and Pintz [141] get an asymptotic formula for $S(n)$
with $\theta_1 \geq \frac{7}{12} + \epsilon$, $\theta_2 \geq \frac{1}{3} + \epsilon$. It is no coincidence, as will become apparent, that
$\frac{7}{12}$ is the exponent in Theorem 7.1 while $\frac{1}{3} = 2 \times \frac{1}{6}$, where $\frac{1}{6}$ is the exponent in
Theorem 9.1. We shall take

$$\theta_1 = \frac{11}{20} + \epsilon, \qquad \theta_2 = 2\left(\frac{1}{18} + \epsilon\right).$$

Here and below, ϵ is a sufficiently small positive absolute constant. The reader
will note that $\frac{1}{18}$ is essentially the same exponent as appears in Theorems 9.2 and
9.3, whereas the $\frac{11}{20}$ corresponds to an exponent for the primes in short intervals
problem for which our upper and lower bounds are very close to their expected
values. Obviously Theorem 10.1 follows from (10.2.2) with these θ_1 and θ_2. In our
proof of (10.2.2) we assume (as we may) that $x - \frac{1}{2}$ is an integer.

To prove (10.2.2) we use (10.1.5). It then suffices to establish that

$$S_{i,j} = u_i v_j \frac{Y}{\mathcal{L}\mathcal{L}'} \, \mathfrak{S}(2n)(1 + O(\mathcal{L}^{-1}))$$

for almost all even $2n$ in $[x, x + x^{\theta_1 \theta_2}]$, with $\mathcal{L} = \log x$, $\mathcal{L}' = \log Y$,

$$u_0 > 0.99, \; u_1 < 1.01, \; v_0 > 0.05, \; v_1 < 2.2. \qquad (10.2.3)$$

Here

$$\mathfrak{S}(2n) = 2 \prod_p \left(1 - \frac{1}{(p-1)^2}\right) \prod_{\substack{p \mid n \\ p > 2}} \left(\frac{p-1}{p-2}\right) \qquad (10.2.4)$$

is the *singular series*. Clearly it is always positive; indeed, it is bounded below by
the positive constant $2 \prod(1 - (p-1)^{-2})$. From (10.2.3) we thus obtain the crucial
inequality

$$u_0 v_1 + u_1 v_0 - u_1 v_1 > 0.$$

Since our choice of functions a_0, a_1, b_0, b_1 is based on the sieve method we
have been developing, we need to establish classes of sequences $a(k)$ ($k \in \mathcal{B}_1$),
$b(m)$ ($m \in \mathcal{B}_2$) for which an asymptotic formula

$$\sum_{k+m=2n} a(k)b(m) = uv \frac{Y}{\mathcal{L}\mathcal{L}'} \, \mathfrak{S}(2n)(1 + O(\mathcal{L}^{-1})) \qquad (10.2.5)$$

holds for almost all $2n$ in $[x, x + x^{\theta_1\theta_2}]$. We apply the Hardy-Littlewood circle method and follow [141, 11] quite closely to obtain the following result. We write

$$H = Y^{\theta_2}, \quad Q = \left[\frac{H^{\frac{1}{2}}}{2}\right], \quad P(z) = \prod_{p<z} p.$$

$$\delta_\chi = \begin{cases} 1 & \text{if } \chi \text{ is the principal character } \chi_0 \ (\mathrm{mod}\, q), \\ 0 & \text{if } \chi \text{ is nonprincipal } (\mathrm{mod}\, q). \end{cases}$$

By B we denote an absolute constant (not always the same one); ϵ is chosen so that $B\epsilon$ is sufficiently small (whenever necessary).

Theorem 10.2. *Suppose that the sequences $a(k)(k \in \mathcal{B}_1)$ and $b(m)(m \in \mathcal{B}_2)$ satisfy the following, for every $A > 0$ and $N > C_2(A)$.*

   (i) *We have*

$$\sum_{k \in \mathcal{B}_1,\, k \le t} \left(a(k)\chi(k) - \frac{\delta_\chi u}{\mathcal{L}}\right) \ll Y\mathcal{L}^{-A} \tag{10.2.6}$$

   for $t \le x$ and any character $\chi \ (\mathrm{mod}\, q)$, $q \le \mathcal{L}^A$.

   (ii) *We have*

$$a(k) = O(\tau(k)^B), \quad b(m) = O(\tau(m)^B),$$

   where τ is the divisor function and $a(k) = 0$ unless $(k, P(\mathcal{L}^A)) = 1$.

   (iii) *We have*

$$\sum_{m \in \mathcal{B}_2,\, m \le t} \left(b(m)\chi(m) - \frac{\delta_\chi v}{\mathcal{L}'}\right) \ll Y\mathcal{L}^{-A} \tag{10.2.7}$$

   for $t \le 2Y$ and any character $\chi \ (\mathrm{mod}\, q)$, $q \le \mathcal{L}^A$.

   (iv) *For any $q \le Q$ and any $z \in [qQ/6Y, 6qQ/Y]$, we have*

$$\sum_{\substack{\chi \ (\mathrm{mod}\, q) \\ \chi \notin E_q}} \int_{Y/2}^{3Y} \left|\sum_{n \in J_y} b(m)\chi(m) - \frac{\delta_\chi v}{\mathcal{L}'}\right|^2 dy \ll (qQ)^2 Y\mathcal{L}^{-A}. \tag{10.2.8}$$

     Here $J_y = [y, y + yz]$ and E_q is a set of $O(q^{1/2}\mathcal{L}^{-2A})$ characters $(\mathrm{mod}\, q)$.
Then (10.2.5) holds for almost all even integers $2n$ in $[N, N + N^{\theta_1\theta_2}]$.

Remark. Here and later, implied constants depend at most on A.

We shall prove Theorem 10.2 in Section 10.3. In Section 10.5, we shall find two sequences $a_0(k), a_1(k)$ that satisfy (i) and (ii), with associated constants u_0, u_1, in the role of u, such that (10.1.1) and the first two inequalities in (10.2.3) hold. In Section 10.6 we shall find sequences $b_0(m), b_1(m)$ that satisfy (ii), (iii), (iv), with associated constants v_0, v_1 in the role of v, such that (10.1.2) and the last two inequalities in (10.2.3) hold. This will establish Theorem 10.1.

We note the following consequence of our construction.

Theorem 10.3. *Let* $x^{0.55+\epsilon} \le M \le x\mathcal{L}^{-1}$. *For all* $q \le \mathcal{L}^A$ *and* $N > C_3(A)$,

$$\frac{0.99M}{\phi(q)\mathcal{L}} < \pi(x;q,a) - \pi(x-M;q,a) < \frac{1.01M}{\phi(q)\mathcal{L}}$$

whenever $(a,q) = 1$.

It is also worth remarking that our previous work on primes in all and in almost all short intervals leads to the following.

Theorem 10.4. *For* $x > C_4$, *the interval* $[x, x + x^\mu]$ *contains Goldbach numbers. Here* $\mu = 0.0292$.

A result like Theorem 10.3 with $q = 1$ was obtained by D. R. Heath-Brown in unpublished work using the sieve identities described in Section 2.5 here. To prove Theorem 10.3, we simply use the fact (already explained above) that Theorem 10.2(i) holds for sequences $a_0(k), a_1(k)$, lying below and above $\rho(k)$, with associated constants in $(0.99, 1.01)$. We then pick out integers congruent to $a \pmod q$ using characters in standard fashion.

To prove Theorem 10.4 we argue as in Montgomery and Vaughan [136], employing Theorem 9.2 in conjunction with the lower bound

$$\pi(x) - \pi(x-y) \gg \frac{y}{\log x}$$

for $x^{0.525} \le y \le x$ that we established in Chapter 7. The constant 0.0292 is larger than 0.525/18. The argument goes like this. Let $2N$ be a large, even integer with $2N \in [X, X + Y^{1/18}] = \mathcal{J}$, say, where $Y = (X)^{0.525}$. Consider $M(N,p) = 2N - p$, where p denotes a prime in the interval $[X - Y, X - \frac{1}{2}Y]$. There are $\gg Y/\log x$ such p by Theorem 7.2 (the additional factor $\frac{1}{2}$ making no difference to the result, of course). Now assume that $M(N,p)$ is not prime for any p for $2N \in \mathcal{J}$. There are thus $\gg Y/\log x$ intervals $[h, h + Y^{1/18}]$ with $h = X - p \in [Y/2, Y]$ containing no prime numbers. However, this contradicts Theorem 9.2, and so $\mathcal{J}$ does indeed contain Goldbach numbers.

10.3 PROOF OF THEOREM 10.2

We write

$$S_1(\alpha) = \sum_k a(k)e(k\alpha), \qquad S_2(\alpha) = \sum_m b(m)e(m\alpha).$$

Thus, by the orthogonality relation

$$\int_0^1 e(n\alpha)\, d\alpha = \begin{cases} 0 & \text{if } n \in \mathbb{Z} \setminus \{0\}, \\ 1 & \text{if } n = 0, \end{cases}$$

we have

$$\sum_{k+m=2n} a(k)b(m) = \int_{1/Q}^{1+1/Q} S_1(\alpha)S_2(\alpha)e(-2n\alpha)\, d\alpha.$$

The basic principle of the circle method is that the main contribution to such an integral should come from the set of α close to rationals with a small denominator (the *major arcs*), while the contribution from the remaining α (the *minor arcs*) should give a smaller order term. We thus divide up the interval $[1/Q, 1 + 1/Q]$ into Farey arcs of order Q, writing $I_{q,r}$ for the arc with centre at r/q. Hence

$$I_{q,r} \subset \left[\frac{r}{q} - \frac{1}{qQ}, \frac{r}{q} + \frac{1}{qQ} \right]$$

for $q \leq Q$, $1 \leq r \leq q$, $(r,q) = 1$. Let

$$I'_{q,r} = \left[\frac{r}{q} - \frac{\mathcal{L}^{4A}}{qY}, \frac{r}{q} + \frac{\mathcal{L}^{4A}}{qY} \right] \quad \text{and} \quad I''_{q,r} = \begin{cases} I_{q,r} \setminus I'_{q,r} & \text{if } q \leq \mathcal{L}^{2A}, \\ I_{q,r} & \text{if } q > \mathcal{L}^{2A}. \end{cases}$$

The major and minor arcs are defined by

$$\mathfrak{M} = \bigcup_{q \leq \mathcal{L}^{2A}} \bigcup_{r=1}^{q} {}^{*} I'_{q,r}, \qquad \mathfrak{m} = \left[\frac{1}{Q}, 1 + \frac{1}{Q} \right] \setminus \mathfrak{M},$$

respectively. Here the asterisk denotes a restriction to those r coprime to q.

Now write

$$c_q(m) = \sum_{r=1}^{q} {}^{*} e \left(\frac{mr}{q} \right), \qquad \tau(\chi) = \sum_{r=1}^{q} {}^{*} \chi(r) e \left(\frac{r}{q} \right).$$

The connection between $\mathfrak{S}(2n)$ and Ramanujan's sum $c_q(m)$ is well known (see [161, (3.24)]), namely,

$$\mathfrak{S}(2n) = \sum_{q=1}^{\infty} \frac{\mu^2(q)}{\phi^2(q)} c_q(-2n).$$

Also, we recall the well-known results ([27, pp. 66–67], for example) for the Gauss sum:

$$|\tau(\chi)| \leq q^{\frac{1}{2}}, \qquad \tau(\chi_0) = \mu(q). \tag{10.3.1}$$

To prove Theorem 10.2 it suffices to show that

$$\sum_{2n \in K} \left| \int_{1/Q}^{1+1/Q} S_1(\alpha) S_2(\alpha) e(-2n\alpha) \, d\alpha - \frac{uvY}{\mathcal{L}\mathcal{L}'} \mathfrak{S}(2n) \right|^2 \ll HY^2 \mathcal{L}^{-A-7}.$$

First we establish that

$$\sum_{2n \in K} \left| \int_{\mathfrak{M}} S_1(\alpha) S_2(\alpha) e(-2n\alpha) \, d\alpha - \frac{uvY}{\mathcal{L}\mathcal{L}'} \mathfrak{S}(2n) \right|^2 \ll HY^2 \mathcal{L}^{-A-7} \tag{10.3.2}$$

and then prove that

$$\sum_{2n \in K} \left| \int_{\mathfrak{m}} S_1(\alpha) S_2(\alpha) e(-2n\alpha) \, d\alpha \right|^2 \ll HY^2 \mathcal{L}^{-A-7}. \tag{10.3.3}$$

In order to tackle (10.3.2) we first replace the S_j by suitable approximations. For $q \leq \mathcal{L}^{2A}$ we have $(k, q) = 1$ whenever $a(k) \neq 0$. Guided by hypothesis (i) we rearrange S_1 as follows:

$$
\begin{aligned}
S_1\left(\frac{r}{q} + \eta\right) &= \sum_k a(k)e(k\eta)\frac{1}{\phi(q)}\sum_\chi \tau(\bar\chi)\chi(kr) \\
&= \frac{1}{\phi(q)}\sum_\chi \tau(\bar\chi)\chi(r)\sum_k a(k)\chi(k)e(k\eta) \\
&= S_1'\left(\frac{r}{q} + \eta\right) \\
&\quad + \sum_\chi \frac{\tau(\bar\chi)}{\phi(q)}\chi(r)\sum_k \left(\left(a(k)\chi(k) - \frac{u\delta_\chi}{\mathcal{L}}\right)e(k\eta)\right).
\end{aligned}
\tag{10.3.4}
$$

Here, using $\tau(\chi_0) = \mu(q)$, for $r/q + \eta \in I_{q,r}'$, we have

$$
S_1'\left(\frac{r}{q} + \eta\right) = \frac{\mu(q)}{\phi(q)}\frac{u}{\mathcal{L}}\sum_k e(k\eta).
$$

We now consider the inner sum over k in the final term of (10.3.4). We first write

$$
v(t) = \sum_{\substack{k \in \mathcal{B}_1 \\ k < t}} \left(a(k)\chi(k) - \frac{u\delta_\chi}{L}\right).
$$

Then hypothesis (i) yields (with $I = (x - 2Y, x]$)

$$
\sum_k \left(a(k)\chi(k) - \frac{u\delta_\chi}{\mathcal{L}}\right)e(k\eta) = \int_I e(t\eta)\,dv(t).
$$

Integration by parts turns this into

$$
\left[e(t\eta)v(t)\right]_{x-2Y}^x - 2\pi i\eta\int_I e(t\eta)v(t)\,dt \ll (1 + Y|\eta|)\max_{t\in I}|v(t)|
$$
$$
\ll \mathcal{L}^{4A}Y\mathcal{L}^{-13A} \ll Y\mathcal{L}^{-9A}.
$$

Here and below we may suppose that $x > C_2(A')$, where A' is sufficiently large in terms of A. We thus obtain, after summing over $\chi \pmod q$,

$$
S_1 - S_1' \ll Y\mathcal{L}^{-8A}.
$$

Now $\mathfrak{M}$ has Lebesgue measure $\ll \mathcal{L}^{6A}Y^{-1}$. Thus, since hypothesis (ii) yields $|S_2| \ll \sum_m |b(m)| \ll Y\mathcal{L}^B$, we obtain

$$
\int_{\mathfrak{M}} |(S_1 - S_1')S_2|\,d\alpha \ll \left(\mathcal{L}^{6A}Y^{-1}\right)\left(Y\mathcal{L}^{-8A}\right)\left(Y\mathcal{L}^B\right) \ll Y\mathcal{L}^{-A}. \tag{10.3.5}
$$

The bound (10.3.5) allows us to replace S_1 by S_1' in proving (10.3.2). Similarly, we may replace $S_1'S_2$ by $S_1'S_2'$, where

$$
S_2'\left(\frac{r}{q} + \eta\right) = \frac{\mu(q)}{\phi(q)}\frac{v}{\mathcal{L}'}\sum_m e(m\eta) \quad \text{for} \quad \frac{r}{q} + \eta \in I_{q,r}'.
$$

Hence we now need only prove the analogue of (10.3.2) for $S_1' S_2'$.

Let $\eta_0 = \mathcal{L}^{4A}/qY$. We then rewrite the integral

$$\int_{\mathfrak{M}} S_1'(\alpha) S_2'(\alpha) e(-2n\alpha) \, d\alpha$$

as

$$uv(\mathcal{L}\mathcal{L}')^{-1} \sum_{q \leq \mathcal{L}^{2A}} \sum_{r=1}^{q}{}^* \frac{\mu^2(q)}{\phi^2(q)} \, e\left(-\frac{2nr}{q}\right) \int_{-\eta_0}^{\eta_0} \sum_k \sum_m e((k+m-2n)\eta) \, d\eta.$$

$$(10.3.6)$$

Using

$$\left| \sum_{t \leq T} e(t\eta) \right| \leq \eta^{-1}$$

we can replace the domain of integration in (10.3.6) by $\left[-\frac{1}{2}, \frac{1}{2}\right]$, with an additional error

$$\ll \sum_{q \leq \mathcal{L}^{2A}} \frac{1}{\phi(q)} \int_{\eta_0}^{1/2} \eta^{-2} d\eta \ll \sum_{q \leq \mathcal{L}^{2A}} \frac{qY}{\phi(q)\mathcal{L}^{4A}} \ll Y\mathcal{L}^{-A}.$$

Since

$$\int_{-1/2}^{1/2} \sum_k \sum_m e((k+m-2n)\eta) \, d\eta = \sum_{\substack{k,m \\ k+m=2n}} 1,$$

this replaces the left-hand side of (10.3.6) by

$$\frac{uv}{\mathcal{L}\mathcal{L}'} \sum_{q \leq \mathcal{L}^{2A}} \frac{\mu^2(q)}{\phi^2(q)} c_q(-2n) \sum_{\substack{k,m \\ k+m=2n}} 1 = \frac{uv}{\mathcal{L}\mathcal{L}'} \sum_{q \leq \mathcal{L}^{2A}} \frac{\mu^2(q)}{\phi^2(q)} c_q(-2n)(Y + O(1))$$

for $2n \in K$. Thus,

$$\sum_{2n \in K} \left| \int_{\mathfrak{M}} S_1'(\alpha) S_2'(\alpha) e(-2n\alpha) \, d\alpha - \frac{uvY}{L^2} \mathfrak{S}(2n) \right|^2$$

$$\ll HY^2 \mathcal{L}^{-A-7} + \sum_{2n \in K} Y^2 \left| \sum_{q > \mathcal{L}^{2A}} \frac{\mu^2(q)}{\phi^2(q)} c_q(-2n) \right|^2.$$

Using known results on sums of Ramanujan sums [161, (3.23)] this last expression is

$$\ll HY^2 \mathcal{L}^{-A-7} + Y^2 \sum_{2n \in K} \left(\sum_{d|2n} \frac{1}{\phi(d)} \min\left(\frac{d}{\mathcal{L}^{2A}}, 1\right) \right)^2.$$

Using the crude inequality $d \ll \mathcal{L}^{1/2}\phi(d)$ we have

$$\sum_{2n \in K} \left(\sum_{\substack{d|2n \\ d \leq \mathcal{L}^{2A}}} \frac{1}{\phi(d)} \frac{d}{\mathcal{L}^{2A}} \right)^2 \ll \mathcal{L}^{-4A+1} \sum_{j \in K} \tau^2(j) \ll H\mathcal{L}^{-A-7}$$

and

$$\sum_{\substack{2n\in K}} \left(\sum_{\substack{d|2n \\ d>\mathcal{L}^{2A}}} \frac{1}{\phi(d)} \right)^2 \ll \mathcal{L}^{-4A+1} \sum_{j\in K} \tau^2(j) \ll HL^{-A-7}.$$

The analogue of (10.3.2) for $S'_1 S'_2$ now follows, and so (10.3.2) itself is established.

In some applications of the circle method the next move would be to apply Bessel's inequality to the left-hand side of (10.3.3), but this would only yield a bound $\ll Y^3 \mathcal{L}^{-A}$ rather than the form $\ll Y^2 H \mathcal{L}^{-A}$ which we need. We thus work more carefully and begin by multiplying out the modulus squared on the left-hand side of (10.3.3) to get

$$\sum_{2n\in K} \int_{\mathfrak{m}} S_1(\zeta) S_2(\zeta) e(-2n\zeta) \int_{\mathfrak{m}} \overline{S_1(\alpha)}\, \overline{S_2(\alpha)} e(2n\alpha)\, d\alpha\, d\zeta$$

$$\ll \int_{\mathfrak{m}} |S_1(\zeta) S_2(\zeta)| \int_{\mathfrak{m}} |S_1(\alpha) S_2(\alpha)| \min\left(H, \frac{1}{\|2(\alpha-\zeta)\|} \right) d\alpha\, d\zeta.$$

We now apply the Cauchy-Schwarz inequality, the bound

$$\int_0^1 |S_1(\alpha)|^2\, d\alpha = \sum_k |a(k)|^2 \ll \sum_k \tau(k)^{2B} \ll Y\mathcal{L}^B,$$

and the corresponding bound for S_2 to the last expression to obtain a bound

$$\ll \left(Y\mathcal{L}^B \right)^{\frac{3}{2}} \sup_{\zeta\in[0,1]} \Xi(\zeta),$$

where

$$\Xi^2(\zeta) = \int_{\mathfrak{m}} |S_2(\alpha)|^2 \min\left(H, \frac{1}{\|2(\alpha-\zeta)\|} \right)^2 d\alpha.$$

Now, if we write $\mathfrak{J}(\xi) = (\xi - 1/H,\ \xi + 1/H) \cap \mathfrak{m}$, we have

$$\Xi^2(\zeta) \ll \sum_\xi \int_{\mathfrak{J}(\xi)} |S_2(\alpha)|^2 \min\left(H, \frac{1}{\|2(\xi-\zeta)\|} \right)^2 d\alpha.$$

Here ξ takes on values $2jH^{-1}$, $j = 0, 1, \ldots, H$. Once we have shown that

$$\sup_\zeta \int_{\mathfrak{J}(\zeta)} |S_2(\alpha)|^2\, d\alpha \ll Y\mathcal{L}^{-3A}, \tag{10.3.7}$$

this gives

$$\Xi^2(\zeta) \ll Y\mathcal{L}^{-3A} \sup_\zeta \sum_\xi \min\left(H, \frac{1}{\|2(\xi-\zeta)\|} \right)^2 \ll H^2 Y\mathcal{L}^{-3A}$$

and completes the proof of (10.3.3).

Since $I_{q,r}$ has length at least $Q^{-2} > 2H^{-1}$, there are at most two punctured intervals $I''_{q,r}$ with $q \le Q$ and $(r,q) = 1$, which intersect $(\zeta - 1/H, \zeta + 1/H)$. Instead of (10.3.7), then, we need only show that

$$\int_{I''_{q,r}} |S_2(\alpha)|^2 d\alpha \ll Y\mathcal{L}^{-3A}$$

for $q \leq Q$, $(r, q) = 1$.

Now the work we did in (10.3.4) can be modified *mutatis mutandis* for S_2. We then need only show that

$$\int_{\eta + r/q \in I''_{q,r}} \left| S'_2 \left(\frac{r}{q} + \eta \right) \right|^2 d\eta \ll Y \mathcal{L}^{-3A} \qquad (10.3.8)$$

and

$$\frac{q}{\phi^2(q)} \int_{-1/qQ}^{1/qQ} \left(\sum_\chi |W(\chi, \eta)| \right)^2 d\eta \ll Y \mathcal{L}^{-3A}. \qquad (10.3.9)$$

In this case

$$W(\chi, \eta) = \sum_m \left(b(m) \chi(m) - \frac{v \delta_\chi}{\mathcal{L}'} \right) e(m\eta).$$

Let

$$I(\eta) = \begin{cases} [\mathcal{L}^{4A}/qY, \frac{1}{2}] & (q \leq \mathcal{L}^{2A}), \\ [0, \frac{1}{2}] & (q > \mathcal{L}^{2A}). \end{cases}$$

Then the left-hand side of (10.3.8) is

$$\ll \frac{1}{\phi^2(q)} \frac{1}{\mathcal{L}^2} \int_{I(\eta)} \min(Y^2, \eta^{-2}) \, d\eta \ll Y \mathcal{L}^{-3A}, \qquad (10.3.10)$$

as desired to establish (10.3.8).

To consider (10.3.9) we split the sum over χ into sums over E_q and $E_q^c = \{\chi \,(\mathrm{mod}\, q) : \chi \notin E_q\}$. We apply Cauchy's inequality to each subsum; it then suffices to prove in each case that

$$\frac{q|E|}{\phi^2(q)} \int_{-1/qQ}^{1/qQ} \sum_{\chi \in E} |W(\chi, \eta)|^2 \, d\eta \ll Y L^{-3A}. \qquad (10.3.11)$$

For the case $E = E_q$ we use hypotheses (ii) and (iv) and Parseval's inequality. In this way we find that the left-hand side of (10.3.11) is

$$\ll \frac{q}{\phi^2(q)} |E_q| \int_{-1/2}^{1/2} \sum_{\chi \in E_q} |W(\chi, \eta)|^2 \, d\eta$$

$$\ll \frac{q}{\phi^2(q)} |E_q|^2 \sum_m (|b(m)|^2 + 1)$$

$$\ll q^{-1} \mathcal{L} |E_q|^2 Y \mathcal{L}^B \ll Y \mathcal{L}^{-3A}.$$

For $E = E_q^c$ we use Gallagher's lemma [44, Lemma 1] (a quick use of Fourier transforms and Plancherel's theorem), which gives

$$\int_{-\frac{1}{qQ}}^{\frac{1}{qQ}} |W(\chi, \eta)|^2 \, d\eta \ll (qQ)^{-2} \int_{Y/2}^{3Y} \left| \sum_{m \in J'_y} \left(b(m) \chi(m) - \frac{v \delta_\chi}{\mathcal{L}'} \right) \right|^2 dy,$$

$$(10.3.12)$$

where

$$J_y' = \left[y - \frac{qQ}{2}, y + \frac{qQ}{2} \right].$$

Since $J_y = [y, y + yz]$, for some $z \in [qQ/6Y, 6qY/Y]$, the right-hand side of (10.3.12) is

$$\ll U_\chi := (qQ)^{-2} \int_{Y/2}^{3Y} \left| \sum_{m \in J_y} \left(b(m)\chi(m) - \frac{v\delta_\chi}{\mathcal{L}} \right) \right|^2 dy.$$

The reader may verify this using the argument given in [153, p. 25]. By hypothesis (iv)

$$\sum_{\chi \in E_q^c} U_\chi \ll Y\mathcal{L}^{-4A}.$$

The bound (10.3.11) with $E = E_q^c$ then follows, and the proof of Theorem 10.2 is complete.   □

10.4 DIRICHLET POLYNOMIALS

In this section we adapt the work we did in Chapter 7 to derive results about mean and large values of Dirichlet polynomials that now include a Dirichlet character. In general we write

$$F(s, \chi) = \sum_{n \asymp F} a_n \chi(n) n^{-s},$$

where χ is a Dirichlet character (mod q) with $q \leq x$. As in Chapter 7, we write $m \asymp M$ to mean $B^{-1}M < m < BM$, where B is a positive absolute constant that need not have the same value at each occurrence. As usual, the same letter is used for a Dirichlet polynomial, $F(s, \chi)$, and its length. We suppose that $2 \leq F \leq x$ for any such polynomial. All the Dirichlet polynomials that arise here will be *divisor-bounded* (recall (7.2.4)), and so we shall not always mention this fact in subsequent theorem and lemma hypotheses, and several will be *prime-factored* in the language of Chapter 7. To transfer information between Chapter 7 and the following it is usually only necessary to replace T by qT.

Let $2 < T \leq x/4$. Let $\mathcal{S}$ be a set of pairs (t, χ), with $t \in [-T, T]$ and χ a character (mod q). We say that $\mathcal{S}$ is *well spaced* if $|t - t'| \geq 1$ whenever $(t, \chi), (t', \chi) \in \mathcal{S}$. Suppressing dependence on $\mathcal{S}$, we write, for $p \geq 1$,

$$\|F\|_p = \left\| F\left(\tfrac{1}{2} + it, \chi \right) \right\|_p = \left(\sum_{(t,\chi) \in \mathcal{S}} \left| F\left(\tfrac{1}{2} + it, \chi \right) \right|^p \right)^{\frac{1}{p}}.$$

We also put

$$\|F\|_\infty = \sup_{(t,\chi) \in \mathcal{S}} \left| F\left(\tfrac{1}{2} + it, \chi \right) \right|.$$

Our prime-factored condition (7.2.3) now becomes

$$\|F\|_\infty \ll F^{\frac{1}{2}} \mathcal{L}^{-A} \quad \text{(all } A > 0\text{)}. \tag{10.4.1}$$

The following lemma encapsulates both the mean value theorem for Dirichlet polynomials as well as Huxley's large-values result, except that we are now averaging over characters as well as the parameter t.

Lemma 10.2. *Let $\mathcal{S}$ be well spaced; then*

$$\|F\|_2^2 \ll \mathcal{L}(qT + F)G, \tag{10.4.2}$$

where

$$G = G(F) = \sum_{n \asymp F} |a_n|^2 n^{-1}.$$

If

$$\left| F \left(\tfrac{1}{2} + it, \chi \right) \right| \geq V$$

for $(t, \chi) \in \mathcal{S}$, then the cardinality of $\mathcal{S}$ is bounded by

$$|\mathcal{S}| \ll \mathcal{L}^2 (GFV^{-2} + G^3 FqTV^{-6}). \tag{10.4.3}$$

Proof. The inequality (10.4.2) is established in the appendix (A.4). To obtain (10.4.3) we begin by proving

$$\|F\|_2^2 \ll \mathcal{L}(F + (qT)^{\frac{1}{2}} |\mathcal{S}|)G$$

using the same reasoning as in the proof in [131, Theorem 8.3]. We may then follow Huxley's argument in [97, Section 2], to complete the proof of (10.4.3). $\quad\square$

In the following when X occurs, it denotes *one* of x, Y^2. We now state the equivalent of Lemma 7.3 in our current terminology where we take $g = 2$ or $g = 4$ in (7.2.13). The proof is, of course, identical.

Lemma 10.3. *Let $M(s, \chi)$, $N(s, \chi)$, $L(s, \chi)$ be Dirichlet polynomials with*

$$MNL = X. \tag{10.4.4}$$

Suppose that L satisfies (10.4.1). Let $qT = X^{1-\theta}$ where $\frac{1}{2} + \epsilon < \theta < \frac{7}{12}$. Suppose that $M = X^{\alpha_1}$, $N = X^{\alpha_2}$,

$$|\alpha_1 - \alpha_2| < 2\theta - 1 - \epsilon \tag{10.4.5}$$

and

$$1 - (\alpha_1 + \alpha_2) < \gamma(\theta) - \epsilon, \tag{10.4.6}$$

where either

$$\gamma(\theta) = \min\left(4\theta - 2, \frac{12\theta - 5}{7}, \frac{48\theta - 25}{7} \right) \tag{10.4.7}$$

or

$$\gamma(\theta) = \min\left(4\theta - 2, \frac{28\theta - 13}{15}, \frac{96\theta - 49}{15} \right). \tag{10.4.8}$$

Then

$$\|MNL\|_1 \ll X^{\frac{1}{2}} \mathcal{L}^{-A} \quad \text{(all } A > 0\text{)}. \tag{10.4.9}$$

Remarks. When we use (10.4.7), we have $\theta = 0.55$, and so $\gamma(\theta) = \frac{1}{5}$. When we use (10.4.8), we have $\theta = \frac{1}{2} + \frac{1}{36}$ and so $\gamma(\theta) = \frac{1}{9}$.

The next lemma is the analogue of Lemma 7.21 in the present context.

Lemma 10.4. *Let $M(s, \chi)$ be a Dirichlet polynomial and suppose that $qT \leq M^\gamma$ where $1 \leq \gamma < B$. Suppose also that $M(s, \chi)$ is prime-factored. Then*

$$\|M\|_\beta \ll M^{\frac{1}{2}} \mathcal{L}^{-A} \tag{10.4.10}$$

for every $A > 0$, provided that

$$\beta \geq 4\gamma - 2h + \epsilon, \tag{10.4.11}$$

where h is an integer satisfying $h \ll 1$,

$$2h - \epsilon < \beta < 6h - \epsilon. \tag{10.4.12}$$

We may use the above result, just as we did its analogue in Chapter 7, to establish the following lemma corresponding to Lemma 7.22.

Lemma 10.5. *Let $L(s, \chi), M(s, \chi), N(s, \chi), R(s, \chi)$ be Dirichlet polynomials with*

$$LMNR = X.$$

Suppose that

$$qT \leq X^{0.45 - \frac{1}{2}\epsilon}, \tag{10.4.13}$$

$$M \geq qT. \tag{10.4.14}$$

Suppose that L, N, and R are all prime-factored. Then each of the following sets of conditions implies

$$\|LMNR\|_1 \ll X^{\frac{1}{2}} (\log X)^{-A} \quad (all\ A > 0) :$$

(i) $N \gg X^{0.1125}, R \gg X^{0.225}, L \gg X^{9/70}$,

(ii) $N \gg X^{0.225}, R \gg X^{0.15}, L \gg X^{0.1}$,

(iii) $N \gg X^{0.15}, R^2 L \gg X^{0.45}, L \gg X^{0.18}$,

(iv) $N \ll X^{0.15}, R \ll x^{0.15}, NR \gg X^{9/35}, LNR \gg X^{63/130 + B\epsilon}$.

The following lemma is key to the strength of our result for sieving $\mathcal{B}_1$. It is a variant of Lemma 2 in [85]. In that paper Heath-Brown and Iwaniec require $L(s)$ to be a Type I sum and consequently do not need the bound $L \leq X^{8/35}$ that we must impose. The reader should note that the proof breaks down at $\theta = 0.55$ (this was the reason Heath-Brown and Iwaniec stopped at this exponent), and this is why we have not considered using shorter intervals. We quote the lemma and its proof almost verbatim from [11].

Lemma 10.6. *Let $M(s, \chi), N(s, \chi), L(s, \chi)$ be Dirichlet polynomials satisfying $MNL = X$. Suppose that L satisfies (10.4.1), (10.4.13) holds, and*

$$\max(M, N) \leq X^{0.46 + \frac{1}{2}\epsilon}, \quad L \leq X^{\frac{8}{35}}. \tag{10.4.15}$$

Then (10.4.9) holds.

Proof. We follow the proof of Lemma 7.3 as far as Case 1 and treat the other cases as follows.

   Case 2. $M^{c(\sigma_1)} < qT$, $N^{c(\sigma_2)} < qT$. From (7.2.16) with $g = 2$ (L here corresponds to R in Lemma 7.3), and the mean value theorem we have

$$V \ll \min\{qTM^{f(\sigma_1)}, qTN^{f(\sigma_2)}, L^{4-4\sigma_3}\}M^{\sigma_1-\frac{1}{2}}N^{\sigma_2-\frac{1}{2}}L^{\sigma_3-\frac{1}{2}}$$
$$+ \min\{qTM^{f(\sigma_1)}, qTN^{f(\sigma_2)}, qTL^{2f(\sigma_3)}\}M^{\sigma_1-\frac{1}{2}}N^{\sigma_2-\frac{1}{2}}L^{\sigma_3-\frac{1}{2}}.$$

Now in the proof of Lemma 7.3 we only used three parameters to try to optimise the minima. Since each $f(\sigma_j)$ involves a minimum of two functions, there are five parameters we can choose here. The reader will see those chosen by inspecting the exponents in the following and will note that these lead to a bound independent of all σ_j. We write

$$V_1 = (qTM^{1-2\sigma_1})^{\frac{5}{16}}(qTM^{4-6\sigma_1})^{\frac{1}{16}}(qTN^{1-2\sigma_2})^{\frac{5}{16}}(qTN^{4-6\sigma_2})^{\frac{1}{16}} \times$$
$$\times (L^{4-4\sigma_3})^{\frac{1}{4}}M^{\sigma_1-\frac{1}{2}}N^{\sigma_2-\frac{1}{2}}L^{\sigma_3-\frac{1}{2}},$$

$$V_2 = \{(qTM^{1-2\sigma_1})^{\frac{5}{16}}(qTM^{4-6\sigma_1})^{\frac{1}{16}}(qTN^{1-2\sigma_2})^{\frac{5}{16}}(qTN^{4-6\sigma_2})^{\frac{1}{16}} \times$$
$$\times (qTL^{2-4\sigma_3})^{\frac{1}{4}}M^{\sigma_1-\frac{1}{2}}N^{\sigma_2-\frac{1}{2}}L^{\sigma_3-\frac{1}{2}},$$

and

$$V_3 = (qTM^{1-2\sigma_1})^{\frac{7}{16}}(qTM^{4-6\sigma_1})^{\frac{1}{48}}(qTN^{1-2\sigma_2})^{\frac{7}{16}}(qTN^{4-6\sigma_2})^{\frac{1}{48}} \times$$
$$\times (qTL^{8-12\sigma_3})^{\frac{1}{12}}M^{\sigma_1-\frac{1}{2}}N^{\sigma_2-\frac{1}{2}}L^{\sigma_3-\frac{1}{2}}\}.$$

We then have

$$V \ll V_1 + \min(V_2, V_3)$$
$$\ll (qT)^{\frac{3}{4}}(MN)^{\frac{1}{16}}L^{\frac{1}{2}} + qT(MN)^{\frac{1}{16}}\min(1, L^{\frac{1}{6}}(MN)^{-\frac{1}{24}})$$
$$\ll (qT)^{3/4}X^{\frac{1}{16}}L^{\frac{7}{16}} + qT(MN)^{\frac{1}{16}}(L^{\frac{1}{6}}(MN)^{-\frac{1}{24}})^{\frac{3}{10}}$$
$$\ll (qT)^{\frac{3}{4}}X^{\frac{13}{80}} + qTX^{\frac{1}{20}} \ll X^{\frac{1}{2}}\mathcal{L}^{-A}$$

by (10.4.13) and (10.4.15). The reader should see that we required (10.4.13) (the equivalent of $\theta > 0.55$ for primes in short intervals) for both terms on the last line.

   Case 3. $M^{c(\sigma_1)} \geq qT$, $N^{c(\sigma_2)} < qT$. We work similarly to Case 2, letting

$$V_4 = (M^{2-2\sigma_1})^{\frac{1}{2}}(qTN^{1-2\sigma_2})^{\frac{1}{8}}(qTN^{4-6\sigma_2})^{\frac{1}{8}}(L^{4-4\sigma_3})^{\frac{1}{4}},$$
$$V_5 = \{(M^{2-2\sigma_1})^{\frac{1}{2}}(qTN^{1-2\sigma_2})^{\frac{1}{8}}(qTN^{4-6\sigma_2})^{\frac{1}{8}}(qTL^{2-4\sigma_3})^{\frac{1}{4}},$$

and

$$V_6 = (M^{2-2\sigma_1})^{\frac{1}{2}}(qTN^{1-2\sigma_2})^{\frac{3}{8}}(qTN^{4-6\sigma_2})^{\frac{1}{24}}(qTL^{8-12\sigma_3})^{\frac{1}{12}}.$$

Thus

$$V \ll (V_4 + \min(V_5, V_6)) \times M^{\sigma_1-\frac{1}{2}}N^{\sigma_2-\frac{1}{2}}L^{\sigma_3-\frac{1}{2}}.$$

This gives

$$\begin{aligned}
V &\ll (qT)^{\frac{1}{4}} M^{\frac{1}{2}} N^{\frac{1}{8}} L^{\frac{1}{2}} + (qT)^{\frac{1}{2}} M^{\frac{1}{2}} \min(N^{\frac{1}{8}}, N^{\frac{1}{24}} L^{\frac{1}{6}}) \\
&\ll (qT)^{\frac{1}{4}} X^{\frac{1}{8}} (ML)^{\frac{3}{8}} + (qT)^{\frac{1}{2}} M^{\frac{1}{2}} N^{\frac{1}{16} + \frac{1}{48}} L^{\frac{1}{12}} \\
&\ll (qT)^{\frac{1}{4}} X^{\frac{1}{8}} (ML)^{\frac{3}{8}} + (qT)^{\frac{1}{2}} X^{\frac{1}{12}} M^{\frac{5}{12}} \\
&\ll X^{\frac{1}{2}} \mathcal{L}^{-A}
\end{aligned}$$

by (10.4.13) and (10.4.15).

Since Case 4 reduces to Case 3 on interchanging M and N, this establishes Lemma 10.6.   □

We now need to consider Dirichlet polynomials of the form

$$\sum_{n \asymp N} \chi(n) n^{-s}, \qquad \sum_{\substack{n \asymp N \\ (n, P(z)) = 1}} \chi(n) n^{-s}$$

where

$$z \geq \exp(\mathcal{L}^{\frac{9}{10}}). \tag{10.4.16}$$

Now, we appealed to Siegel's theorem (in the form given in [27]) in Chapter 2 to consider character sums over primes, but this only worked for small q. To apply our sieve method we would like something like Vinogradov's zero-free region for $\zeta(s)$ in order to get Type II information. To do this we treat the possible exceptional character separately and average over q. We also need more precise results on L-function zeros, which can be found in Prachar [144] (see Satz 6.2 in Chapter 8 in particular). From these, we know that there is at most one character χ in the set $\Gamma = \{\chi \pmod{q} : q \leq x\}$ for which $L(s, \chi)$ has a zero $\beta + i\gamma$ with $\beta > 1 - \epsilon/\mathcal{L}$, $|\gamma| \leq x$. If χ exists, say $\chi = \chi_1$, then χ_1 is real and primitive with conductor $q \gg \mathcal{L}^A$ for all A. What is more, for $\chi \in \Gamma \backslash \{\chi_1\}$, we have

$$\beta < 1 - \frac{\epsilon}{\max(\log q, \mathcal{L}^{\frac{4}{5}})}.$$

Let E_q consist of those characters $\chi \pmod{q}$ for which $L(s, \chi)$ has a zero $\beta + i\gamma$ with

$$\beta \geq 1 - \frac{\epsilon}{\mathcal{L}^{\frac{4}{5}}}, \quad |\gamma| \leq x.$$

Then, if $\chi \in E_q$, we have either $\chi = \chi_1$ or $\log q > \mathcal{L}^{4/5}$. Thus

$$|E_q| \begin{cases} = 0 & \text{if } q \leq \mathcal{L}^{2A}, \\ \leq 1 & \text{if } \mathcal{L}^{2A} < q \leq \exp(\mathcal{L}^{\frac{4}{5}}). \end{cases}$$

We still have not got all the information required to adapt the techniques from Chapter 7 to primes in arithmetic progressions. However, for $q > \exp(\mathcal{L}^{4/5})$, we can appeal to a zero-density theorem given by Montgomery [131, Theorem 12.1]:

$$\sum_{\chi} N(\sigma, x, \chi) \ll (qx)^{3(1-\sigma)} \mathcal{L}^{14},$$

where $N(\sigma, x, \chi)$ denotes the number of zeros of $L(s, \chi)$ in $[\sigma, 1] \times [-x, x]$. Taking $1 - \sigma = \epsilon/\mathcal{L}^{4/5}$ we get $|E_q| \ll \exp(7\epsilon\mathcal{L}^{1/5})$. In all cases we therefore obtain the bound $|E_q| \ll q^{1/2}\mathcal{L}^{-A}$ claimed in Theorem 10.2(iv).

We can now assume that any well-spaced set $\mathcal{S}$ occurring will obey the restriction

$$\chi \notin E_q \quad \text{for} \quad (t, \chi) \in \mathcal{S}. \tag{10.4.17}$$

Lemma 10.7. *Suppose that* $\min\{|t| : (t, \chi) \in \mathcal{S}\} \gg \mathcal{L}^A \delta_\chi$ *for all* $A > 0$. *Let*

$$M(s, \chi) = \sum_{\substack{m \asymp M \\ (m, P(z)) = 1}} \chi(m) m^{-s},$$

where z satisfies (10.4.16). *Then M satisfies* (10.4.1).

Proof. Suppose that $N \geq \exp\left(\mathcal{L}^{9/10}\right)$ and $N' \in (N, 2N]$. Then, by familiar techniques, we have

$$N(s, \chi) = \sum_{N < n \leq N'} \Lambda(n)\chi(n)n^{-it} \ll N\mathcal{L}^{-A}.$$

By partial summation we then obtain

$$\sum_{N < p \leq N'} \chi(p)p^{-it} \ll N\mathcal{L}^{-A}.$$

The idea of proof is very simple: If one fixes all variables but one, that one runs over a long enough range to get cancellation by the above bound. We thus decompose $M(s, \chi)$ into $O(\mathcal{L})$ sums

$$M_r(s, \chi) = \sum_{\substack{p_1 \geq \cdots \geq p_r \geq z \\ p_1 \ldots p_r \asymp M}} \chi(p_1) \ldots \chi(p_r)(p_1 \ldots p_r)^{-s}.$$

The lemma follows since, by a partial summation,

$$\left| M_r \left(\tfrac{1}{2} + it, \chi \right) \right| \leq \sum_{p_1 \geq \cdots \geq p_{r-1} \geq z} M^{-\frac{1}{2}} \sup_u \left| \sum_{\substack{z \leq p_r \leq u \\ p_1 \ldots p_r \asymp M}} \chi(p_r)p_r^{-it} \right|$$

$$\ll M^{-\frac{1}{2}} \sum_{\substack{p_1 \geq \cdots \geq p_{r-1} \geq z \\ p_1 \ldots p_{r-1} \ll M}} \left(\frac{M}{p_1 \ldots p_{r-1}} \right)^{\frac{1}{2}} \mathcal{L}^{-A}$$

$$\ll \mathcal{L}^{-A} \sum_{n \ll M} n^{-\frac{1}{2}} \ll M^{\frac{1}{2}}\mathcal{L}^{-A}.$$

$\square$

We now need results for Dirichlet L-functions $L(s, \chi)$ and its partial sums in order to obtain the required Type I information.

Lemma 10.8. *Let* $x^\epsilon \leq N \leq x$, $N \leq N_1 \leq 2N$, $q \leq \mathcal{L}^A$. *Then, for* $t \leq x$,

$$\sum_{N < n \leq N_1} \frac{\chi(n)}{n^{\frac{1}{2}+it}} \ll N^{\frac{1}{2}-\eta} + N^{\frac{1}{2}}(1 + |t|)^{-1}, \tag{10.4.18}$$

where $\eta = \eta(\epsilon) > 0$.

Proof. Much more precise bounds can be given for these sums when the parameters are in restricted ranges. The result we need can be readily obtained via Perron's formula from the bounds given in [98]. Basically the final term on the right side of (10.4.18) covers the case $\chi = \chi_0$ with small t. For $\chi \neq \chi_0$, or for larger t, known bounds for $L(s, \chi)$ lead to the first term on the right side of (10.4.18). $\square$

Lemma 10.9. *For $q \leq x$, $T \leq x$, we have*

$$\sum_{\chi \pmod q} \int_{-T}^{T} \left| L\left(\tfrac{1}{2} + it, \chi\right)\right|^4 dt \ll qT\mathcal{L}^B.$$

Proof. The fourth-power moment in this form has been proved by Ramachandra [146]. $\square$

Lemma 10.10. *Let χ be a character $\pmod q$; then*

$$L(\sigma + it, \chi) \ll \left(q(|t| + 1)\right)^{1+\epsilon-\sigma} + \frac{1}{|\sigma + it - 1|}.$$

Proof. See Rademacher's bound [145]. $\square$

Lemma 10.11. *Let*

$$N(s, \chi) = \sum_{n \leq N} \chi(n)n^{-s}.$$

Then for $2 \leq T \leq x$, $q \leq x$, $N \leq x$,

$$\|N\|_4^4 \ll qT\mathcal{L}^B + \mathcal{L}^4|\mathcal{S}|(q^2T^{-2} + N^2T^{-4}) + \delta_\chi N^2 \sum_{(t,\chi) \in \mathcal{S}} (1 + |t|)^{-4}.$$

Proof. The reader will be able to obtain this result using Perron's formula and the previous results. The full proof is given in [11, Lemma 9]. $\square$

For the rest of this chapter we write

$$T_0 = \exp(\mathcal{L}^{\frac{1}{3}}), \quad T_1 = x^{0.45 - \frac{2}{3}\epsilon}.$$

Lemma 10.12. *Let $L(s, \chi), M(s, \chi), N(s, \chi)$ be Dirichlet polynomials satisfying $LMN = x$, with $L \leq x^{1/2}$, and*

$$L(s, \chi) = \sum_{\ell \asymp L} \chi(\ell)\ell^{-s}.$$

Suppose that $q \leq \mathcal{L}^A$, that $|t| \in [\tfrac{1}{2}T, T]$ for all $(t, \chi_0) \in \mathcal{S}$, and that

$$T_0 \leq T \leq T_1. \tag{10.4.19}$$

Then (10.4.9) holds if either

$$\max(M, x^{0.45}) \max(N, x^{0.225}) \leq x^{0.775 + \frac{1}{8}\epsilon} \tag{10.4.20}$$

or

$$\max(M, N) \leq x^{0.46 + \frac{1}{8}\epsilon}. \tag{10.4.21}$$

Proof. The case when (10.4.20) holds is essentially Lemma 7.4, so we assume that (10.4.21) holds. (The reader will note that (10.4.20) is precisely the condition (7.3.1).) We have added in the condition $L \le x^{1/2}$ so that the result given by Lemma 10.11 has the right form. To be precise, there is a possible unwanted factor $(1 + L^2 T^{-3})$ occurring, which is clearly < 2 when $L \le x^{1/2}$.

Now suppose that (10.4.21) holds. This case is then essentially Lemma 2 in [85]. If $L \le x^{8/35}$, then the result follows from Lemma 10.6. If $L > x^{8/35}$, then assume first that $L \ge qT$. In this case (assuming without loss of generality that $M \ge N$)

$$N < (x/qT)^{\frac{1}{2}}, \quad M \le x^{0.46} \Rightarrow (10.4.20) \text{ holds.}$$

Now we can suppose that $L < qT$, so that $\|L\|_4^4 \ll \mathcal{L}^{3A} qT$ from Lemma 10.11. Thus, when (10.4.21) holds, we may replace $L^{4-4\sigma_3}$ by $\mathcal{L}^{3A} qT L^{2-4\sigma_3}$ where it occurs in the proof of Lemma 10.6. Since $\mathcal{L}^{3A} qT < x^{16/35}$, it may readily be verified that the proof goes through as before. □

10.5 SIEVING THE INTERVAL $\mathcal{B}_1$

We begin by restating Lemma 7.2. To be precise, we let c_k in that result be $c_k \chi(k)$.

Lemma 10.13. *Let*

$$F(s, \chi) = \sum_{k \sim x} c_k \chi(k) k^{-s}$$

be a divisor-bounded Dirichlet polynomial. Suppose that

$$\int_{T_0}^{T_1} \left| F\left(\frac{1}{2} + it, \chi \right) \right| dt \ll x^{\frac{1}{2}} \mathcal{L}^{-A} \quad (\text{all } A > 0). \tag{10.5.1}$$

Then

$$\sum_{k \in I} c_k \chi(k) = \frac{Y}{y_1} \sum_{k \in I_1} c_k \chi(k) + O(Y \mathcal{L}^{-A}) \quad (\text{all } A > 0), \tag{10.5.2}$$

where $y_1 = x \exp(-3(\log x)^{1/3})$ and $I_1 = (x - y_1, x]$.

In the following, when we say "(10.5.1) holds" it is tacitly assumed that we mean it holds for all $q \le \mathcal{L}^A$. We remark that (10.5.2) is elementary when $L \ge x^{1/2}, q \le \mathcal{L}^A$ if

$$c_k = \sum_{\substack{n\ell=k \\ \ell \sim L}} a_n.$$

Hence we can suppose in the following that Type I sums have length $L < x^{1/2}$.

To construct the sequence $a_0(k)$ using our sieve method we must obtain a supply of Dirichlet polynomials satisfying (10.5.2). Lemmas 10.3, 10.5, and 10.6 start off the process. We now give the sieve fundamental lemma in the present context, which follows from our Type I information. Let $w = \exp((\log x)^{9/10})$. For $n \ge 1, z \ge 2$ let

$$\rho(n, z) = \begin{cases} 1 & \text{if } (n, P(z)) = 1, \\ 0 & \text{otherwise.} \end{cases} \tag{10.5.3}$$

Lemma 10.14. *Let*

$$c_k = \sum_{\substack{mn\ell=k \\ m \sim M,\, n \sim N}} a_m b_n \rho(\ell, w). \tag{10.5.4}$$

Suppose that a_m, b_n are divisor-bounded and either

$$\max(x^{0.45}, M) \max(x^{0.225}, N) \leq x^{0.775} \tag{10.5.5}$$

or

$$\max(M, N) \leq x^{0.46 + \frac{1}{8}\epsilon}. \tag{10.5.6}$$

Then (10.5.2) holds.

Proof. This result corresponds to Lemma 7.4 and the proof of (7.5). The improvement we have made comes from the alternative possibility furnished by (10.5.6) coming from Lemma 10.12. $\qquad\square$

Working as in Chapter 7 we can use Lemma 10.14 to establish the following result.

Lemma 10.15. *Let*

$$c_k = \sum_{\substack{mn\ell=k \\ m \sim M,\, n \sim N}} a_m b_n \rho(\ell, z),$$

where a_m and b_n are divisor-bounded. Suppose that

$$x^{0.4} \leq M \leq x^{0.46}, \ N \leq x^{0.46}, \ z \ll x/MN; \tag{10.5.7}$$

then (10.5.2) holds.

Lemma 10.15 covers all sums that arise below where one variable lies in the interval $[x^{0.4}, x^{0.46}]$. Using familiar techniques we can extend this result to the case of one variable lying in the interval $[1, x^{1/2}]$. As in the work we did in Chapter 7 (see Lemma 7.15, for example), we must break the range down into subranges and the maximum value for z varies continuously (in a familiar zigzag pattern) over each subrange. The result we obtain is as follows.

Lemma 10.16. *Lemma 10.15 holds if in place of (10.5.7) we suppose that*

$$M \in \left[x^{\frac{1}{10}\alpha}, x^{\frac{1}{10}(\alpha+1)}\right], \ \text{where} \ \alpha \in \{0, 1, 2, 3, 4\},$$

$$x^{0.1} < z \ll \left(\frac{x^{1.1}}{M^2}\right)^{\frac{1}{9-2\alpha}}, \ MN \leq x^{0.775} z^{\alpha-4}. \tag{10.5.8}$$

There is one additional asymptotic formula that we need that is not covered by the above, when the product MN can be over $x^{0.775}$.

Lemma 10.17. *Let c_k be as in Lemma 10.15 with $z = x^{0.1}$. Then, if $M \leq N \leq x^{0.4}$, (10.5.2) holds.*

Proof. We use our standard procedure for getting variables in the right range. Our Type I information comes from Lemma 10.14. Our working is therefore valid when winding Buchstab's identity around when our two "outer" variables do not exceed $x^{0.46}$. We are alright if $x^{0.4} < mp_1 \ldots p_j \leq x^{0.46}$ by Lemma 10.15: there is no need for a further iteration. For $mp_1 \ldots p_j > x^{0.46}$ we note that $mp_1 \ldots p_j < x^{1/2}$. It is then clear that we are in the asymptotic formula region furnished by Lemma 10.3.   □

We now construct the sequence $a_0(k)$ by starting with $\rho(k)$, using Buchstab's identity (in the form (3.1.1)) up to six times and discarding some non-negative terms from the resulting identity. The first two applications of (3.1.1) are as follows. Let $k \in \mathcal{B}_1$. Then we have, with $z_0 = x^{11/90}$,

$$\rho(k) = \rho(k, x^{\frac{1}{2}}) = \rho(k, z_0) - \sum_{\substack{z_0 \leq p_1 < x^{1/2} \\ p_1 n_2 = k}} \rho(n_2, p_1)$$

$$= \rho(k, z_0) + \sum_{\substack{z_0 \leq p_1 < x^{1/2} \\ p_1 n_2 = k}} (-\rho(n_2, z_1)) + \sum_{\substack{p_1 p_2 n_3 = k \\ z_1 \leq p_2 < p_1 \\ z_0 \leq p_1 < x^{1/2}}} \rho(n_3, p_2). \qquad (10.5.9)$$

$$= c_k(1) + c_k(2) + c_k(3).$$

Using the role-reversal process, when necessary, we may continue applying Buchstab's identity to generate sums of the form

$$\sum_{\substack{(n_1, \ldots, n_j) \in \mathcal{C}_j \\ n_1 \ldots n_{j+1} = k}} (-1)^j \rho(n_1, z_1) \ldots \rho(n_{j+1}, z_{j+1}) \qquad (10.5.10)$$

where $n_j \geq x^{0.1}$ in the region $\mathcal{C}_j$ and z_i is a function of $\{n_t : t \neq i\}$. When applying the role-reversal trick, it may not be obvious which variable to choose in applying the next step,

$$\rho(n_{j+1}, z_{j+1}) = \rho(n_{j+1}, w_{j+1}) - \sum_{\substack{w_{j+1} \leq p_{j+1} < z_{j+1} \\ p_{j+1} n_{j+2} = n_{j+1}}} \rho(n_{j+2}, p_{j+1}),$$

but we must be sure that the *first term* on the right yields a sum over $\mathcal{C}_j$ satisfying (10.5.2). In [11] this step is described as "decomposing n_{j+1} to reach $p_{j+1} n_{j+2}$." In general (exceptions being described below), we decompose the *largest variable* permitted by Lemma 10.16, such that there is *no decomposition for any region $\mathcal{C}_j$ where* (10.5.2) *can be shown to hold.* We never go beyond the stage $j = 6$.

We shall show below that it is always possible to decompose one of $n_1, \ldots, n_{j+1}$ for $j \leq 4$. In accordance with our usual philosophy, however, we must ensure that no $\mathcal{C}_5$ is generated for which no further decomposition can be made, and this forces us to discard certain regions $\mathcal{C}_4$. Similarly, we must also discard any $\mathcal{C}_6$ for which (10.5.2) cannot be shown to hold.

Now suppose that

$$\rho(k) = \rho(k, z_0) + \cdots + \sum_{\mathcal{C}_6} \sum_{\substack{(n_1, \ldots, n_6) \in \mathcal{C}_6 \\ n_1 \ldots n_7 = k}} \rho(n_1, z_1) \ldots \rho(n_7, z_7)$$

is the identity obtained by the above algorithm. Discarding regions $\mathcal{C}_4$ and $\mathcal{C}_6$ as explained above gives an inequality $\rho(k) \geq a_0(k)$ in which $a_0(k)$ satisfies (10.5.2).

From the zero-free region (compare Lemma 2.7) we have

$$\sum_{k \in \mathcal{B}_1} a_0(k)\chi(k) = \frac{y_1(\delta_\chi u_0 + O(\mathcal{L}^{-A}))}{\log x}. \tag{10.5.11}$$

As in previous applications of our method, the constant u_0 is found by subtracting, from 1, integrals in four and six dimensions corresponding to the discarded regions of $\mathcal{C}_4$ and $\mathcal{C}_6$. The reader should note that we get $(\log x)^{-1}$ in (10.5.11) with a $O(\mathcal{L}^{-A})$ error because we are working in a short interval. The constant here is, of course, ineffective owing to the usual difficulty with a possible Siegel zero.

We spend the rest of this section demonstrating that the regions of $\mathcal{C}_4$ that must be discarded correspond to (α_1, α_2) lying in a very small part of the plane. This is illustrated graphically in [11, p. 33].

Lemma 10.18. *In the decomposition procedure described above,*

(a) if a subproduct m of $n_1 \ldots n_{j+1}$ satisfies $x^{0.4} \ll m \ll x^{0.46}$,
  then (10.5.2) holds for the sum (10.5.10);

(b) if $j \leq 4$, one of the variables may be decomposed;

(c) if $j \geq 5$, any variable that is at least $x^{0.225}$ may be decomposed.

Proof. (a) Suppose that

$$n_1 \ldots n_\ell \in [x^{0.4}, x^{0.46}].$$

We can use Lemma 10.15 if $n_{j+1} > x^{0.14}$ (in this case, $x/(n_1 \ldots n_\ell n_{j+1}) < x^{0.46}$) and Lemma 10.3 if $x^{0.1} \leq n_{j+1} < x^{0.14}$.

(b) If $j = 2$, we have variables p_1, p_2, n_3 with $p_2 \leq p_1, p_2 \leq n_3$. By (a), the case where p_1 or n_3 is in $[x^{0.4}, x^{0.46}]$ is already covered. Since both variables cannot exceed $x^{0.46}$, at least two of p_1, p_2, n_3 do not exceed $\leq x^{0.4}$ and so we may use Lemma 10.17.

If $j = 3$, with $n_1 \leq n_2 \leq n_3 \leq n_4$, then $n_3 \leq x^{0.4}$, $n_1 n_2 \leq x^{0.5}$, $n_1 n_2 n_3 \leq x^{0.75}$. If (10.5.2) does not hold, either Lemma 10.16 or 10.17 allows us to decompose n_4.

If $j = 4$, $n_1 \leq n_2 \leq n_3 \leq n_4 \leq n_5$, and $n_5 \leq x^{0.225}$, we have $x^{0.4} \leq n_4 n_5 \leq x^{0.45}$ and (a) applies. If $n_5 > x^{0.225}$, we may decompose $n_1 \ldots n_4$ into two products $\leq x^{0.5}$. (This is obvious if $n_4 > x^{0.275}$ or $n_1 n_2 > x^{0.275}$. If $\max(n_1 n_2, n_4) \leq x^{0.275}$ then $n_2 \leq x^{0.175}$ and $\max(n_1 n_4, n_2 n_3) \leq x^{0.45}$.) We may now finish the proof using case (a) and Lemmas 10.17 and 10.16.

(c) For $j = 5$ we have $n_1 n_2 n_3 \leq (x^{0.775})^{3/5} < x^{0.5}$, $n_4 n_5 < x^{0.775-0.3} < x^{0.5}$, and the final part of case (b) applies. The general case follows similarly. $\square$

Remark. Part (a) of the above lemma will be crucial in future proofs, for we can immediately assume that no combination of the variables lies in the range $[x^{0.4}, x^{0.46}]$. Hence, if we have a combination of variables $> M > x^{0.4}$, we can assume $x > x^{0.46}$. Similarly, if the combination is $< M < x^{0.46}$, we can presume it is $< x^{0.4}$.

In the rest of this section we decompose, as we may, only regions $\mathcal{C}_j$ that are not known to satisfy (10.5.2) and (in particular) do not have property (a) of Lemma 10.18. We follow [11] closely. The many regions under consideration are illustrated in Diagram 1 of that work. The reader might like, as an exercise, to draw this for themselves to be satisfied that we have dealt with all regions. I remember while writing [11] that this was the only way to be certain!

Lemma 10.19. *We may decompose the subsum of* $c_k(3)$ *with* $x^{0.36} \leq p_1 < x^{0.4}$, $p_1 p_2 \leq x^{0.54}$ *to reach* $\mathcal{C}_6$.

Proof. Noting that $\max(p_1, p_2 p_3) \leq x^{0.4}$ (and so Lemma 10.17 is applicable), we may decompose successively to get $p_1 p_2 n_3$, $p_1 p_2 p_3 n_4$, $p_1 p_2 p_3 p_4 n_5$. Once we have obtained asymptotic formulae where (10.5.2) already holds, there remain two cases to consider.

(i) $p_2 \geq x^{0.15}$. In this case $p_1 p_2 \geq x^{0.51}$, and so $p_3 p_4 n_5 \leq x^{0.49}$. We can thus decompose p_1 into $p_5 n_6$ (using Lemma 10.16 with $\alpha = 4$). We have $n_6 > x^{0.18}$, $n_6 p_5^2 \geq x^{0.36+0.1} > x^{0.45}$, $p_3 p_4 n_5 \gg x^{0.46}$. Lemma 10.5(iii) then yields (10.5.2).

(ii) $p_2 < x^{0.15}$. We thus have $p_2 p_3 p_4 < x^{0.45}$. Since we have assumed (10.5.2) does not hold, this means that $p_2 p_3 p_4 < x^{0.4}$, and so $\max(p_1, p_2 p_3 p_4) < x^{0.4}$. We can therefore apply Lemma 10.17 to decompose n_5 to get $p_1 p_2 p_3 p_4 p_5 n_6$ and finish by applying Lemma 10.18(c) in view of the size of p_1. $\square$

Lemma 10.20. *Lemma 10.16 holds if in place of* (10.5.8) *we suppose that*
$$N \geq x^{0.18}, x^{0.3} \leq M \leq x^{0.36}, MN \leq x^{0.54}, MNz \leq x^{0.775}.$$

Proof. If $p_3 \leq x^{0.15}$, then
$$n_5 \gg \frac{x}{mnp_3 p_4} \gg \frac{x}{MNp_3^2} \gg x^{0.15}.$$
Hence one of p_3, n_5 always exceeds $x^{0.15}$. We can restrict n, p_3, p_4, n_5 to dyadic intervals with left endpoints N_1, N_2, N_3, N_4 arranged so that
$$x^{0.1} \leq N_1 \leq N_2 \leq N_3 \leq N_4,$$
$$N_3 > x^{0.15}, N_4 > x^{0.18}, N_1 N_2 N_3 N_4 \asymp xM^{-1}.$$
We have $MN_1 \geq x^{0.4}$, and so we can assume that $MN_1 \gg x^{0.46}$. This then forces $N_3 N_4 \ll x^{0.44}$, and so we can assume that $N_3 N_4 < x^{0.4}$. We cannot then have $N_1 N_2 N_4 < x^{0.4}$ since this forces $N_3 > x^{0.24}$, which is absurd. Thus $N_1 N_2 N_4 \geq x^{0.46}$. We are then only left to deal with the case
$$N_4 \geq x^{0.18}, N_3 > x^{0.15}, MN_1 \gg x^{0.46}, N_2^2 N_4 \geq N_1 N_2 N_4 > x^{0.46},$$
but now Lemma 10.5(iii) applies to complete the proof. $\square$

It follows from the above result that (10.5.2) holds for the subsum of $c_k(3)$ with $x^{0.3} \leq p_1 \leq x^{0.36}, x^{0.18} \leq p_2 \leq x^{0.54} p_1^{-1}$. The hypotheses of Lemma 10.20 are immediately satisfied if $p_1 p_2^2 \leq x^{0.775}$. Otherwise we have
$$p_1 = \frac{p_1^2 p_2^2}{p_1 p_2^2} \leq x^{0.305}, \quad p_2 \geq x^{0.235}, \quad p_1 p_2^3 \geq x^{1.05}$$

and so the variable n_3 must be prime. Since $MN\,(x/p_1p_2)^{1/2} \ll x^{0.77}$, Lemma 10.20 gives the desired result.

Lemma 10.21. *Lemma 10.15 holds if in place of (10.5.8) we suppose that*

$$N \geq x^{0.15}, MNz^2 \leq x^{0.82}, x^{0.3} \leq M \leq x^{0.36}, MN \leq x^{0.54}.$$

Proof. First we note that $N \leq x^{0.24}$, $N^2z^2 \leq x^{1.06-0.3}$, $np_3 < x^{0.4}$. After two applications of Buchstab's identity we count numbers $mnp_3p_4n_5$ with

$$n_5 \gg \frac{x}{(MNz^2)} \gg x^{0.18}, \quad p_3^2 n_5 \geq p_3p_4n_5 \gg x^{0.46}, \quad p_4m \geq x^{0.4}.$$

Thus Lemma 10.15 is applicable unless $p_4m \geq x^{0.46}$. In that case Lemma 10.5(iii) holds and the proof is complete. $\square$

Lemma 10.22. *We may decompose the subsum of $c_k(3)$ for which $x^{0.3} \leq p_1 \leq x^{0.36}$, $p_2 \leq x^{0.15}$, to reach $\mathcal{C}_6$.*

Proof. The argument of case (ii) in the proof of Lemma 10.19 remains valid for this result. $\square$

Lemma 10.23. *Let*

$$c_k = \sum_{\substack{p_1 \sim M,\, p_2 \sim M \\ p_1p_2n_3=k}} \rho(n_3, p_2)$$

and suppose that $x^{0.23} \leq M \leq x^{0.3}, MN \geq x^{0.46}$. Then c_k satisfies (10.5.2) if either

$$\text{(a)} \quad MN^2 \leq x^{0.7}, N \geq x^{0.18} \quad \text{or} \quad \text{(b)} \quad MN^3 \leq x^{0.82}. \tag{10.5.12}$$

Proof. In either case we can apply Buchstab's identity in a straightforward fashion to reach $p_1p_2p_3p_4n_5$. When (10.5.12)(a) holds, we have

$$p_1p_2 = (p_1p_1p_2^2)^{\frac{1}{2}} \leq x^{0.5}, \qquad p_1p_2p_3 \leq p_1p_2^2 < x^{0.775}.$$

By Lemma 10.16, we may therefore assume that $p_4 \geq x^{1.1}(p_1p_2)^{-2}$. Thus $p_1p_4 \geq x^{1.1}p_1^{-1}p_2^{-2} \geq x^{0.4}$; indeed, $p_1p_4 \geq x^{0.46}$ and $p_4 \geq x^{0.16}$, so $p_3 \geq x^{0.16}$, $n_5 \geq x^{0.16}$, and Lemma 10.5(iii) applies.

Now suppose that (10.5.12)(b) holds. Since $p_1p_2^3 \leq x^{0.82}$, we have $n_5 \geq x^{0.18}$. Now

$$n_5p_4^2 = \frac{p_1p_2p_3p_4n_5}{p_1p_2p_3}\,p_4 \gg \frac{p_4x}{p_1p_2^2} \gg \frac{x^{1.1}}{(p_1)^{\frac{1}{3}}(p_1p_2^3)^{\frac{2}{3}}} \gg x^{1.1-0.1-\frac{1.64}{3}} > x^{0.45}.$$

Lemma 10.5(iii) applies unless $p_3 < x^{0.15}$. In this case, $p_1p_3 < x^{0.45}$, and so we may assume that $p_1p_3 < x^{0.4}$. This leads to

$$n_5p_3 = \frac{p_1p_2p_3p_4n_5}{p_1p_2p_4} \gg \frac{x}{p_1p_2p_3} \gg \frac{x^{0.6}}{p_2} > x^{0.4}.$$

Since $n_5p_3 \leq x^{0.54}/p_4 \leq x^{0.44}$, this case can be handled. $\square$

We shall skip the proofs of subsequent lemmas that consist merely of simple rearrangements of the variables and given conditions. The reader has seen plenty of these by now and could supply all the details themselves. Full proofs are given in [11].

Lemma 10.24. *We may decompose the subsum of $c_k(3)$ with $x^{0.23} \leq p_1 \leq x^{0.3}, x^{0.7} \leq p_1 p_2^2, x^{0.46} \leq p_1 p_2 \leq x^{0.54}$ to reach $\mathcal{C}_6$.*

Let us write $(p_1, p_2) = (x^{\alpha_1}, x^{\alpha_2})$ and consider the decomposition algorithm for $\alpha_1 \leq 0.3$. Lemma 10.15 covers $\alpha_1 + \alpha_2 \geq 0.54$ since then $x^{0.4} \ll x/(p_1 p_2) \ll x^{0.46}$; and it also covers $\alpha_1 + \alpha_2 \in [0.4, 0.46]$. Appealing to Lemmas 10.23, 10.24, we see that within the region $\{(\alpha_1, \alpha_2) : \alpha_1 \leq 0.3, \quad 0.46 \leq \alpha_1 + \alpha_2 \leq 0.54,\}$ no $\mathcal{C}_4$ is discarded except when (α_1, α_2) lies in the small triangle

$$\triangle_1 : \quad \alpha_1 \leq 0.3, \ \alpha_1 + 3\alpha_2 \geq 0.82, \ \alpha_2 \leq 0.18.$$

This region is contained in a larger triangle described below as $\triangle_3$.

(For part of the region covered by Lemma 10.24, (10.5.2) holds:

$$p_2 \geq x^{0.225}, p_1^4 p_2^3 \geq x^{1.8}, p_1 p_2 \leq x^{0.54}.$$

Here $p_1 n_3 \leq x^{0.775}, x^{0.46} < n_3 \leq x^{0.5}$, permitting us to take $z = x^{1.1} n_3^{-2} > x^{-0.9} p_1^2 p_2^2 > p_2^{1/2}$ in Lemma 10.16.)

For $\alpha_1 \leq 0.3$ it remains to consider $\alpha_1 + \alpha_2 \leq 0.4$. If $\alpha_1 + \alpha_2 \geq 0.36$, we may reach $\mathcal{C}_6$. (We easily reach $p_1 p_2 p_3 p_4 p_5 n_6$. If $p_3 \geq x^{0.15}$, we have $p_1 \geq x^{0.18}$, $p_1 p_2^2 \geq x^{0.46}$, $p_4 p_5 n_6 \geq x^{0.4}$, and indeed $p_4 p_5 n_6 \geq x^{0.46}$; so Lemma 10.5(iii) applies. If $p_3 < x^{0.15}$, we argue as in case (ii) of Lemma 10.19.)

If $0.3 \leq \alpha_1 + \alpha_2 \leq 0.36$ and $\alpha_2 \leq 0.15$, we repeat the argument just used. We also reach $\mathcal{C}_6$ for $0.3 \leq \alpha_1 + \alpha_2 \leq 0.36, \alpha_2 > 0.15, \alpha_1 + 4\alpha_2 \leq 0.82$. To see this, we easily reach $p_1 p_2 p_3 p_4 p_5 n_6$, with $n_6 \geq 0.18$. If $p_3 p_4 p_5 > x^{0.4}$, then $p_3 p_4 p_5 > x^{0.46}$ and Lemma 10.5(iii) applies. If $p_3 p_4 p_5 \leq x^{0.4}$, then we decompose n_6. Thus, within the strip $0.3 \leq \alpha_1 + \alpha_2 \leq 0.4$, we are left with the triangle

$$\triangle_2 : \quad \alpha_1 + 4\alpha_2 \geq 0.82, \ \alpha_2 \leq \alpha_1, \ \alpha_1 + \alpha_2 \leq 0.36,$$

for which certain $\mathcal{C}_4$ are discarded.

As for $\alpha_1 + \alpha_2 \leq 0.3$, we can reach $\mathcal{C}_6$ by the argument of Lemma 10.19(ii).

We turn our attention to $0.3 \leq \alpha_1 \leq 0.4$. First take

$$0.3 \leq \alpha_1 \leq 0.4, \quad \alpha_1 + \alpha_2 \leq 0.54. \tag{10.5.13}$$

If $\alpha_1 \geq 0.36$ or $\alpha_2 \leq 0.15$, we reach $\mathcal{C}_6$ (Lemmas 10.19 and 10.22). If $\alpha_1 < 0.36$, $\alpha_1 + \alpha_2 \leq 0.54$, $\alpha_1 + 3\alpha_2 \leq 0.82$, $\alpha_2 \geq 0.15$, then Lemma 10.21 yields (10.5.2); while if $\alpha_2 \geq 0.18$, we may appeal to the remark following Lemma 10.20. Within (10.5.13), we are left with a region $\triangle_3 \backslash \triangle_1$, where $\triangle_3$ is the triangle

$$\triangle_3 : \quad \alpha_2 \leq 0.18, \ \alpha_1 \leq 0.36, \ \alpha_1 + 3\alpha_2 \geq 0.82,$$

for which regions $\mathcal{C}_4$ with (10.5.2) unavailable are discarded.

Next we consider

$$0.3 \leq \alpha_1 \leq 0.4, \quad \alpha_1 + \alpha_2 \geq 0.6, \quad \alpha_1 + 2\alpha_2 \leq 1. \tag{10.5.14}$$

We can dismiss the case $\alpha_2 \leq 0.225$ immediately, using Lemma 10.6.

Lemma 10.25. *Lemma 10.23 remains true if* (10.5.12) *is replaced by either*

$$M \geq x^{0.3}, \; N \geq x^{0.3}, \; MN \leq x^{0.64} \tag{10.5.15}$$

or

$$x^{0.36} \leq M \leq x^{0.4}, \; N \geq x^{0.3}. \tag{10.5.16}$$

Remark. Note that in Lemma 10.25, k is a product of three primes any of which can be decomposed.

Lemma 10.26. *For the subsums of* $c_k(3)$ *for which either*

$$x^{0.3} \ll p_1 \ll x^{0.4}, \; x^{0.225} \ll p_2 \ll x^{0.3}, \; x^{0.6} \ll p_1 p_2 \ll x^{0.64} \tag{10.5.17}$$

or

$$x^{0.36} \ll p_1 \ll x^{0.4}, \; x^{0.225} \ll p_2 \ll x^{0.3}, \; p_1 p_2 \gg x^{0.6}, \tag{10.5.18}$$

we may decompose to reach $\mathcal{C}_6$.

Returning to the region (10.5.14), we are left with a quadrilateral Γ,

$$\Gamma : \quad \alpha_2 \leq \alpha_1 \leq 0.36, \; \alpha_1 + 2\alpha_2 \leq 1, \; \alpha_1 + \alpha_2 \geq 0.64,$$

for which regions $\mathcal{C}_4$ with (10.5.2) unavailable are discarded.

It remains to consider $0.46 \leq \alpha_1 \leq 0.5$. Lemma 10.20 with a role-reversal gives a region,

$$0.46 \leq \alpha_1 \leq 0.5, \; 0.64 \leq \alpha_1 + \alpha_2 \leq 0.7, \; \alpha_2 \geq 0.18, \tag{10.5.19}$$

where (10.5.2) holds. For $0.46 \leq \alpha_1 \leq 0.5$, $\alpha_2 \geq 0.225$, $4\alpha_1 + \alpha_2 \leq 2.2$, we may deduce (10.5.2) from Lemma 10.16 since

$$n_3 p_1 \ll x^{0.775}, \; \frac{x^{1.1}}{p_1^2} > p_2^{\frac{1}{2}}.$$

Lemma 10.27. *Lemma 10.23 remains true if* (10.5.12) *is replaced by*

$$x^{0.46} \leq M \leq x^{0.5}, \; N \geq x^{0.15}, \; MN^2 \leq x^{0.82}.$$

Lemma 10.28. *For the subsum of* $c_k(3)$ *with either* $\alpha_2 \geq 0.18$ *or* $\alpha_2 \leq 0.15$ *we may decompose to reach* $\mathcal{C}_6$.

In our discussion of the region

$$0.46 \leq \alpha_1 \leq 0.5, \; \alpha_1 + 2\alpha_2 \leq 1, \; \alpha_2 \geq 0.1,$$

there remains only the triangle

$$\triangle_4 : \quad \alpha_2 \leq 0.18, \; \alpha_1 \leq 0.5, \; \alpha_1 + 2\alpha_2 \geq 0.82,$$

for which regions $\mathcal{C}_4$ with (10.5.2) unavailable are discarded.

A numerical calculation shows that the four-dimensional integrals corresponding to $\triangle_2, \triangle_3, \triangle_4$, and Γ are less than 0.0003, 0.002, 0.002, and 0.0003, respectively. The total of all six-dimensional integrals is less than 0.0001, and so

$$u_0 > 0.9953.$$

We now consider the construction of $a_1(k)$. This is done by an analogous decomposition in which the final identity is

$$\rho(k) = \rho(k, x^{\frac{11}{90}}) + \cdots - \sum_{\substack{(n_1,\ldots,n_7)\in\mathcal{C}_7 \\ n_1\ldots n_8=k}} \rho(n_1, z_1)\cdots\rho(n_8, z_8).$$

Lemma 10.18(c) shows that we always reach $\mathcal{C}_5$. We then discard certain $\mathcal{C}_5$ in order to avoid generating $\mathcal{C}_6$ that do not allow one more decomposition, and any $\mathcal{C}_7$ for which (10.5.2) is unavailable, to get an inequality

$$\rho(k) \le b_0(k).$$

The following lemma informs our choice as to which parts of $\mathcal{C}_5$ must be discarded.

Lemma 10.29. *In* (10.5.10) *suppose that* $j = 5$, $n_1 \le \cdots \le n_6$, *and* $n_1 n_2 n_3 \le x^{0.46}$; *then we may decompose to reach* $\mathcal{C}_7$.

Proof. We may suppose $n_1 n_2 n_3 < x^{0.4}$, so $n_5 n_6 \ge x^{0.4}$ and indeed $n_5 n_6 \ge x^{0.46}$. If both n_5, n_6 are $\ge x^{0.225}$, then we use Lemma 10.18(c) to decompose both in turn. Assume now that $n_5 < x^{0.225} < n_6$. Plainly we may suppose also that $n_1 n_2 n_3 n_4 \ge x^{0.46}$. Note that $n_2 n_3 n_4 \le x^{0.44}$.

If $n_6 < x^{0.3}$, then we group variables as $n_1 n_6, n_2 n_3 n_4, n_5$ and apply Lemma 10.6. Thus we may suppose that $n_6 \ge x^{0.3}$ and so arrive at the inequality $n_1 n_6 \ge x^{0.46}$. It follows that $n_3 n_4 n_5 \le x^{0.44}$ and indeed $n_3 n_4 n_5 \le x^{0.4}$.

We may now decompose n_6 to reach $n_1 n_2 n_3 n_4 n_5 p_6 n_7$. If $n_7 \ge x^{0.225}$, we may decompose again. Otherwise, Lemma 10.6 applies to $n_3 n_4 n_5, n_1 n_2 p_6, n_7$. To see this we observe that $n_1 n_2 n_3 n_4 n_5 \ge x^{0.57}$ and indeed $n_1 n_2 n_3 n_4 n_5 \ge x^{0.6}$, so that $p_6 \le x^{0.2}$. If $n_1 n_2 \le x^{0.26}$, we see that $\max(n_3 n_4 n_5, n_1 n_2 p_6) \le x^{0.4}$. If $n_1 n_2 > x^{0.26}$, then $n_1 n_2 n_3 n_4 n_5 > x^{0.65}$, $p_6 < x^{0.175}$, and

$$n_1 n_2 p_6 \le x^{0.4\times\frac{2}{3}+0.175},$$

leading to the same conclusion. This completes the proof. $\square$

Note that in a region $\mathcal{C}_5$ with $n_1 n_2 n_3 \ge x^{0.46}$ we have $n_4 \ge x^{0.15}$. We can therefore give an asymptotic formula if $n_6 \ge x^{0.18}$. Thus the variables counted in the discarded regions are products of five primes, and so the corresponding integrals take the form

$$\int \frac{d\alpha_1 \ldots d\alpha_5}{\alpha_1 \ldots \alpha_5(1 - \alpha_1 - \alpha_2 - \alpha_3 - \alpha_4 - \alpha_5)},$$

where

$$\alpha_1 \le \alpha_2 \le \alpha_3 \le \alpha_4 \le \alpha_5 \le 1 - \alpha_1 - \cdots - \alpha_5 < 0.18$$

with

$$0.46 \le \alpha_1 + \alpha_2 + \alpha_3 \le 0.5.$$

This integral is less than 10^{-6}. Allowing for the number of possible decompositions starting with $p_1 p_2 n_3$ (certainly less than a dozen) we will have to count this integral a few times. Nevertheless, the seven-dimensional integrals are so small that we may readily conclude that

$$u_1 < 1.0001.$$

10.6 SIEVING THE INTERVAL $\mathcal{B}_2$

We begin with an analogue of Lemma 10.13. Let $\mathcal{E}(q) = E_q^c \backslash \{\chi_0\}$. Throughout this section we have $1 \leq q \leq Q$.

Lemma 10.30. *Write* $T_2 = T_2(q) = Y/(qQ)$ *and let*

$$G(s, \chi) = \sum_{m \asymp Y} b_m \chi(m) m^{-s}$$

be a divisor-bounded Dirichlet polynomial. Suppose that

$$M_0(G) = \sum_{\chi \in \mathcal{E}(q)} \int_{-Y}^{Y} \frac{\left| G\left(\frac{1}{2} + it, \chi\right) \right|^2}{T_2^2 + t^2} \, dt \ll \frac{Y}{T_2^2} \mathcal{L}^{-A} \quad (all \ A > 0) \quad (10.6.1)$$

and

$$M_1(G) = \int_{[-Y,Y] \backslash [-T_0, T_0]} \frac{\left| G\left(\frac{1}{2} + it, \chi_0\right) \right|^2}{T_2^2 + t^2} \, dt \ll \frac{Y}{T_2^2} \mathcal{L}^{-A} \quad (all \ A > 0).$$

$$(10.6.2)$$

Write

$$\|H\|_E = \left(\sum_{\chi \in E} \int_{Y/2}^{3Y/2} |H(y, \chi)|^2 dy \right)^{\frac{1}{2}},$$

$$W(u, v) = \frac{1}{2\pi i} \int_{1/2 - iT_0}^{1/2 + iT_0} G(s, \chi_0) \frac{(u + uv)^s - u^s}{s} \, ds.$$

Then

$$\left\| \sum_{m \in J_y} b(m) \chi(m) - \delta_\chi W(y, z) \right\|_{E_q^c}^2 \ll (qQ)^2 Y \mathcal{L}^{-A} \quad (all \ A > 0).$$

Proof. Write

$$\mathcal{G}(\chi, y, z) = \frac{1}{2\pi i} \int_{1/2 - iY}^{1/2 + iY} G(s, \chi) \frac{(y + yz)^s - y^s}{s} \, ds$$

and put $\mathcal{G}'(\chi, y, z) = \mathcal{G}(\chi, y, z) - \delta_\chi W(u, v)$. By Perron's formula

$$\sum_{m \in J_y} b(m) \chi(m) = \mathcal{G}(\chi, y, z) + O(Y^\epsilon).$$

Hence we need only show that

$$\|\mathcal{G}(\chi, y, z)\|_{\mathcal{E}(q)}^2 \ll (qQ)^2 Y \mathcal{L}^{-A} \quad (10.6.3)$$

and

$$\|\mathcal{G}'(\chi, y, z)\|_{\{\chi_0\}}^2 \ll (qQ)^2 Y \mathcal{L}^{-A}. \quad (10.6.4)$$

By Lemma 9.1 the left-hand side of (10.6.3) is

$$\ll Y^2 \mathcal{L} \sum_{\chi \in \mathcal{E}(q)} \int_{-Y}^{Y} \frac{\left| G\left(\frac{1}{2} + it, \chi\right) \right|^2}{T_2^2 + t^2} \, dt \ll \frac{Y^2}{T_2^2} Y \mathcal{L}^{-A} \ll (qQ)^2 Y \mathcal{L}^{-A},$$

where we used (compare (9.1.5))

$$\left| \frac{(1+z)^{\frac{1}{2}+it} - 1}{\frac{1}{2} + it} \right|^2 \ll \min(z^2, |t|^{-2}) \ll \frac{1}{T_2^2 + t^2}$$

in the first step and (10.6.1) in the second step. Similarly (10.6.2) yields (10.6.4).

□

We can quickly finish the proof now by appealing to the results of Chapter 9 with just one complication. To see the relation between what we have left to establish — (10.6.1), (10.6.2) — and the problem of primes in almost all short intervals, note that by dividing up the integration range into dyadic blocks,

$$\sum_{\chi \in \mathcal{E}(q)} \int_{U \leq |t| \leq Y} \frac{|G\left(\frac{1}{2} + it, \chi\right)|^2}{T_2^2 + t^2} \, dt \ll \mathcal{L} \frac{1}{U^2} (Uq + Y) \mathcal{L}^B$$

for some fixed B. Hence, if we choose $U = T_2 \mathcal{L}^{A+B}$, we obtain the required bound (and similarly for (10.6.2)). It thus suffices to establish

$$\sum_{\chi \in \mathcal{E}(q)} \int_{-U}^{U} \left| G\left(\tfrac{1}{2} + it, \chi\right) \right|^2 \, dt \ll Y \mathcal{L}^{-A} \tag{10.6.5}$$

and

$$\int_{[-U,U] \setminus [-T_0, T_0]} \left| G\left(\tfrac{1}{2} + it, \chi_0\right) \right|^2 \, dt \ll Y \, \mathcal{L}^{-A} \tag{10.6.6}$$

for our upper- and lower-bound functions $b_j(m)$. We can translate the work of Chapter 9 into the present context by noting that, with the exception of Watt's mean value theorem, all the results from Chapter 9 have an exact analogue here with T replaced by qT. Assuming for the moment that we did not need Watt's mean value result in Chapter 9, we can thus obtain (10.6.5), (10.6.6) for $Uq \approx Y^{17/18}$. Ignoring log powers, this means

$$qT_2 = \frac{Y}{Q} \approx Y^{1 - \frac{1}{2}\theta_2}.$$

At long last the reader can see that the exponent $\frac{1}{9}$ we have been using for θ_2 is double the exponent we obtained in Chapter 9.

The above argument would complete the proof — indeed it would furnish us with much better upper- and lower-bound constants than we have quoted — but for the complication of not having the required generalization of Watt's mean value result. We can mitigate this problem by appealing to [70], where a partial analogue of this result was obtained. There we proved the following.

Lemma 10.31. *Let $\epsilon > 0$ be given and q a positive integer. Then, for all $M \geq 1, T \geq q^{3/5}$ we have*

$$\frac{1}{\phi(q)} \sum_{\chi \pmod q} \int_{-T}^{T} \left| L\left(\tfrac{1}{2} + it, \chi\right) \right|^4 \left| \sum_{m \leq M} a_m \chi(m) m^{-it} \right|^2 \, dt$$

$$\tag{10.6.7}$$

$$\ll (TM)^{1+\epsilon} q^\epsilon \left(1 + M^2 T^{-\frac{1}{2}} q^{-\frac{25}{64}} \right) \max_{m \leq M} |a_m|^2.$$

If the exponent for q had been $\frac{1}{2}$ in (10.6.7), then the analogue of Watt's result would have been exact. However, we can redo the working for Chapter 9 and obtain the quoted bounds

$$\frac{Y}{20\mathcal{L}'} \leq \sum_{m \in \mathcal{B}_2} b_0, \quad \sum_{m \in \mathcal{B}_2} b_1 \leq \frac{11Y}{5\mathcal{L}'}.$$

The only difference from our previous work is that the Type I information is slightly weakened, reducing our capability of applying Buchstab's identity twice more. To be precise, where we would have had a factor $M^2 Y^{-0.47222\cdots}$, we now obtain $M^2 Y^{-0.4661\cdots}$. This completes the proof of Theorem 10.1.

10.7 FURTHER APPLICATIONS

We close this chapter by pointing out two further applications of the vector sieve in the context of the Hardy-Littlewood circle method. In [10] we proved the following result.

Theorem 10.5. *Suppose that $\frac{4}{7} \leq \theta < 1$. Then every sufficiently large odd N is the sum of three primes from the interval*

$$\left[\tfrac{1}{3}N - N^\theta, \tfrac{1}{3}N + N^\theta\right].$$

The above theorem uses several of the results we have established in this chapter. The following result from [64] has a different flavour since it uses the variant of the circle method given in [28] for dealing with inequalities.

Theorem 10.6. *Let $\lambda_1, \lambda_2, \lambda_3$ be nonzero real numbers, not all negative. Suppose that λ_1/λ_2 is irrational and algebraic. Let $\mathcal{V}$ be a well-spaced sequence in the sense that $u, v \in \mathcal{V}, u \neq v \Rightarrow |u - v| > c > 0$. Let $\delta > 0$ and write $E_2(\mathcal{V}, X, \delta)$ for the number of $v \in \mathcal{V}, v \leq X$ such that the inequality*

$$|\lambda_1 p_1^2 + \lambda_2 p_2^2 + \lambda_3 p_3^2 - v| < v^{-\delta}$$

has no solution in primes p_1, p_2, p_3. Then

$$E_2(\mathcal{V}, X, \delta) \ll X^{\frac{6}{7} + 2\delta + \epsilon}$$

for any $\epsilon > 0$.

Exercise

The following is not an exercise on the vector sieve, but rather an examination of the problem of primes in arithmetic progressions. The reader may have wondered if the methods we have evolved are applicable to this question. If the reader works through the required ideas, they should find that a good zero-free region for L-functions is necessary to adapt our working; indeed, a region nearly as good as that known for the Riemann zeta-function must be assumed. That prevents the consideration of primes in *all* progressions unless q is very small, as occurs in Theorem 10.3 above and Theorem 10.8 below. In our work on Carmichael numbers [65] we established the following result, which works for "most" progressions by the zero density results in [44].

Theorem 10.7. *Given $\epsilon > 0$, there are constants $K(\epsilon) \geq 2$ and $c > 0$ such that if $q > K$ and for every $d|q$ with χ a primitive character (mod d) we have*

$$L(s, \chi) \neq 0 \text{ for } \operatorname{Re} s > 1 - \frac{1}{(\log q)^{\frac{3}{4}}}, \quad |t| \leq \exp\left(\epsilon(\log q)^{\frac{3}{4}}\right), \qquad (10.7.1)$$

then, for any a with $(a, q) = 1$, we have

$$\pi(x; q, a) \geq \frac{cx}{\phi(q) \log x} \qquad (10.7.2)$$

whenever $x^{0.472} > q$.

Readers should attempt to prove this for themselves. All the ideas necessary have been used so far in this book, except that the integers counted (say n) must be weighted with a factor such as $(1 - n/x)$ in order to nullify the effects of integration over t (since we are no longer in short intervals, it is only the summation over characters that is significant). The numerical work is the same as would arise for primes in short intervals if there was no appeal to Watt's theorem. The work to obtain the exponent 0.472 (=$1 - \theta$ in the primes in short-intervals notation) is therefore the same as that required in the next chapter for the problem of Gaussian primes in small regions.

Remarks. The reader can verify that the hypothesis (10.7.1) can be weakened to

$$\operatorname{Re} s > 1 - \frac{J \log \log q}{\log q},$$

where J is another (very large) constant (and with a corresponding reduction in upper bound for $|t|$), but that would make the proof more complicated and not all the arguments follow verbatim from results we have established to date.

The reader should also note in this regard the hybrid problem of Bombieri's theorem in short intervals [117]. This leads to identical numerical work to the above problem when dealing with the case of short intervals with a small range for q. That is to say, we can prove that

$$\pi(x + y; q, a) - \pi(x; q, a) \geq c\frac{y}{q \log x}$$

for $y = x^{0.53}$ on average over $q < x^{0.002}$. A small improvement on this is noted in [70].

Finally, we remark that if one does not need tight constants as in Theorem 10.3, we can use the method of Chapter 7 (with Lemma 10.31 replacing Watt's mean value theorem there) to obtain the following.

Theorem 10.8. *Let $x^{0.525} \leq M \leq x$. For all $q \leq \mathcal{L}^A$ and $N > C(A)$ we have*

$$\pi(x; q, a) - \pi(x - M; q, a) > \frac{9M}{100\phi(q)\mathcal{L}}$$

whenever $(a, q) = 1$.

The value $C(A)$ is, of course, ineffective in view of the appeal to Siegel's theorem required to obtain a sufficient zero-free region.

Chapter Eleven

Generalizing to Algebraic Number Fields

11.1 INTRODUCTION

We begin in this chapter to generalize our basic method to algebraic number fields. This chapter is based on joint work with Lewis and Kumchev [68, 69]. First we consider, in an abstract setting, what is needed to make the method work. Let $\mathcal{B}$ be a set we want to sieve. This implies that it is a subset of some set with a multiplicative structure. The most general situation would be a free semigroup with the "primes" as basis elements (see [143], Sections 2.5 and 2.6, for analytic number theory in this abstract setting). We will want to remove members of $\mathcal{B}$ having small prime (or irreducible) factors. That already implies that we have some notion of the size of an element. There must be some way of ordering the primes. Often there will be a natural "norm" function to do this. We do not want there to be infinitely many elements with a given norm (unless we have some other measure of size by which we can restrict ourselves to finitely many objects). Our applications will all be to *Dedekind domains*, that is, integral domains where the set of integral ideals enjoys the unique factorization property. Indeed, all our examples will be rings of integers $\mathfrak{O}$ in an algebraic number field K, and in this case there is a natural norm for an element or ideal. The simplest generalization of $\mathbb{Z}$ with these properties is $\mathbb{Z}[i]$, the set of Gaussian integers. Here we have a Euclidean ring with norm $N(z) = |z|^2$. In this case primes and irreducibles are one and the same and their definition is well known. We need only concern ourselves with Gaussian integers in the first quadrant (the positive real line being taken as lying in this quadrant, but not the positive imaginary axis), which we denote by $\mathbb{Z}[i]^*$. The primes are then p if $p \equiv 3 \bmod 4$, p a prime in $\mathbb{N}$, $1 + i$ and $a + bi$ where $a^2 + b^2 = p \equiv 1 \pmod 4$, p a prime in $\mathbb{N}$. Since for $p \equiv 1 \pmod 4$ there are two Gaussian primes $a + bi, b + ai$, we can order them by argument. That is, in general we can write $p_1 < p_2$ to mean either $N(p_1) < N(p_2)$ or $N(p_1) = N(p_2)$ with $\arg(p_1) < \arg(p_2)$. We can define the Möbius function on $\mathbb{Z}[i]$ in an analogous way to its original definition on $\mathbb{N}$. This gives us enough background to prove the analogue of our Fundamental Theorem in $\mathbb{Z}[i]$. We note that we have the following Prime Number Theorem:

$$\sum_{\substack{p \in \mathbb{Z}[i]^* \\ N(p) \leq x}} 1 = \operatorname{Li}(x) + O\left(x \exp\left(-(\log x)^{\frac{3}{5}-\epsilon}\right)\right). \tag{11.1.1}$$

This can be obtained either by noting that the sum is

$$1 + 2\pi(x, 4, 1) + \pi(x^{\frac{1}{2}}, 4, 3), \tag{11.1.2}$$

and using the known results for primes in arithmetic progressions $(\mathrm{mod}\,4)$, or by appealing to the Prime Ideal Theorem.

In order to make progress on natural multiplicative problems, we need to replace the Riemann zeta-function in our arguments with the Dedekind zeta-function (see [106, Section 5.10]). For $\mathbb{Z}[i]$ this takes the form (here we write $K = \mathbb{Q}[i]$)

$$\zeta_K(s) = \sum_{n\in\mathbb{Z}[i]^*} \frac{1}{N(n)^s}.$$

We have the usual relations:

$$(\zeta_K(s))^{-1} = \sum_{n\in\mathbb{Z}[i]^*} \frac{\mu(n)}{N(n)^s}, \qquad \frac{\zeta_K'}{\zeta_K}(s) = -\sum_{n\in\mathbb{Z}[i]^*} \frac{\Lambda(n)}{N(n)^s}.$$

Here $\mu(n), \Lambda(n)$ are the natural generalizations of the Möbius and von Mangoldt function to K. There are many interesting questions that can be asked about Gaussian primes. These are clearly related to problems on the distribution of rational primes. For example, the question: Are there infinitely many Gaussian primes with imaginary part equal to 1? is equivalent to the question asked on page 1 of this book: Is the quadratic $n^2 + 1$ infinitely often prime? Unfortunately many of the most interesting problems are not amenable to study! Nevertheless, we are able to tackle two natural analogues of our results on primes in short intervals that are worthy of consideration:

Problem 1. Are Gaussian primes well distributed in narrow sectors $n \in \mathbb{Z}[i], \beta \le \arg(n) < \beta + \gamma$?

Problem 2. What is the maximum size of the gaps between Gaussian primes? That is, if p is a Gaussian prime, how big must θ be so that we can find another, say q, with $|p - q| < p^\theta$?

We shall answer both of these questions in this chapter. Clearly Problem 2 looks very similar to the problem of primes in short intervals. There is a conjecture that one can "walk to infinity" on the Gaussian primes with bounded steps. That is, there is a path from $1 + i$ taking bounded steps that includes Gaussian primes of arbitrarily large modulus. Such a conjecture is obviously false for rational primes, but no one has yet been able to disprove the Gaussian prime case.

Before turning to a specific problem, we should note that what we have done for $\mathbb{Z}[i]$ holds for all imaginary quadratic fields with class number 1. We can extend this further to treat all imaginary quadratic fields in a standard fashion by using ideals, and we shall consider the second problem in this more general context. The reader should immediately see that such a generalization also holds in the case of Problem 1.

11.2 GAUSSIAN PRIMES IN SECTORS

We now begin the simplest application of the alternative sieve to an algebraic number field situation given in [69]. The value of θ in the results and working below can be improved to 0.118, as was shown in [119]. One consequence of our main theorem (Theorem 11.2) is the following result on rational primes.

Theorem 11.1. *There exist infinitely many primes p with*

$$p = m^2 + n^2, \quad m, n \in \mathbb{Z}, \ n < p^\theta \tag{11.2.1}$$

and

$$\theta < 0.119. \tag{11.2.2}$$

Kubilius [114] and Ankeny [2] have shown that there exist infinitely many primes p with

$$p = m^2 + n^2, \quad m, n \in \mathbb{Z}, \ n \ll \log p, \tag{11.2.3}$$

assuming the Riemann Hypothesis for Hecke L-functions over $\mathbb{Q}(i)$. The best unconditional previous result in this area was due to Coleman [24] (where references to earlier work may be found) who obtained (11.2.1) with $\theta < 0.1631$. His work was the first application of a sieve method to this problem, but he only used the analogue of the linear Rosser-Iwaniec sieve in $\mathbb{Z}[i]$. We shall use essentially the same arithmetical information as Coleman, although it should be noted that one of his important lemmas (Lemma 9 in [24]) does not work for $\theta < \frac{1}{8}$. Our main result on Gaussian primes in sectors is as follows.

Theorem 11.2. *Let $X > X_0$. Then, for any given β, γ with*

$$0 \le \beta \le \frac{\pi}{2}, \quad X^{-0.381} \le \gamma \le \frac{\pi}{2}, \tag{11.2.4}$$

the number of Gaussian primes p satisfying

$$\beta \le \arg p < \beta + \gamma, \quad |p|^2 \le X, \tag{11.2.5}$$

is

$$> \frac{cX\gamma}{\log X}, \tag{11.2.6}$$

where c is an absolute positive constant.

11.3 NOTATION AND OUTLINE OF THE METHOD

We let $\gamma = X^{\theta - 1/2}$, where $\frac{1}{8} > \theta > \frac{1}{9}$. As should be expected, the proof for larger γ becomes progressively easier, but some arguments need to be changed as γ increases. As usual we will obtain a lower bound that reduces with decreasing θ, and our ultimate goal is to prove the lower-bound constant is positive for $\theta = 0.119$. The reader will notice that the proof could be described as "primes in short intervals with one hand tied behind your back." The difficulties in our current problem are discussed at the end of this section. For now the reader should note that τ below corresponds to θ in the problem of primes in short intervals. Also, the parameter

T plays essentially the same role as it does in Chapter 7, although here it is the "length" of a Fourier sum. We write

$$\sigma = \frac{1}{2} - \theta, \quad \tau = \frac{1}{2} + \theta, \quad \rho = \frac{1 + 22\theta}{12},$$
$$\zeta = \frac{1 + 4\theta}{3}, \quad \nu = \frac{10\theta + 7}{12}, \quad \chi = \frac{34\theta + 7}{24}.$$

For $\theta = 0.119$ the values of these parameters will be

$$\sigma = 0.381, \quad \tau = 0.619, \quad \rho = 0.3015, \quad \zeta = 0.492, \quad \nu = 0.6825, \quad \chi = 0.46025.$$

We write $\eta = (\log X)^{-A}$, where A can be made as large as we like (as usual, it arises from a term that in reality is $\exp(-(\log X)^{\delta})$ for some $\delta > 0$). We write C for fixed exponents of logarithms, which might arise as in previous chapters from an averaged divisor function (now defined on $\mathbb{Z}[i]$, of course). We may thus write, for example,

$$\eta(\log X)^C = O(\eta), \qquad (\log X)^C (\log X)^C = O((\log X)^C).$$

Implied constants will depend at most on the A implicit in this notation. We put

$$W = X^{\sigma}, \qquad T = W\eta^{-1}.$$

Also, we write (script letters now denoting sets of Gaussian integers)

$$\mathcal{A} = \{n \in \mathbb{Z}[i] : \beta \leq \arg n < \beta + \gamma, \tfrac{1}{2}X \leq |n|^2 < X\},$$
$$\mathcal{B} = \{n \in \mathbb{Z}[i] : 0 \leq \arg n < \tfrac{\pi}{2}, \tfrac{1}{2}X \leq |n|^2 < X\},$$
$$S(\mathcal{E}, z) = \{n \in \mathcal{E} : p|n \Rightarrow |p|^2 \geq z\} \text{ when } z \text{ is a positive real},$$
$$S(\mathcal{E}, q) = \{n \in \mathcal{E} : p|n \Rightarrow |p| \geq |q|\} \text{ if } q \in \mathbb{Z}[i],$$
$$\mathcal{E}_m = \{n \in \mathcal{E} : m|n\}.$$

The number of solutions to (11.2.5) with $|p|^2 > \frac{1}{2}X$ is then $S(\mathcal{A}, X^{1/2})$. As in previous chapters, our philosophy is to relate $S(\mathcal{A}, X^{1/2})$ to $S(\mathcal{B}, X^{1/2})$. From (11.1.2) we obtain

$$S(\mathcal{B}, X^{\frac{1}{2}}) \sim \frac{X}{\log X} \quad \text{as } X \to \infty. \tag{11.3.1}$$

We would expect that

$$S(\mathcal{A}, X^{\frac{1}{2}}) \sim \delta S(\mathcal{B}, X^{\frac{1}{2}}) \quad \text{as } X \to \infty, \tag{11.3.2}$$

where $\delta = \frac{2}{\pi}\gamma$. This follows for $\gamma > X^{-0.3}$ by the work of Ricci [148]. Although we are unable to establish (11.3.2) for smaller values of γ, nevertheless we can derive formulae of the type

$$\sum_{p \sim P \, q \sim Q} S(\mathcal{A}_{pq}, z) = \delta(1 + \eta) \sum_{p \sim P \, q \sim Q} S(\mathcal{B}_{pq}, z) \tag{11.3.3}$$

for various ranges of the parameters P, Q. Throughout this chapter we modify our usual $\sim$ notation so that $b \sim B$ indicates the restriction $B \leq |b|^2 < 2B$ with $0 \leq \arg b < \pi/2$. Henceforth all Gaussian integers will be assumed to lie in

$\mathbb{Z}[i]^*$. We thus interpret the inequality $\beta \leq \arg n < \beta + \gamma \pmod{\pi/2}$, so that, for example,

$$1 \leq \arg n < 2 \quad \text{becomes} \quad 0 \leq \arg n < 2 - \pi/2 \quad \text{or} \quad 1 \leq \arg n < \pi/2.$$

This is permissible since the angular distribution of Gaussian primes has period $\pi/2$.

To obtain (11.3.3) we now adapt the alternative sieve to $\mathbb{Z}[i]$. For this we require arithmetical information which is obtained from results on Hecke characters. To motivate the introduction of these characters and their associated L-functions, we first note that our condition on $\arg n$ may be expressed as

$$\left\{ \frac{2}{\pi} \arg n \right\} \in \mathfrak{I},$$

with $\{\cdot\}$ denoting the fractional part and $\mathfrak{I}$ representing an interval (mod 1) of length δ. We can thus pick out the condition on n using Fourier analysis. We note that

$$e \left(\frac{2}{\pi} m \arg n \right) = \left(\frac{n}{|n|} \right)^{4m}.$$

The reader can now see why we introduce the Hecke characters for $n \in \mathbb{Z}[i]$:

$$\lambda^m(n) = \left(\frac{n}{|n|} \right)^{4m}.$$

We may then use the Fourier analysis as given in Lemma 2.1 to show that, for any set $\mathcal{E} \subset \mathbb{Z}[i]$, $T \geq 1$, and any weights a_n,

$$\sum_{\substack{n \in \mathcal{E} \\ \beta \leq \arg n < \beta + \gamma}} a_n = \delta \sum_{n \in \mathcal{E}} a_n + O\left(\frac{1}{T} \sum_{n \in \mathcal{E}} |a_n| \right) + O\left(\gamma \sum_{m=1}^{T} \left| \sum_{n \in \mathcal{E}} a_n \lambda^m(n) \right| \right).$$

(11.3.4)

It is now clear why the L-functions of importance here are the Hecke L-functions, defined for $\mathrm{Re}\, s > 1$ by

$$L(s, \lambda^m) = \sum_{n \in \mathbb{Z}[i]^*} |n|^{-2s} \lambda^m(n). \tag{11.3.5}$$

These may be continued to the whole complex plane, as was first shown by Hecke [90]. The function is entire except when $m = 0$ when there is a simple pole at $s = 1$. See also [106, Sections 3.8, 5.10].

Once we have obtained results like (11.3.3) we can utilize Buchstab's identity in $\mathbb{Z}[i]$ just as we used it in $\mathbb{N}$. The formula now reads

$$S(\mathcal{E}, z) = S(\mathcal{E}, w) - \sum_{w \leq |p|^2 < z} S(\mathcal{E}_p, p). \tag{11.3.6}$$

Modifying the notation in the introduction to this chapter, we write $|q|^2 < |p|^2$ to indicate that

$$|q| \leq |p| \quad \text{and} \quad \arg q < \arg p \quad \text{if} \quad |p| = |q|.$$

Applying (11.3.6) twice to both $S(\mathcal{A}, X^{1/2})$ and $S(\mathcal{B}, X^{1/2})$, we obtain that

$$S(\mathcal{B}, X^{\frac{1}{2}}) = S(\mathcal{B}, z_1) - \sum_{z_1 \le |p|^2 < X^{1/2}} S(\mathcal{B}_p, z_2(p)) + \sum_{\substack{|p|^2 < X^{1/2} \\ z_2(p) \le |q|^2 < |p|^2}} S(\mathcal{B}_{pq}, q)$$

(11.3.7)

and

$$S(\mathcal{A}, X^{\frac{1}{2}}) = S(\mathcal{A}, z_1) - \sum_{z_1 \le |p|^2 < X^{1/2}} S(\mathcal{A}_p, z_2(p)) + \sum_{\substack{|p|^2 < X^{1/2} \\ z_2(p) \le |q|^2 < |p|^2}} S(\mathcal{A}_{pq}, q).$$

(11.3.8)

Here z_1 and $z_2(p)$ could be any values with $z_1 < X^{1/2}$, $z_2(p) < |p|^2$. The appropriate choices will be made later. Now, if we can show that

$$S(\mathcal{A}, z_1) = \delta S(\mathcal{B}, z_1)(1 + O(\eta))$$

and

$$\sum_{z_1 \le |p|^2 < X^{1/2}} S(\mathcal{A}_p, z_2(p)) = \delta(1 + O(\eta)) \sum_{z_1 \le |p|^2 < X^{1/2}} S(\mathcal{B}_p, z_2(p)),$$

then it follows as in the previous chapters that

$$S(\mathcal{A}, X^{\frac{1}{2}}) = \delta(1 + O(\eta))S(\mathcal{B}, X^{\frac{1}{2}}) - \sum_{z_2(p) \le |q|^2 < |p|^2 < X^{1/2}}{}' (\delta S(\mathcal{B}_{pq}, q) - S(\mathcal{A}_{pq}, q))$$

$$\ge \delta(1 + O(\eta))S(\mathcal{B}, X^{\frac{1}{2}}) - \delta \sum_{z_2(p) \le |q|^2 < |p|^2 < X^{1/2}}{}' S(\mathcal{B}_{pq}, q),$$

where $'$ indicates that the summation is only over those regions of p and q for which we have been unable to establish (11.3.3). If $|p|^2 = X^{\alpha}$, $|q|^2 = X^{\beta}$, and the regions for which we have not obtained (11.3.3) correspond to $(\alpha, \beta) \in \mathfrak{I}$, then, working as in previous chapters,

$$\sum_{z_2(p) \le |q|^2 < |p|^2 < X^{1/2}}{}' S(\mathcal{B}_{pq}, q) = (1 + O(\eta))S(\mathcal{B}, X^{1/2})\mathfrak{I}_1,$$

(11.3.9)

where

$$\mathfrak{I}_1 = \int_{(\alpha, \beta) \in \mathfrak{I}} \frac{1}{\alpha \beta^2} \omega\left(\frac{1 - \alpha - \beta}{\beta}\right) d\beta \, d\alpha.$$

As usual, $\omega(u)$ is Buchstab's function (we do not need a new Buchstab function in this context!). In fact we will apply Buchstab's identity twice more to certain sums and so we obtain (just as in earlier work)

$$S(\mathcal{A}, X^{\frac{1}{2}}) \ge \delta S(\mathcal{B}, X^{\frac{1}{2}})(1 + O(\eta)), (1 - \mathfrak{I}_1 - \mathfrak{I}_2)$$

(11.3.10)

where $\mathfrak{I}_2$ is a similar four-dimensional integral.

Before we introduce the technical details for the current problem we shall explain the relation of the auxiliary results we obtain to those of previous chapters. We have a mean-value estimate, a zero-free region, and a mean-square L-function estimate that correspond to results used for rational primes in short intervals.

We lack an analogue of the Halász-Montgomery-Huxley large-values estimate, the fourth power moment (or what is often equivalent in this context — the "reflection principle"), and Watt's mean value theorem. Except for the last-mentioned of these results, our problems stem from only having the "Fourier variable" in play for our auxiliary estimates. The reader should compare our work later in this chapter when we consider Problem 2 and it will be only Watt's result that we lack.

11.4 THE ARITHMETICAL INFORMATION

We now collect the auxiliary results that give the arithmetical information required. The first of these corresponds to the mean value theorem for Dirichlet polynomials as used in Chapter 7.

Lemma 11.1. *Suppose that $T \geq 1, N \geq 1$. Then, for any coefficients $c(n)$, we have*

$$\sum_{m=1}^{T} \left| \sum_{|n|^2 \leq N} c(n) \lambda^m(n) \right|^2 \ll (T+N) \sum_{|n|^2 \leq N} {}' |c_1(n)|^2, \tag{11.4.1}$$

where

$$c_1(n) = \sum_{\substack{\arg \nu = \arg n \\ |\nu|^2 \leq N}} c(\nu),$$

and the $'$ on the right-hand side of (11.4.1) indicates that the summation is only over primitive Gaussian integers n, that is, $n = x + iy, (x, y) = 1$.

Proof. This is a large sieve type inequality for $\mathbb{Z}[i]$ (more accurately, a dual form of the large sieve). We give a proof in the appendix (Section A.3). $\quad\square$

To use the above result we need the following elementary lemma whose proof we leave to the reader. When we quote Lemma 11.1 in the future, we may tacitly assume that the following result has also been applied.

Lemma 11.2. *In the notation of Lemma 11.1 we have*

$$\sum_{|n|^2 \leq N} {}' |c_1(n)|^2 \ll N \sum_{|n|^2 \leq N} \frac{|c(n)|^2 \tau(n)}{|n|^2}.$$

In particular, if $c(n)$ is bounded by a divisor function, then

$$\sum_{|n|^2 \leq N} {}' |c_1(n)|^2 \ll N (\log N)^C.$$

The next result corresponds to the mean square of the Riemann zeta-function (not explicitly used in Chapter 7 because we had the reflection principle and the fourth-power moment to hand).

Lemma 11.3. *Suppose that* $T \geq 2, U \geq 2$. *Then*

$$\sum_{m=1}^{T} \int_{-U}^{U} |L(\tfrac{1}{2} + it, \lambda^m)|^2 \, dt \ll TU(\log TU)^C. \qquad (11.4.2)$$

Proof. This follows from Lemma 10 in [24]. □

It is important in our situation that the right-hand side of (11.4.2) has a factor TU, not $T^2 + U^2$. That is why the known fourth-power moment result is not quite suitable for our purposes. We state this result in the next lemma and use it to obtain a weaker fourth-power result in the following lemma that suffices for our purposes.

Lemma 11.4. *Suppose that* $T \geq 2, U \geq 2$. *Then*

$$\sum_{m=1}^{T} \int_{-U}^{U} |L(\tfrac{1}{2} + it, \lambda^m)|^4 \, dt \ll (T^2 + U^2)(\log TU)^C. \qquad (11.4.3)$$

Proof. See [22, Theorem 6.3]. □

Lemma 11.5. *Suppose that* $T \geq 2, T^{5/3} \geq U \geq 2$. *Then*

$$\sum_{m=1}^{T} \int_{-U}^{U} |L(\tfrac{1}{2} + it, \lambda^m)|^4 \, \frac{dt}{1 + |t|} \ll T^{\frac{5}{3}}(\log T)^C. \qquad (11.4.4)$$

Proof. This follows from Lemmas 11.3 and 11.4 along with the bound

$$L(\tfrac{1}{2} + it, \lambda^m) \ll (|t| + |m|)^{\frac{1}{3}} \log^4(|t| + |m|)$$

(see (6.16) in [24]). □

Lemma 11.6. *Say* $R, U \geq 2, T \geq S \geq R + \exp((\log UT)^{4/5}), 1 \leq m \leq T, |t| \leq U$. *Then*

$$\left| \sum_{R \leq |p|^2 < S} |p|^{2it} \lambda^m(p) \right| \ll S \exp\left(-\frac{\log S}{2(\log TU)^{\frac{7}{10}}} \right). \qquad (11.4.5)$$

Proof. This result may be obtained in the same way as the corresponding results for rational primes (compare Lemma 1.5) since the Hecke L-functions have a similar zero-free region (see [23]). □

In the following we write

$$\mu = \exp\left((\log X)^{\frac{4}{5}} \right).$$

The next lemma has already been seen for rational primes. It corresponds to the triangle T in Diagram 7.1 (compare (7.2.12) with (11.4.7)). This is all we can do to obtain Type II information for the current problem. Since we have no corresponding large-values result we cannot get the larger parallelogram P that was obtained in Chapter 7.

Lemma 11.7. *Suppose that* $U, V \geq W, UV \leq X\mu^{-1}$, *and the coefficients* a_u, b_v *are bounded by divisor functions. Then*

$$\sum_{m=1}^{T} \left| \sum_{u \sim U} \sum_{v \sim V} \sum_{puv \in \mathcal{B}} a_u b_v \lambda^m (uvp) \right| \ll X\eta. \tag{11.4.6}$$

Proof. Using Perron's formula it suffices to establish that, for $|t| \leq X$,

$$\sum_{m=1}^{T} \left| \sum_{u \sim U} \sum_{v \sim V} \sum_{|p|^2 \leq 4X/(UV)} a_u b_v \lambda^m (uvp) |uvp|^{2it} \right| \ll X\eta.$$

Using Cauchy's inequality with Lemmas 11.1 and 11.6, the left-hand side above is

$$\ll \frac{X}{UV} \left(U(U+T)(\log X)^C \right)^{\frac{1}{2}} \left(V(V+T)(\log X)^C \right)^{\frac{1}{2}}$$

$$\exp \left(-\frac{(\log X)^{\frac{4}{5}}}{(\log X)^{\frac{7}{10}}} \right) \ll X\eta \tag{11.4.7}$$

as required. $\square$

We now consider the Type I information that can be obtained.

Lemma 11.8. *Suppose that* $U \leq X^\tau \mu^{-1}$. *Then, if* a_u *is bounded by a divisor function, we have*

$$\sum_{m=1}^{T} \left| \sum_{u \sim U} a_u \sum_{uv \in \mathcal{B}} \lambda^m (uv) \right| \ll X\eta. \tag{11.4.8}$$

Proof. We had results like this in Chapter 7, but there we appealed to the fourth-power moment of the Riemann zeta-function. The details appear sufficiently different here to warrant a complete proof. Let $M = X^{1/2}$. Write

$$S_m = \sum_{u \sim U} a_u \sum_{uv \in \mathcal{B}} \lambda^m (uv).$$

Using Perron's formula with $c = 1 + (\log X)^{-1}$ we then have

$$S_m = \sum_{u \sim U} a_u \lambda^m (u) \frac{1}{2\pi i} \int_{c-Mi}^{c+Mi} L(s, \lambda^m) \left(\frac{X}{|u|^2} \right)^s (1 - 2^{-s}) \frac{ds}{s}$$

$$+ O\left(\sum_{u \sim U} |a_u| \frac{X^{\frac{1}{2}}}{|u|^2} \log X \right)$$

$$= \mathfrak{I}_m + O\left(X^{\frac{1}{2}} (\log X)^C \right), \quad \text{say.}$$

We now shift the contour of integration back to $\mathrm{Re}\, s = \frac{1}{2}$. Writing $\mathfrak{I}'_m$ for the new integral and using the bound $|L(s, \lambda^m)| \leq (|m| + |s|)^{1-\sigma/2}$, we get

$$\mathfrak{I}_m = \mathfrak{I}'_m + O\left(\max_{1/2 \leq \sigma \leq 1} \sum_{u \sim U} |a_u| \left(\frac{X}{|u|^2} \right)^\sigma \frac{X^{\frac{1-\sigma}{4}}}{M} \right).$$

Since $U \leq X^\tau$, the error term above is no more than $X^\phi (\log X)^C$, where

$$\phi = \max_{\sigma \leq 1} \left(\tau(1 - \sigma) + \sigma + \frac{1 - \sigma}{4} - \frac{1}{2} \right) = \frac{1}{2}.$$

Thus

$$\sum_{m=1}^{T} |S_m| = \sum_{m=1}^{T} |\mathcal{I}'_m| + O(X\eta).$$

Also,

$$\sum_{m=1}^{T} |\mathcal{I}'_m| \leq \sum_{m=1}^{T} X^{\frac{1}{2}} \int_{-M}^{M} |L(\tfrac{1}{2} + it, \lambda^m)| \left| \sum_{u \sim U} |u|^{-1-2it} \lambda^m (u) a_u \right| \frac{dt}{1 + |t|}$$

$$\leq (XT)^{\frac{1}{2}} (T + U)^{\frac{1}{2}} (\log X)^C$$

using Cauchy's inequality with Lemmas 11.1 and 11.3. The bound (11.4.8) then follows since $\max(XTU, XT^2) \leq X^2 \mu^{-1}$. ∎

The next result corresponds to the Type I/II information in Chapter 7. The reader can compare this to Lemma 7.4 and note the present weakness caused by our inability to acquire the required fourth-power moment.

Lemma 11.9. *Suppose that $U \leq X^\zeta \mu^{-1}, V \leq X^\rho \mu^{-1}, \theta > \frac{1}{14}, UV \leq X^\nu \mu^{-1}$. Then, if a_u, b_v are bounded by divisor functions,*

$$\sum_{m=1}^{T} \left| \sum_{u \sim U} a_u \sum_{v \sim V} b_v \sum_{uvr \in \mathcal{B}} \lambda^m (uvr) \right| \ll X\eta. \tag{11.4.9}$$

Proof. Working as in the previous lemma leads to terms of a satisfactory size together with the expression

$$\sum_{m=1}^{T} \int_{-M}^{M} |L(\tfrac{1}{2} + it, \lambda^m)| \left| \sum_{u \sim U} |u|^{-s} \lambda^m (u) a_u \right| \left| \sum_{v \sim V} |v|^{-s} \lambda^m (v) b_v \right| \frac{dt}{1 + |t|},$$

where $s = 1 + 2it$. We may then apply Hölder's inequality and Lemmas 11.1 and 11.5 to give an upper bound

$$\ll (\log X)^C X^{\frac{1}{2}} T^{\frac{5}{12}} (U + T)^{\frac{1}{2}} (V^2 + T)^{\frac{1}{4}}.$$

The conditions on U, V, θ ensure that each of the following four conditions are satisfied:

$$T^{\frac{5}{6}} UV \ll X\eta, \quad T^{\frac{11}{6}} V \ll X\eta, \quad T^{\frac{7}{3}} \ll X\eta, \quad T^{\frac{4}{3}} U \ll X\eta.$$

The bound (11.4.9) then follows. ∎

11.5 ASYMPTOTIC FORMULAE FOR PROBLEM 1

Some asymptotic formulae follow immediately from (11.3.4) with Lemma 11.7. For example, we could count numbers of the form

$$p_1 \ldots p_j \in \mathcal{A} \quad \text{with } p_i \sim P_i,$$

where
$$P_1 \geq \mu, \quad P_2 \ldots P_\ell \geq W, \quad P_{\ell+1} \ldots P_j \geq W.$$
Thus the hypotheses of Lemma 11.7 are satisfied, and so the number of such numbers, say S', is $\delta S + O(X\gamma\eta)$. Here S represents the number of solutions to
$$p_1 \ldots p_j \in \mathcal{B} \quad \text{with } p_i \sim P_i.$$
It is straightforward to obtain an asymptotic formula for S from (11.3.1) and thus deduce a formula for S'. For example, if a_r, b_s are bounded by divisor functions, we have
$$\sum_{\substack{rsp\in\mathcal{A} \\ |r|^2 > W, |s|^2 > W \\ |rs|^2 \mu \leq X}} a_r b_s = \delta \sum_{\substack{rsp\in\mathcal{B} \\ |r|^2 > W, |s|^2 > W \\ |rs|^2 \mu \leq X}} a_r b_s + O(X\gamma\eta). \tag{11.5.1}$$
We now state the result that is the generalization to $\mathbb{Z}[i]$ of our Fundamental Theorem.

Lemma 11.10. *Suppose that $W \ll Y \ll X^{\tau-\epsilon}$ for some $\epsilon > 0$. Let $\ell(n)$ be a given function of $n \in \mathbb{Z}[i], n \sim Y$, of the form $\ell(n) = R|n|^k$, where $R > 0, -2 \leq k \leq 2$, and*
$$\mu \leq \ell(n) \ll X^\tau Y^{-1}.$$
Then, assuming that a_n is bounded by a divisor function, we have
$$\sum_{n\sim Y} a_n S(\mathcal{A}_n, \ell(n)) = \delta \sum_{n\sim Y} a_n S(\mathcal{B}_n, \ell(n)) + O(X\gamma\eta). \tag{11.5.2}$$

Proof. By Buchstab's identity
$$\sum_{n\sim Y} a_n S(\mathcal{A}_n, \ell(n)) = \sum_{n\sim Y} a_n S(\mathcal{A}_n, \mu) - \sum_{n\sim Y} a_n \sum_{\mu \leq |q|^2 < \ell(n)} S(\mathcal{A}_{nq}, q)$$
$$= S_1 - S_2, \quad \text{say.}$$
From (11.5.1) we obtain a formula for S_2 after an appeal, if necessary, to Perron's formula to remove the condition $|q|^2 < \ell(n)$. Here we need $|nq|^2 \ll X^\tau, |n|^2 \geq W, |q|^2 \geq \mu$. The required formula for S_1 may be obtained by appealing to the fundamental lemma in Chapter 4. Heath-Brown's result (Lemma 4.1) generalizes to the following:
$$\sum_{d|Q} \mu(d) = \sum_{d|Q, |d|^2 < E} \mu(d) + O\left(\sum_{d|Q, E \leq |d|^2 < E\mu} 1 \right), \tag{11.5.3}$$
where Q is now the product of the Gaussian primes p with $|p|^2 < \mu$. To apply this result we need the arithmetical information from Lemma 11.8, which leads to the inequality
$$\sum_{n\sim Y} |a_n| \sum_{|d| \leq X^{\epsilon/4}} \left| |\mathcal{A}_{dn}| - \frac{|\mathcal{A}|}{|dn|^2} \right| \ll X\gamma\eta.$$
We can thus obtain the formula
$$\sum_{n\sim Y} a_n S(\mathcal{A}_n, \mu) = \delta \sum_{n\sim Y} a_n S(\mathcal{B}_n, \mu) + O(X\gamma\eta)$$
by modifying the arguments of Chapter 5. $\qquad \square$

We now extend the result of Lemma 11.10 to smaller values of Y. We have worked in a similar way from Chapter 7 onward. We get the by now familiar zigzag fluctuation in z as Y varies. We thus write

$$z(Y) = \begin{cases} X^\tau/Y & \text{if } Y \geq W, \\ (X^\tau/Y)^{\frac{1}{j+2}} & \text{if } WX^{-(j+1)\theta} \leq Y < WX^{-j\theta}, \ j = 0, 1, 2, \ldots . \end{cases}$$

Lemma 11.11. *Suppose that $X^\theta \ll Y \ll W$. Let $\ell(n)$ be a given function of $n \in \mathbb{Z}[i], n \sim Y$, with $\ell(n) = R|n|^k$, where $R > 0, -2 \leq k \leq 2$, and*

$$\ell(n) \ll z(Y).$$

Then, assuming that a_n is bounded by a divisor function,

$$\sum_{n \sim Y} a_n S(\mathcal{A}_n, \ell(n)) = \delta \sum_{n \sim Y} a_n S(\mathcal{B}_n, \ell(n)) + O(X\gamma\eta). \tag{11.5.4}$$

Proof. The proof is similar to earlier work (compare Lemmas 7.15 and 10.15). The reader will be able to supply all the details if required. $\square$

Lemma 11.12. *We have*

$$S(\mathcal{A}, X^\theta) = \delta S(\mathcal{B}, X^\theta) + O(X\gamma\eta).$$

Proof. This follows using the same procedures as needed to prove Lemma 11.11 $\square$

Lemma 11.13. *Suppose that $W \leq P \ll X^\chi$ and*

$$PX^{-2\theta} \ll Q \ll X^\nu/(P\mu).$$

Then we can give an asymptotic formula for

$$\sum_{p \sim P, q \sim Q} S(\mathcal{A}_{pq}, q). \tag{11.5.5}$$

Proof. First we note that $\chi < \zeta$ and

$$P > W \ \Rightarrow \ \frac{X^\nu}{P\mu} \leq \frac{X^\rho}{\mu}.$$

It follows, using Lemma 11.9 and a fundamental lemma, that we can give an asymptotic formula for

$$\sum_{p \sim P, q \sim Q} S(\mathcal{A}_{pq}, \mu).$$

Applying Buchstab's identity to (11.5.5) leaves us to estimate

$$\sum_{p \sim P, q \sim Q} \sum_{\mu \leq |r|^2 < |q|^2} S(\mathcal{A}_{pqr}, r). \tag{11.5.6}$$

We note that we can assume

$$|r|^2 \ll \left(\frac{X}{PQ}\right)^{\frac{1}{2}},$$

for otherwise the inner sum in (11.5.6) is empty. Thus

$$|pr|^2 \ll X^{\frac{1}{2}} \left(\frac{P}{Q}\right)^{\frac{1}{2}} \leq X^\tau.$$

Hence (11.5.6) counts numbers $pqrs \in \mathcal{A}$ that satisfy

$$|qs|^2 \gg W, |p|^2 > W, |r|^2 > \mu,$$

and so a formula may be obtained from (11.5.1). Here we have tacitly applied Perron's formula to decouple r and s. $\qquad \square$

Lemma 11.14. *Suppose that* $W \leq PQ \leq X^\zeta \mu^{-1}, PQR \leq X^\nu \mu^{-1}, P \geq Q \geq R \geq X^\theta$. *Then we can give an asymptotic formula for*

$$\sum_{\substack{p \sim P, q \sim Q \\ r \sim R}} S(\mathcal{A}_{pqr}, X^\tau/PQ).$$

Proof. From Buchstab's identity and a fundamental lemma result as in Lemma 11.13, it suffices to estimate

$$\sum_{\substack{p \sim P, q \sim Q \\ r \sim R}} \sum_{\mu \leq |s|^2 < X^\tau/PQ} S(\mathcal{A}_{pqrs}, s).$$

This counts numbers $pqrst$ where

$$|s|^2 \geq \mu, \quad |pq|^2 \geq W, \quad \text{and} \quad |rt|^2 \gg \frac{X}{|pqs|^2} \gg \frac{X}{PQX^\tau/PQ} = W.$$

After removing the interdependence between s and t in a standard fashion we can apply (11.5.1) to complete the proof. $\qquad \square$

Lemma 11.15. *Suppose that* $P \geq Q \geq X^\theta, W \leq PQ \leq X^\zeta \mu^{-1}, PQ^2 \leq X^\nu \mu^{-1}$. *Then, if* $Q^2 P < X^\tau$, *we can give an asymptotic formula for*

$$S = \sum_{p \sim P, q \sim Q} S(\mathcal{A}_{pq}, q).$$

Otherwise we have

$$S = \sum_{p \sim P, q \sim Q} S(\mathcal{A}_{pq}, X^\tau/PQ) - \sum_{\substack{p \sim P, q \sim Q \\ X^\tau/PQ \leq |r|^2 < |q|^2}} S(\mathcal{A}_{pqr}, X^\tau/PQ)$$

$$+ \sum_{\substack{p \sim P, q \sim Q \\ X^\tau/PQ \leq |s|^2 < |r|^2 < |q|^2}} S(\mathcal{A}_{pqrs}, s), \qquad (11.5.7)$$

and we can give an asymptotic formula for the first two terms on the right-hand side of (11.5.7).

Proof. This follows from Buchstab's identity and Lemmas 11.10 and 11.14. $\qquad \square$

Lemma 11.16. *Suppose that* $P \geq Q \geq X^\theta, WX^{-\theta} \leq PQ < W$. *Then, if* $PQ^3 < X^\tau$, *we can give an asymptotic formula for*

$$S = \sum_{p \sim P, q \sim Q} S(\mathcal{A}_{pq}, q).$$

Otherwise we have

$$S = \sum_{p \sim P, q \sim Q} S(\mathcal{A}_{pq}, (X^\tau/PQ)^{\frac{1}{2}})$$

$$- \sum_{\substack{p \sim P, q \sim Q \\ (X^\tau/PQ)^{1/2} \leq |r|^2 < |q|^2}} S(\mathcal{A}_{pqr}, X^\tau/PQ|r|^2) + \sum_{\nabla} S(\mathcal{A}_{pqrs}, s) \quad (11.5.8)$$

with ∇ *representing*

$$p \sim P, \quad q \sim Q \quad \left(\frac{X^\tau}{PQ}\right)^{\frac{1}{2}} \leq |r|^2 < |q|^2, \quad \frac{X^\tau}{PQ|r|^2} \leq |s|^2 < |r|^2,$$

and we can give asymptotic formulae for the first two sums on the right-hand side of (11.5.8).

Proof. We combine Buchstab's identity with Lemmas 11.10 and 11.11. It should be noted that the variable s may take values with modulus squared smaller than X^θ. Indeed, for $\theta = 0.119$ the value of $|s|^2$ can reduce down to $X^{0.0475}$. □

11.6 THE FINAL DECOMPOSITION FOR PROBLEM 1

We must now apply Buchstab's identity to $S(\mathcal{A}, X^{1/2})$ in such a way that (11.3.10) holds with $\mathfrak{I}_1 + \mathfrak{I}_2 < 1$. With a first move that the reader has seen many times before, we have

$$S(\mathcal{A}, X^{\frac{1}{2}}) = S(\mathcal{A}, X^\theta) - \sum_{X^\theta \leq |p|^2 < X^{1/2}} S(\mathcal{A}_p, z(|p|^2)) + \sum_{\mho} S(\mathcal{A}_{pq}, q). \quad (11.6.1)$$

Here $\mho$ represents the conditions

$$X^\theta \leq |p|^2 < X^{\frac{1}{2}} \quad z(|p|^2) \leq |q|^2 < \min\left(|p|^2, \frac{X^{\frac{1}{2}}}{|p|}\right).$$

We can give asymptotic formulae for the first two terms on the right-hand side of (11.6.1) using Lemmas 11.10–11.12. We now split the consideration of the final sum into six cases as follows. We here write

$$d\boldsymbol{\alpha}_j = \frac{d\alpha_j \dots d\alpha_1}{\alpha_1 \dots \alpha_{j-1}\alpha_j^2}.$$

1. $W \leq |p|^2 < X^{1/2}$. We can apply Lemma 11.13 to part of this sum and discard the remainder. We thus incur a loss amounting to

$$\phi_1 \delta S(\mathcal{B}, X^{\frac{1}{2}}),$$

where

$$\phi_1 = \int^* \omega\left(\frac{1 - \alpha_1 - \alpha_2}{\alpha_2}\right) d\alpha_2.$$

Here

$$\int^* = \int_\sigma^X \int_{\nu-\alpha_1}^{(1-\alpha_1)/2} + \int_{1/4+3\theta/2}^X \int_{\tau-\alpha_1}^{\alpha_1-2\theta} + \int_X^{1/2} \int_{\tau-\alpha_1}^{(1-\alpha_1)/2}.$$

Numerical integration gives, for $\theta = 0.119$,

$$\phi_1 < 0.275.$$

2. $X^{1/3} \le |p|^2 < W$. If $|pq|^2 < X^\varsigma$, then $|pq^2|^2 = |pq|^4|p|^{-2} < X^{2/3} < X^\nu$. We thus apply Lemma 11.15 to part of the sum and throw away what remains. This gives a loss from a two-dimensional integral

$$\phi_2 = \int_{1/3}^\sigma \int_{\varsigma-\alpha_1}^{(1-\alpha_1)/2} \omega\left(\frac{1-\alpha_1-\alpha_2}{\alpha_2}\right) d\alpha_2,$$

and from a four-dimensional integral

$$\phi_3 = \int_{\mathcal{F}} \omega\left(\frac{1-\alpha_1-\alpha_2-\alpha_3-\alpha_4}{\alpha_4}\right) d\alpha_4.$$

Here $\mathcal{F}$ is the region bounded by the inequalities

$$\tfrac{1}{3} \le \alpha_1 \le \sigma, \qquad \tfrac{1}{2}(\tau - \alpha_1) \le \alpha_2 \le \varsigma - \alpha,$$
$$\tau - \alpha_1 - \alpha_2 \le \alpha_3 \le \min(\alpha_2, \alpha_1 + \alpha_2 - 2\theta)$$
$$\tau - \alpha_1 - \alpha_2 \le \alpha_4 \le \min\left(\alpha_3, \tfrac{1}{2}(1 - \alpha_1 - \alpha_2 - \alpha_3)\right).$$

At $\theta = 0.119$ we have

$$\phi_2 < 0.341, \quad \phi_3 < 0.001.$$

3. $X^\rho \le |p|^2 < X^{1/3}$. Again $|pq|^2 < X^\varsigma \Rightarrow |pq^2|^2 < X^\nu$, so the analysis of this section is identical to that in the previous one, except that the upper limit for $|q|$ is now $|p|$. Thus corresponding losses are now

$$\phi_4 = \int_\rho^{1/3} \int_{\varsigma-\alpha_1}^{\alpha_1} \omega\left(\frac{1-\alpha_1-\alpha_2}{\alpha_2}\right) d\alpha_2,$$

and

$$\phi_5 = \int_\rho^{1/3} \int_{(\tau-\alpha_1)/2}^{\varsigma-\alpha_1} \int_{\tau-\alpha_1-\alpha_2}^{\alpha_2} \int_{\tau-\alpha-\alpha_2}^{\alpha_3} \omega\left(\frac{1-\alpha_1-\alpha_2-\alpha_3-\delta}{\alpha_4}\right) d\alpha_4.$$

At $\theta = 0.119$ we have

$$\phi_4 < 0.156, \quad \phi_5 < 0.001.$$

4. $WX^{-\theta} \le |p|^2 < X^\rho$. Now the constraint $|pq^2|^2 < X^\nu$ is more stringent than $|pq| < X^\varsigma$. For this case we have a two-dimensional loss,

$$\phi_6 = \int_{1/2-2\theta}^\rho \int_{(\nu-\alpha_1)/2}^{\alpha_1} \omega\left(\frac{1-\alpha-\alpha_2}{\alpha_2}\right) d\alpha_2,$$

and a four-dimensional loss

$$\phi_7 = \int_{1/2-2\theta}^{\rho} \int_{(\tau-\alpha_1)/2}^{(\nu-\alpha_1)/2} \int_{\tau-\alpha_1-\alpha_2}^{\alpha_2} \int^{*} \omega\left(\frac{1-\alpha_1-\alpha_2-\alpha_3-\alpha_4}{\alpha_4}\right) d\boldsymbol{\alpha}_4.$$

Here the $*$ indicates the conditions

$$\tau - \alpha_1 - \alpha_2 \leq \alpha_4 \leq \alpha_3, \quad \text{and} \quad \alpha_4 > 1 - \alpha_1 - \alpha_3 - \sigma \text{ if } \alpha_1 + \alpha_3 \geq \sigma.$$

At $\theta = 0.119$ we have

$$\phi_6 < 0.112, \quad \phi_7 < 0.001.$$

5. $|p|^2 < WX^{-\theta}, |pq|^2 > WX^{-\theta}$. In view of Lemma 11.16, we can always give an asymptotic formula for the part of the sum with $|p|^2 < X^{\tau/4}$, and a formula for that part of the remainder with $|pq^3|^2 < X^\tau, |pq|^2 < W$. When $|pq|^2 > W$, we have a two-dimensional loss

$$\phi_8 = \int_{\nu/3}^{\sigma-\theta} \int_{(\nu-\alpha_1)/2}^{\alpha_1} \omega\left(\frac{1-\alpha-\alpha_2}{\alpha_2}\right) d\boldsymbol{\alpha}_2$$

and a four-dimensional loss

$$\phi_9 = \int_{\mathcal{G}} \omega\left(\frac{1-\alpha_1-\alpha_2-\alpha_3-\alpha_4}{\alpha_4}\right) d\boldsymbol{\alpha}_4,$$

with $\mathcal{G}$ defined by

$$\tfrac{1}{3}\tau \leq \alpha_1 \leq \sigma-\theta, \qquad \tfrac{1}{2}(\tau-\alpha_1) \leq \min\left(\tfrac{1}{2}(\nu-\alpha_1),\alpha_1\right),$$

$$\tau - \alpha_1 - \alpha_2 \leq \alpha_3 \leq \min(\alpha_2, \alpha_1 + \alpha_2 - 2\theta)$$

$$\tau - \alpha_1 - \alpha_2 \leq \alpha_4 \leq \min\left(\alpha_3, \tfrac{1}{2}(1-\alpha_1-\alpha_2-\alpha_3)\right).$$

(We omit here a couple of extra conditions on the integral ranges corresponding to other ways of combining variables to get asymptotic formulae.) At $\theta = 0.119$ we have

$$\phi_8 < 0.036, \quad \phi_9 < 0.001.$$

When $|pq|^2 < W$, there is just a four-dimensional loss, say ϕ_{10}. We leave the reader to deduce the integration ranges. At $\theta = 0.119$ we have

$$\phi_{10} < 0.017.$$

6. $|pq|^2 < WX^{-\theta}$. Here we have just a four-dimensional loss, say ϕ_{11}. We have $\phi_{11} < 0.001$ at $\theta = 0.119$.

Conclusion of Proof. We have

$$\sum_{j=1}^{11} \phi_j < 0.943.$$

Thus

$$S(\mathcal{A}, X^{\frac{1}{2}}) > \delta 0.057 S(\mathcal{B}, X^{\frac{1}{2}}) + O(X\gamma\eta).$$

This completes the proof of Theorem 11.2. By the above method the lower bound constant ceases to be positive for some value of θ between 0.117 and 0.118 (see [119] for more calculations). However, the more interesting challenge would be to improve the auxiliary results: the fourth-power moment or large-values results we lack. □

11.7 PRIME IDEALS IN SMALL REGIONS

In the following sections we consider Problem 2 in a more general setting. First we note that the Hecke L-functions allow us to pick out numbers in the intersection of an annulus and a sector. This type of region is called a *polar box* by some authors. Our statement of Problem 2 looked as though circles $|p - q| < |p|^\theta$ were the natural objects of study. Instead we look at regions (in the case of $\mathbb{Z}[i]$) of the form

$$X \leq |n|^2 < X + Y, \qquad \alpha \leq \arg(n) < \alpha + \beta, \qquad (11.7.1)$$

where $Y/X \approx \beta$. We shall have similar regions arising in the next chapter.

Let d be a square-free negative integer and write $K = \mathbb{Q}(\sqrt{d})$ for the imaginary quadratic field with discriminant

$$D = \begin{cases} 4d & \text{if } d \equiv 2, 3 \, (\text{mod } 4), \\ d & \text{if } d \equiv 1 \, (\text{mod } 4). \end{cases}$$

We use the standard notation $\mathfrak{O} = \mathfrak{O}_K$ for its ring of integers. In the rest of this chapter, we study the distribution of the prime ideals of $\mathfrak{O}$ in small regions corresponding to (11.7.1). A reader who does not like to think in such generality can always substitute Gaussian integers (the case $d = -1$, $D = -4$) at every stage. The main result is Theorem 11.4, which first appeared in [68] and improved on earlier work by Coleman (see [22, Theorem 2.1] and [25, Theorem 4]). In the special case of the Gaussian integers we get the following answer to Problem 2 as a corollary.

Theorem 11.3. *There is an absolute constant C such that for every $z \in \mathbb{C}$ one can find a Gaussian prime $\pi \neq z$ satisfying*

$$|z - \pi| \leq C \left(|z|^{0.53} + 1 \right).$$

The reader should compare the above result with the results of Chapter 7. The exponent $\theta = \frac{11}{20}$ had been obtained by Coleman [25, Theorem 4], who adapted the work of Heath-Brown and Iwaniec [85] to algebraic number fields. The exponent 0.53 was first published in [68], and a small improvement to 0.528 appears in [119]. Our result is not quite as strong as in Chapter 7 because we do not have an analogue for the Hecke L-functions of Watt's mean value theorem [164]. The reader will, however, be able to see that having the t variable and the Fourier variable both in play for the Hecke L-functions improves the quality of all other auxiliary results to the same status as that enjoyed in Chapter 7.

Before stating our main result we must introduce some notation and assume that the reader has some knowledge of algebraic number theory. We denote by $\mathfrak{a}, \mathfrak{b}, \ldots$ the integral ideals of $\mathfrak{O}$ and reserve $\mathfrak{p}$ (with or without subscripts) for prime ideals; also, we write $\hat{\alpha}, \hat{\beta}, \ldots$ for the ideal numbers of K, and $\hat{\pi}$ for the prime ideal numbers. Thus, given any ideal $\mathfrak{a}$, there exists an algebraic integer $\hat{\alpha}$ such that, upon writing $\mathfrak{O}'$ for the ring of integers in $K(\hat{\alpha})$, we have $\mathfrak{a} = (\mathfrak{O}'\hat{\alpha}) \cap \mathfrak{O}$ (see [156, Chapters 5 and 9]). In $\mathbb{Z}[i]$ all ideals are already principal, and the ideal numbers are simply the elements of $\mathbb{Z}[i]$. Let w be the number of roots of unity contained in the field K, so that

$$w = \begin{cases} 4 & \text{if } D = -4, \\ 6 & \text{if } D = -3, \\ 2 & \text{if } D < -4. \end{cases}$$

The Hecke Grössencharaktere $\lambda(-)$ is defined on the multiplicative group of the ideal numbers of K by

$$\lambda(\hat{\mu}) = \left(\frac{\hat{\mu}}{|\hat{\mu}|} \right)^w.$$

This is consistent with our notation in $\mathbb{Z}[i]$ where $w = 4$. If $\mathcal{C}$ is an ideal class, let $\mathcal{C}^{-1}$ denote its inverse in the ideal class group (since $\mathbb{Z}[i]$ has class number 1, this step was not needed in our previous discussion). For each class $\mathcal{C}$, we choose and fix an ideal $\mathfrak{a}_0 \in \mathcal{C}^{-1}$. Then, given an ideal $\mathfrak{a} \in \mathcal{C}$, we can find an algebraic integer $\xi_\mathfrak{a} \in \mathfrak{a}_0$ with $(\xi_\mathfrak{a}) = \mathfrak{a}\mathfrak{a}_0$. Here (ξ) denotes the principal ideal generated by ξ. Now, $\xi_\mathfrak{a}$ is unique up to multiplication by units, and so $\arg \lambda(\xi_\mathfrak{a})$ is unique mod 2π. Therefore, if $\mathcal{C}$ is an ideal class and x, y, ϕ_0, ϕ are real numbers subject to the restrictions $0 < \phi < 2\pi$ and $0 < y < x$, the following set of ideals is well-defined:

$$\mathcal{A} = \{\mathfrak{a} \in \mathcal{C} : x - y \le N\mathfrak{a} < x, \ \phi_0 \le \arg \lambda(\xi_\mathfrak{a}) \le \phi_0 + \phi\}. \qquad (11.7.2)$$

The following is the general result we shall prove in the remainder of this chapter.

Theorem 11.4. *There is an $x_0 > 0$ such that if $x \ge x_0$, $y \ge x^{0.765}$, $\phi \ge x^{-0.235}$, and $\mathcal{A}$ is the collection of ideals defined by* (11.7.2), *then*

$$\sum_{\mathfrak{p} \in \mathcal{A}} 1 \gg \frac{\phi y}{\log x}. \qquad (11.7.3)$$

Remark. We could also restrict the ideals to lie in arithmetic progressions, but the modulus would have to be small ($\ll (\log x)^A$ for some A) to avoid problems from any possible Siegel zeros.

11.8 FIRST STEPS

We write $\tau(\mathfrak{a})$ for the number of ideals dividing the ideal $\mathfrak{a}$, and $\mu(\mathfrak{a})$ is the analogue for ideals of the Möbius function:

$$\mu(\mathfrak{a}) = \begin{cases} 1 & \text{if } \mathfrak{a} = \mathfrak{O}, \\ (-1)^k & \text{if } \mathfrak{a} \text{ is the product of } k \text{ distinct prime ideals}, \\ 0 & \text{if } \mathfrak{p}^2 | \mathfrak{a} \text{ for some } \mathfrak{p}. \end{cases}$$

Suppose that $\mathcal{A}$ is defined by (11.7.2) with $y \ge x^{(\theta+1)/2+\epsilon}$ and $\phi \ge x^{(\theta-1)/2+\epsilon}$. Our goal will be to establish (11.7.3) when $\theta \ge 0.53 - 2\epsilon$. Write

$$y_1 = x \exp(-3 \log^{\frac{1}{3}} x)$$

and let $\mathcal{C}$ be the ideal class appearing in the definition of the set $\mathcal{A}$. Our comparison set will be

$$\mathcal{B} = \{\mathfrak{a} \in \mathcal{C} : x - y_1 \le N\mathfrak{a} < x\}. \qquad (11.8.1)$$

Our working will be very similar to that in Chapters 7 and 10, but here we must use Dirichlet polynomials of the form

$$F(s, \lambda^m) = \sum_{\mathfrak{a}} c(\mathfrak{a}) \lambda^m(\mathfrak{a})(N\mathfrak{a})^{-s}, \tag{11.8.2}$$

where λ is the Hecke Grössencharaktere defined above, the summation is over integral ideals $\mathfrak{a} \in \mathcal{C}$ with $N\mathfrak{a} \asymp x$, and the coefficients $c(\mathfrak{a})$ satisfy

$$|c(\mathfrak{a})| \le \tau(\mathfrak{a})^B. \tag{11.8.3}$$

The condition (11.8.3) corresponds exactly to our earlier description of being divisor-bounded, and we shall henceforth assume that all expressions of the form (11.8.2) satisfy this requirement. Sometimes, when $m = 0$ in (11.8.2), we will write simply $F(s)$. The $N\mathfrak{a}^{-s}$ terms detect the condition $x - y \le N\mathfrak{a} < x$, which is a direct analogue of the problem of finding primes in short intervals, while the restriction $\phi_0 \le \arg \lambda(\xi_\mathfrak{a}) \le \phi_0 + \phi$ is picked out by the Hecke characters as in our work on Problem 1. Now we could just combine Perron's formula and (11.3.4) to do this. Although this approach will work, it will be simpler to adopt a different way of approximating the characteristic functions corresponding to these conditions. This relieves problems caused by the error terms when they are combined. Our aim is thus to replace the sum on the left-hand side of (11.7.3) by $\sum_{\mathfrak{p}} \Psi(\mathfrak{p})$, where $\Psi(\cdot)$ is a smooth function that approximates the characteristic function of $\mathcal{A}$ from below. The precise definition of this function is given by (11.8.12). Clearly, it suffices to show that

$$\sum_{\mathfrak{p}} \Psi(\mathfrak{p}) \gg \frac{\phi y}{\log x}. \tag{11.8.4}$$

If $z \ge 2$, we define the quantities

$$S(\mathcal{A}, z) = \sum_{(\mathfrak{a}, \mathfrak{P}(z))=1} \Psi(\mathfrak{a}) \quad \text{and} \quad S(\mathcal{B}, z) = \sum_{\substack{\mathfrak{a} \in \mathcal{B} \\ (\mathfrak{a}, \mathfrak{P}(z))=1}} 1,$$

where $\mathfrak{P}(z)$ is the ideal

$$\mathfrak{P}(z) = \prod_{N\mathfrak{p} < z} \mathfrak{p}.$$

The left-hand side of (11.8.4) can then be expressed as $S(\mathcal{A}, x^{1/2})$. We generalize another function we have found useful in our sieve method by writing

$$\rho(\mathfrak{a}, z) = \begin{cases} 1 & \text{if } (\mathfrak{a}, \mathfrak{P}(z)) = 1, \\ 0 & \text{otherwise.} \end{cases}$$

We will also need to generalize the ordering we used for the Gaussian primes. Here we use the convention employed by Coleman [24], namely, $\mathfrak{p}_2 < \mathfrak{p}_1$ if $N\mathfrak{p}_2 < N\mathfrak{p}_1$, or if $N\mathfrak{p}_2 = N\mathfrak{p}_1$ and $\arg(\hat{\pi}_2) < \arg(\hat{\pi}_1)$ where $\hat{\pi}_j$ is the ideal number of $\mathfrak{p}_j$. With multiple sums it will become very inconvenient to have conditions of the form

$$A < N\mathfrak{p}_1, \; N\mathfrak{p}_2 < B, \; \mathfrak{p}_2 < \mathfrak{p}_1. \tag{11.8.5}$$

We therefore write, for real A, B, $A < \mathfrak{p}_1 < B$ to mean $A < N\mathfrak{p}_1 < B$. We will continue to use $N\mathfrak{p}$ for less complicated expressions and in places where its use is unavoidable (for example, later we will need to write $(x/N(\mathfrak{p}_1\mathfrak{p}_2\mathfrak{p}_3))^{1/2}$). Also, we will use $\mathfrak{P}(\mathfrak{p}_1)$ to mean

$$\prod_{\mathfrak{p}_2 < \mathfrak{p}_1} \mathfrak{p}_2,$$

and we can then define $\rho(\mathfrak{a}, \mathfrak{p})$, $S(\mathcal{A}, \mathfrak{p})$, and $S(\mathcal{B}, \mathfrak{p})$ in an analogous fashion to $\rho(\mathfrak{a}, z)$, $S(\mathcal{A}, z)$, and $S(\mathcal{B}, z)$. Using the above notation we then have as one version of Buchstab's identity:

$$\rho(\mathfrak{l}, z_1) = \rho(\mathfrak{l}, z_2) - \sum_{\substack{\mathfrak{p}\mathfrak{b}=\mathfrak{l} \\ z_2 \leq N\mathfrak{p} < z_1}} \rho(\mathfrak{b}, \mathfrak{p}). \tag{11.8.6}$$

Here z_2 is any value with $2 \leq z_2 < z_1$).

We can follow the same programme used before. We decompose $S(\mathcal{A}, x^{1/2})$ by means of Buchstab's identity to obtain

$$S(\mathcal{A}, x^{\frac{1}{2}}) = \sum_{j=1}^{k} S_j - \sum_{j=k+1}^{\ell} S_j. \tag{11.8.7}$$

Here, $S_j \geq 0$, and for $j \leq k' < k$ and $j > k$ we will be able to prove asymptotic formulae of the form

$$S_j = \frac{\phi y}{2\pi y_1} S_j^*(1 + o(1)), \tag{11.8.8}$$

where S_j^* is a sifting function analogous to S_j in which the set $\mathcal{B}$ is being sifted instead. Using our standard philosophy, that is, discarding the sums S_j, $k' < j \leq k$, we obtain the lower bound

$$S(\mathcal{A}, x^{\frac{1}{2}}) \geq \frac{\phi y}{2\pi y_1} \left(S(\mathcal{B}, x^{\frac{1}{2}}) - \sum_{k' < j \leq k} S_j^* \right)(1 + o(1)).$$

As in previous applications of our method, it remains to ensure that the right-hand side of this inequality is non-negative, that is, that not too many sums S_j are discarded. Since most of our work is exactly analogous to that in Chapter 7, we can spare the readers many technical details.

We shall write $N\mathfrak{a} \sim X$ for the condition $X \leq N\mathfrak{a} < 2X$, and $N\mathfrak{a} \asymp X$ for the condition $c_1 X \leq N\mathfrak{a} < c_2 X$, where c_1 and c_2 are some absolute constants. We shall use the letter B to denote a sufficiently large, positive constant not necessarily the same in each occurrence; η will stand for a fixed positive number sufficiently small in terms of ϵ.

We now demonstrate that asymptotic formulae of the form

$$\sum_{\mathfrak{a}} c(\mathfrak{a})\Psi(\mathfrak{a}) = \frac{\phi y}{2\pi y_1} \sum_{\mathfrak{a} \in \mathcal{B}} c(\mathfrak{a}) + \text{error terms} \tag{11.8.9}$$

can be derived from mean-value estimates for Dirichlet polynomials of the form (11.8.2). We start by defining the function $\Psi(\mathfrak{a})$. The basic idea is to replace the characteristic function of our set with a smooth lower bound before replacing it with a Fourier series and a complex integral. Let $\Delta_1 = yx^{-\eta}$ and let $\psi_1(t)$ be an infinitely differentiable function on $\mathbb{R}$ such that $\psi_1(t) \in [0, 1]$ for all t, satisfying

$$\psi_1(t) = \begin{cases} 1 & \text{if } x - y + \Delta_1 \leq t \leq x - \Delta_1, \\ 0 & \text{if } t \notin (x - y, x), \end{cases}$$

with

$$\psi_1^{(j)}(t) \ll_j \Delta_1^{-j} \quad \text{for } j = 1, 2, \dots .$$

We show how to construct such a function in the appendix (Section A.5). Also, let $\Delta_2 = \phi x^{-\eta}$, let $r = [2/\eta] + 1$, and let $\psi_2(t)$ be a 2π-periodic function $\in [0, 1]$ for all t, satisfying

$$\psi_2(t) = \begin{cases} 1 & \text{if } \phi_0 + \Delta_2 \leq t \leq \phi_0 + \phi - \Delta_2, \\ 0 & \text{if } \phi_0 + \phi \leq t \leq \phi_0 + 2\pi, \end{cases}$$

with the Fourier expansion

$$\psi_2(t) = \widehat{\psi}_2(0) + \sum_{m \neq 0} \widehat{\psi}_2(m) e^{imt}, \tag{11.8.10}$$

where $\widehat{\psi}_2(0) = (\phi - \Delta_2)/2\pi$ and, for $m \neq 0$,

$$|\widehat{\psi}_2(m)| \leq \min\left(\phi, \frac{2}{\pi|m|}, \frac{c(r)}{\pi|m|}\left(|m|\Delta_2\right)^{-r}\right). \tag{11.8.11}$$

Again, we give the explicit construction in the appendix for this function. We can then define $\Psi(\mathfrak{a})$ by

$$\Psi(\mathfrak{a}) = \begin{cases} \psi_1(N\mathfrak{a})\,\psi_2(\arg\lambda(\xi_\mathfrak{a})), & \mathfrak{a} \in \mathcal{C}, \\ 0, & \mathfrak{a} \notin \mathcal{C}. \end{cases} \tag{11.8.12}$$

We now turn to establish (11.8.9). From (11.8.10) and the definition of $\xi_\mathfrak{a}$, we get

$$\psi_2(\arg\lambda(\xi_\mathfrak{a})) = \sum_{m \in \mathbb{Z}} \widehat{\psi}_2(m)\lambda^m(\mathfrak{a}\mathfrak{a}_0).$$

Thus, choosing $M = [\Delta_2^{-1}x^\eta] + 1$ and using (11.8.11) to estimate the tail of the last series, we obtain

$$\psi_2(\arg\lambda(\xi_\mathfrak{a})) = \sum_{|m| \leq M} \widehat{\psi}_2(m)\lambda^m(\mathfrak{a}\mathfrak{a}_0) + O(x^{-2}).$$

Hence,

$$\sum_\mathfrak{a} c(\mathfrak{a})\Psi(\mathfrak{a}) = \sum_{|m| \leq M} \widehat{\psi}_2(m)\lambda^m(\mathfrak{a}_0) \sum_{\mathfrak{a} \in \mathcal{C}} c(\mathfrak{a})\lambda^m(\mathfrak{a})\psi_1(N\mathfrak{a}) + O(1).$$

Let

$$\widehat{\psi}_1(s) = \int_0^\infty \psi_1(t)t^{s-1}\,dt$$

be the Mellin transform of $\psi_1(t)$ (the reader unfamiliar with this transform might like to note that if $s = \sigma + i\tau$, the change of variable $t = e^y$ converts this to the *Fourier* transform of the function $\psi_1(\exp(y))\exp(y\sigma)$). By Mellin's inversion formula (which is effectively the Fourier inversion formula with a shift of the line of integration by complex analysis), we have

$$\psi_1(t) = \frac{1}{2\pi i} \int_{\frac{1}{2}-i\infty}^{\frac{1}{2}+i\infty} \widehat{\psi}_1(s)t^{-s}\, ds.$$

Thus, since $\psi_1(N\mathfrak{a}) = 0$ unless $N\mathfrak{a} \asymp x$, we derive the formula

$$\sum_{\mathfrak{a}} c(\mathfrak{a})\Psi(\mathfrak{a}) = \frac{1}{2\pi i} \sum_{|m|\leq M} \widehat{\psi}_2(m)\lambda^m(\mathfrak{a}_0) \int_{1/2-i\infty}^{1/2+i\infty} F(s,\lambda^m)\widehat{\psi}_1(s)\, ds + O(1),$$

where $F(s,\lambda^m)$ is the Dirichlet polynomial (11.8.2). Integrating by parts r times yields the bound

$$\widehat{\psi}_1(\sigma + it) \ll_r \Delta_1 x^{\sigma-1} \left(\frac{x}{|t|\Delta_1}\right)^r \quad (r = 1, 2, \dots).$$

Applying this estimate with $r = [2/\eta] + 1$, we obtain

$$\sum_{\mathfrak{a}} c(\mathfrak{a})\Psi(\mathfrak{a}) = \frac{1}{2\pi i} \sum_{|m|\leq M} \widehat{\psi}_2(m)\lambda^m(\mathfrak{a}_0) \int_{1/2-iT_1}^{1/2+iT_1} F(s,\lambda^m)\widehat{\psi}_1(s)\, ds + O(1),$$

$$(11.8.13)$$

where $T_1 = x^{1+\eta}\Delta_1^{-1}$.

We note that

$$\widehat{\psi}_1(s) - \widehat{\psi}_1(1)x^{s-1} = \int_{x-y}^{x} \psi_1(t)(t^{s-1} - x^{s-1})\, dt$$

$$= -(s-1) \int_{x-y}^{x} \psi_1(t) \int_{t}^{x} v^{s-2}\, dv\, dt$$

$$\ll |s|y^2 x^{\sigma-2}.$$

We thus have a formula analogous to the one for $(x^s - (x-Y)^s)/s$ used in Lemma 7.2:

$$\widehat{\psi}_1(s) = \widehat{\psi}_1(1)x^{s-1} + O\big(|s|y^2 x^{\sigma-2}\big). \quad (11.8.14)$$

We now have the ingredients to establish the analogue of Lemma 7.2 in the present context. Since the expression $\exp\big(-B(\log x)^{1/4}\big)$ will occur frequently, we will replace it with δ in the following. The B in the definition of δ may vary from line to line but will always be positive.

Lemma 11.17. *Let the Dirichlet polynomial $F(s,\lambda^m)$ be defined by (11.8.2) and suppose that its coefficients satisfy (11.8.3). Also, suppose that the following mean-value bounds hold for $F(s,\lambda^m)$:*

$$\sum_{0<|m|\leq M} \int_{-T_1}^{T_1} \left|F\big(\tfrac{1}{2}+it,\lambda^m\big)\right| dt \ll x^{\frac{1}{2}}\delta, \quad (11.8.15)$$

and

$$\int_{T_0}^{T_1} \left| F\left(\tfrac{1}{2} + it, \lambda^0\right) \right| dt \ll x^{\frac{1}{2}} \delta, \tag{11.8.16}$$

where in (11.8.16) $T_0 = \exp(\log^{1/3} x)$. *Then,*

$$\sum_{\mathfrak{a}} c(\mathfrak{a}) \Psi(\mathfrak{a}) = \frac{\phi y}{2\pi y_1} \sum_{\mathfrak{a} \in \mathcal{B}} c(\mathfrak{a}) + O(\phi y \delta). \tag{11.8.17}$$

Proof. Since

$$\left| \widehat{\psi}_1(\sigma + it) \right| \le y x^{\sigma - 1} \quad \text{and} \quad \left| \widehat{\psi}_2(m) \right| \le \phi,$$

we conclude from (11.8.13)–(11.8.16) that

$$\sum_{\mathfrak{a}} c(\mathfrak{a}) \Psi(\mathfrak{a}) = \frac{\widehat{\psi}_2(0)}{2\pi i} \int_{1/2 - iT_0}^{1/2 + iT_0} F(s) \widehat{\psi}_1(s) \, ds + O(\phi y \delta). \tag{11.8.18}$$

Also, we have (11.8.14) and $F(\tfrac{1}{2} + it) \ll x^{1/2}(\log x)^B$, so (11.8.18) gives

$$\sum_{\mathfrak{a}} c(\mathfrak{a}) \Psi(\mathfrak{a}) = \frac{\phi y}{4\pi^2 i} \int_{1/2 - iT_0}^{1/2 + iT_0} F(s) x^{s-1} ds + O(\phi y \delta). \tag{11.8.19}$$

Now we invoke the comparison principle as used in Lemma 7.2. In our context that means we must write $y_1 = x \exp(-3 \log^{1/3} x)$, $\Delta_3 = y_1 x^{-\eta}$, and let $\psi_3(t)$ be a function analogous to $\psi_1(t)$ with y_1 and Δ_3 in place of y and Δ_1. Then, by a simpler version of the above argument, we obtain

$$\sum_{\mathfrak{a} \in \mathcal{C}} c(\mathfrak{a}) \psi_3(N\mathfrak{a}) = \frac{y_1}{2\pi i} \int_{1/2 - iT_0}^{1/2 + iT_0} F(s) x^{s-1} \, ds + O(y_1 \delta)$$

since (11.8.16) holds. On the other hand,

$$\sum_{\mathfrak{a} \in \mathcal{C}} c(\mathfrak{a}) \psi_3(N\mathfrak{a}) = \sum_{\mathfrak{a} \in \mathcal{B}} c(\mathfrak{a}) + O(x^{1-\eta}),$$

and so

$$\sum_{\mathfrak{a} \in \mathcal{B}} c(\mathfrak{a}) = \frac{y_1}{2\pi i} \int_{1/2 - iT_0}^{1/2 + iT_0} F(s) x^{s-1} \, ds + O(y_1 \delta). \tag{11.8.20}$$

Combining (11.8.19) and (11.8.20) then completes the proof. $\quad\square$

Lemma 11.17 will be our main tool for proving asymptotic formulae of the form (11.8.9), but there is one case in which we will not be able to satisfy its hypotheses. The alternative method required is provided by the following lemma.

Lemma 11.18. *Define the Dirichlet polynomial*

$$F(s, \lambda^m) = \sum_{\mathfrak{a}, \mathfrak{b}} c(\mathfrak{a}) \lambda^m(\mathfrak{a}\mathfrak{b}) N(\mathfrak{a}\mathfrak{b})^{-s}, \tag{11.8.21}$$

where the coefficients $c(\mathfrak{a})$ satisfy (11.8.3) *and the summation is over integral ideals $\mathfrak{a}$ and $\mathfrak{b}$ subject to*

$$\mathfrak{a}\mathfrak{b} \in \mathcal{C}, \quad N\mathfrak{a} \le x^{\eta}, \quad N(\mathfrak{a}\mathfrak{b}) \asymp x. \tag{11.8.22}$$

If (11.8.15) *holds, then*

$$\sum_{\substack{a,b \\ Na \leq x^\eta}} c(a)\Psi(ab) = \frac{\phi y}{2\pi y_1} \sum_{\substack{ab \in \mathcal{B} \\ Na \leq x^\eta}} c(a) + O(\phi y \delta). \qquad (11.8.23)$$

Proof. In the following we use $*$ to denote the conditions (11.8.22). As above, we obtain that the left-hand side of (11.8.23) equals

$$\sum_{|m| \leq M} \widehat{\psi}_2(m) \lambda^m(a_0) \sum_{a,b}^* c(a) \lambda^m(ab) \psi_1(N(ab)) + O(1),$$

where $M = [\Delta_2^{-1} x^\eta] + 1$. As before, we can estimate the contribution from $|m| > 0$ using the properties of the Mellin transform $\widehat{\psi}_1(s)$ and hypothesis (11.8.15). We obtain

$$\sum_{a,b}^* c(a)\Psi(ab) = \widehat{\psi}_2(0) \sum_{a,b}^* c(a)\psi_1(N(ab)) + O(\phi y \delta). \qquad (11.8.24)$$

From [138, Corollary 1 on p. 417]) we have the formula

$$\sum_{\substack{a \in \mathcal{C} \\ Na \leq X}} 1 = C_K \cdot X + O\left(X^{\frac{2}{3}+\eta}\right), \qquad (11.8.25)$$

where $\mathcal{C}$ can be any ideal class and C_K is a constant depending only on the field K (and not on the class $\mathcal{C}$). Using (11.8.24) and (11.8.25) we arrive at

$$\sum_{a,b}^* c(a)\Psi(ab) = \frac{\phi}{2\pi} \sum_{\substack{a,b \\ x-y \leq N(ab) < x}}^* c(a) + O(\phi y \delta). \qquad (11.8.26)$$

Now suppose that $\mathcal{C}_1, \ldots, \mathcal{C}_h$ are the distinct ideal classes of K. Then, from (11.8.25) again,

$$\sum_{\substack{a,b \\ x-y \leq N(ab) < x}}^* c(a) = \sum_{j=1}^{h} \sum_{\substack{a \in \mathcal{C}_j \\ Na \leq x^\eta}} c(a) \sum_{\substack{b \in \mathcal{C}\mathcal{C}_j^{-1} \\ x-y \leq N(ab) < x}} 1$$

$$= C_K y \sum_{j=1}^{h} \sum_{\substack{a \in \mathcal{C}_j \\ Na \leq x^\eta}} \frac{c(a)}{Na} + O\left(x^{\frac{2}{3}+2\eta}\right). \qquad (11.8.27)$$

Substituting this back into (11.8.26), we have

$$\sum_{\substack{a,b \\ Na \leq x^\eta}} c(a)\Psi(ab) = C_K \cdot \frac{\phi y}{2\pi} \sum_{j=1}^{h} \sum_{\substack{a \in \mathcal{C}_j \\ Na \leq x^\eta}} \frac{c(a)}{Na} + O(\phi y \delta). \qquad (11.8.28)$$

The result follows from (11.8.28) and (11.8.27) with y_1 in place of y. $\qquad \square$

11.9 ESTIMATES FOR DIRICHLET POLYNOMIALS

As in previous chapters, we will use the same letter for a Dirichlet polynomial and for its length; for example, the summation in polynomial (11.8.2) will be over ideals with $N\mathfrak{a} \asymp F$. We must now obtain analogues of the results used in Chapters 7 and 10, namely, estimates for mean and large values of Dirichlet polynomials of the form (11.8.2). We will use these estimates to satisfy the hypotheses (11.8.15) and (11.8.16) of Lemma 11.17.

We will use a similar notation to that given in Chapter 10. We need to be careful because we want to state some results for $m = 0$ separately in order to give (11.8.16). Let $2 < T, M \le x/4$. If S_1 is a set of pairs (t, m), with $|t| \le T$, $|m| \le M$, we say that S_1 is *well-spaced* if $|t - t'| \ge 1$ whenever $(t, m), (t', m) \in S_1$. Let $2 < T \le x/4$ and let S be a set of real numbers t, $|t| \le T$. Correspondingly, we say that S_0 is *well-spaced* if $|t - t'| \ge 1$ whenever $t, t' \in S_0$. We define the norm

$$_1\|F\|_n = \begin{cases} \left(\sum_{(t,m)\in S_1} \left|F\left(\tfrac{1}{2} + it, \lambda^m\right)\right|^n \right)^{\frac{1}{n}} & \text{if } 1 \le n < \infty, \\ \sup_{(t,m)\in S_1} \left|F\left(\tfrac{1}{2} + it, \lambda^m\right)\right| & \text{if } n = \infty. \end{cases}$$

Also, we write

$$G_1 = G_1(F) = \sum_{N\mathfrak{a}\asymp F} |c(\mathfrak{a})|^2 (N\mathfrak{a})^{-1}.$$

To deal with the case $m = 0$ needed for (11.8.16) we put

$$_0\|F\|_n = \begin{cases} \left(\sum_{(t,m)\in S_0} \left|F\left(\tfrac{1}{2} + it, \lambda^m\right)\right|^n \right)^{\frac{1}{n}} & \text{if } 1 \le n < \infty, \\ \sup_{(t,m)\in S_0} \left|F\left(\tfrac{1}{2} + it, \lambda^m\right)\right| & \text{if } n = \infty, \end{cases}$$

with

$$G_0 = G_0(F) = \sum_{N\mathfrak{a}\asymp F} d(\mathfrak{a})|c(\mathfrak{a})|^2 (N\mathfrak{a})^{-1}.$$

Here $d(\mathfrak{a})$ denotes the number of distinct ideals with norm equal to $N\mathfrak{a}$. We also write

$$U_j = \begin{cases} T^2 + M^2 & \text{if } j = 1 \\ T & \text{if } j = 0. \end{cases}$$

Sometimes we will need to impose the condition we have often called *prime-factored*:

$$_j\|F\|_\infty \ll F^{\frac{1}{2}}\delta. \tag{11.9.1}$$

The next lemma combines the mean value and large-values results required in the present context and is exactly analogous to Lemma 10.2. It follows from the work of Coleman ([22, Theorem 6.2, Theorem 7.3] in exactly the same way.

Lemma 11.19. *For either $j = 0$ or $j = 1$ let $\mathcal{S}_j$ be well-spaced; then*

$$_j\|F\|_2^2 \ll (U_j + F)G_j \log x. \tag{11.9.2}$$

Furthermore, if

$$\left|F\left(\tfrac{1}{2} + it, \lambda^m\right)\right| \geq V$$

for all $(t, m) \in \mathcal{S}_1$, or

$$\left|F\left(\tfrac{1}{2} + it\right)\right| \geq V$$

for all $t \in \mathcal{S}_0$, then

$$|\mathcal{S}_j| \ll \left(\frac{G_j F}{V^2} + \frac{G_j^3 F U_j}{V^6}\right) \log^4 x. \tag{11.9.3}$$

The next two lemmas are analogues of Lemma 7.3 (with $g = 4$) and Lemma 7.22. The proofs are, of course, essentially identical since we have the same "input" of results for the mean and large values.

Lemma 11.20. *In the case $j = 1$, let $P(s, \lambda^m)$, $Q(s, \lambda^m)$, $F(s, \lambda^m)$ be divisor-bounded Dirichlet polynomials. Suppose that $PQF = x$ and that $F(s, \lambda^m)$ satisfies (11.9.1). For the case $j = 0$ we remove the Hecke character λ^m. Let $U_j \leq x^{1-\theta-\eta}$, where $\frac{1}{2} + \eta < \theta \leq \frac{7}{12}$. Suppose that $P = x^{\sigma_1}$, $Q = x^{\sigma_2}$, with*

$$|\sigma_1 - \sigma_2| \leq 2\theta - 1, \tag{11.9.4}$$

$$1 - (\sigma_1 + \sigma_2) \leq \gamma(\theta), \tag{11.9.5}$$

where

$$\gamma(\theta) = \min\left(4\theta - 2, \frac{28\theta - 13}{15}, \frac{96\theta - 49}{15}\right).$$

Then,

$$_j\|FPQ\|_1 \ll x^{\frac{1}{2}}\delta. \tag{11.9.6}$$

The implied constant depends only on η.

Remark. We shall henceforth suppose that $\frac{19}{36} \leq \theta \leq \frac{17}{32}$ (which includes the value $\theta = 0.528$ or 0.53), and thus we have $\gamma(\theta) = 4\theta - 2$.

Lemma 11.21. *In the case $j = 1$, let $F(s, \lambda^m)$, $P(s, \lambda^m)$, $Q(s, \lambda^m)$, $R(s, \lambda^m)$ be divisor-bounded Dirichlet polynomials with $FPQR = x$. For the case $j = 0$ we remove the Hecke character λ^m. Suppose that F, Q, R satisfy (11.9.1). Also, suppose that $U_j \leq x^{1-\theta-\eta}$, where $\frac{1}{2} + \eta < \theta \leq \frac{7}{12}$. Then,*

$$_j\|FPQR\|_1 \ll x^{\frac{1}{2}}\delta, \tag{11.9.7}$$

provided that any of the following sets of conditions hold:

(i) $P \gg x^{1-\theta}$, $Q \gg x^{(1-\theta)/2}$, $R \gg x^{(1-\theta)/4}$, $F \gg x^{2(1-\theta)/7}$;

(ii) $P \gg x^{1-\theta}$, $Q \gg x^{(1-\theta)/2}$, $R \gg x^{(1-\theta)/3}$, $F \gg x^{2(1-\theta)/9}$;

(iii) $P \gg x^{1-\theta}$, $Q \gg x^{(1-\theta)/3}$, $R^2 F \gg x^{1-\theta}$, $F \gg x^{2(1-\theta)/5}$;

(iv) $P \gg x^{1-\theta}$, $Q, R \ll x^{(1-\theta)/3}$, $QR \gg x^{4(1-\theta)/7}$, $FQR \gg x^{14(1-\theta)/13}$.

Our next result is the fourth-power moment result we need for our Type I information. The reader may compare this result with Lemma 10.11, whose proof is nearly identical, and so we leave it out here (it is given in full in [68, Lemma 6]).

Lemma 11.22. *Let*

$$F(s, \lambda^m) = \sum_{\substack{\mathfrak{a} \in \mathcal{C} \\ N\mathfrak{a} \asymp F}} \lambda^m(\mathfrak{a})(N\mathfrak{a})^{-s}. \tag{11.9.8}$$

Let $M \leq x$, $T \leq x$, $F \leq x$ *and suppose that* $\mathcal{S}_1$ *contains no pairs* (t, m) *with* $m = 0$. *Then,*

$${}_1 \|F\|_4^4 \ll U_j (\log x)^B + |\mathcal{S}_1| F^2 U_j^{-4} (\log x)^B, \tag{11.9.9}$$

where $V = M + T$. *The corresponding result in the case* $j = 0$ *is*

$${}_0 \|F\|_4^4 \ll T^2 (\log x)^B + |\mathcal{S}_0| F^2 T^{-4} (\log x)^B + F^2 \sum_{t \in \mathcal{S}} \frac{1}{1 + |t|^4}. \tag{11.9.10}$$

The following lemma give us Type I/II information. It is an exact analogue of Lemma 7.4, and the proof follows similarly so we shall omit it here. However, we should mention that an approximate functional equation for Hecke L-functions will be required. In [106, Theorem 5.3] a very general result is proved. The specific theorem needed here can be found in [96, Theorem 2], although the statement of the result there appears quite formidable.

Lemma 11.23. *Let* $P(s, \lambda^m)$ *and* $Q(s, \lambda^m)$ *be Dirichlet polynomials satisfying* (11.8.3) *and let* $F(s, \lambda^m)$ *be defined by* (11.9.8). *Suppose that* $PQF = x$, $(M^2 + T^2) \leq x^{1-\theta-6\eta}$, *where* $\frac{1}{2} + \eta < \theta \leq \frac{7}{12}$, *and that*

$$\max(P, x^{1-\theta}) \max\left(Q, x^{\frac{1}{2}(1-\theta)}\right) \leq x^{\frac{1}{2}(1+\theta)+2\eta}.$$

Then (11.9.6) *holds.*

The next two lemmas provide bounds of the form (11.9.1). The first is a generalization of Lemma 11.6, and the proof follows familiar lines.

Lemma 11.24. *Let* $z \geq \exp\left(\log^{7/10} x\right)$ *and suppose that for all* $(t, m) \in \mathcal{S}_1$, $|t| \ll x^B$, $0 < |m| \ll x^B$. *Then, the Dirichlet polynomial*

$$F(s, \lambda^m) = \sum_{\substack{\mathfrak{a} \in \mathcal{C}, N\mathfrak{a} \asymp F \\ (\mathfrak{a}, \mathfrak{P}(z))=1}} \lambda^m(\mathfrak{a})(N\mathfrak{a})^{-s}$$

satisfies (11.9.1). *The case* $j = 0$ *also holds (with the Hecke character removed) if* $|t| \gg \exp\left((\log x)^{1/3}\right)$.

Lemma 11.25. *Let* $x^\epsilon \leq F \leq x, 0 < |m| \leq x, |t| \leq x.$ *Then,*

$$\sum_{\substack{\mathfrak{a}\in\mathcal{C}\\N\mathfrak{a}\asymp F}} \lambda^m(\mathfrak{a})(N\mathfrak{a})^{-\frac{1}{2}-it} \ll F^{\frac{1}{2}-\eta}, \tag{11.9.11}$$

where $\eta = \eta(\epsilon) > 0.$ *In the case* $m = 0$ *this becomes*

$$\sum_{\substack{\mathfrak{a}\in\mathcal{C}\\N\mathfrak{a}\asymp F}} (N\mathfrak{a})^{-\frac{1}{2}-it} \ll F^{\frac{1}{2}-\eta} + \frac{F^{\frac{1}{2}}}{1+|t|}.$$

Proof. This is an analogue of (5.6.3) and can be deduced from [23, Theorem 6]. The case $j = 0$ needs to take into account the contribution from small values of t caused by the pole of $\zeta_K(s, \mathcal{C})$ at $s = 1$. ▢

11.10 ASYMPTOTIC FORMULAE FOR PROBLEM 2

In this section we prove asymptotic formulae of the form (11.8.8) for various choices of the coefficients $c(\mathfrak{a})$ that arise from our sieve method. We write $w = \exp\left(\log^{7/10} x\right)$ and $y_1 = x \exp\left(-3\log^{1/3} x\right)$. The first lemma will be very familiar in appearance to the reader; we give details of the proof since we now have to discuss ideals and ideal classes. We will use it as we used similar results (Lemma 10.13, for example) to provide formulae for sums corresponding to $S(\mathcal{A}, z)$.

Lemma 11.26. *Let* $\frac{19}{36} \leq \theta \leq \frac{17}{32}$ *and define*

$$c(\mathfrak{a}) = \sum_{\substack{\mathfrak{m}\mathfrak{n}\mathfrak{l}=\mathfrak{a}\\N\mathfrak{m}\asymp P, N\mathfrak{n}\asymp Q}} a(\mathfrak{m})b(\mathfrak{n})\rho(\mathfrak{l}, w).$$

Suppose that $a(\mathfrak{m})$, $b(\mathfrak{n})$ *satisfy* (11.8.3) *and that* P *and* Q *satisfy the condition*

$$\max(P, x^{1-\theta})\max(Q, x^{\frac{1}{2}(1-\theta)}) \leq x^{\frac{1}{2}(1+\theta)}.$$

Then, (11.8.17) *holds.*

Proof. For clarity, the conditions $N\mathfrak{m} \asymp P, N\mathfrak{n} \asymp Q$ will be tacitly assumed in the following. We have

$$c(\mathfrak{a}) = \sum_{\mathfrak{m},\mathfrak{n}} \sum_{\substack{\mathfrak{m}\mathfrak{n}\mathfrak{l}=\mathfrak{a}\\\mathfrak{d}|\mathfrak{l},\mathfrak{d}|\mathfrak{P}(w)}} a(\mathfrak{m})b(\mathfrak{n})\mu(\mathfrak{d}).$$

We write $z = x^\eta$ and apply the sieve identity

$$\sum_{\mathfrak{d}|\mathfrak{l},\mathfrak{d}|\mathfrak{P}(w)} \mu(\mathfrak{d}) = \sum_{\substack{\mathfrak{d}|\mathfrak{l},\mathfrak{d}|\mathfrak{P}(w)\\N\mathfrak{d}<z}} \mu(\mathfrak{d}) + O\left(\sum_{\substack{\mathfrak{b}|\mathfrak{l},\mathfrak{d}|\mathfrak{P}(w)\\z\leq N\mathfrak{d}<wz}} 1\right),$$

which is the natural generalization of Heath-Brown's fundamental lemma given in Chapter 4, and which can be proved in the same manner. Hence we derive

$$\sum_{\mathfrak{a}} c(\mathfrak{a})\Psi(\mathfrak{a}) = \sum_{\mathfrak{a}} c_1(\mathfrak{a})\Psi(\mathfrak{a}) + O\left(\sum_{\mathfrak{a}} c_2(\mathfrak{a})\Psi(\mathfrak{a})\right),$$

where

$$c_1(\mathfrak{a}) = \sum_{\mathfrak{m},\mathfrak{n}} \sum_{\substack{\mathfrak{d}|\mathfrak{P}(w) \\ \mathfrak{d}<z}} \sum_{\substack{\mathfrak{m}\mathfrak{n}\mathfrak{l}=\mathfrak{a} \\ \mathfrak{l}\equiv 0\,(\mathrm{mod}\,\mathfrak{d})}} a(\mathfrak{m})b(\mathfrak{n})\mu(\mathfrak{d}),$$

$$c_2(\mathfrak{a}) = \sum_{\mathfrak{m},\mathfrak{n}} \sum_{\substack{\mathfrak{d}|\mathfrak{P}(w) \\ z\le\mathfrak{d}<wz}} \sum_{\substack{\mathfrak{m}\mathfrak{n}\mathfrak{l}=\mathfrak{a} \\ \mathfrak{l}\equiv 0\,(\mathrm{mod}\,\mathfrak{d})}} |a(\mathfrak{m})b(\mathfrak{n})|.$$

It thus suffices to demonstrate the following three formulae:

$$\sum_{\mathfrak{a}} c_1(\mathfrak{a})\Psi(\mathfrak{a}) = \frac{\phi y}{2\pi y_1} \sum_{\mathfrak{a}\in\mathcal{B}} c_1(\mathfrak{a}) + O(\phi y\delta), \qquad (11.10.1)$$

$$\sum_{\mathfrak{a}} c_2(\mathfrak{a})\Psi(\mathfrak{a}) = \frac{\phi y}{2\pi y_1} \sum_{\mathfrak{a}\in\mathcal{B}} c_2(\mathfrak{a}) + O(\phi y\delta), \qquad (11.10.2)$$

$$\sum_{\mathfrak{a}\in\mathcal{B}} c_2(\mathfrak{a}) = O(y_1\delta). \qquad (11.10.3)$$

To prove (11.10.1), we consider the Dirichlet polynomial

$$P_1(s,\lambda^m)Q(s,\lambda^m)F_1(s,\lambda^m),$$

where

$$P_1(s,\lambda^m) = \sum_{\mathfrak{m}} \sum_{\substack{\mathfrak{d}|\mathfrak{P}(w) \\ N\mathfrak{d}\sim D}} a(\mathfrak{m})\mu(\mathfrak{d})\lambda^m(\mathfrak{m}\mathfrak{d})(N(\mathfrak{m}\mathfrak{d}))^{-s},$$

$$Q(s,\lambda^m) = \sum_{\mathfrak{n}} b(\mathfrak{n})\lambda^m(\mathfrak{n})(N\mathfrak{n})^{-s}, \quad F_1(s,\lambda^m) = \sum_{\mathfrak{l}\asymp FD^{-1}} \lambda^m(\mathfrak{l})(N\mathfrak{l})^{-s},$$

with D of the form $D = z2^{-j}$, $0 \le j \ll \log x$, and $\mathfrak{d}$, $\mathfrak{l}$, $\mathfrak{m}$, $\mathfrak{n}$ belonging to fixed ideal classes. We can divide the sum on the left-hand side of (11.10.1) into subsums with $N\mathfrak{d} \sim D$ and with $\mathfrak{d}$, $\mathfrak{l}$, $\mathfrak{m}$, $\mathfrak{n}$ in fixed ideal classes with $\mathfrak{d}\mathfrak{l}\mathfrak{m}\mathfrak{n} \in \mathcal{C}$. If $PQD \ge x^\eta$, we can apply Lemma 11.17 to the corresponding subsum. Mean-value estimates of the forms (11.8.15) and (11.8.16) are provided by Lemma 11.23. If $PQD \le x^\eta$, we use Lemmas 11.18 and 11.23 to satisfy (11.8.15). Summing the resulting asymptotic formulae over D and the appropriate ideal classes, we complete the proof of (11.10.1). We can obtain (11.10.2) in a similar manner.

To complete the proof it remains to establish (11.10.3). Using (11.8.25), we get

$$\sum_{\mathfrak{a}\in\mathcal{B}} c_2(\mathfrak{a}) \ll y_1 \sum_{\mathfrak{m},\mathfrak{n}} \frac{|a(\mathfrak{m})b(\mathfrak{n})|}{N(\mathfrak{m}\mathfrak{n})} \sum_{\substack{\mathfrak{d}|\mathfrak{P}(w) \\ N\mathfrak{d}>z}} \frac{1}{N\mathfrak{d}}. \qquad (11.10.4)$$

Working as in Lemma 4.2 we obtain

$$\sum_{\substack{\mathfrak{d}|\mathfrak{P}(w) \\ N\mathfrak{d}\ge z}} \frac{1}{N\mathfrak{d}} \ll \exp\left(-\tfrac{1}{10}\eta(\log x)^{\frac{3}{10}}\log\log x\right),$$

and so (11.10.3) follows from (11.10.4) and the properties of the coefficients $a(\mathfrak{m})$, $b(\mathfrak{n})$. $\qquad\square$

The next step, as previously, is to use this lemma to obtain a similar result in which w is replaced by a larger quantity. The reader will not be at all surprised to see the zigzag nature of the function $\nu(\alpha)$ below. Indeed this function and the related quantity α^* are identical to their definitions in Chapter 7. We therefore omit the proof; it is given in full in [68, Lemma 18]. The reader should by now be convinced that all details from the rational case carry over analogously; the only weakness present here comes from the lack of Watt's mean-value result to boost the Type I/II estimates.

Lemma 11.27. *Let $\frac{19}{36} \leq \theta \leq \frac{17}{32}$ and define*

$$c(\mathfrak{a}) = \sum_{\substack{\mathfrak{m}\mathfrak{n}\mathfrak{l}=\mathfrak{a} \\ N\mathfrak{m}\sim P, N\mathfrak{n}\sim Q}} a(\mathfrak{m})b(\mathfrak{n})\rho(\mathfrak{l},z),$$

where $a(\mathfrak{m})$, $b(\mathfrak{n})$ satisfy (11.8.3). Let $2P = x^\alpha$, $Q = x^\beta$ and let h be the positive integer satisfying

$$\tfrac{1}{2} - h(2\theta - 1) \leq \alpha < \tfrac{1}{2} - (h-1)(2\theta - 1). \tag{11.10.5}$$

Define

$$\alpha^* = \max\left(\frac{2h(1-\theta) - \alpha}{2h-1}, \frac{2(h-1)\theta + \alpha}{2h-1}\right)$$

and suppose that

$$0 \leq \alpha \leq \tfrac{1}{2}, \quad 0 \leq \beta \leq \frac{1+\theta}{2} - \alpha^* = \beta^*(\alpha), \quad say.$$

Then (11.8.17) holds whenever $z \leq x^{\nu(\alpha)}$, where

$$\nu(\alpha) = \min\left(\frac{2}{2h-1}(\theta - \alpha), \gamma(\theta)\right) \tag{11.10.6}$$

and $\gamma(\theta)$ is the function appearing on the right side of (11.9.5).

11.11 THE FINAL DECOMPOSITION FOR PROBLEM 2

The reader might try to write this section themselves as an exercise and then compare our presentation. We are simply working as in Chapter 7 but are slightly hampered by the lack of Type I/II information, making two further Buchstab decompositions unattainable at present for certain sums. Several of the arguments below are identical to those in Section 7.9 with prime ideals replacing rational primes. Write

$$z(\mathfrak{n}) = x^{\nu(\alpha)} \quad \text{where } 2^j \leq N\mathfrak{n} < 2^{j+1} = x^\alpha.$$

Also, put $\nu_0 = 2\theta - 1$, $z_0 = x^{\nu_0}$. By a standard opening move we have

$$S(\mathcal{A}, x^{1/2}) = S(\mathcal{A}, z_0) - \sum_{z_0 \leq \mathfrak{p} < x^{1/2}} S(\mathcal{A}_\mathfrak{p}, z(\mathfrak{p})) + \sum_{z(\mathfrak{p}_1) \leq \mathfrak{p}_2 < \mathfrak{p}_1 < x^{1/2}} S(\mathcal{A}_{\mathfrak{p}_1\mathfrak{p}_2}, \mathfrak{p}_2)$$

$$= S_1 - S_2 + S_3, \quad \text{say.}$$

By Lemma 11.27 we can give an asymptotic formula for S_1 and S_2. We can also give an asymptotic formula for those parts of S_3 with $N\mathfrak{p}_2 < z(\mathfrak{p}_1\mathfrak{p}_2)$. To consider the remainder of S_3 we work in our usual fashion. We start by writing $N\mathfrak{p}_j = x^{\alpha_j}$. Then we may convert sums over prime ideals to integrals in our usual manner (with the prime ideal theorem replacing the PNT). For example,

$$\sum_{\substack{z_0 \leq \mathfrak{p}_2 < \mathfrak{p}_1 < x^{1/2} \\ \mathfrak{p}_1 \mathfrak{p}_2^2 < x}} S(\mathcal{B}_{\mathfrak{p}_1 \mathfrak{p}_2}, \mathfrak{p}_2) = S(\mathcal{B}, x^{\frac{1}{2}})(I + o(1)),$$

where

$$I = \int_{\nu_0}^{1/2} \int_{\nu_0}^{\min(\alpha_1, (1-\alpha_1)/2)} \omega\left(\frac{1 - \alpha_1 - \alpha_2}{\alpha_2}\right) \frac{d\alpha_2}{\alpha_1 \alpha_2^2}$$

and $\omega(t)$ once again is Buchstab's function.

The expressions that will usually arise in the following have the form

$$\sum_{m_j \sim x^{\alpha_j}} S(\mathcal{A}_{m_1 \dots m_n}, m_n), \tag{11.11.1}$$

where all the prime factors of each m_j exceed z_0 and there may be certain interdependencies between the variables. We write (G standing for *good*, as we have written before)

$$G_n = \{(\alpha_1, \dots, \alpha_n) \in \mathbb{R}^n : \text{an asymptotic formula can be obtained for (11.11.1)}\}.$$

Also, we let

$$D = \{(\alpha, \beta) : \nu_0 \leq \alpha \leq \tfrac{1}{2}, \nu_0 \leq \beta \leq \beta^*(\alpha)\}.$$

The significance of D (for *decomposable*, just as in Chapter 7) is that if we have a sum of the form (11.11.1) and we can partition $\alpha_1, \dots, \alpha_n$ into two sets whose sums are α, β with $(\alpha, \beta) \in D$, then we can apply Buchstab's identity followed by Lemma 11.27. Usually we will need to do this twice, so we want to be able to partition $\alpha_1, \dots, \alpha_n, \gamma$ into two sets whose sums are $(\alpha, \beta) \in D$ for all $\gamma \in [\nu_0, \alpha_n]$. Henceforth we may write, as in previous chapters, $\boldsymbol{\alpha}_n = (\alpha_1, \dots, \alpha_n)$.

We treat S_3 by considering α_2 belonging to the following disjoint sets whose union is $\{\alpha_2 : \nu_0 \leq \alpha_1 \leq \frac{1}{2}, \nu(\alpha_1) \leq \alpha_2 \leq \alpha_1\}$:

$T_1 = \{\alpha_2 : \frac{1}{3} \leq \alpha_1 \leq \frac{1}{2}, (1 - \alpha_1)/2 \leq \alpha_2 \leq \alpha_1\}$;

$T_2 = \{\alpha_2 : \theta/2 \leq \alpha_1 \leq 1 - \theta, \theta/2 \leq \alpha_2 \leq \min(\alpha_1, (1 - \alpha_1)/2)\}$;

$A_1 = \{\alpha_2 : \nu_0 \leq \alpha_1 \leq \frac{1}{4}, \nu(\alpha_1) \leq \alpha_2 \leq \alpha_1\}$;

$A_2 = \{\alpha_2 : \frac{1}{4} < \alpha_1 \leq \frac{2}{5}, \nu(\alpha_1) \leq \alpha_2 \leq (1 - \alpha_1)/3\}$;

$A_3 = \{\alpha_2 : \frac{2}{5} < \alpha_1 \leq \frac{1}{2}, \nu(\alpha_1) \leq \alpha_2 \leq \alpha_1/2\}$;

$B_1 = \{\alpha_2 : \frac{1}{4} \leq \alpha_1 \leq \frac{2}{5}, (1 - \alpha_1)/3 < \alpha_2 \leq \min(\theta/2, \alpha_1, 1 - 2\alpha_1)\}$;

$B_2 = \{\alpha_2 : \frac{1}{4}(2 - \theta) \leq \alpha_1 \leq \frac{1}{2}, \max(2 - 4\alpha_1, \alpha_1) < \alpha_2 \leq \min(\theta, 1 - \alpha_1)\}$.

Region T_1 corresponds to products $\mathfrak{p}_1\mathfrak{p}_2$ with $\mathfrak{p}_2 < \mathfrak{p}_1 < x^{1/2}$. As the trivial estimate for the number of such products in $\mathcal{B}$ is $O(y_1(\log x)^{-2})$, discarding T_1

will not affect the final bound. Although it is possible to apply Buchstab's identity twice more to a large part of region T_2 (using a two-dimensional sieve as in Chapter 7), we lose practically nothing in discarding the whole set. Indeed, calculations show that the four-dimensional loss one would incur exceeds the two-dimensional loss because many of the new sums we generate cannot be given an asymptotic formula. We therefore discard a sum of size

$$\frac{\phi y}{2\pi} K S(\mathcal{B}, x^{\frac{1}{2}})(1 + o(1)),$$

where

$$K = \int_{\alpha_2 \in T_2} \frac{d\alpha_2}{\alpha_1 \alpha_2 (1 - \alpha_1 - \alpha_2)} < 0.2 \text{ at } \theta = 0.53. \qquad (11.11.2)$$

The reader might like to draw a diagram of the above regions (with T_1 removed) and compare and contrast it with Diagram 7.3.

In like manner to earlier work we note that both the regions B_1 and B_2 correspond to products of three prime ideals and $\alpha_2 \in B_1 \Leftrightarrow (1 - \alpha_1 - \alpha_2, \alpha_2) \in B_2$. It therefore suffices to consider only one of these regions during the calculations.

We now show that it is always possible to apply Buchstab's identity twice more to a sum corresponding to a subset R of A_j or B_k:

$$\sum_{\alpha_2 \in R} S(\mathcal{A}_{\mathfrak{p}_1 \mathfrak{p}_2}, \mathfrak{p}_2). \qquad (11.11.3)$$

Since $\alpha_2 \leq \theta/2$ and $\alpha_1 \leq \frac{1}{2}$, we can decompose (11.11.3) as

$$\sum_{\alpha_2 \in R} S(\mathcal{A}_{\mathfrak{p}_1 \mathfrak{p}_2}, z(\alpha_2)) - \sum_{\substack{\alpha_2 \in R \\ \nu(\alpha_2) \leq \alpha_3 \leq \alpha_2}} S(\mathcal{A}_{\mathfrak{p}_1 \mathfrak{p}_2 \mathfrak{p}_3}, \mathfrak{p}_3). \qquad (11.11.4)$$

Here

$$\nu(\alpha_2) = \max(\nu(\alpha_1), \nu(\alpha_1 + \alpha_2)), \quad z(\alpha_2) = x^{\nu(\alpha_2)},$$

with the convention that $\nu(\gamma) = 0$ for $\gamma > \frac{1}{2}$. Now Lemma 11.27 gives an asymptotic formula for the first sum in (11.11.4). To consider the second sum we split into two cases. Write

$$F_1 = \{\alpha_3 : \alpha_1 + \alpha_3 \leq \tfrac{1}{2} \text{ or } \alpha_2 + \alpha_3 \leq \beta^*(\alpha(1))\},$$
$$F_2 = \{\alpha_3 : \alpha_1 + \alpha_3 > \tfrac{1}{2} \text{ and } \alpha_2 + \alpha_3 > \beta^*(\alpha(1))\}.$$

For the sum over F_1 we can apply Buchstab's identity once more in a straightforward fashion. However, for the sum over F_2 we reverse the roles of two of the ideals. We have already seen this technique for rational primes, and its use here is essentially identical. We note that $S(\mathcal{A}_{\mathfrak{p}_1 \mathfrak{p}_2 \mathfrak{p}_3}, \mathfrak{p}_3)$ counts products of ideals $\mathfrak{p}_1 \mathfrak{p}_2 \mathfrak{p}_3 m$ with $\mathfrak{p}|m \Rightarrow \mathfrak{p} > \mathfrak{p}_3$. We rewrite this as

$$S(\mathcal{A}_{\mathfrak{p}_2 \mathfrak{p}_3 m}, (x/N(\mathfrak{p}_2 \mathfrak{p}_3 m))^{\frac{1}{2}}).$$

The advantage we have gained is that $N(\mathfrak{p}_2 m) < x^{1/2}$ and $N\mathfrak{p}_3 \leq x^{\theta/2}$. Thus it is now possible to apply Buchstab's identity and use Lemma 11.27. Hence

$$
\sum_{\substack{\alpha_2 \in R \\ \nu(\alpha_2) \le \alpha_3 \le \alpha_2}} S(\mathcal{A}_{\mathfrak{p}_1 \mathfrak{p}_2 \mathfrak{p}_3}, \mathfrak{p}_3) \tag{11.11.5}
$$

$$
= \sum_{\alpha_3 \in F_1} S(\mathcal{A}_{\mathfrak{p}_1 \mathfrak{p}_2 \mathfrak{p}_3}, \mathfrak{p}_3) + \sum_{\alpha_3 \in F_2} S(\mathcal{A}_{\mathfrak{m} \mathfrak{p}_2 \mathfrak{p}_3}, (N\mathfrak{p}_2 \mathfrak{p}_3 \mathfrak{m})^{\frac{1}{2}})
$$

$$
= \sum_{\alpha_3 \in F_1} S(\mathcal{A}_{\mathfrak{p}_1 \mathfrak{p}_2 \mathfrak{p}_3}, z_0) + \sum_{\alpha_3 \in F_2} S(\mathcal{A}_{\mathfrak{m} \mathfrak{p}_2 \mathfrak{p}_3}, z_0)
$$

$$
- \sum_{\substack{\alpha_3 \in F_1 \\ \nu_0 \le \alpha_4 \le \alpha_3}} S(\mathcal{A}_{\mathfrak{p}_1 \mathfrak{p}_2 \mathfrak{p}_3 \mathfrak{p}_4}, \mathfrak{p}_4) - \sum_{\substack{\alpha_3 \in F_2 \\ \nu_0 \le \alpha_4 \le \alpha_1/2}} S(\mathcal{A}_{\mathfrak{m} \mathfrak{p}_2 \mathfrak{p}_3 \mathfrak{p}_4}, \mathfrak{p}_4).
$$

In the above $N\mathfrak{m} \sim x^{1 - \alpha_1 - \alpha_2 - \alpha_3}$. Lemma 11.27 supplies asymptotic formulae for the first two sums on the right-hand side of (11.11.5), and possibly for some part of the final two sums we have $\alpha_4 \in G_4$. As before, we need to allow the possibility that the losses from the final sums in (11.11.5) may be larger than the losses from the original two-dimensional sum. Hence the corresponding loss is

$$
\int_R \min\left(\omega\left(\frac{1 - \alpha_1 - \alpha_2}{\alpha_2} \right) \frac{1}{\alpha_2}, I_1(\alpha_2) + I_2(\alpha_2) \right) \frac{d\alpha_2}{\alpha_1 \alpha_2}, \tag{11.11.6}
$$

where

$$
I_1(\alpha_2) = \int_{R_1} \omega\left(\frac{1 - \alpha_1 - \alpha_2 - \alpha_3 - \alpha_4}{\alpha_4} \right) \frac{d\alpha_4 \, d\alpha_3}{\alpha_3 \alpha_4^2},
$$

$$
I_2(\alpha_2) = \alpha_1 \int_{R_2} \omega\left(\frac{\alpha_1 - \alpha_4}{\alpha_4} \right) \omega\left(\frac{1 - \alpha_1 - \alpha_2 - \alpha_3}{\alpha_3} \right) \frac{d\alpha_4 \, d\alpha_3}{\alpha_3^2 \alpha_4^2},
$$

with

$$
R_1 = \{ \alpha_4 \notin G_4 : \alpha_3 \in F_1, \nu_0 \le \alpha_4 \le \alpha_3 \},
$$

$$
R_2 = \{ \alpha_4 \notin G_4 : \alpha_3 \in F_2, \nu_0 \le \alpha_4 \le \alpha_1/2 \}.
$$

We can now use exact analogues of ideas used in previous chapters to reduce the size of the integrals that have to be discarded. The reader has already seen these devices, so we can be brief with the details.

1) *Further decompositions.* So long as $(\alpha_1, \ldots, \alpha_4, \gamma) \in D$ for all $\gamma \in [\nu_0, \alpha_4]$ we can perform two more decompositions to arrive at a six-dimensional integral. Indeed, we can perform even more Buchstab decompositions for parts of the resulting sums. The contribution from eight or more decompositions is extremely small just as we saw in Sections 3.5 and 7.4 since $(8!\nu_0)^{-1} \approx 4.13 \times 10^{-4}$ for $\theta = 0.53$.

2) *Making visible the almost-primes.* Just as in earlier chapters, we can make explicit the almost-primes counted by the sifting function $S(\mathcal{A}_{\mathfrak{p}_1\ldots\mathfrak{p}_n}, \mathfrak{p}_n)$ using Buchstab's identity in reverse:

$$S(\mathcal{A}_{\mathfrak{p}_1\ldots\mathfrak{p}_n}, \mathfrak{p}_n) = S(\mathcal{A}_{\mathfrak{p}_1\ldots\mathfrak{p}_n}, (x/N\mathfrak{p}_1\ldots\mathfrak{p}_n)^{\frac{1}{2}})$$

$$+ \sum_{\mathfrak{p}_n < \mathfrak{p}_{n+1} < (x/N\mathfrak{p}_1\ldots\mathfrak{p}_n)^{1/2}} S(\mathcal{A}_{\mathfrak{p}_1\ldots\mathfrak{p}_{n+1}}, \mathfrak{p}_{n+1}). \qquad (11.11.7)$$

As in previous applications of the method, this process may be further iterated.

Numerical calculations give the results shown in Table 11.1 for $\theta = 0.53$ and for $\theta = 0.528$ (this value is discussed in detail in [119]).

Region	Loss at $\theta = 0.53$	Loss at $\theta = 0.528$
A_1	0.063	0.116
A_2	0.183	0.167
A_3	0.25	0.296
B_1	0.12	0.107
B_2	0.12	0.107

Table 11.1 Upper bounds for losses

Recalling (11.11.2) this gives an overall loss no more than 0.936 for $\theta = 0.53$. (note that this rises to 0.994 at $\theta = 0.528$). As usual, the loss is a continuous function of θ, and so we obtain

$$\sum_{\mathfrak{p} \in \mathcal{A}} 1 \geq \frac{\phi y}{20 \log x}$$

for all $\theta \geq 0.53 - 2\epsilon$ and for all $x > x_0(\epsilon)$, which completes the proof. $\square$

Chapter Twelve

Variations on Gaussian Primes

12.1 INTRODUCTION

In the previous chapter we considered the distribution of Gaussian primes in terms of sectors or polar boxes. In this chapter we consider what happens if we study Gaussian primes of the form $p = u + iv$ where u is unrestricted but v is constrained to a given set. The material we present here is based on the work of Fouvry and Iwaniec [36] who considered the case of v belonging to a fairly dense sequence, in particular treating the case that v is a rational prime, and on the work of Friedlander and Iwaniec [38] who considered the case that v is a square. As consequences of these works, we know that there are infinitely many rational primes $p = u^2 + v^2$ with v prime, and infinitely many rational primes $p = u^2 + v^4$. The two main theorems we prove here are as follows.

Theorem 12.1. (Fouvry-Iwaniec) *Let λ_ℓ be complex numbers with $|\lambda_\ell| \leq 1$. Then, for all $A > 1$,*

$$\sum_{\ell^2 + m^2 \leq x} \lambda_\ell \Lambda(\ell^2 + m^2) = \sum_{\ell^2 + m^2 \leq x} \lambda_\ell \psi(\ell) + O\left(x(\log x)^{-A}\right), \qquad (12.1.1)$$

where

$$\psi(\ell) = \prod_{p \nmid \ell} \left(1 - \frac{\chi(p)}{p-1}\right) \qquad (12.1.2)$$

and χ is the non-principal character $(\mathrm{mod}\,4)$. The constant implied in the error term depends only on A.

Remark. Write

$$\theta(\ell) = \prod_{p \mid \ell} \left(1 - \frac{\chi(p)}{p-1}\right)^{-1}.$$

Then we note that

$$\psi(\ell) = \theta(\ell)\psi(1)$$

and

$$\psi(1) = L(1,\chi)^{-1}C = \frac{4C}{\pi}$$

with

$$C = \prod_p \left(1 - \frac{\chi(p)}{(p-1)(p-\chi(p))}\right). \qquad (12.1.3)$$

Theorem 12.2. (Friedlander-Iwaniec) *We have*

$$\sum_{\substack{a,b\in\mathbb{N} \\ a^2+b^4\leq x}} \Lambda(a^2+b^4) = \frac{4\kappa}{\pi} x^{\frac{3}{4}}\left(1+O\left(\frac{\log\log x}{\log x}\right)\right), \qquad (12.1.4)$$

where

$$\kappa = \int_0^1 (1-t^4)^{\frac{1}{2}}\, dt = \frac{\Gamma(\frac{1}{4})^2}{6\sqrt{2\pi}}. \qquad (12.1.5)$$

The most striking deduction from Theorem 12.1 comes by taking

$$\lambda_\ell = \frac{\Lambda(\ell)}{\log x}.$$

In this case $\theta(\ell) = 1+O(\ell^{-1})$ for those ℓ that make a significant contribution to the right-hand side of (12.1.1), that is, prime values of ℓ. Hence, with the elementary integral

$$\int_{-\sqrt{x}}^{\sqrt{x}} \sqrt{x-u^2}\, du = \frac{\pi x}{2},$$

we obtain the following.

Corollary. *We have, with C given by (12.1.3),*

$$\sum_{\ell^2+m^2\leq x} \Lambda(\ell)\Lambda(\ell^2+m^2) = 2Cx + O\left(x(\log x)^{-A}\right). \qquad (12.1.6)$$

12.2 OUTLINE OF THE FOUVRY-IWANIEC METHOD

In contrast to the previous chapter, our sieving will take place in $\mathbb{Z}$ and not in $\mathbb{Z}[i]$. We shall use the idea mentioned at the end of Chapter 2, namely, that we can obtain the main term from the Type I information available, but use Type II information to detect cancellations in sums involving the Möbius function. We need to work a little more carefully than in the outline presented there. If we go back to our original derivation of Vaughan's identity (2.1.2) with $a_4(n)$ given by (2.1.3) we obtain, for $n > U$,

$$\Lambda(n) = \sum_{\substack{hd=n \\ d\leq V}} \mu(d)\log h - \sum_{\substack{dm|n \\ d\leq V, m\leq U}} \mu(d)\Lambda(m) + \sum_{\substack{dm|n \\ d>V, m>U}} \mu(d)\Lambda(m). \quad (12.2.1)$$

Now, for a given sequence of complex numbers λ_ℓ, $1 \leq \ell \leq x^{1/2}$, write

$$a_n = \sum_{\ell^2+m^2=n} \lambda_\ell, \qquad P(x) = \sum_{n\leq x} a_n\Lambda(n),$$

$$A_d(x) = \sum_{\substack{n\leq x \\ n\equiv 0\,(\mathrm{mod}\,d)}} a_n, \qquad M_d(x) = \frac{1}{d}\sum_{n\leq x} a_n(d),$$

with

$$a_n(d) = \sum_{\ell^2+m^2=n} \lambda_\ell \rho_\ell(d),$$

and $\rho_\ell(d)$ denotes the number of solutions to $\nu^2 + \ell^2 \equiv 0 \pmod d$. Here we expect $M_d(x)$ to be a good approximation, at least on average, to $A_d(x)$. Hence we define remainder terms

$$R_d(x) = A_d(x) - M_d(x), \quad R(x, D) = \sum_{d \le D} |R_d(x)|.$$

Also, we put $a'_n = a_n \log n$ and define A'_d as A_d with a_n replaced by a'_n. Then we have

$$P(x) = A(x; V, U) + B(x; V, U) + P(U), \qquad (12.2.2)$$

where

$$A(x; V, U) = \sum_{d \le V} \mu(d) \left(A'_d(x) - A_d(x) \log d - \sum_{m \le U} \Lambda(m) A_{md}(x) \right) \qquad (12.2.3)$$

and

$$B(x; V, U) = \sum_{\substack{dh \le x \\ d > V}} \mu(d) a_{dh} \left(\sum_{m|h, m>U} \Lambda(m) \right). \qquad (12.2.4)$$

We thus have

$$|B(x; V, U)| \le \sum_{U < h < x/V} (\log h) \left| \sum_{V < d \le x/h} \mu(d) a_{dh} \right|. \qquad (12.2.5)$$

The reader should compare the above with the end of Chapter 2; $B(x; V, U)$ corresponds to R there with $U = N/V$. We have gone back to the full form of Vaughan's identity because the choice $U = N/V$ would cause the method to break down when we try to estimate the Type II sums.

By partial summation

$$A'_d(x) = (\log x) A_d(x) - \int_1^x A_d(t) \frac{dt}{t}.$$

Hence we can replace $A_d(x)$ by $M_d(x) + R_d(x)$ in (12.2.3) to obtain

$$A(x; V, U) = M(x; V, U) + R(x; V, U) \qquad (12.2.6)$$

with

$$M(x; V, U) = \sum_{n \le x} \sum_{d \le x} \frac{\mu(d)}{d} \left(a_n(d) \log(n/d) - \sum_{m \le U} \frac{\Lambda(m)}{m} a_n(dm) \right) \qquad (12.2.7)$$

and

$$R(x; V, U)$$

$$= \sum_{d \le V} \mu(d) \left(R_d(x) \log(x/d) - \int_1^x R_d(t) \frac{dt}{t} - \sum_{m \le U} \lambda(m) R_{md}(x) \right).$$

We then have (an expression we shall estimate in the next section)

$$|R(x; V, U)| \leq R(x, UV) \log x + \int_1^x R(t, V) \frac{dt}{t}. \qquad (12.2.8)$$

Now write

$$\lambda_\ell(n) = \begin{cases} \lambda_\ell & \text{if } \exists\, m \text{ with } \ell^2 + m^2 = n, \\ 0 & \text{otherwise.} \end{cases}$$

Also, put

$$\sigma_\ell(n; V, U) = \sum_{d \leq V} \frac{\mu(d)}{d} \left(\rho_\ell(d) \log(n/d) - \sum_{m \leq U} \frac{\Lambda(m)}{m} \rho_\ell(dm) \right).$$

Then (12.2.7) becomes

$$M(x; V, U) = \sum_{n \leq x} \sum_\ell \lambda_\ell(n) \sigma_\ell(n; V, U). \qquad (12.2.9)$$

We show later how this, apart from a smaller order error, becomes the main term

$$\sum_{\ell^2 + m^2 \leq x} \lambda_\ell \psi(\ell).$$

It now remains to obtain Type I information to estimate (12.2.8) and use the cancellation introduced by the $\mu(d)$ factor to obtain a satisfactory bound for (12.2.5). For the former of these we shall need to consider the distribution of the roots of $\nu^2 + 1 \equiv 1 \pmod{d}$. The latter problem involves the use of bilinear forms in Gaussian integers (see Section 12.5).

12.3 SOME PRELIMINARY RESULTS

The Poisson summation formula is an important tool in analytic number theory and is used to prove many of the auxiliary results we have quoted in previous chapters. We state it in its own right now. Given a function $f \in L^1(\mathbb{R})$, we denote its Fourier transform by $\widehat{f}$. That is,

$$\widehat{f}(t) = \int_{-\infty}^\infty f(x) e(-xt) \, dt.$$

Lemma 12.1. (Poisson Summation Formula) *Suppose that both f and $\widehat{f}$ belong to $L^1(\mathbb{R})$ and have bounded variation. Then*

$$\sum_{m \in \mathbb{Z}} f(m) = \sum_{n \in \mathbb{Z}} \widehat{f}(n). \qquad (12.3.1)$$

In particular, if $d > 0$ and $u \in \mathbb{R}$, then

$$\sum_{m \in \mathbb{Z}} f(md + u) = \sum_{k \in \mathbb{Z}} \widehat{f}\left(\frac{k}{d}\right) e\left(\frac{ku}{d}\right). \qquad (12.3.2)$$

Proof. See [106, pp. 69 and 70]. □

The following is a form of the large-sieve inequality proved in [36, p. 252]. The proof depends on the fact that the fractions ν/d arising from solutions to $\nu^2 + 1 \equiv 0 \,(\mathrm{mod}\, d)$ are well-spaced modulo 1. The result follows from Lemma A.4 in the appendix once one establishes that, for $8J < d \le 9J$, the distinct roots of $\nu^2 + 1 \equiv 0 \,(\mathrm{mod}\, d)$ satisfy

$$\left\| \frac{\nu_1}{d_1} - \frac{\nu_2}{d_2} \right\| > \frac{1}{36J}.$$

We leave that as an exercise for the reader. We write $\mathcal{S}(d)$ for a representative set of such ν for each d.

Lemma 12.2. *For any complex numbers α_n we have*

$$\sum_{d \sim D} \sum_{\nu \in \mathcal{S}(d)} \left| \sum_{n \le N} \alpha_n e\left(\frac{\nu n}{d}\right) \right|^2 \ll (D + N) \sum_{n \le N} |\alpha_n|^2. \qquad (12.3.3)$$

Now write

$$\rho_{k,\ell}(d) = \sum_{\nu^2 + \ell^2 \equiv 0 \,(\mathrm{mod}\, d)} e\left(\frac{\nu k}{d}\right) \qquad (12.3.4)$$

and put $\rho(d) = \rho_{0,1} = |\mathcal{S}(d)|$, that is, the number of solutions to $\nu^2 + 1 \equiv 0 \,(\mathrm{mod}\, d)$. For a given sequence of complex numbers $\alpha_{k,\ell}$ we write

$$A^2 = \sum_{k,\ell} |\alpha_{k,\ell}|^2 \quad \text{and} \quad \tilde{A}^2 = \sum_{k,\ell} \tau(k\ell) |\alpha_{k,\ell}|^2.$$

Lemma 12.3. *Let $D, K, L \ge 1$ be given. For any complex numbers $\alpha_{k,\ell}$ we have*

$$\sum_{d \le D} \left| \sum_{k \le K} \sum_{\ell \le L} \alpha_{k,\ell} \rho_{k,\ell}(d) \right| \ll (\log 3D)^3 D^{\frac{1}{2}} (D + KL)^{\frac{1}{2}} \tilde{A}. \qquad (12.3.5)$$

Proof. First we note that $\rho_{k,\ell}(d) = \rho_{k\ell,1}(d)$ when $(d, \ell) = 1$. Otherwise, we put $(d, \ell^2) = ab^2$ where a is square-free, so that $d = ab^2 d_1$, $\ell = ab\ell_1$ with $(d_1, a\ell_1) = 1$. If we substitute $\nu = ab\nu_1$ into $\nu^2 + \ell^2 \equiv 0 \,(\mathrm{mod}\, d)$ and divide throughout by $a^2 b^2$ we obtain $\nu_1^2 + \ell_1^2 \equiv 0 \,(\mathrm{mod}\, d_1)$. Thus we obtain

$$\rho_{k,\ell}(d) = \sum_{\substack{\nu_1 \,(\mathrm{mod}\, bd_1) \\ \nu_1^2 + \ell_1^2 \equiv 0 \,(\mathrm{mod}\, d_1)}} e\left(\frac{\nu_1 k}{bd_1}\right).$$

If $b \nmid k$, then the above sum is zero since each root ν of $\nu^2 + \ell_1^2 \equiv 0 \,(\mathrm{mod}\, d_1)$ leads to terms

$$\sum_{g=1}^{b} e\left(\frac{(\nu + gd_1)k}{bd_1}\right) = 0.$$

We may thus suppose that $k = bk_1$ and so obtain

$$\rho_{k,\ell}(d) = b \sum_{\substack{\nu_1 \,(\mathrm{mod}\, d_1) \\ \nu_1^2 + \ell_1^2 \equiv 0 \,(\mathrm{mod}\, d_1)}} e\left(\frac{\nu_1 k_1}{d_1}\right) = b\rho_{k_1,\ell_1,1}(d_1).$$

Here we have changed ν_1 to $\nu_1\ell_1 \,(\mathrm{mod}\, d_1)$ and divided the resulting congruence by ℓ_1^2.

We have thus shown that the left-hand side of (12.3.5) can be bounded above by

$$\sum_{ab^2 d \le D} \sum_{\nu \in \mathcal{S}(d)} \left| \sum_{k \le K/b} \sum_{\substack{\ell \le L/(ab) \\ (\ell, d) = 1}} \alpha_{bk, ab\ell} e\left(\frac{\nu k \ell}{d}\right) \right|. \qquad (12.3.6)$$

Since

$$\sum_{ab^2 \le D} b(ab^2)^{-1} < (\log 3D)^2,$$

it now suffices to establish that

$$\sum_{d \le D} \sum_{\nu \in \mathcal{S}(d)} \left| \sum_{k \le K} \sum_{\substack{\ell \le L \\ (\ell, d) = 1}} \alpha_{k, \ell} e\left(\frac{\nu k \ell}{d}\right) \right| \ll (\log 3D) D^{\frac{1}{2}} (D + KL)^{\frac{1}{2}} \tilde{A}. \quad (12.3.7)$$

From (1.3.1) we have, for any sequence q_ℓ,

$$\sum_{(\ell, d) = 1} q_\ell = \sum_{b|d} \mu(b) \sum_{b|\ell} q_{\ell b}.$$

Thus the left-hand side of (12.3.7) can be bounded by

$$\sum_{b \le D} \sum_{d \le D/b} \sum_{\nu \in \mathcal{S}(db)} \left| \sum_{k \le K} \sum_{\ell \le L/b} \alpha_{k, \ell} e\left(\frac{\nu k \ell}{d}\right) \right|$$

$$\le \sum_{b \le D} \rho(b) \sum_{d \le D/b} \sum_{\nu \in \mathcal{S}(d)} \left| \sum_{k \le K} \sum_{\ell \le L/b} \alpha_{k, \ell} e\left(\frac{\nu k \ell}{d}\right) \right|$$

$$\ll \left(\sum_{b \le D} \frac{\rho(b)}{b} \right) D^{\frac{1}{2}} (D + KL)^{\frac{1}{2}} \tilde{A},$$

by (12.3.3). Writing $\chi(n)$ for the nontrivial character $(\mathrm{mod}\, 4)$, we have

$$\sum_{b \le D} \frac{\rho(b)}{b} \le \sum_{d \le D} \frac{1}{d} \sum_{c \le D/d} \frac{\chi(c)}{c} < \sum_{d \le D} \frac{1}{d} < \log 3D.$$

This completes the proof. $\qquad \qquad \qquad \qquad \qquad \qquad \qquad \square$

Our final preliminary result, which we will need to deal with (12.2.5), is a mean value theorem for Gaussian integers in an arithmetic progression. It will enable us to pass from "long" arithmetic progressions to short ones for which a Siegel-Walfisz result is known to hold.

Lemma 12.4. *Let $A \geq D \geq 1$ and suppose that f is a function defined on $\mathbb{Z}[i]$ and supported on the disc $|z| \leq A$. Write, for any $E \geq 1$,*

$$\Sigma(E) = \sum_{d \leq E} d^2 \sum_{a \,(\mathrm{mod}\, d)} \left| \sum_{z \equiv a \,(\mathrm{mod}\, d)} f(z) \right|^2. \tag{12.3.8}$$

Here a runs over the d^2 residue classes $(\mathrm{mod}\, d)$ in $\mathbb{Z}[i]$. Let $\epsilon > 0$ be given. Then, for any $G \geq 1$ we have

$$\Sigma(D) \leq 2D\Sigma(G) + O\big(AD(D^{1+\epsilon} + AG^{1-\epsilon})\|f\|^2\big), \tag{12.3.9}$$

where

$$\|f\|^2 = \sum_z |f(z)|^2.$$

Proof. Write

$$\Sigma(d, a) = \sum_{z \equiv a \,(\mathrm{mod}\, d)} f(z)$$

and put

$$Q^2(d) = d^2 \sum_{a \,(\mathrm{mod}\, d)} |\Sigma(d, a)|^2.$$

For Gaussian integers s, z we write $s \cdot z = \mathrm{Re}\, s\bar{z}$. Applying additive characters modulo d in a standard way then yields

$$Q^2(d) = \sum_{s \,(\mathrm{mod}\, d)} \left| \sum_z f(z) e\left(\frac{s \cdot z}{d}\right) \right|^2.$$

Now put $s = g + hi$, $z = m + ni$ and write the fractions $a/d, b/d$ in their lowest terms to obtain

$$Q^2(d) = \sum_{\substack{u|d \ g \,(\mathrm{mod}\, u) \\ v|d \ h \,(\mathrm{mod}\, v)}} {}^{*} \left| f(z) e\left(\frac{gm}{u} + \frac{hn}{v}\right) \right|^2,$$

where $*$ indicates summation over primitive residue classes only. Thus

$$\Sigma(D) \leq D \sum_{[u,v] \leq D} [u, v]^{-1} \sum_{\substack{g \,(\mathrm{mod}\, u) \\ h \,(\mathrm{mod}\, v)}} {}^{*} \left| f(z) e\left(\frac{gm}{u} + \frac{hn}{v}\right) \right|^2, \tag{12.3.10}$$

with $[u, v]$ denoting the least common multiple of u and v.

Now let $t = [u, v]$. We split the argument according to the size of t. When t is small, we can write $g/u, h/v$ over a common denominator and let the outer summation run over all residue classes $(\mathrm{mod}\, t)$. This gives

$$\sum_{\substack{g \,(\mathrm{mod}\, u) \\ h \,(\mathrm{mod}\, v)}} {}^{*} \left| f(z) e\left(\frac{gm}{u} + \frac{hn}{v}\right) \right|^2 \leq t^2 \sum_{z_1 \equiv z_2 \,(\mathrm{mod}\, t)} f(z_1)\overline{f}(z_2) = Q^2(t).$$

Hence the terms with $t \leq G$ contribute to $\Sigma(D)$ at most

$$D \sum_{t \leq G} \frac{1}{t} \left(\sum_{[u,v]=t} 1 \right) Q^2(t) \leq 2D\Sigma(G). \tag{12.3.11}$$

Here we have noted the elementary fact that

$$\sum_{[u,v]=t} 1 < 2t.$$

For larger values of t we must resort to large-sieve inequalities. From the one-dimensional large sieve (see Lemma A.4), for any coefficients c_n, we have

$$\sum_{\ell \leq L} \sum_{b \,(\mathrm{mod}\, h\ell)}^{*} \left| \sum_{n \leq N} c_n e\left(\frac{bn}{h\ell}\right) \right|^2 \leq (HL^2 + N) \sum_n |c_n|^2.$$

We apply this with

$$c_n = \sum_m c_{mn} e\left(\frac{am}{hk}\right).$$

We then apply the one-dimensional large-sieve inequality in the form (this follows immediately from Lemma A.4)

$$\sum_{h \leq H} \sum_{k \leq K} \tau(hk)^{-1} \sum_{a \,(\mathrm{mod}\, hk)}^{*} \left| \sum_m c_{mn} e\left(\frac{an}{hk}\right) \right|^2 \leq (H^2 K^2 + M) \sum_m |c_{mn}|^2$$

to obtain

$$\sum_{\substack{h \leq H \\ \ell \leq L}} \sum_{k \leq K} \tau(hk)^{-1} \sum_{\substack{a \,(\mathrm{mod}\, hk) \\ h \,(\mathrm{mod}\, h\ell)}}^{*} \left| \sum_{\substack{m \leq M \\ n \leq N}} c_{mn} e\left(\frac{am}{hk} + \frac{bn}{h\ell}\right) \right|^2$$

$$\leq (HK^2 + M)(HL^2 + N) \sum_{mn} |c_{mn}|^2.$$

Using the inequality $\tau(hk) \ll (hk)^\epsilon$ then gives the following result.

For any $X > Y \geq 1, M \geq 1, N \geq 1, \epsilon > 0$, and any complex numbers c_{mn} we have

$$\sum_{Y < hk\ell \leq X} (hk\ell)^{-1} \sum_{\substack{a \,(\mathrm{mod}\, hk) \\ h \,(\mathrm{mod}\, h\ell)}}^{*} \left| \sum_{\substack{m \leq M \\ n \leq N}} c_{mn} e\left(\frac{am}{hk} + \frac{bn}{h\ell}\right) \right|^2$$

$$\leq \left((M + N + X)X^{1+\epsilon} + MNY^{\epsilon-1}\right) \sum_{mn} |c_{mn}|^2.$$

We apply this with $Y = G, X = D, M = N = A, c_{mn} = f(m + ni)$ to those terms in (12.3.10) with $D \geq t \geq G$. This gives a contribution

$$O\left(AD(D^{1+\epsilon} + AG^{1-\epsilon})\|f\|^2\right),$$

which completes the proof of (12.3.9).   □

12.4 FOUVRY-IWANIEC TYPE I INFORMATION

Our aim now is to estimate (12.2.8). For a given sequence of complex numbers $\lambda_\ell, 1 \leq \ell \leq x^{1/2}$, write

$$||\lambda||^2 = \sum_\ell |\lambda_\ell|.$$

First we need to introduce a smoothing function into the sums to enable us to apply Poisson summation in a satisfactory manner. As in Chapter 11, we can find a non-negative function g on the positive reals such that

$$g(u) = \begin{cases} 1 & \text{if } 0 < u \leq x - y, \\ 0 & \text{if } u \geq x, \end{cases}$$

and

$$g^{(j)}(u) \ll y^{-j},$$

where the implied constant depends only on j. See Section A.5 for an explicit construction. We write

$$A_d(g) = \sum_{n \equiv 0 \, (\text{mod } d)} a_n g(n).$$

Thus

$$A_d(g) = \sum_\ell \sum_{\nu^2 + \ell^2 \equiv 0 \, (\text{mod } d)} \sum_{m \equiv \nu \, (\text{mod } d)} g(\ell^2 + m^2).$$

An application of (12.3.2) (and recalling the definition of $\rho_{k,\ell}(d)$: (12.3.4)) yields

$$A_d(g) = \frac{1}{d} \sum_k \sum_\ell \lambda_\ell \rho_{k,\ell}(d) G_\ell(k/d), \tag{12.4.1}$$

where

$$G_\ell(u) = \int_{-\infty}^\infty g(\ell^2 + t^2) e(-tu) \, dt.$$

The smoothed version of $M_d(x)$ that we would expect to come from the term $k = 0$ in (12.4.1) is thus

$$M_d(g) = \frac{1}{d} \sum_\ell \lambda_\ell \rho_\ell(d) G_\ell(0).$$

We write $R_d(g) = A_d(g) - M_d(g)$. We will be able to deduce our required Type I estimate from the following result.

Lemma 12.5. *Let λ_ℓ and g be as above. Let $\epsilon > 0$ and suppose that $1 \leq D \leq x$. Then we have*

$$\sum_{d \leq D} |R_d(g)| \ll ||\lambda|| y^{-1} D^{\frac{1}{2}} x^{\frac{5}{4} + \epsilon}, \tag{12.4.2}$$

where the implied constant depends only on ϵ.

Proof. First suppose that $\frac{1}{2}E < d \le E \le D$. We have

$$R_d(g) = \frac{2}{d} \operatorname{Re} \sum_{k=1}^{\infty} \sum_{\ell} \lambda_\ell \rho_{k,\ell}(d) G_\ell(k/d). \tag{12.4.3}$$

We now consider how quickly $G_\ell(u)$ decays as a function of u since we shall want to discard all large values of k in (12.4.3).

Integrating by parts j times we arrive at

$$G_\ell(u) = (2\pi i u)^{-j} \int_{-\sqrt{x}}^{\sqrt{x}} e(-ut) \frac{\partial^j}{\partial t^j} g(\ell^2 + t^2)\, dt.$$

Now, for certain positive integers c_{hj}, we have

$$\frac{\partial^j}{\partial t^j} g(\ell^2 + t^2) = \sum_{0 \le 2h \le j} c_{hj} t^{j-2h} g^{(j-h)}(\ell^2 + t^2) \ll \left(\frac{\sqrt{x}}{y} \right)^j,$$

where we have used $y \le x$ to show that $(\sqrt{x})^{j-2i} y^{i-j} \le (\sqrt{x} y^{-1})^j$. It follows that

$$G_\ell(u) \ll \sqrt{x} \left(\frac{\sqrt{x}}{yu} \right)^j$$

for all $u > 0$ and for any $j \ge 0$. Hence, if $k \ge K = E x^{1/2+\epsilon} y^{-1}$ and $\frac{1}{2}E \le d \le E$, we obtain

$$G_\ell(k/d) \ll k^{-2} E^{-1} x^{-\frac{1}{2}}$$

by choosing j sufficiently large in terms of ϵ. The contribution to (12.4.3) from terms with $k \ge K$ is thus

$$\frac{y\rho(d)}{dx^{1+\epsilon}} \sum_{\ell} |\lambda_\ell| \le ||\lambda|| \frac{x^{\frac{5}{4}-\epsilon} \rho(d)}{dy}, \tag{12.4.4}$$

using Cauchy's inequality and $y \le x$. Summing this over $d \le D$ then gives a term

$$||\lambda|| y^{-1} x^{\frac{5}{4}-\epsilon} \sum_{d \le D} \frac{\rho(d)}{d},$$

which is of smaller order than the right-hand side of (12.4.2).

We must now treat the remaining terms with $1 \le k \le K$ so that Lemma 12.3 is applicable. We cannot put $\alpha_{k,\ell} = \lambda_\ell G_\ell(k/d)$ in view of the dependence on d. However, by the change of variable $t = \sqrt{x} v k^{-1}$ we obtain

$$G_\ell(k/d) = \frac{2\sqrt{x}}{k} \int_0^\infty g\left(\ell^2 + xv^2 k^{-2}\right) \cos(2\pi v \sqrt{x}/d)\, dv.$$

Since the integrand vanishes for $k \le v$ and $\ell \ge \sqrt{x}$, we conclude that

$$d|R_d(g)| \le 4\sqrt{x} \int_0^K \left| \sum_{\substack{v < k < K \\ \ell \le \sqrt{x}}} \lambda_\ell k^{-1} g\left(\ell^2 + xv^2 k^{-2}\right) \rho_{k,\ell}(d) \right| dv + O(J),$$

where J corresponds to (12.4.4), which we have already seen is suitably small. We can now apply Lemma 12.3 with $\alpha_{k,\ell} = \lambda_\ell k^{-1} g \left(\ell^2 + xv^2 k^{-2} \right)$. We then have

$$\tilde{A}^2 \leq \sum_\ell \tau(\ell) |\lambda_\ell|^2 \sum_{k>v} \tau(k) k^{-2} \ll ||\lambda||^2 x^{\frac{1}{2}\epsilon} \frac{\log(v+2)}{v+1},$$

using the standard bound for the divisor function. Also,

$$\int_0^K \left(\frac{\log(v+2)}{v+1} \right)^{\frac{1}{2}} dv \ll (K \log(K+1))^{\frac{1}{2}}.$$

From (12.3.5) we thereby obtain

$$\sum_{E/2 \leq d \leq E} d |R_d(g)| \ll x^{\frac{1}{4}\epsilon} ||\lambda|| (E + K\sqrt{x})^{\frac{1}{2}} (EKx)^{\frac{1}{2}} (\log x)^4,$$

and so

$$\sum_{E/2 \leq d \leq E} |R_d(g)| \ll x^{\frac{1}{4}\epsilon} ||\lambda|| (1 + K\sqrt{x}/E)^{\frac{1}{2}} (Kx)^{\frac{1}{2}} (\log x)^4.$$

On substituting the value for K, summing over $E = 2^{-j} D, j = 0, 1, \ldots$, and using $(\log x)^5 \ll x^{\epsilon/4}$, we thus obtain (12.4.2) as required. $\qquad\square$

Lemma 12.6. *Let λ_ℓ be a sequence of complex numbers for $1 \leq \ell \leq \sqrt{x}$. Then, for $1 \leq D \leq x, \epsilon > 0$ we have*

$$\sum_{d \leq D} |R_d(x)| \ll ||\lambda|| D^{\frac{1}{4}} x^{\frac{1}{2}+\epsilon}. \tag{12.4.5}$$

Here the implied constant depends only on ϵ.

Proof. We have

$$E_d(x) = A_d(x) - A_d(g) = \sum_{\substack{n \equiv 0 \,(\mathrm{mod}\, d) \\ x-y < n \leq x}} (1 - g(n)).$$

Hence

$$\sum_d |E_d(x)| \leq \sum_{x-y < n \leq x} |a_n| \tau(n) \ll x^\epsilon \sum_{x-y < \ell^2+m^2 \leq x} |\lambda_\ell|$$

$$\ll yx^\epsilon \sum_{\ell \leq \sqrt{x}} |\lambda_\ell| (x + y - \ell^2)^{-\frac{1}{2}}$$

$$\ll yx^\epsilon ||\lambda|| \left(\sum_{\ell \leq \sqrt{x}} (x + y - \ell^2)^{-1} \right)^{\frac{1}{2}}$$

$$\ll ||\lambda|| \left(y^{\frac{1}{2}} + yx^{-\frac{1}{4}} \right) x^\epsilon.$$

The same estimate can be obtained for the sum of $|M_d(x) - M_d(g)|$ by a similar argument. From (12.4.2) we thus derive the bound

$$\sum_{d \leq D} |R_d(x)| \ll ||\lambda|| \left(y^{-1} D^{\frac{1}{2}} x^{\frac{5}{4}} + y^{\frac{1}{2}} + yx^{-\frac{1}{4}} \right) x^\epsilon.$$

The choice $y = D^{1/4} x^{3/4}$ then leads to (12.4.5) and completes the proof. $\qquad\square$

We now consider the strength of Lemma 12.6. If λ_ℓ represents the characteristic function of the primes, then we would expect to get an asymptotic formula on average for $A_d(x)$ if we could show that

$$\sum_{d \leq D} |R_d(x)| \ll x(\log x)^{-A}$$

for some sufficiently large A. According to the Fouvry-Iwaniec result we have presented, this holds for D as large as $x^{1-\epsilon}$. In the notation of Section 12.2 that means we can take $UV = x^{1-\epsilon}$ and still obtain

$$|R(x; V, U)| \ll x(\log x)^{-A}.$$

We thus have extremely good Type I information. From our previous work we know that we only require a modicum of Type II information to obtain an asymptotic formula for the number of primes p represented as $m^2 + \ell^2$ in this situation.

12.5 REDUCING THE BILINEAR FORM PROBLEM

Here we turn our attention to (12.2.5). Our aim is to show that

$$|B(x; V, U)| \ll \Delta x(\log x)^5, \tag{12.5.1}$$

where $\Delta = (\log x)^{-A}$ for any fixed $A > 0$. Previously we have used Perron's formula to remove conditions like $dh \leq x$ in sums. In our present context we do not want to introduce an h^{it} factor, so we shall use an elementary range-splitting argument instead. To this end we write

$$\mathbf{B}(M, N) = \sum_{m \sim M} \left| \sum_{N < n \leq N'} \mu(n) a_{mn} \right| \tag{12.5.2}$$

where $N' = e^\Delta N$. We then sum over $M = 2^j U$ and $N = e^{k\Delta}V$ with $\Delta x \leq MN <$ to obtain

$$|B(x; V, U)| \leq (\log x) \sum_{M,N} \mathbf{B}(M, N) + O\left(\Delta x(\log x)^2\right). \tag{12.5.3}$$

Here the error term $O\left(\Delta x(\log x)^2\right)$ comes from estimating trivially those terms that may be omitted in the summation, namely, $dh \leq 2\Delta x$ and $e^{-2\Delta}x < dh \leq x$. Since there are no more than $2\Delta^{-1}(\log x)^2$ sums $\mathbf{B}(M, N)$ to consider, the bound (12.5.1) will follow from

$$\mathbf{B}(M, N) \ll \Delta^2 x(\log x)^2. \tag{12.5.4}$$

The trivial bound for such a sum is just ΔMN, of course.

Now it will be important to have m and n coprime in the following. To achieve this we note that

$$\mathbf{B}(M, N) \leq \sum_{d < \Delta^{-1}} \mathbf{B}_d(M, N) + O(\Delta^2 x),$$

where $\mathbf{B}_d(M, N)$ denotes the sum (12.5.2) except that the variables are now restricted by $(m, n) = d$. We have

$$\sum_{\substack{m \sim M}} \left| \sum_{\substack{N < n \leq N' \\ (m,n)=d}} \mu(n)a_{mn} \right| = \sum_{\substack{m \sim M}} \left| \sum_{\substack{N < n \leq N' \\ (n,m/d)=1}} \mu(nd)a_{mnd} \right|$$

$$\leq \sum_{\substack{m \sim Md}} \left| \sum_{\substack{N < n \leq N' \\ (n,m)=1}} \mu(n)a_{mn} \right|.$$

To obtain the last inequality, note that if $d|m$, then

$$(m, n) = 1 \Rightarrow (n, d) = 1 \quad \text{and} \quad (n, m/d) = 1.$$

Thus $\mu(dn) = \mu(d)\mu(n)$ in this case. Our final sum then counts all the terms in the previous sum (these correspond to $d^2|m$) with many more besides (of course, those terms with $(n, d) > 1$ are not counted in the previous sum since then $\mu(nd) = 0$). We thus have

$$\mathbf{B}_d(M, N) \leq \mathbf{B}_1(Md, N/d),$$

and so it suffices to show that

$$\mathbf{B}_1(M, N) \ll \Delta^3 x (\log x)^2 \tag{12.5.5}$$

for any M, N with $M \geq U, N \geq \Delta V$, and $\Delta x < MN < x$.

The condition $(m, n) = 1$ enables us to switch from the bilinear form in m to one in z, w where $z, w \in \mathbb{Z}[i]$, as we now show. We henceforth write (as in the proof of Lemma 12.4) $z \cdot w = \operatorname{Re} z\overline{w}$. Since this notation would otherwise appear rather awkwardly in a subscript, we write $\lambda(\ell) = \lambda_\ell$. Also, we shall write $(w, z) = 1$ to mean $(|w|^2, |z|^2) = 1$. Since $(m, n) = 1$, every representation of mn as the sum of two squares arises precisely four times from each representation of m and n. We can thus rewrite our expression for a_{mn} as

$$a_{mn} = \frac{1}{4} \sum_{|w|^2=m} \sum_{|z|^2=n} \lambda(z \cdot w). \tag{12.5.6}$$

Hence

$$\mathbf{B}_1(M, N) \leq \frac{1}{4} \sum_{\substack{|w|^2 \sim M}} \left| \sum_{\substack{N < |z|^2 \leq N' \\ (z,w)=1}} \mu(|z|^2)\lambda(z \cdot w) \right|. \tag{12.5.7}$$

Paradoxically, now that we have transferred our bilinear form to be over $\mathbb{Z}[i]$, we remove the coprimality condition we had imposed. We use the familiar (1.3.1) to do this. The inner sum in (12.5.7) thus becomes

$$\sum_{r \,|\, |w|^2} \mu(r) \sum_{\substack{N < |z|^2 \leq N' \\ |z|^2 \equiv 0 \,(\mathrm{mod}\, r)}} \mu(|z|^2)\lambda(z \cdot w). \tag{12.5.8}$$

Now $|z|^2$ is only counted when it is square-free. So, if $|z|^2 \equiv 0 \,(\mathrm{mod}\, r)$, there exists $\zeta \,|\, z$ with $|\zeta|^2 = r$ and ζ is unique up to associates. It follows that all solutions to $|z|^2 = n$ with $r \mid n$ are counted exactly four times by the solutions to the system of equations

$$|z|^2 = n, \quad |\zeta|^2 = r \quad \text{with} \quad \zeta \mid z.$$

Hence we can replace (12.5.8) with

$$\frac{1}{4} \sum_{r \mid \mid w \mid^2} \mu(r) \sum_{|\zeta|^2 = r} \sum_{N < r|z|^2 \leq N'} \mu(r|z|^2) \lambda(\zeta z \cdot w).$$

Now $\zeta z \cdot w = z \cdot (\bar{\zeta} w)$. So we can replace w with $\bar{\zeta} w$, and it then follows that r^2 divides $|w|^2$. We then estimate the number of ζ for each r as no more than $4\rho(r)$. Ignoring the condition $\bar{\zeta} \mid w$ we arrive at

$$\mathbf{B}_1 \leq \frac{1}{4} \sum_r \rho(r) \sum_{\substack{|w|^2 \sim rM \\ r^2 \mid \mid w \mid^2}} \left| \sum_{N < r|z|^2 \leq N'} \mu(r|z|^2) \lambda(z \cdot w) \right|.$$

The terms with $r \geq \Delta^{-2}$ contribute

$$\ll \Delta M N \sum_{r > \Delta^{-2}} \rho(r)^2 r^{-2} \ll \Delta^3 x (\log x)^2,$$

which is a suitably small error. Now put

$$\mathbf{C}_r(M, N) = \sum_{|w|^2 \sim M} \left| \sum_{N < |z|^2 \leq N'} \mu(r|z|^2) \lambda(z \cdot w) \right|. \qquad (12.5.9)$$

If we show that

$$\mathbf{C}_r(M, N) \ll \Delta^5 x (\log x)^2 \qquad (12.5.10)$$

for every r, M, N with $r < \Delta^{-2}, M \geq U, N \geq \Delta^3 V$, and $\Delta x < MN < x$, then (12.5.5) will follow.

Now we shall want to reduce the summation over w to primitive Gaussian integers, that is to say, those $w = u + iv$ with $(u, v) = 1$. To this end we write

$$\mathbf{C}_{r,c}(M, N) = \sum_{|w|^2 \sim M}^{*} \left| \sum_{N < |z|^2 \leq N'} \mu(r|z|^2) \lambda(cz \cdot w) \right|, \qquad (12.5.11)$$

where $*$ indicates summation over primitive Gaussian integers only. Thus we have

$$\mathbf{C}_r(M, N) = \sum_{c \geq 1} \mathbf{C}_{r,c}(M/c^2, N)$$

$$= \sum_{1 \leq c \leq \Delta^{-4}} \mathbf{C}_{r,c}(M/c^2, N) + O(\Delta^5 x)$$

since $\mathbf{C}_{r,c}(M, N) \ll \Delta M N$. We have thus reduced the problem to showing that

$$\mathbf{C}_{r,c}(M, N) \ll \Delta^5 M N \qquad (12.5.12)$$

for every c, r, m, N satisfying

$$c < \Delta^{-4}, \quad r < \Delta^{-2}, \quad M \geq \Delta^4 U, \quad N > \Delta^3 V, \quad \Delta^5 x < MN < x.$$

In the next section we prove the following result, which completes the demonstration of (12.5.1).

Lemma 12.7. *In the above notation, if $N^\epsilon < M < N^{1-\epsilon}$ and $c, r < N < N' \leq 2N$, then, for every $j > 0$,*

$$\mathbf{C}_{r,c}(M, N) \ll MN(\log N)^{-j}. \tag{12.5.13}$$

Here the implied constant depends only on ϵ and j.

12.6 CATCHING THE CANCELLATION INTRODUCED BY μ

Fouvry and Iwaniec [36] stated their results in more general terms than we shall do here. We shall simplify the previous notation for clarity. Write $\beta(n) = \lambda(nc)$ for $n \in \mathbb{N}$ and put

$$\alpha(z) = \begin{cases} \mu(r|z|^2) & \text{if } N < |z|^2 < N', \\ 0 & \text{otherwise.} \end{cases}$$

Let

$$H(w) = \sum_z \alpha(z)\beta(z \cdot w), \qquad H'(w) = \frac{\overline{H(w)}}{|H(w)|}$$

and put

$$\gamma(w) = \begin{cases} H'(w) & \text{if } |w|^2 \sim M, \\ 0 & \text{otherwise.} \end{cases}$$

Hence

$$\begin{aligned}
\mathbf{C}_{c,r}(M, N) &= \sum_w \sum_z \alpha(z)\beta(z \cdot w) \\
&= \sum_\ell \beta(\ell) \sum_w \gamma(w) \sum_{\substack{z \\ z \cdot w = \ell}} \alpha(z).
\end{aligned}$$

Hence, by Cauchy's inequality,

$$\mathbf{C}_{c,r}(M, N) \leq \left(\sum_\ell |\beta(\ell)|^2 \right)^{\frac{1}{2}} \left(\sum_w |\gamma(w)|^2 \right)^{\frac{1}{2}} \left(\sum_w \sum_\ell \left| \sum_{\substack{z \\ z \cdot w = \ell}} \alpha(z) \right|^2 \right)^{\frac{1}{2}}$$

$$\ll \|\lambda\| M^{\frac{1}{2}} \mathbf{D}(\alpha)^{\frac{1}{2}}, \quad \text{say,}$$

where

$$\mathbf{D}(\alpha) = \sum_w^* g(w) \sum_\ell \left| \sum_{\substack{z \\ z \cdot w = \ell}} \alpha(z) \right|^2. \tag{12.6.1}$$

Here g could be any non-negative function with $g(w) \geq 1$ for $|w|^2 \sim M$. We shall use another function that can be constructed in the manner of Section A.5. We thus take $g(w) = G(|w|^2)$, where $G(t) \in [0, 1]$, $G(t) = 1$ if $t \sim M$, with

$$G(t) > 0 \Rightarrow \tfrac{1}{2}M < t < 4M, \qquad \frac{d^j}{dt^j} G(t) \ll M^{-j}. \tag{12.6.2}$$

We write

$$A(z) = \sum_{z_1 - z_2 = x} \alpha(z_1)\alpha(z_2).$$

(This is written as $\alpha * \alpha(z)$ in [36] to bring out the fact that it is a convolution.) We note that

$$A(0) = \mu^2(r) \sum_{\substack{N < n \le N' \\ (n,r)=1}} \mu^2(n) \sum_{\substack{z \in \mathbb{Z}[i] \\ |z|^2 = n}} 1 \ll N.$$

Squaring out $\mathbf{D}(\alpha)$ we obtain

$$\mathbf{D}(\alpha) = \sum_{w}^{*} g(w) \sum_{z} A(z). \tag{12.6.3}$$

Now, since w is primitive, the condition $z \cdot w = 0$ forces z to be a rational integer multiple of iw. Hence

$$\mathbf{D}(\alpha) = \sum_{c \in \mathbb{Z}} \sum_{w}^{*} g(w)A(cw) = \mathbf{D}_0(\alpha) + 2\mathbf{D}'(\alpha), \quad \text{say}, \tag{12.6.4}$$

where $\mathbf{D}_0(\alpha)$ covers the term $c = 0$, and $\mathbf{D}'(\alpha)$ is the contribution from $c > 0$. Thus

$$\mathbf{D}_0(\alpha) = A(0) \sum_{w}^{*} g(w) \ll MN \tag{12.6.5}$$

and

$$\mathbf{D}'(\alpha) = \sum_{z \ne 0} g(z^*)A(z), \tag{12.6.6}$$

where z^* is the *primitive kernel* of z; that is,

$$z^* = \frac{z}{(x,y)} \quad \text{when} \quad z = x + iy.$$

We now remove the primitivity condition by our familiar formula (1.3.1) to get

$$\mathbf{D}'(\alpha) = \sum_{b,c>0} \mu(b)\mathbf{D}(\alpha; bc), \tag{12.6.7}$$

where

$$\mathbf{D}(\alpha; b, c) = \sum_{z \equiv 0 \pmod{bc}} g(z/c)A(z). \tag{12.6.8}$$

Now we only count values of z with $|z|^2 \le N' < 2N$ and $\frac{1}{2}M \le |z/c|^2 \le 4M$. It follows that there is no contribution from values of $c > 2\sqrt{N/M} = \Gamma_0$, say. Now let Γ be a parameter to be chosen optimally later satisfying $1 \le \Gamma \le \Gamma_0$ and write $C = \Gamma_0/\Gamma$. Trivially

$$\mathbf{D}(\alpha; b, c) \ll MNb^{-2}.$$

Hence the contribution to $\mathbf{D}'(\alpha)$ from those terms with $b \ge \Gamma$ or $c \le C$ is $O(N^{3/2}M^{1/2}/\Gamma)$. We thus have

$$\mathbf{D}'(\alpha) = \sum_{b \le \Gamma} \mu(b) \sum_{C < c < \Gamma_0} \mathbf{D}(\alpha; b, c) + O\left(\frac{N^{\frac{3}{2}}M^{\frac{1}{2}}}{\Gamma}\right). \tag{12.6.9}$$

Now it can be easy to lose sight of the fact that we expect to be able to make some saving on the trivial bound for our sums by detecting cancellation caused by the sign changes in $\alpha(z) = \mu(r|z|^2)$. We need to make further transformations of the sum in order to do this — in particular we need to isolate the summation over z in some sense. We use Fourier transforms on $\mathbb{R}^2$ to this end. We regard the complex variable w (until now always lying in $\mathbb{Z}[i]$) as a vector in $\mathbb{R}^2$ and introduce another variable ω with the same ambiguity. The notation $w \cdot \omega$ has the same meaning whether we regard it as $\operatorname{Re} w\bar{\omega}$ or as the standard scalar product in $\mathbb{R}^2$. So we write

$$f(\omega) = \int_{\mathbb{R}^2} g(w)e(-\omega \cdot w)\, dw,$$

which gives, by Fourier inversion,

$$g(w) = \int_{\mathbb{R}^2} f(w)e(\omega \cdot w)\, dw.$$

Since g is radial, so is f, and we can express f as $F(\omega) = F(|\omega|^2)$ with

$$F(s) = \pi \int_0^\infty J_0(2\pi\sqrt{st})G(t)\, dt,$$

and where $J_0(x)$ is the standard Bessel function (see [155, Theorem 3.3 in Chapter 4]). Clearly $F(s) \ll M$. If $sM > 1$, we can use (12.6.2) along with the following facts about Bessel functions,

$$\int u^\nu J_{\nu-1}(u)\, du = u J_\nu(u), \qquad |J_\nu(u)| \ll u^{-\frac{1}{2}},$$

with repeated integration by parts, to show that $F(s) \ll M(sM)^{-3/2}$. Thus we obtain

$$F(s) \ll M(1 + sM)^{-\frac{3}{2}}. \tag{12.6.10}$$

By a simple change of variable we have

$$g(z/c) = \int_{\mathbb{R}^2} c^2 f(c\omega)e(\omega \cdot z)\, d\omega.$$

Thus (12.6.8) becomes

$$\mathbf{D}(\alpha; b, c) = \int_{\mathbb{R}^2} c^2 f(c\omega)S_{bc}(\omega)\, d\omega,$$

where

$$S_d(\omega) = \sum_{z \equiv 0 \,(\mathrm{mod}\, d)} A(z)e(\omega \cdot z).$$

Recalling the definition of $A(z)$ this gives

$$S_d(\omega) = \sum_{z_1 \equiv z_2 \,(\mathrm{mod}\, d)} \alpha(z_1)\alpha(z_2)e(\omega \cdot (z_1 - z_2))$$

$$= \sum_{a \,(\mathrm{mod}\, d)} \left| \sum_{z \equiv a \,(\mathrm{mod}\, d)} \alpha(z)e(\omega \cdot z) \right|^2.$$

Here a runs over the d^2 residue classes $(\mathrm{mod}\ d)$ in $\mathbb{Z}[i]$. The final representation given of $S_d(\omega)$ shows that it is real and non-negative. We now make use of this fact by replacing f by an explicit upper bound that is easier to handle. By (12.6.10)

$$c^2 f(c\omega) = c^2 F(c^2|\omega|^2) \ll \frac{c^2 M}{(1 + c^2 M|\omega|^2)^{\frac{3}{2}}},$$

and this last expression is a decreasing function of c, so we may replace c by $C = \Gamma_0/\Gamma = 2\sqrt{N/M}/\Gamma$ to obtain an upper bound

$$\ll \frac{\Gamma^{-2}N}{(1 + N\Gamma^{-2}|\omega|^2)^{\frac{3}{2}}} \ll \Gamma N h(\omega)$$

with $h(\omega) = (1 + |\omega|^2 N)^{-3/2}$. Thus

$$\mathbf{D}(\alpha; b, c) \ll \Gamma N \int_{\mathbb{R}^2} h(\omega) S_{bc}(\omega)\, d\omega. \qquad (12.6.11)$$

We let $d = bc$ and note that $C < d < \Gamma_0\Gamma = D$, say. We now write

$$\Sigma(D, \omega) = \sum_{d \le D} d^2 S_d(\omega)$$

and recognize that this is the expression we estimated in Lemma 12.4 with $f(z)$ there equal to $\alpha(z)e(\omega \cdot z)$ here. It follows that

$$\Sigma(D, \omega) \le 2D \sum_{d \le G} d^2 S_d(\omega) + O(N^2 DG^{\epsilon-1}), \qquad (12.6.12)$$

where the choice of G is at our disposal subject to $(DG)^2 < N^{1-\epsilon}$. Assembling our results so far and noting that

$$\int_{\mathbb{R}^2} h(\omega)\, d\omega \ll \frac{1}{N}$$

then gives

$$\mathbf{D}'(\alpha) \ll \Gamma^3 M \int_{\mathbb{R}} h(\omega)\Sigma(D)\, d\omega + \frac{N(MN)^{\frac{1}{2}}}{\Gamma}$$

$$\ll \Gamma^3 MD \sum_{d \le G} d^2 \mathbf{D}_d(\alpha) + N^{\frac{3}{2}} M^{\frac{1}{2}}\left(\Gamma^3 G^{\epsilon-1} + \frac{1}{\Gamma}\right),$$

where

$$\mathbf{D}_d(\alpha) = \int_{\mathbb{R}^2} h(\omega) S_d(\omega)\, d\omega.$$

Assuming that $M > \Gamma^{14} N^\epsilon$ we may choose $G = \Gamma^6$ to give $(\Gamma^3 G^{\epsilon-1} + 1/\Gamma) < 2/\Gamma$. Now

$$\mathbf{D}_d(\alpha) = \sum_{z \equiv 0\ (\mathrm{mod}\ d)} A(z)\widehat{h}(z),$$

where $\widehat{h}(z)$ is the Fourier transform of $h(\omega)$, namely,

$$\int_{\mathbb{R}^2} h(\omega)e(\omega \cdot z)\, d\omega = \frac{2\pi}{N}\exp\left(-\frac{2\pi}{\sqrt{N}}\right).$$

Thus

$$\mathbf{D}_d(\alpha) = \frac{2\pi}{N} \sum_{z_1 \equiv z_2 \,(\mathrm{mod}\, d)} \alpha(z_1)\alpha(z_2) \exp\left(\frac{-2\pi|z_1 - z_2|}{\sqrt{N}}\right).$$

All told we have therefore established that

$$\mathbf{D}(\alpha) \ll \Gamma^4 \sqrt{MN} \sum_{d \le \Gamma^6} d^2 \mathbf{D}_d(\alpha) + \sqrt{N}(M + \sqrt{MN}\Gamma^{-1}). \tag{12.6.13}$$

In conclusion we therefore obtain

$$\mathbf{C}_{c,r}(M,N) \ll (MN)^{\frac{3}{4}}||\lambda||\Gamma^{-\frac{1}{2}} + ||\lambda||M^{\frac{3}{4}}N^{\frac{1}{4}}\Gamma^2 \left(\sum_{d \le \Gamma^6} d^2 \mathbf{D}_d(\alpha)\right)^{\frac{1}{2}}. \tag{12.6.14}$$

Since $||\lambda|| \le x^{1/4}$, the first term on the right-hand side of (12.6.14) is

$$\ll MN\left(x^{\frac{1}{4}}(MN)^{-\frac{1}{4}}\Gamma^{-\frac{1}{2}}\right) \ll MN\Delta^{\frac{5}{4}}\Gamma^{-\frac{1}{2}}.$$

This term will therefore give a suitable contribution to (12.5.13) if Γ is taken as a large enough power of $\log x$, to be precise, $\Gamma = \Delta^{5/2}(\log x)^{2j}$. With this choice for Γ, the proof of (12.5.13) is completed by establishing the following result.

Lemma 12.8. *Let $B > 0, A > 0$ be given and suppose $N \ge x^{1/2}$. Then, for all $d \le (\log x)^B$, we have*

$$\mathbf{D}_d(\alpha) \ll N(\log x)^{-A}. \tag{12.6.15}$$

The implied constant here depends only on A and B.

Proof. Let $\nu = \sqrt{N}$ and put $\theta = (\log x)^{-A}$. First we note that (and the reader should remember that here $a, d \in \mathbb{Z}[i]$)

$$\mathbf{D}_d(\alpha) \ll \max_{a \,(\mathrm{mod}\, d)} \max_{|y| \le 2\nu} \mathbf{S}(a, d, y)$$

with

$$\mathbf{S}(a, d, y) = \sum_{\substack{z \equiv a \,(\mathrm{mod}\, d) \\ |z| \le \nu}} \mu(r|z|^2) \exp\left(-\frac{2\pi|z - y|}{\nu}\right).$$

We can suppose that $(a, d) = 1$ henceforth, since if $(a, d) = t$ we can replace r by $r|t^2|$, which will cause no problems. Our main concern is to remove the factor

$$\exp\left(-\frac{2\pi|z - y|}{\nu}\right).$$

We shall split up the summation over z in $\mathbf{S}(a, d, y)$ into small regions. Clearly values with $|z| < \nu\theta$ contribute $\ll N(\log x)^{-2A}$ and so can be neglected. The remaining values of z are confined to non-overlapping regions of the form

$$R(Z, \xi) = \{z \in \mathbb{Z}[i] : Z \le |z| \le Z + \nu\theta, \xi \le \arg(z) \le \xi + \theta\}.$$

There are $\ll (\log x)^{2A}$ such regions to consider. For $z \in R(Z, \xi)$ we have

$$\exp\left(-\frac{2\pi|z - y|}{\nu}\right) = \exp\left(-\frac{2\pi|Z - y|}{\nu}\right) + O(\theta),$$

and so it suffices to show that

$$\sum_{\substack{z \equiv a \,(\mathrm{mod}\, d) \\ z \in R(Z,\xi)}} \mu(r|z|^2) \ll \nu\theta^3.$$

Now the region $R(Z, \xi)$ is a polar box: the type of region considered for Problem 2 in the previous chapter. We can use Perron's formula and Fourier analysis as previously to bound the sum

$$\sum_{\substack{z \equiv a \,(\mathrm{mod}\, d) \\ z \in R'(Z,\xi)}} \mu(z),$$

where now μ is the Möbius function defined on $\mathbb{Z}[i]$ and

$$R'(Z, \xi) = \{z \in R(Z, \xi) : (z, r) = 1, \ (z, \overline{z}) = 1 \text{ or } 1 + i\}.$$

We do this using Hecke L-functions as before, except that we now have to introduce Dirichlet characters to pick out the condition $z \equiv a \,(\mathrm{mod}\, d)$. There may now be an exceptional real zero for one of these L-functions. However, since $d < (\log x)^B$, we can obtain a Siegel-Walfisz-type result (see [31] for the required zero-free region; the result is then analogous to Lemma 2.7), which completes the proof. $\qquad \square$

12.7 THE MAIN TERM FOR THEOREM 12.1

We recall that it only remains to prove (see (12.2.9)) that

$$M(x; V, U) = \sum_{n \le x} \sum_{\ell} \lambda_\ell(n) \sigma_\ell(n; V, U) = \sum_{\ell^2 + m^2 \le x} \lambda_\ell \psi(\ell) + E,$$

where E is a smaller order error. We note that, for primes p,

$$\rho_\ell(p) = \begin{cases} 1 + \chi(p) & \text{if } p \nmid \ell, \\ 1 & \text{if } p \mid \ell. \end{cases}$$

Hence

$$\psi(\ell) = \prod_{p \nmid \ell} \left(1 - \frac{\chi(p)}{p - 1}\right) = \prod_p \left(1 - \frac{\rho_\ell(p)}{p}\right)\left(1 - \frac{1}{p}\right)^{-1}.$$

This last expression equals

$$-\sum_b \frac{\mu(b)}{b} \rho_\ell(b) \log b.$$

Thus

$$\psi(\ell) - \sigma_\ell(n; V, U) = \sum_{d > V} \frac{\mu(d)}{d} \left(\rho_\ell(d) \log(n/d) - \sum_{m \leq U} \frac{\Lambda(m)}{m} \rho_\ell(dm) \right)$$

$$= \delta_\ell(n; V, U), \quad \text{say.}$$

Here we have used

$$\sum_d \frac{\mu(d)\rho(d)}{d} = \sum_d \frac{\mu(d)\rho(dm)}{d} = 0.$$

The proof is finally completed by recalling that $\rho_\ell(d) = (r(d), \ell)\rho(d/(d, \ell^2))$, from which it follows that, for any A,

$$\delta_\ell(n; V, U) \ll_A \tau(\ell)(\log x)(\log V)^{-2-A}.$$

Thus

$$\sum_{\ell^2 + m^2 = n \leq x} \lambda_\ell \delta_\ell(n; V, U) \ll x(\log x)^{-A},$$

as required.

12.8 THE FRIEDLANDER-IWANIEC OUTLINE FOR $a^2 + b^4$

Unfortunately, the working so far is not strong enough to consider the case

$$\lambda_\ell = \begin{cases} 1 & \text{if } \ell = b^2, \\ 0 & \text{otherwise.} \end{cases}$$

However, the basic idea of using Vaughan's identity with the Type II information given by estimating a bilinear form like (12.2.5) (where the cancellations produced by $\mu(n)$ are vital) is still the starting point. Also, we shall use essentially the same Type I information. For technical reasons it will be more convenient to use

$$\lambda_\ell = \begin{cases} 2 & \text{if } \ell = b^2 > 0, \\ 1 & \text{if } \ell = 0, \\ 0 & \text{otherwise.} \end{cases}$$

Thus we write

$$a_n = \sum_{\ell^2 + a^2 = n} \lambda_\ell, \qquad A_d(x) = \sum_{\substack{n \leq x \\ n \equiv 0 \,(\mathrm{mod}\, d)}} a_n,$$

$$M_d(x) = \frac{1}{d} \sum_{n \leq x} a_n(d), \quad \text{with} \quad a_n(d) = \sum_{\ell^2 + a^2 = n} \lambda_\ell \rho_\ell(d),$$

and put $R_d(x) = A_d(x) - M_d(x)$. We then have the following variant of Lemma 12.6.

Lemma 12.9. *For* $1 \leq D \leq x, \epsilon > 0$, *and with* λ_ℓ *as above, we have*

$$\sum_{d \leq D} |R_d(x)| \ll D^{\frac{1}{4}} x^{\frac{9}{16}+\epsilon}. \tag{12.8.1}$$

Here the implied constant depends only on ϵ.

Proof. The only difference from our previous working is the estimate of the error involved in replacing $A_d(x)$ by the smoothed sum $A_d(g)$. Before, we had

$$\sum_d |A_d(x) - A_d(g)| \ll ||\lambda|| \left(y^{\frac{1}{2}} + yx^{-\frac{1}{4}} \right) x^\epsilon.$$

Working as in Section 12.4 but making use of the new properties of λ, we obtain

$$\sum_d |A_d(x) - A_d(g)| \ll yx^\epsilon \sum_{b^4 \leq x} \frac{1}{x + y - b^4} \ll yx^{-\frac{1}{4}} x^\epsilon.$$

We can then apply (12.4.2) with $y = D^{1/4} x^{13/16}$, and (12.8.1) follows. $\qquad\square$

The choice for λ_ℓ allows us to give a very neat expression for $M_d(x)$, as we now show.

Lemma 12.10. *For* d *cube-free we have*

$$M_d(x) = \kappa g(d) x^{\frac{3}{4}} + O\left(h(d) x^{\frac{1}{2}} \right), \tag{12.8.2}$$

where κ *is given by* (12.1.5) *and* $G(d), h(d)$ *are the multiplicative functions given by*

$$g(p)p = 1 + \chi(p) \left(1 - \frac{1}{p} \right), \quad g(p^2)p^2 = 1 + \rho(p) \left(1 - \frac{1}{p} \right),$$

$$h(p)p = 1 + 2\rho(p), \qquad h(p^2)p^2 = p + 2\rho(p),$$

with the exception that $g(4) = \frac{1}{4}$.

Proof. We follow the proof in [37] closely. We have

$$M_d(x) = \frac{2}{d} \sum_{|c| \leq x^{1/4}} \rho_{c^2}(d) \left((x - c^4)^{\frac{1}{2}} + O(1) \right).$$

Since d is cube-free, we can write $d = d_1 d_2^2$ with $d_1 d_2$ square-free. Thus

$$\rho_{c^2}(d) = (c, d_2)\rho\left(\frac{d_1 d_2}{(c, d_1 d_2)} \right),$$

except when d_2 is even and c is odd, in which case we have $\rho_{c^2}(d) = 0$. Hence, for d not divisible by 4 we have

$$M_d(x) = \frac{2}{d} \sum_{\nu_j | d_j} \nu_2 \rho\left(\frac{d_1 d_2}{\nu_1 \nu_2} \right) \sum_{\substack{c^4 \leq x \\ (c, d_1 d_2) = \nu_1 \nu_2}} \left((x - c^4)^{\frac{1}{2}} + O(1) \right)$$

$$= \frac{2}{d} \sum_{\nu_j | d_j} \nu_2 \rho\left(\frac{d_1 d_2}{\nu_1 \nu_2} \right) \left(\phi\left(\frac{d_1 d_2}{\nu_1 \nu_2} \right) \frac{2\kappa x^{\frac{3}{4}}}{d_1} d_2 + O\left(\tau\left(\frac{d_1 d_2}{\nu_1 \nu_2} \right) x^{\frac{1}{2}} \right) \right).$$

The reader can then verify that (12.8.2) then follows in this case. For d cube-free but divisible by 4, then the above working is applicable, except that c and hence ν_2 must be restricted to even integers. This makes the value of $g(4)$ exceptional. $\qquad\square$

We now define the error term

$$r_d(x) = A_d(x) - g(d)A(x).$$ (12.8.3)

Combining Lemmas 12.9 and 12.10 then gives the following result.

Lemma 12.11. *Given $\epsilon > 0$, we have, for all $t \leq x$,*

$$\sum_{d \leq D} |r_d(t)| \ll D^{\frac{1}{4}} x^{\frac{9}{16} + \epsilon},$$

where the summation is over cube-free values of d only.

Remark. From this result we see that we have a "level of distribution" $x^{3/4 - \epsilon}$ for this problem; that is, the variable with the "unknown" coefficient can go up to this value in our Type I sums.

12.9 THE FRIEDLANDER-IWANIEC ASYMPTOTIC SIEVE

The Type I information provided by Lemma 12.11 would be sufficient to produce primes provided that we had Type II information for sums where one variable, say n, ranges between N_1 and N_2 with $N_1 = x^{3/8 - \epsilon}$ and $N_2 = x^{5/8 + \epsilon}$, by (2.1.4) with $\beta = \frac{3}{8} - \epsilon$. The result proved by Friedlander and Iwaniec (see Theorem 12.6) essentially gives such a result except for n in the range $[x^{1/2}\delta, \delta^{-1}x^{1/2}]$, where $\delta = (\log x)^{-J}$ for some (large) J. Of course, we have become accustomed to dealing with situations where there is a "missing interval" and we would assume that we can get an asymptotic formula for the number of primes in this situation since the "gap" is so small. To be precise, such a gap should only entail the addition of an error term of size $(\log \log x)/(\log x)$ times the main term. The innovation in the Friedlander-Iwaniec asymptotic sieve is that this was achieved by using (essentially) Vaughan's identity and not Buchstab's identity as has become familiar in this book so far. The Type II information available in their application of the sieve is also more specialized than we have required, and this makes the combinatorial arguments needed to prove their result rather more convoluted. However, this specialised information is only required in *half* the usual interval (in this case from $x^{3/8 - \epsilon}$ to $x^{1/2}\delta$). Our intrepid explorers in this previously unchartered jungle of prime number theory proved the following three results in [38]. The final one is required for our current application.

Theorem 12.3. *Let x be a large positive number and suppose that $\delta = \delta(x), \eta = \eta(x)$ are positive functions, both of which tend to zero with increasing x. Let D be a parameter satisfying*

$$x^{\frac{2}{3}} < D < x.$$ (12.9.1)

Let a_n be a real sequence of non-negative numbers supported on the square-free integers (so that $a_n = \mu^2(n)a_n$) and write

$$A_d(x) = \sum_{\substack{n \leq x \\ n \equiv 0 \,(\mathrm{mod}\, d)}} a_n.$$

Assume that

$$A_1(x) \gg A_1(x^{\frac{1}{2}})(\log x)^2 \tag{12.9.2}$$

and that uniformly for $d \leq x^{1/3}$,

$$A_d(x) \ll d^{-1}\tau(d)^8 A_1(x). \tag{12.9.3}$$

Also, for all $d \leq D$,

$$A_d(x) = g(d)A_1(x) + r_d(x), \tag{12.9.4}$$

where $r_d(x)$ is a remainder term satisfying

$$\max_{t \leq x} \sum_{d \leq D} \mu^2(d) |r_d(t)| \leq A_1(x)(\log x)^{-2^{22}}. \tag{12.9.5}$$

Here $g(d)$ is a multiplicative function satisfying

$$0 \leq g(p) < 1, \qquad g(p) \ll p^{-1}, \tag{12.9.6}$$

and

$$\sum_{p \leq y} g(p) = \log \log y + c + O\big((\log y)^{-10}\big) \tag{12.9.7}$$

for some constant c (depending at most on g), for all $y \geq e^e$. Suppose, in addition, that

$$\max_{1 \leq C \leq x/D} \sum_m \left| \sum_{\substack{n \sim N \\ mn \leq x}} \gamma(n; C)\mu(mn)a_{mn} \right| \ll A_1(x)(\log x)^{-2^{22}} \tag{12.9.8}$$

for every N with

$$\eta D^{\frac{1}{2}} < N < \delta x^{\frac{1}{2}}, \tag{12.9.9}$$

where the coefficients $\gamma(n; C)$ are given by

$$\gamma(n; C) = \sum_{\substack{d \leq C \\ d|n}} \mu(d). \tag{12.9.10}$$

Then we have

$$\sum_{p \leq x} a_p \log p = H A_1(x) \left(1 + O\left(\frac{\log \delta}{\log \eta}\right)\right). \tag{12.9.11}$$

Here

$$H = \prod_p (1 - g(p)) \left(1 - \frac{1}{p}\right)^{-1}. \tag{12.9.12}$$

Remark. The conditions imposed on g ensure that H defined by (12.9.12) is a convergent product.

Theorem 12.4. *Given the hypotheses of Theorem 12.3 except that a_n is no longer assumed to be supported on the square-free integers, and so we have the additional hypotheses*

$$0 \le g(p^2) \le g(p), \qquad g(p^2) \ll p^{-2}, \tag{12.9.13}$$

$$\sum_{n \le x} a_n^2 \le A_1(x)^2 x^{-\frac{2}{3}}, \tag{12.9.14}$$

and, in place of (12.9.5), we have

$$\max_{t \le x} \sum_{d \le DL^2} \mu^2(d) \, |r_d(t)| \le A_1(x) L^{-2}, \tag{12.9.15}$$

where $L = (\log x)^{2^{24}}$ and d only takes cube-free values. Then (12.9.11) remains valid.

Theorem 12.5. *Let the hypotheses of Theorem 12.4 be given with the exception of (12.9.8). Let P be a parameter satisfying*

$$2 \le P \le \exp\left(\frac{\log(1/\eta)}{2^{35} \log \log x} \right) \tag{12.9.16}$$

and put

$$\Pi = \prod_{p < P} p.$$

Suppose that

$$\max_{1 \le C \le x/D} \sum_m \left| \sum_{\substack{n \sim N \\ (n,\Pi)=1 \\ mn \le x}} \gamma(n; C)\mu(mn)a_{mn} \right| \ll A_1(x)(\log x)^{-2^{26}}. \tag{12.9.17}$$

Then (12.9.11) remains valid.

Remark. In our present context we can take $D = x^{3/4-\epsilon}, \delta = (\log x)^{-J}$ for some large J, η can be as small as $x^{-1/24+\epsilon}$, and this will suffice to prove Theorem 12.2. We note that the hypothesis (12.9.8) is of the same type, albeit slightly more complicated, as we derived for Theorem 12.1 (see Section 12.5).

Proof. (Theorem 12.3) The large negative powers of the logarithm in (12.9.5) and (12.9.8) may appear rather bizarre. In fact, the actual results required are given as (12.9.18) and (12.9.19). It is shown in [38, pp. 1046–1047] that these are consequences of (12.9.5), (12.9.8), and (12.9.3), and these last three formulae then make no further appearance in the proof. The bounds actually required in the proof are as follows:

$$\sum_{d \le D} \mu(d)^2 \tau_5(d) \, |r_d(t)| \ll A(x)(\log x)^{-3}, \tag{12.9.18}$$

$$\sum_m \tau_5(m) \left| \sum_{\substack{n \sim N \\ mn \le x}} \gamma(n; C)\mu(mn)a_{mn} \right| \ll A(x)(\log x)^{-3}. \tag{12.9.19}$$

In a further technical reduction at the start of the proof, it is shown that the formula (12.9.7) implies that

$$\sum_{\substack{d \leq y \\ (d,\nu)=1}} \mu(d)g(d) \ll \sigma_\nu (\log y)^{-6} \tag{12.9.20}$$

uniformly in $\nu \geq 1, y \geq 2$, where

$$\sigma_\nu = \prod_{p|\nu} \left(1 + \frac{1}{\sqrt{p}}\right) \tag{12.9.21}$$

(see [38, pp. 1048–1049]).

The starting point is again Vaughan's identity as we presented it in Chapter 2. For $n > z$ we have, by (2.1.2) with $U = z, V = y$, that

$$\Lambda(n) = \sum_{\substack{b|n \\ b \leq y}} \mu(b) \log(n/b) - \sum_{\substack{bc|n \\ b \leq y, c \leq z}} \mu(b)\Lambda(c) + \sum_{\substack{bc|n \\ b > y, c > z}} \mu(c)\Lambda(c). \tag{12.9.22}$$

This basic identity must now be modified in three ways. First an extra parameter s is introduced to split up the final term in (12.9.22) into three subsums:

$$\sum_{\substack{bc|n \\ b > sy \\ c > sz}} \mu(b)\Lambda(c) + \sum_{\substack{bc|n \\ sy \geq b > y \\ c > z}} \mu(b)\Lambda(c) + \sum_{\substack{bc|n \\ b > sy \\ sz \geq c > z}} \mu(b)\Lambda(c)$$

$$= f_1(n; y, z) + f_2(n; y, z) + f_2(n; y, z), \quad \text{say.}$$

The choice for s is taken to be the smallest power of 2 satisfying

$$1 < \frac{8s\sqrt{D}\eta}{\delta\sqrt{x}}.$$

Next, for technical reasons (to separate variables later), the parameters y, z are subject to some averaging. To be precise, we integrate our decomposition of $\Lambda(n)$ with respect to y and z as follows:

$$\int_Y^{eY} \int_Z^{eZ} \quad * \quad \frac{dy\, dz}{yz}. \tag{12.9.23}$$

Here $*$ represents both sides of (12.9.22). If $n > eZ$, the left-hand side remains $\Lambda(n)$, of course, but the sums on the right-hand side will have been smoothed somewhat. We shall write $I(*)$ to denote the application of (12.9.23).

Our final refinement of Vaughan's identity at this stage is to multiply both sides of (12.9.22) by a function $\rho(n)$ supported on almost-primes in some sense. We will need

$$\rho(n) \begin{cases} = 1 & \text{if } n \text{ is prime,} \\ \geq 0 & \text{otherwise.} \end{cases}$$

So long as we are only concerned with square-free values of n, this has not altered the left-hand side of (12.9.22), although it will help estimate one of the terms (now modified) on the right-hand side. We can construct a suitable function ρ using the techniques from Chapter 4. To be precise, we can use the upper-bound sieve construction to give

$$\rho(n) = \sum_{\nu \mid n} \lambda_\nu,$$

where $\lambda_\nu = 0$ for $\nu > \eta^{-1}$ or $\nu \nmid P(\eta^{-1})$, and for other values of ν either $\lambda_\nu = \mu(\nu)$ or $\lambda_\nu = 0$. Since the construction gives $\lambda_1 = 1$, we have $\rho(n) = 1$ if n is a prime exceeding η^{-1}. Indeed, $\rho(mn) = \rho(m)$ whenever n is a prime exceeding η^{-1}.

Now

$$\sum_{p \leq x} a_p \log p = \sum_{Z < n \leq x} a_n \Lambda(n) + O\left(\frac{A_1(x)}{\log x}\right)$$

$$= S(x, Z) + O\left(\frac{A_1(x)}{\log x}\right), \quad \text{say,}$$

by (12.9.2). We then write

$$T(x; Y) = \sum_{n \leq x} a_n I\left(\sum_{\substack{b \mid n \\ b \leq y}} \mu(b) \log(n/b)\right),$$

$$T(x; Y, Z) = \sum_{n \leq x} a_n I\left(\sum_{\substack{bc \mid n \\ b \leq y, c \leq z}} \mu(b) \Lambda(c)\right)$$

and put

$$S_j(x; Y, Z) = \sum_{n \leq x} a_n I(f_j(n; y, z)).$$

We thus have

$$S(x, Z) = T(x; Y) - T(x; y, Z) + \sum_{j=1}^{3} S_j(x; Y, Z), \qquad (12.9.24)$$

and we can complete the proof of Theorem 12.3 by obtaining suitable formulae or bounds for the five terms on the right-hand side of (12.9.24).

Evaluation of $T(x; Y)$. Writing $\log(n/b) = \log n - \log b$ and ignoring the averaging over z and y (which is unnecessary for this term), we get

$$T(x; y) = T_1(x; y) - T_2(x; y), \quad \text{say.}$$

We shall find that T_1 gives a suitable error term, while T_2 provides the main term in (12.9.11). Not surprisingly, the working is very similar to that used to obtain the main term for Theorem 12.1; only the presence of the factors $\rho(n)$ complicates matters slightly.

We have

$$T_1(x; y) = \sum_{b \leq y} \mu(b) \sum_{\substack{n \leq x \\ n \equiv 0 \,(\text{mod } b)}} a_n \rho(n) \log n. \qquad (12.9.25)$$

Now write

$$V_b(x) = \sum_{\substack{n \le x \\ n \equiv 0 \,(\mathrm{mod}\, b)}} a_n \sum_{\nu \mid n} \lambda_\nu$$

$$= \sum_\nu \lambda_\nu A_{[\nu,b]}(x) = \sum_\nu \lambda_\nu \left(g([\nu,b]) A_1(x) + r_{[\nu,b]}(x) \right)$$

$$= \sum_\nu \left(V_b(\nu;1) + V_b(\nu;2) \right), \quad \text{say.}$$

Then, by partial summation, we have

$$T_1(x;y) = \sum_{b \le y} \left(V_b(x) - \int_1^x V_b(t)\, \frac{dt}{t} \right). \tag{12.9.26}$$

Since $g([\nu,b]) = g(\nu)g(b/(\nu,b))$ we have

$$\sum_{b \le y} \mu(b) g(b/(\nu,b)) = \sum_{d \mid \nu} \mu(d) \sum_{\substack{b \le y/d \\ (b,\nu)=1}} \mu(b) \ll \tau(\nu)\sigma_\nu (\log x)^{-6}$$

by (12.9.20). Hence

$$\sum_{b \le y} \sum_\nu V_b(\nu;1) \ll A_1(x)(\log x)^{-6} \sum_\nu g(\nu)\tau(\nu)\sigma_\nu \ll A_1(x)(\log x)^{-4}.$$

Also,

$$\sum_{b \le y} \sum_\nu V_b(\nu;2) \ll \sum_\nu \sum_{b \le y} \left| \lambda_\nu \mu(b) r_{\nu,b} \right| \ll A_1(x)(\log x)^{-3}$$

using (12.9.18) since $y\eta^{-1} < D$. Since the same bounds also hold using $V_b(t)$ for $1 \le t \le x$, this gives $T_1(x;y) \ll A_1(x)(\log x)^{-2}$ as required.

Now

$$T_2(x;y) = \sum_{b \le y} \mu(b) \log b \sum_\nu \lambda_\nu A_{[\nu,b]}(x) = M + R, \quad \text{say,}$$

where

$$M = \sum_{b \le y} \mu(b) \log b \sum_\nu \lambda_\nu g([\nu,b]) A_1(x),$$

and R can be estimated in the same manner as the terms involving $V_b(\nu;2)$ above. As before, we can write $g([\nu,b]) = g(\nu)g(b/(\nu,b))$. Using (12.9.20) we can extend the summation range for b in the expression for M to infinity with an additional error $\ll A_1(x)(\log x)^{-3}$. We then note that

$$\sum_b \mu(b) g(b/(\nu,b)) \log b = \sum_{n \mid \nu} \mu(n) \sum_{(b,\nu)=1} \mu(b)g(b) \log(bn)$$

$$= \sum_{n \mid \nu} \mu(n) \sum_{(b,\nu)=1} \mu(b)g(b) \log(b)$$

$$= \begin{cases} -H & \text{if } \nu = 1, \\ 0 & \text{otherwise.} \end{cases}$$

Combining our results then gives

$$T(x; y) = HA_1(x) + O\left(x(\log x)^{-2}\right). \tag{12.9.27}$$

Estimation of $T(x; Y, Z)$. The estimation of this term is very similar to the working for $T_1(x; y)$. The reader should quickly verify, using (12.9.20), (12.9.6), and (12.9.18), that

$$T(x; y, z) \ll A_1(x)(\log x)^{-2}. \tag{12.9.28}$$

Estimation of $S_1(x; Y, Z)$. Again, we do not need to make use of the averaging over z and y. However, this is the term for which we needed to introduce $\rho(n)$. We have

$$S_1(x; y, z) = \sum_{\substack{bcd \le x \\ b > sy, c > sz}} \mu(b)\Lambda(c)\rho(bcd)a_{bcd}.$$

The satisfactory estimate of this sum depends crucially on the short range to which c is constrained, namely,

$$\delta < \frac{c}{\sqrt{x}} < \delta^{-1}, \tag{12.9.29}$$

and the small size of the variable d, that is, $d < \delta^{-2}$. We throw away the possibility of detecting any cancellation with the $\mu(b)$ factor and so obtain

$$|S_1(x; y, z)| \le \sum_c \sum_d \sum_{\substack{n \le x \\ cd|n}} \rho(n)a_n$$

$$= \sum_c \sum_d \sum_\nu \lambda_\nu A_{[cd,\nu]}(x).$$

Using (12.9.4) and (12.9.18) we arrive at the bound

$$|S_1(x; y, z)| \le A_1(x)L(x)M(x) + O\left(A_1(x)(\log x)^{-2}\right), \tag{12.9.30}$$

with

$$L(x) = \sum_c \Lambda(c)g(c), \qquad M(x) = \sum_d \sum_\nu \lambda_{nu}g([d, \nu]).$$

Since $g(p) \ll p^{-1}$, we deduce from (12.9.29) that $L(x) \ll \log(\delta^{-1})$.

We recall that $g([d, \nu]) = g(d)g(\nu/(\nu, d))$ and use this observation to tackle $M(x)$. Now, for *any* multiplicative function f we have

$$\sum_\nu \lambda_\nu f(\nu/(\nu, d)) \ge 0$$

(this follows from the construction of the λ_{nu} to give an upper-bound sieve). We then use Rankin's trick (compare the proof of Lemma 4.2) to infer that

$$M(x) \le \delta^{-2\epsilon} \sum_\nu \lambda_\nu g(\nu) \sum_d g(d/(\nu, d))d^{-\epsilon}$$

$$= \delta^{-2\epsilon} \prod_p \left(1 + \frac{g(p)}{p^\epsilon}\right) \sum_\nu \lambda_\nu g(\nu)h(\nu),$$

where $h(\nu)$ is the multiplicative function given by

$$h(p) = \left(1 + p^{-\epsilon}\right)\left(1 + g(p)p^{-\epsilon}\right)^{-1}.$$

We then have $h(p) < 2(1 + g(p))^{-1}$ and $g(p)h(p) < 1$. It follows that

$$\sum_{\nu} \lambda_{\nu} g(\nu) h(\nu) \ll \prod_{p < \eta^{-1}} (1 - g(p)h(p)) = \prod_{p < \eta^{-1}} (1 - g(p))(1 + g(p)p^{-\epsilon})^{-1}.$$

If we choose $\epsilon = (\log(1/\eta))^{-1}$, then

$$\prod_{p \geq \eta^{-1}} (1 + g(p)P^{-\epsilon}) \ll 1, \qquad \delta^{-2\epsilon} \ll 1.$$

Thus

$$M(x) \ll \prod_{p < \eta^{-1}} (1 - g(p)) \ll (\log(1/\eta))^{-1}. \qquad (12.9.31)$$

Combining our estimates then furnishes us with the bound

$$S_1(x; y, z) \ll A_1(x) \frac{\log \delta}{\log \eta}, \qquad (12.9.32)$$

as required.

Estimation of $S_2(x; Y, Z)$. For this term we must make use of the averaging over y but do not need the integration over z. For the present we assume that $a_n = 0$ for $n > x$ to simplify summation conditions. Since

$$S_2(x; y, z) = \sum_{e} \sum_{y < b \leq sy} \sum_{c > z} \mu(b)\Lambda(c)\rho(ebc)a_{ebc},$$

we have

$$|S_2(x; y, z)| \leq (\log x) \sum_{k} \left| \sum_{y < b \leq sy} \mu(b)\rho(bk)a_{bk} \right|$$

$$= (\log x) \sum_{\nu_1, \nu_2} |\lambda_{\nu_1 \nu_2}| \sum_{k} \left| \sum_{y < b\nu_2 \leq sy} \mu(b)a_{bk\nu_1\nu_2} \right|.$$

We use the integration over y to remove the ν_2 from the range of summation for b as follows:

$$\int_{Y}^{eY} \left| \sum_{y < b\nu_2 \leq sy} \cdots \right| \frac{dy}{y} \leq \int_{\eta Y}^{eY} \left| \sum_{y < b \leq sy} \cdots \right| \frac{dy}{y}, \qquad (12.9.33)$$

where we have used the fact that $\eta\nu_2 < 1$. The reader should note that, as with previous methods for removing joint summation conditions, this only introduces an extra logarithmic factor. Thus

$$S_2(x; Y, z) \leq (\log x) \int_{\eta Y}^{eY} \sum_{\ell} \tau_3(\ell) \left| \sum_{y < b \leq sy} \mu(b)a_{b\ell} \right| \frac{dy}{y}.$$

Using (12.9.19) we thereby obtain

$$S_2(x; Y, z) \ll A_1(x)(\log x)^{-1}. \tag{12.9.34}$$

Estimation of $S_3(x; Y, Z)$. For this final sum we need the integration over z, and we will need to detect cancellations involving both the c and b variables after transforming the sum. Now

$$S_3(x; y, z) = \sum_e \sum_{b > sy} \mu(b) \sum_{z < c \le sz} \Lambda(c)\rho(ebc)a_{ebc}.$$

First we note that $\rho(enc) = \rho(en)$ since c is a prime. This would now be the same sum as the last one discussed were it not for the fact that we have $\Lambda(c)$ where we would want $\mu(c)$. We therefore begin by splitting up $\Lambda(c)$. Write

$$\lambda(\zeta, \xi) = \begin{cases} \log(\zeta/\xi) & \text{if } \zeta \ge \xi, \\ 0 & \text{otherwise.} \end{cases}$$

We also note that $\Lambda(c) = -\mu(c)\Lambda(c)$ for the values of c counted. Starting to develop Vaughan's identity again for $\Lambda(c)$ we have, for any value C,

$$\begin{aligned} \mu(c)\Lambda(c) &= \sum_{m|c} \mu(m) \log m \\ &= \sum_{m|c} \mu(m) \log(m/C) \\ &= \lambda^+(c) - \lambda^-(c), \end{aligned}$$

where

$$\lambda^+(c) = \sum_{m|c} \lambda(m, C), \qquad \lambda^-(c) = \sum_{m|c} \lambda(C, m).$$

We put $C = x/D$ and then we have

$$\lambda^-(c) = \int_1^C \gamma(c, t) \frac{dt}{t}.$$

We can therefore apply (12.9.10) to this part of the sum to give a bound

$$\sum_\ell \tau_4(\ell) \left| \sum_{z < c \le sz} \mu(c)\lambda^-(c)a_{c\ell} \right| \ll A_1(x)(\log x)^{-1}. \tag{12.9.35}$$

For the remainder of the sum we observe that

$$\sum_e \sum_{b > sy} \mu(b)\rho(eb) \sum_{z < \ell m \le sz} \mu(\ell)\lambda(m, C)a_{eb\ell m}$$

$$= \int_C^x \left(\sum_{b > sy} \mu(b) \sum_{ebz < x} \rho(eb) \sum_{\substack{\ell t < sz \\ eb\ell t < x}} \mu(\ell) \sum_{\substack{m > t \\ z < \ell m \le sz}} a_{eb\ell m} \right) \frac{dt}{t}.$$

Now write $d = eb\ell$. Then $d < D$, and d is only counted when it is square-free by our condition on a_n. The inner sum above thus equals

$$A_d(\min(x, ebsz)) - A_d(\max(dt, ebz)).$$

Resorting once more to (12.9.4) the inner sum becomes

$$g(d) \sum_{\substack{dt<n\leq x \\ ebz<n\leq ebsz}} a_n + r_d(\min(s, ebsz)) - r_d(\max(dt, ebz)). \qquad (12.9.36)$$

For that part of the sum corresponding to the first term in (12.9.36) we can catch the cancellation caused by the $\mu(b)$ term, having rewritten $d = eb\ell$. Write

$$b_1 = \max\left(sy, \frac{n}{esz}\right), \qquad b_2 = \min\left(\frac{n}{ez}, \frac{n}{e\ell t}\right).$$

Then our range of summation for b is $b_1 < b < b_2$, and we obtain

$$\sum_{\substack{(b,e\ell)=1 \\ b_1<b<b_2}} \mu(b)\rho(eb)g(eb\ell) = g(e\ell) \sum_{\nu_1} g(\nu_1) \sum_{\nu_2} \lambda_{\nu_1\nu_2} \sum_{\substack{(b,e\ell\nu_1)=1 \\ b_1<b\nu_1<b_2}} \mu(b)g(b).$$

The inner sum here provides all the cancellation we need since it is

$$\ll \sigma_{e\ell\nu_1}(\log x)^{-6}$$

by (12.9.20). Summing over all the other variables trivially gives a contribution $\ll A_1(x)(\log x)^{-1}$.

The final terms in (12.9.36) can be bounded using (12.9.18) in a similar manner to previous estimates. To be precise, these terms, after integration over z and applying the device (12.9.33) with z in place of y, contribute

$$\leq \int_Z^x \int_C^x \sum_{d<D} \tau_5(d) \left| r_d(\min(x, sz)) - r_d(\max(t, z)) \right| \frac{dt}{t} \frac{dz}{z}.$$

An application of (12.9.18) then completes the estimation of this term to yield, all told, $S_3(x; y, Z) \ll A_1(x)(\log x)^{-1}$.

Completion of the Proof. The reader will see that we have now estimated all the terms from (12.9.24) in a satisfactory manner, and the proof of Theorem 12.3 is finished. $\qquad\square$

We leave it as an exercise for the reader to work out the details necessary to generalize the proof to obtain Theorems 12.4 and 12.5. The solution may be found in [38]! No new ideas are required to obtain these results.

12.10 SKETCH OF THE CRUCIAL RESULT

The most difficult part of the proof by far is to establish (12.9.17) in this context. This occupies more than 70 pages in [37], and so it would have been inappropriate to reproduce it here (the present author not having found any great simplification of the argument!). We content ourselves with just mentioning a few landmarks in a remarkable *tour de force*. When proving the corresponding result for the proof of Theorem 12.1 (see Section 12.6), we had more "room to play with" since we needed only to save an arbitrary power of a logarithm on the "most trivial" estimate (because ℓ ranged over quite a dense set). Now we must not throw anything away connected with ℓ being a square and still save a logarithm power from the cancellation induced by the Möbius function. We state for reference the crucial result [37, Proposition 4.1].

Theorem 12.6. *Let $\gamma > 0$ and $A > 0$. Then*

$$\max_{1 \leq C \leq x/D} \sum_m \left| \sum_{\substack{n \sim N \\ (n,\Pi)=1 \\ mn \leq x}} \gamma(n;C)\mu(mn)a_{mn} \right| \ll A_1(x)(\log x)^{4-A} \qquad (12.10.1)$$

for every N with

$$X^{\frac{1}{4}+\gamma} < N < x^{\frac{1}{2}}(\log x)^{-B}$$

when $1 \leq C \leq N^{1-\gamma}$. Here B and the implied constant in (12.10.1) need to be taken sufficiently large in terms of γ and A.

The initial transformations are similar to those used previously. We have

$$a_{mn} = \frac{1}{4} \sum_{|w|^2=m} \sum_{|z|^2=n} \lambda(z \cdot w).$$

We introduce the condition $(m,n) = 1$ as before. We then have to estimate a bilinear form

$$B^*(M,N) = \sum_{(|w|^2,|z|^2)=1} \alpha(w)\beta(z)\lambda(z \cdot w),$$

with $\alpha(w), \beta(z)$ supported on the intervals $|w|^2 \sim M, |z|^2 \sim N$, respectively. It is also shown in [37] that one can take $z \equiv 1 \pmod{2(1+i)}$ and also lying in a fixed residue class $\pmod 8$. Moreover one needs only deal with values of z for which

$$|\beta(z)| \leq \tau = (\log x)^J$$

for some large value J that needs to be determined later in the proof. We then remove the $(m,n) = 1$ condition to give

$$B^*(M,N) = B(M,N) + O\left(\left(M^{\frac{1}{4}}N^{\frac{5}{4}} + P^{-1}M^{\frac{3}{4}}N^{\frac{3}{4}} \right) \log^3 N \right),$$

where

$$B(M,N) = \sum_{w,z} \alpha(w)\beta(z)\lambda(z \cdot w).$$

Next z is restricted to the familiar type of region amenable to study using Hecke L-functions,

$$\{z : N' < |z|^2 \leq (1+\theta)N', \phi < \arg z < \phi + 2\pi\theta\},$$

where $N' \sim N$ and $\theta = (\log N)^{-A}$ for some (large) A. Using a trivial estimate for small ϕ one can assume that $\phi > \pi\theta \pmod{\pi/2}$. As in Section 12.6, we apply Cauchy's inequality to obtain

$$B^2(M,N) \ll MD(M,N),$$

where

$$D(M,N) = \sum_w f(w) \left| \sum_z \lambda(z \cdot w) \right|^2$$

and $f(w)$ is radial with smooth support in the range $\frac{1}{4}M \leq |w|^2 \leq 4M$. The square in the sum above is multiplied out to give a double sum over z_1, z_2. The next goal is to introduce the condition $(z_1, z_2) = 1$. It is then shown that, apart from an acceptably small error, it suffices to bound

$$D^*(M, N) = \sum_{(z_1, z_2)=1} \beta(z_1)\beta(z_2)C(z_1, z_2),$$

where

$$C(z_1, z_2) = \sum_w f(w)\lambda(z_1 \cdot w)\lambda(z_2 \cdot w).$$

The significance of the last sum is that there is no restriction on w, so that it becomes possible to give an approximate evaluation for this expression. To be precise, Friedlander and Iwaniec show that

$$C(z_1, z_2) = \frac{1}{|z_1 z_2|^{\frac{1}{2}}} \sum_{h_1} \sum_{h_2} F\left(h_1(\Gamma|z_1|)^{-\frac{1}{2}}, h_2(\Gamma|z_2|)^{-\frac{1}{2}}\right) G(h_1, h_2),$$

$$(12.10.2)$$

where

$$\Gamma = |\operatorname{Im} \bar{z}_1 z_2|,$$

$$F(u_1, u_2) = \int\int f\left(\frac{z_2}{|z_2|}t_1^2 - \frac{z_1}{|z_1|}t_2^2\right) e(u_1 t_1 + u_2 t_2) \, dt_1 \, dt_2,$$

and

$$G(h_1, h_2) = \frac{1}{\Gamma} \sum_{g_1^2 z_2 \equiv g_2^2 z_1 \,(\operatorname{mod} \Gamma)} e\left(\frac{g_1 h_1 + g_2 h_2}{\Gamma}\right).$$

Now the terms with $(h_1, h_2) \neq (0,0)$ in (12.10.2) are dealt with relatively easily. One can show that

$$F\left(h_1(\Gamma|z_1|)^{-\frac{1}{2}}, h_2(\Gamma|z_2|)^{-\frac{1}{2}}\right) \ll \left(1 + h_1^2 H^{-2}\right)^{-1} \left(1 + h_2^2 H^{-2}\right)^{-1} M^{\frac{1}{2}} \log N,$$

with $H = M^{-1/4} N^{3/4}$ and

$$|G(h_1, h_2)| \leq \frac{4\tau_3(\Gamma)}{\Gamma} \left(z_1 h_1^2 - z_2 h_2^2, \Gamma\right).$$

Summing over $(h_1, h_2) \neq (0,0)$ then gives a suitably small term. For the term with $(h_1, h_2) = (0, 0)$ it is important to make the dependence on z_1, z_2 explicit. Thus one writes $F_0(z_1, z_2), G_0(z_1, z_2)$ for $F(0,0), G(0,0)$, respectively. This leaves

$$D_0(M, N) = \sum_{(z_1, z_2)=1} \beta(z_1)\bar{\beta}(z_2)\frac{1}{|z_1 z_2|^{\frac{1}{2}}} F_0(z_1, z_2)G_0(z_1, z_2) \qquad (12.10.3)$$

to consider.

Now it is shown that

$$F_0(z_1, z_2) = \widehat{f}(0)2 \log\left(2\frac{|z_1 z_2|}{\Gamma}\right) + O\left(\Gamma^2 M^{\frac{1}{2}} N^{-2} \log N\right),$$

where

$$\widehat{f}(0) = \int_0^\infty f(u)\, du \ll M^{\frac{1}{2}},$$

and

$$G_0(z_1, z_2) = 2 \sum_{4d|\Gamma} \frac{\phi(d)}{d} \left(\frac{z_2/z_1}{d} \right),$$

where $\left(\frac{a}{b} \right)$ is an extension of the Jacobi symbol to even moduli (see [37, (8.15)]). The most difficult part of the Friedlander-Iwaniec proof is the treatment of (12.10.3). The trivial bound for this expression only misses the desired result by a logarithmic power; so we have kept in all the necessary savings from ℓ being a square. The aim is to use the cancellation induced by the Möbius function still implicit in $\beta(z_j)$ to save an arbitrary power of a logarithm. The most problematic part is the influence of the term $G_0(z_1, z_2)$, in particular, the quadratic character it contains.

The next step is to show that

$$D_0(M, N) = 2\widehat{f}(0)N^{-\frac{1}{2}}T(\beta) + O\left((\tau^{-1} + \theta)Y(\beta)M^{\frac{1}{2}}N^{-\frac{1}{2}}\log N \right),$$

with

$$T(\beta) = \sum_{z_1, z_2)=1} \beta(z_1)\bar{\beta}(z_2)G_0(z_1, z_2)\log\left(2|z_1 z_2/\Gamma|\right),$$

and

$$Y(\beta) = \sum_{z_1, z_2)=1} |\beta(z_1)\beta(z_2)|\tau\left(|z_1|^2\right)\tau\left(|z_2|^2\right)\tau_3(\Gamma).$$

There is not much difficulty in obtaining the bound

$$Y(\beta) \ll \theta^4 N^2 (\log N)^{2^{19}}.$$

All the difficult work is now concentrated in $T(\beta)$, and the remainder of [37] is devoted to establishing the following result.

Proposition. *Let $\sigma > 0$ be given. Then in the above terminology we have*

$$T(\beta) \ll N^2(\log N)^{-\sigma} + P^{-1}N^2 \log N.$$

Rewriting, $T(\beta)$ becomes

$$2 \sum_d \frac{\phi(d)}{d} \sum_{\substack{(z_1, z_2)=1 \\ \Gamma \equiv 0 \,(\mathrm{mod}\, 4d)}} \beta(z_1)\bar{\beta}(z_2) \left(\frac{z_1/z_2}{d} \right) \log\left(2\frac{|z_1 z_2|}{\Gamma} \right).$$

This is split up into three terms corresponding to the size of d. With the introduction of smooth functions to facilitate range splitting, we have $T(\beta) = U(\beta) + V(\beta) + W(\beta)$. In $U(\beta)$ we have

$$d \ll \frac{\Gamma}{(\log x)^J}, \qquad d < (\log x)^J,$$

for some J. In $V(\beta)$ we have

$$d > (\log x)^J, \qquad d \ll \frac{\Gamma}{(\log x)^J}.$$

Finally, in $W(\beta)$ we have

$$d \gg \frac{\Gamma}{(\log x)^J}.$$

Different techniques are required for each of the three ranges.

Friedlander and Iwaniec next tackle what they call "Jacobi-twisted sums over arithmetic progressions." They establish the following large-sieve-type result.

Lemma 12.12. *Let $D, R, S \geq 1$ be given. Let the complex numbers $\alpha_{r,s}$ be supported on the set*

$$\{(r, s) : (r, 2s) = 1,\ r \sim R, s \sim S\}.$$

Then

$$\sum_{d \sim D} \sum_{a \,(\mathrm{mod}\, d)} \left| \sum_{\bar{r}s \equiv a \,(\mathrm{mod}\, d)} \alpha_{r,s} \left(\frac{r}{d}\right) \right|^2 \leq \Xi(D, R, S) \sum_{r,s} \tau(r) |\alpha_{r,s}|^2,$$

where

$$\Xi(D, R, S) \ll D + RSD^{-\frac{1}{2}} + D^{\frac{1}{3}}(RS)^{\frac{2}{3}}(\log 2RS)^4 + (R + S)^{\frac{1}{12}}(RS)^{\frac{11}{12}+\epsilon}.$$

This result is used to estimate $V(\beta)$ in a satisfactory manner. For $U(\beta)$ one can use the Hecke L-functions in a similar manner to the proof of Lemma 12.8. As in that result, it is the small size of d in this sum that is crucial, enabling us to get a satisfactory estimate even should there be "exceptional" zeros. The final term $W(\beta)$ is bounded via the following result.

Proposition. *Let*

$$S_\chi^k(\beta') = \sum_{(z,\Pi)=1} \beta(z)\chi(z)\psi(z)(z/|z|)^k,$$

where

$$\psi(z) = i^{(r-1)/2}\left(\frac{s}{|r|}\right), \quad when\ z = r + is,$$

and χ is a character $(\mathrm{mod}\, 4d)$. *Then there is a constant $\delta > 0$ such that*

$$S_\chi^k(\beta') \ll N^{1-\delta} \tag{12.10.4}$$

uniformly in $(|k| + 1)d \leq N^\delta$.

This result takes up most of the final 20 pages of [37]. The source of the cancellation in this final sum is the term $\psi(z)$. Among the ingredients necessary to establish (12.10.4) we mention the bounds

$$\sum_{\substack{w,z \\ w \equiv z \equiv 1 \,(\mathrm{mod}\, 2(1+i))}} a(w)b(w)\psi(wz) \ll (M + N)^{\frac{1}{12}}(MN)^{\frac{11}{12}+\epsilon},$$

where $|a(w)|, |b(z)| \leq 1$ and these coefficients are supported on the discs $|w|^2 \leq M, |z|^2 \leq N$;

$$\sum_{z \in \mathcal{B}} \chi(z)\psi(z)(z/|z|)^k \ll d(|k|+1)|w|N^{\frac{3}{4}} \log(N|w|);$$

and

$$\sum_{\substack{z \in \mathcal{B} \\ (z,w)=1}} \chi(z)\psi(z)(z/|z|)^k \ll d(|k|+1)|w|\tau\left(|w|^2\right)N^{\frac{3}{4}} \log(N|w|).$$

In the above, $\mathcal{B}$ consists of those $z \in \mathbb{Z}[i]$ with $z \equiv 1 \pmod{2(1+i)}$ and belonging to a polar box that lies in the region $|z|^2 \leq N$. Analogous results are then demonstrated for

$$\sum_{m,n} a(m)b(n)\Theta(cmn) \quad \text{and} \quad \sum_{n \leq N} \Theta(mn), \qquad (12.10.5)$$

with or without the additional condition $(m,n) = 1$, where

$$\Theta(n) = \sum_{|z|^2=n} \psi(z)\chi(z)(z/|z|)^k.$$

The final stages involve a combinatorial identity for a sum of an arithmetical function over square-free integers and the following result, which is a consequence of Vaughan's identity and the bounds obtained for the sums in (12.10.5).

Lemma 12.13. *For any $c \geq 1$ we have*

$$\sum_{n \leq x} \Lambda(n)\Theta(cn) \ll c(|k|+1)dx^{\frac{76}{77}}.$$

Remark. The reader may be amused to see that the final task was to bound a sum over primes, albeit a rather different animal from the one we started with! This last result may have some intrinsic interest, but there appear to have been no developments on it since the publication of [37].

12.11 AND NOW?

When [37] first appeared, it seemed that there was a possibility to extend the result to proving the existence of infinitely many primes of the form $a^2 + b^6$. The set of numbers of this form has a similar density to the problem considered in Chapter 13. However, there appear to be very serious difficulties in making such an extension of the present results. Nevertheless, Friedlander and Iwaniec have shown in [41] that there are infinitely many primes of the form $a^2 + b^6$ under the assumption that there are infinitely many Siegel zeros in a certain sense. As in Heath-Brown's work [76], this extreme assumption means that there is a character $\chi(n)$ whose behaviour closely mimics $\mu(n)$ on square-free integers n. It would be very desirable to remove this hypothesis! The reader might like to try, as an exercise, to modify the alternative sieve so that it could be used in situations like those described in this chapter.

Chapter Thirteen

Primes of the Form $x^3 + 2y^3$

13.1 INTRODUCTION

The crucial point in the work of the previous chapter was the use that could be made of factorization over $\mathbb{Z}[i]$. Heath-Brown in [82] used the same basic idea to consider the form $x^3 + 2y^3$ that factorizes over $\mathbb{Z}[\rho]$, where here, and henceforth, $\rho = \sqrt[3]{2}$. The details of his work are very different to those of Friedlander and Iwaniec, however. The Type I and II information must be different, of course, but Heath-Brown uses Buchstab iteration in the spirit of our sieve (working in a way similar to Theorem 5.2 here) supplemented by the upper-bound Selberg sieve much as he did in his work on primes in short intervals [80]. He can do this since he generates "genuine" Type II information rather than the special Type II information using cancellation induced by the Möbius function as in [37]. In fact, his Type II information is a little more specialized than what we have used, but it does exactly match the requirements of the sieving process. We shall prove his theorem without involving the Selberg upper-bound sieve. The result he proves is as follows.

Theorem 13.1. *There are infinitely many primes of the form $x^3 + 2y^3$ with integer x, y. Indeed, there is a positive constant c such that if*

$$\eta = \eta(X) = (\log X)^{-c},$$

then the number of such primes with $X < x, y \leq X(1 + \eta)$ is

$$\sigma_0 \frac{\eta^2 X^2}{3 \log X} \{1 + O((\log \log X)^{-\frac{1}{6}})\} \tag{13.1.1}$$

as $X \to \infty$, where

$$\sigma_0 = \prod_p \left(1 - \frac{\nu_p - 1}{p}\right) \tag{13.1.2}$$

and ν_p denotes the number of solutions of the congruence $x^3 \equiv 2 \pmod{p}$.

Remarks. The product in (13.1.2) is conditionally convergent but not absolutely convergent. Heath-Brown remarks that one ought to be able to substitute other ranges for x and y or even take an arbitrary bounded set $\mathcal{R} \subseteq \mathbb{R}^2$ with a positive Jordan content.

This original result has since been extended by Heath-Brown in conjunction with Moroz to arbitrary irreducible cubic forms [88] and even cubic polynomials in two variables [89]. Another way in which one might hope to extend the theorem that has yet to be realized would be to consider incomplete norm forms for fields of higher degree.

The problem considered here can be compared with our work on Gaussian primes. Previously we were looking for such primes of the form $m + ni$ where either the modulus and argument were restricted (Chapter 11) or one of the variables m, n was restricted (Chapter 12). Now we are working in $\mathbb{Z}[\rho]$ and so looking for primes of the form $p = x + y\rho + z\rho^2$. With the natural norm we have $N(p) = x^3 + 2y^3 + 4z^3 - 6xyz$, and we wish this to be a rational prime with $z = 0$. For example, we have $p = 1 + 2\rho$ with $N(p) = 17$ and $p = 3 + 2\rho$ gives $N(p) = 43$. If we use Hecke L-functions with Grössencharacters then, under the appropriate Generalized Riemann Hypothesis, we can get an approximation to this result where we only need $z \ll (|x| + |y|)^\epsilon$ (compare (11.2.3) for the situation in $\mathbb{Z}[i]$).

13.2 OUTLINE OF THE PROOF

Following Heath-Brown, and in a similar manner to previous chapters, we let c denote a positive absolute constant, not necessarily the same at each occurrence. Similarly, given a parameter A, we use $c(A)$ to denote a "constant" depending only on A, again potentially different at each occurrence. The reader should be warned that the parameter A may have different meanings in different places.

We shall progress using similar language to the second problem in Chapter 11. That is, it is more natural to consider the corresponding sieve problem for ideals of the field $K = \mathbb{Q}(\rho)$. The prime ideal theorem then gives

$$\pi_K(x) = \text{Li}(x) + O\big(x \exp(-c\sqrt{\log x}\,)\big) \qquad (13.2.1)$$

for a suitable positive constant c, where $\pi_K(x)$ is the number of prime ideals of norm at most x. An important constant that arises in the following is

$$\gamma_0 = \frac{\pi \log \varepsilon_0}{\sqrt{27}},$$

which is the residue of the pole of the Dedekind zeta-function $\zeta_K(s)$ at $s = 1$, and $\varepsilon_0 = 1 + \rho + \rho^2$ is the fundamental unit of K.

We write $\text{i}(x, y)$ and $\text{i}(\beta)$ for the ideals generated by $x + y\rho$ and β, respectively. These are more usually written $(x + y\rho), (\beta)$, but we felt that curved brackets have too many other meanings in our formulae! We write $\tau(\text{j})$ for the number of divisors of j. For a set $\mathcal{J}$ of integral ideals of K, and any integral ideal $\mathfrak{r}$, we write

$$\mathcal{J}_\mathfrak{r} = \{\text{j} \in \mathcal{J} : \mathfrak{r}|\text{j}\}.$$

We write

$$\mathcal{A} = \{\text{i}(x, y) : x, y \in (X, X(1 + \eta)] \cap \mathbb{N}, (x, y) = 1\}$$

and, with j denoting a general ideal,

$$\mathcal{B} = \{\text{j} : N(\text{j}) \in (3X^3, 3X^3(1 + \eta)]\}.$$

We will follow our usual comparison principle of relating the number of primes (we hope that saying "primes" for "prime ideals" will not confuse the reader — we will try to say "rational primes" when we are back in $\mathbb{N}$) in $\mathcal{A}$ to the number of primes in $\mathcal{B}$. We also set

$$S(\mathfrak{I}, z) = \{\mathfrak{r} \in \mathfrak{I} : \mathfrak{p} | \mathfrak{r} \Rightarrow N(\mathfrak{p}) \geq z\}$$

for any real $z > 1$. We then have that the number of primes in $\mathcal{A}$ is

$$S(\mathcal{A}, 2X^{\frac{3}{2}}).$$

Given an ideal $\mathfrak{j}$, we write

$$S(\mathfrak{I}, \mathfrak{j}) \quad \text{for} \quad S(\mathfrak{I}, N(\mathfrak{j})).$$

Now, Buchstab's identity in K yields

$$S(\mathcal{E}, z) = S(\mathcal{E}, w) - \sum_{w \leq N(\mathfrak{p}) < z} S(\mathcal{E}_{\mathfrak{p}}, \mathfrak{p}). \tag{13.2.2}$$

Later on we shall abbreviate expressions like $w \leq N(\mathfrak{p}) < z$ to $w \leq \mathfrak{p} < z$ for simplicity. When dealing with Gaussian primes in Chapter 11, we had to be careful with primes of the same norm. By Lemma 13.1 this will not arise when sieving $\mathcal{A}$ but can arise for $\mathcal{B}$. However, since there are only finitely many prime ideals with any given norm, a suitable ordering can be chosen to make (13.2.2) well-defined.

We note that if η is sufficiently small, no two values of $x + y\rho$ are associates, so that $\mathcal{A}$ contains distinct ideals. Now we are interested in first-degree prime ideals $\mathfrak{p}$, that is, those with $N(\mathfrak{p}) = p$ where p is a rational prime. We will therefore often need the following elementary result. We leave the proof to the reader (or see [82, Section 4]).

Lemma 13.1. *No prime ideal of degree greater than 1 can divide an element of $\mathcal{A}$, nor can a product of two distinct first-degree prime ideals of the same norm. Thus if a square-free ideal $\mathfrak{r}$ divides an element of $\mathcal{A}$, then $N(\mathfrak{r})$ must be square-free.*

If every prime ideal factor $\mathfrak{p}$ of $\mathfrak{i}(x, y)$ has $N(\mathfrak{p}) \geq 2X^{3/2}$, then $\mathfrak{i}(x, y)$ will be a prime ideal, and so $x^3 + 2y^3$ is a prime, by Lemma 13.1. Writing $\pi(\mathcal{E})$ for the number of prime ideals in $\mathcal{E}$ it then suffices to show that

$$\pi(\mathcal{A}) = \kappa\pi(\mathcal{B}) + O\left(\frac{\eta^2 X^2}{\log X}(\log \log X)^{-\frac{1}{6}}\right), \tag{13.2.3}$$

where

$$\kappa = \sigma_0 \eta (3X)^{-1},$$

in order to prove Heath-Brown's theorem.

The Type I information we need for $\mathcal{A}$ and $\mathcal{B}$ is supplied by the following two results. Here and elsewhere in this chapter, $\mathcal{R}$ is the set of ideals $\mathfrak{j}$ for which $N(\mathfrak{j})$ is square-free.

Lemma 13.2. *Let $\omega_2(\mathfrak{j})$ be the multiplicative function on ideals defined by*

$$\omega_2(\mathfrak{p}^e) = (1 + N(\mathfrak{p})^{-1})^{-1}.$$

Then for any positive integer A there exists a corresponding $c(A)$ such that

$$\sum_{\substack{N(j)\sim Q \\ j\in\mathcal{R}}} \tau(j)^A \left| |\mathcal{A}_j| - \frac{6\eta^2 X^2}{\pi^2 N(j)} \omega_2(j) \right| \tag{13.2.4}$$

$$\ll (Q + XQ^{\frac{1}{2}} + X^{\frac{3}{2}})(\log QX)^{c(A)}.$$

Lemma 13.3. *For any positive integer A there exists a corresponding $c(A)$ such that*

$$\sum_{N(j)\sim Q} \tau(j)^A \left| |\mathcal{B}_j| - \gamma_0 \frac{3\eta X^3}{N(j)} \right| \ll X^2 Q^{\frac{1}{3}} (\log Q)^{c(A)}. \tag{13.2.5}$$

The Type II information we require is given by the following result. The unknown coefficients $c_{\mathfrak{r}}, d_{\mathfrak{s}}$ must have a special form to be specified later, but quoting the Type II result here will enable the reader to see the basic Buchstab decomposition immediately. We now write $\tau = (\log\log X)^{-1/6}$.

Lemma 13.4. *Under conditions on $c_{\mathfrak{r}}, d_{\mathfrak{s}}$ to be specified later, we have*

$$\sum_{\substack{N(\mathfrak{s})\sim V}} \sum_{\substack{\mathfrak{r} \\ rs\in\mathcal{A}}} c_{\mathfrak{r}} d_{\mathfrak{s}} = \kappa \sum_{\substack{N(\mathfrak{s})\sim V}} \sum_{\substack{\mathfrak{r} \\ rs\in\mathcal{B}}} c_{\mathfrak{r}} d_{\mathfrak{s}} + O\left(X^2(\log X)^{-A}\right) \tag{13.2.6}$$

when $X^{1+\tau} \ll V \ll X^{3/2-\tau}$.

From our work in Chapters 3 and 5 we would expect to get an asymptotic formula for

$$\sum_{N(\mathfrak{r})\sim R} c_{\mathfrak{r}} S(\mathcal{A}_{\mathfrak{r}}, z) \tag{13.2.7}$$

for any $z \ll X^{1/2-2\tau}$ if $R < X^{2-\tau}$. Since we are considering objects of "size" X^3, it is natural at this point to put $x = X^3$, $w = X^\tau$, $z_0 = X^{1/2-3\tau}$, $z_1 = X^{1+\tau}$, $z_2 = X^{2-\tau}$, $z_3 = 2X^{3/2}$ (We might have expected to put $z_0 = z_2/z_1 = X^{1/2-2\tau}$, but reducing the value slightly helps with a technical point later). We would then begin our Buchstab decomposition as follows:

$$S(\mathcal{A}, z_3) = S(\mathcal{A}, z_0) - \sum_{z_0 \le \mathfrak{p} < z_3} S(\mathcal{A}_{\mathfrak{p}}, \mathfrak{p}) \tag{13.2.8}$$

$$= S(\mathcal{A}, z_0) - S_1 - S_2 - S_3 + S_4 + S_5 + S_6,$$

where

$$S_1 = \sum_{z_0 \le \mathfrak{p} < z_1} S(\mathcal{A}_{\mathfrak{p}}, z_0), \quad S_2 = \sum_{z_1 \le \mathfrak{p} \le z_2} S(\mathcal{A}_{\mathfrak{p}}, \mathfrak{p}), \quad S_3 = \sum_{z_2 < \mathfrak{p} < z_3} S(\mathcal{A}_{\mathfrak{p}}, z_0),$$

and

$$S_4 = \sum_{z_0 \le \mathfrak{q} < \mathfrak{p} < z_1} S(\mathcal{A}_{\mathfrak{p}\mathfrak{q}}, \mathfrak{q}), \quad S_5 = \sum_{\substack{z_2 < \mathfrak{p} < z_3 \\ z_0 \le \mathfrak{q} < (x/N(\mathfrak{p}))^{1/2}}} S(\mathcal{A}_{\mathfrak{p}\mathfrak{q}}, \mathfrak{q}).$$

We would expect to give asymptotic formulae for S_2 by our Type II information and for S_1 and S_3 since they are of the form (13.2.7). We would discard S_5 in its entirety (in this sum $\mathfrak{q} < z_1$, so we have no immediate hopes of an asymptotic formula). For most of S_4, either $z_1 \leq N(\mathfrak{pq}) \leq z_2$ or $x/z_2 \leq N(\mathfrak{pq}) \leq x/z_1$. The remainder of this sum is then discarded. By our usual method this gives

$$S(\mathcal{A}, z_3) > (\kappa - \epsilon)S(\mathcal{B}, z_3),$$

where

$$\epsilon \ll \log \left(\frac{\frac{1}{2}}{\frac{1}{2} - \tau} \right) \ll \tau,$$

which is the order of size error quoted in Theorem 13.1. Of course, this is only a lower bound, but an upper bound with the same error can be obtained similarly. For the upper bound we only discard after an odd number of Buchstab decompositions, of course. We thus throw away more at the first stage, but there is part of S_4 above that cannot be estimated but for which another Buchstab decomposition is possible. We thus arrive at

$$S(\mathcal{A}, z_3) < (\kappa + \epsilon)S(\mathcal{B}, z_3),$$

which completes the proof.

Although Heath-Brown's argument appears to be more complicated than the above, the reader can compare his work with our alternative presentation to see that in all essentials it is the same. When performing the Buchstab iterations, either "inside" or "outside" the Fundamental Theorem (that is in the proof of the corresponding Fundamental Theorem or in (13.2.8)) we restrict the norms of the prime variables to intervals of the form $J(m) = [X^{m\xi}, X^{(m+1)\xi})$, where

$$\xi = (\log \log X)^{-\frac{5}{6}} = \tau^5. \tag{13.2.9}$$

The reason for this choice of parameter will soon become apparent. We write $\mathcal{J}(m)$ for the set of prime ideals so restricted. Of course we must have $m > \tau^{-4} - 1$ here, and the first interval will have z_0 as its left-hand end-point. We note that the total number of intervals cannot exceed $3\tau^{-1}$. Hence if we have a collection of intervals with corresponding $m_j, 1 \leq j \leq r + 1, m_{j+1} < m_j$ for $j \leq r$, and $X^{m_1} < z_0$, then

$$\sum_{j=1}^{r} m_j < \frac{1 + \tau}{\xi}$$

$$\Rightarrow \quad \xi \sum_{j=1}^{r+1} (m_j + 1) < 1 + \tau + 3\tau^{-1}\xi + (\tfrac{1}{2} - 3\tau) \leq \frac{3}{2} - \tau. \tag{13.2.10}$$

It was to obtain the final inequality here that we reduced the value of z_0 above.

We consider first how the division argument affects the proof of a result for S_1 above that resembles our Fundamental Theorem. Suppose that we have performed r Buchstab decompositions and have $\mathfrak{p}_j \in \mathcal{J}(m_j)$ with $m_t > m_{t+1}$ for all t. Suppose also that

$$\sum_{j=1}^{r} m_j < \frac{1 + \tau}{\xi},$$

so that the Type II information would not be applicable. We then put

$$\sum_{\mathfrak{p}_j \in \mathcal{J}(m_j)} S(\mathcal{A}_{\mathfrak{p}_1\ldots\mathfrak{p}_r}, \mathfrak{p}_r) = \sum_{\mathfrak{p}_j \in \mathcal{J}(m_j)} S(\mathcal{A}_{\mathfrak{p}_1\ldots\mathfrak{p}_r}, w)$$

$$- \sum_{\substack{\mathfrak{p}_j \in \mathcal{J}(m_j), j \leq r \\ w \leq \mathfrak{p}_{r+1} < \mathfrak{p}_r}} S(\mathcal{A}_{\mathfrak{p}_1\ldots\mathfrak{p}_{r+1}}, \mathfrak{p}_{r+1}). \qquad (13.2.11)$$

We then divide up the summation range for $\mathfrak{p}_{r+1}$ as for the other variables. If $r+1$ is even, we discard the sum with $m_{r+1} = m_r$ (of course, $m_{r+1} \leq m_r$ follows from $\mathfrak{p}_{r+1} < \mathfrak{p}_r$). If $r+1$ is odd, we use, for the term $m_{r+1} = m_r$, the possibly crude bound

$$S(\mathcal{A}_{\mathfrak{p}_1\ldots\mathfrak{p}_{r+1}}, \mathfrak{p}_{r+1}) \leq S(\mathcal{A}_{\mathfrak{p}_1\ldots\mathfrak{p}_{r+1}}, w).$$

We shall prove that we can obtain an asymptotic formula for this last sum from our Type I information. We need such a formula for the first term on the right-hand side of (13.2.11) in any case. The required result is stated as Lemma 13.15. For the terms $m_{r+1} < m_r$ we either have

$$\sum_{j=1}^{r+1} m_j < \frac{1+\tau}{\xi},$$

in which case we continue, or, using (13.2.10),

$$\frac{1+\tau}{\xi} \geq \sum_{j=1}^{r+1} m_j \quad \text{and} \quad \sum_{j=1}^{r+1} (m_j + 1) \leq \frac{\frac{3}{2} - \tau}{\xi},$$

in which case we use our Type II information. In this way we obtain an upper bound in place of our usual asymptotic equality for our Fundamental Theorem result. However, we can use exactly the same procedure with the rules for odd and even r reversed to get the corresponding lower bound. It now remains to show that the errors introduced by the sums with $m_{r+1} = m_r$ (which we either discarded or for which we gave a crude upper bound) altogether give a suitably small contribution. Before we do this, we pause to consider the advantages of Heath-Brown's procedure. First, we have removed the $\mathfrak{p}_{r+1} < \mathfrak{p}_r$ condition without using Perron's formula (so we do not have to worry about $N(\mathfrak{p})^{it}$ factors). Indeed, the technique was used in a similar way in the previous chapter. Second, we have ensured that the variable $\mathfrak{s} = \mathfrak{p}_1 \ldots \mathfrak{p}_r$ remains square-free, and this will be important later. Finally, we can at any stage "pull out" the prime variable $\mathfrak{p}_1$ and use its properties (basically we will need an analogue of Siegel's theorem for this variable).

Now we consider the errors introduced by the division argument. We can convert sums over prime ideals into integrals using the following result, which is a corollary of the prime ideal theorem. We have used the analogous result for rational primes throughout this book and assumed the equivalent result in Chapter 11.

Lemma 13.5. *Let* $J_1, \ldots, J_m$ *be intervals of the form* $J_i = [a_i, \rho a_i)$, *with* $\rho > 1$ *and* $a_i \geq A > 1$ *for each* $i \leq m$. *Let* $Y \geq 1$ *be given and define* $\mathcal{J}(Y, m) \subseteq \mathbb{R}^m$

as the set of $(x_1, \ldots, x_m)$ with $x_i \in J_i$ and $\prod x_i \leq Y$. Then there are positive absolute constants c_1 and c_2 such that

$$\sum_{(N(P_1),\ldots,N(P_m))\in \mathcal{J}(Y,m)} \prod_{i=1}^{m} \log N(P_i)$$

$$= |\mathcal{J}(Y,m)| + O\left(mY(c_1 + \log \rho)^{m-1} \exp\left(-c_2(\log A)^{\frac{1}{2}}\right)\right),$$

uniformly in m. Here $||$ denotes m-dimensional Lebesgue measure.

We recall the following notation slightly modified from Chapter 3. Write $\boldsymbol{\alpha}_r = (\alpha_1, \ldots, \alpha_r)$ and put

$$d\boldsymbol{\alpha}_r = \frac{d\alpha_r \ldots d\alpha_1}{\alpha_1 \ldots \alpha_{r-1}\alpha_r^2}, \quad \mathcal{D}_r = \{\boldsymbol{\alpha}_r : \alpha_j < \alpha_{j-1}, 2 \leq j \leq r, \alpha_r > \alpha_{r-1} - \xi\}.$$

The integrals corresponding to discarded sums can then be bounded by

$$I_{2r} = \int_{\mathcal{R}} \omega\left(\frac{1 - \alpha_1 - \cdots - \alpha_{2r}}{\alpha_{2r}}\right) d\alpha_{2r}.$$

Here $\mathcal{R} = (\tau, \frac{1}{3})^{2r} \cap \mathcal{D}_{2r}$. We estimate the integral over α_{2r} as

$$\leq \frac{\xi}{\tau^2}$$

and substitute 1 for $\omega(u)$. By induction and elementary integration we then obtain (as in earlier chapters)

$$I_{2r} \leq \xi \frac{\left(\log\left(\frac{1}{3}/\tau\right)\right)^{2r-1}}{\tau^2(2r-1)!}. \tag{13.2.12}$$

Hence

$$\sum_r I_{2r} \leq \xi\tau^{-3}.$$

The reader will quickly verify that we obtain the same order estimate for the sums given a crude upper bound. Since $\xi\tau^{-3} < \tau$, the errors introduced are of a suitable size.

Similar size errors are introduced by replacing

$$S(\mathcal{A}_{\mathfrak{p}_1\ldots\mathfrak{p}_r}, p_r)$$

with

$$S(\mathcal{A}_{\mathfrak{p}_1\ldots\mathfrak{p}_r}, X^{m_r\xi}),$$

using Buchstab's identity and giving a satisfactory bound for the additional sum. More details can be found in [82], but we have altered the argument there by carrying out the division argument at every application of Buchstab's identity (this saves a factor τ^{-1}) and avoiding "edge effect" sums by our choice of z_0.

If $\mathfrak{s} = \prod_{j=1}^{r} \mathfrak{p}_j$ with the variables constrained as above, we then write

$$d_{\mathfrak{s}} = \prod_{i=j}^{n+1} \frac{\log N(\mathfrak{p}_j)}{m_j \xi \log X}.$$

Since $N(\mathfrak{p}_j) \in J(m_j)$, we find that

$$1 \le \frac{\log N(\mathfrak{p}_j)}{m_j \xi \log X} \le 1 + m_j^{-1} \le 1 + \xi \tau^{-1}.$$

Thus

$$1 \le d_{\mathfrak{s}} \le 1 + O(\xi \tau^{-2}),$$

which leads to another error of the same size as above.

We can deal with S_3 in a similar manner to S_1, and S_2 can be put in the right form almost immediately (with much smaller errors). We have thus shown that we only need to obtain our Type II information for sums of the form

$$\sum_{\mathfrak{r}} c_{\mathfrak{r}} \sum_{\substack{\mathfrak{s} \\ \mathfrak{r}\mathfrak{s} \in \mathcal{A}}} d_{\mathfrak{s}}.$$

The coefficients $c_{\mathfrak{r}}$ will take only the value 1 or 0, and will be supported on ideals $\mathfrak{r} \in \mathcal{R}$, all of whose prime factors $\mathfrak{p}$ satisfy $N(\mathfrak{p}) \ge X^\tau$. Similarly the coefficients $d_{\mathfrak{s}}$ will be supported on ideals $\mathfrak{s}$ as described above.

We shall want to prove, with coefficients as above, that

$$\sum_{\mathfrak{r}} c_{\mathfrak{r}} \sum_{\substack{\mathfrak{s} \\ \mathfrak{r}\mathfrak{s} \in \mathcal{A}}} d_{\mathfrak{s}} = \kappa \sum_{\mathfrak{r}} c_{\mathfrak{r}} \sum_{\substack{\mathfrak{s} \\ \mathfrak{r}\mathfrak{s} \in \mathcal{B}}} d_{\mathfrak{s}} + O(\nabla), \tag{13.2.13}$$

where ∇ is a suitably small error. To do this we use a device that has appeared in Chapters 2 and 12, namely, writing

$$\Lambda(n) = \sum_{d|n} \mu(d)(\log(L/d))$$

for some parameter L and then treating the part of the sum with $d \le L$ as a Type I sum, while the sum with larger d is a genuine Type II sum (with cancellation hopefully coming from the $\mu(d)$ factor). In the present context we write

$$d_{\mathfrak{s}} = e_{\mathfrak{s}} + f_{\mathfrak{s}}, \tag{13.2.14}$$

where the "leading part" $e_{\mathfrak{s}}$ is given by

$$e_{\mathfrak{s}} = \frac{w'(N(\mathfrak{s}))}{\prod_{i=1}^{n+1}(m_i \xi \log X)} \sum_{N(\mathfrak{j})<L} \mu(\mathfrak{j}) \log\left(\frac{L}{N(\mathfrak{j})}\right). \tag{13.2.15}$$

Here $\mu(\ )$ is the analogue of the Möbius function on the ideals of $\mathbb{Z}[\rho]$, we have fixed $L = X^{\tau/2}$ and written

$$w(t) = w(t, \mathbf{m}) = \left|\{\mathbf{x} \in \mathbb{R}^{n+1} : x_j \in J(m_j), \prod x_j \le t\}\right|.$$

In the case $n = 0$ the function $w(t)$ is only piecewise continuously differentiable, and so we define $w'(t)$ to be the right-hand derivative here. Heath-Brown constructs the function $w(t)$ so that the prime ideal theorem shows that

$$\frac{w(t)}{\prod_{i=1}^{n+1}(m_i \xi \log X)}$$

is an approximation to

$$\sum_{N(\mathfrak{s}) \leq t} d_{\mathfrak{s}}.$$

The basic idea is that $e_{\mathfrak{s}}$ can be treated with no problems when L is small, yet it simulates the distribution of $d_{\mathfrak{s}}$ over residue classes. Hence the average of $f_{\mathfrak{s}}$ in residue classes should be small. This is demonstrated by the following lemma. In subscripts we will now write (β) for the ideal generated by β to prevent too clumsy a notation at this point. Also, the quantity $\exp(\pm c\sqrt{\log L})$ will appear frequently, so we shall break with a long-standing convention in this tome and write

$$\mathcal{L} = \exp\left(\sqrt{\log L}\right).$$

Lemma 13.6. *Let $\mathcal{C} \subseteq \mathbb{R}^3$ be a cube of side $S_0 \geq L^2$ and suppose that for every vector $(x, y, z) \in \mathcal{C}$ we have $x, y, z \ll V^{1/3}$ and*

$$x^3 + 2y^3 + 4z^3 - 6xyz \gg V.$$

For each $\beta = a + b\rho + c\rho^2 \in K$ let $\hat{\beta}$ be the vector (a, b, c). Let a constant $A > 0$ be given. Then for any integer $\alpha \in \mathbb{Z}[\rho]$ we have

$$\sum_{\substack{\beta \equiv \alpha \,(\mathrm{mod}\, q) \\ \hat{\beta} \in \mathcal{C}}} f_{(\beta)} \ll V\mathcal{L}^{-c},$$

uniformly for $q \leq (\log X)^A$.

In the above it should be noted that we do not require α to be coprime to q. Also, a phenomenon to which we have become accustomed, we find that the implied constant is ineffective as a result of possible Siegel zeros. The term ∇ in (13.2.13) will come from two errors corresponding to our decomposition (13.2.14), which we write as $\nabla_e + \nabla_f$ in a self-explanatory notation. We further write $\nabla_f(V)$ for the contribution to ∇_f coming from terms with $N(\mathfrak{s}) \sim V$. Our Type II information will then follow from the next two lemmas.

Lemma 13.7. *There is an absolute constant c such that*

$$\nabla_e \ll M^{-1}\eta^{\frac{5}{2}} X^2 (\log X)^c,$$

where

$$M = \prod_{i=1}^{n+1} m_i.$$

Lemma 13.8. *Suppose we have a bound of the form*

$$\sum_{\substack{\beta \equiv \alpha \,(\mathrm{mod}\, q) \\ \hat{\beta} \in \mathcal{C}}} f_{(\beta)} \ll V\mathcal{L}^{-c}, \tag{13.2.16}$$

subject to the conditions of Lemma 13.6, uniformly in a range

$$q \leq Q_1 \leq \exp\left((\log X)^{\frac{1}{3}}\right).$$

Then there exists an absolute positive constant c such that

$$\nabla_f(V) \ll X^2 Q_1^{-\frac{1}{160}} (\log X)^c$$

for $z_1 \ll V \ll z_2$.

This result is the most novel (and difficult!) part of Heath-Brown's argument, and it is here that the structure of the expression $x^3 + 2y^3$ is most crucial.

We conclude this section with the final part of the proof of Theorem 13.1. We have

$$\pi(\mathcal{A}) - \kappa\pi(\mathcal{B}) \ll \tau\frac{\eta^2 X^2}{\log X} + \xi\tau^{-4}\frac{\eta^2 X^2}{\log X} + (\eta^{\frac{5}{2}} + Q_1^{-\frac{1}{160}})X^2(\log X)^c. \quad (13.2.17)$$

Now $\xi = \tau^5$. If we suppose that (13.2.17) holds with the constant $c = c_0$, say, then we can take

$$\eta = (\log X)^{-2c_0 - 2}$$

and

$$Q_1 = (\log X)^{160(c_0+1)}\eta^{-320} = (\log X)^{600(c_0+1)}.$$

We can then apply Lemma 13.6 and so deduce that

$$\pi(\mathcal{A}) - \kappa\pi(\mathcal{B}) \ll \tau\frac{\eta^2 X^2}{\log X}.$$

The result then follows from the quoted value for τ. $\square$

13.3 PRELIMINARY RESULTS

In this section we gather together some basic results, mainly on divisor sums in K but also some elementary facts from the geometry of numbers and a tool for counting points "near" a nonsingular hypersurface. We leave their proof to the reader. Heath-Brown gives explicit proofs to most of them in [82, Section 4].

Lemma 13.9. *The number of integral ideals of $K = \mathbb{Q}(\rho)$ with norm at most x is*

$$\gamma_0 x + O\left(x^{\frac{2}{3}}\right),$$

where γ_0 is as given in Lemma 2.2.

Lemma 13.10. *For any integer $A > 0$ there is a constant $c(A)$ such that*

$$\sum_{n \le x} \tau(n)^A \ll x(\log x)^{c(A)}$$

and

$$\sum_{N(\mathfrak{i}) \le x} \tau(\mathfrak{i})^A \ll x(\log x)^{c(A)}.$$

Indeed there is a positive constant $\delta = \delta(A)$ such that

$$\sum_{x < n \le x+y} \tau(n)^A \ll y(\log x)^{c(A)}$$

and

$$\sum_{x < N(\mathfrak{i}) \le x+y} \tau(\mathfrak{i})^A \ll y(\log x)^{c(A)}$$

for $x^{1-\delta} \le y \le x$.

Lemma 13.11. *Let $\mathcal{C} = (a_1, a_1 + S_0] \times (a_2, a_2 + S_0] \times (a_3, a_3 + S_0]$ be a cube of side S_0 and suppose that $\max |a_i| \le S_0^A$ for some positive constant A. For any $\beta = x + y\rho + z\rho^2 \in K$ write $\hat{\beta} = (x, y, z)$. Then there is a constant $c(A)$ such that*

$$\sum_{\hat{\beta} \in \mathcal{C}} \tau(\beta)^2 \ll S_0^3 (\log S_0)^{c(A)}.$$

Lemma 13.12. *Let $x, y \ge 2$ be given. Then for any positive integer A there exists a positive constant $c(A)$ such that*

$$\sum_{\substack{|m| \le x, \, |n| \le y \\ mn \ne 0}} \tau(m + n\rho)^A \ll xy(\log x)^{c(A)}.$$

The next result is used to count non-singular points near a hypersurface.

Lemma 13.13. *Let $\mathcal{C}_i \subseteq \mathbb{R}^n$ be disjoint hypercubes with parallel edges of length S_0 and contained in a ball of radius R centred on the origin. Let F be a real cubic form in n variables and let F_0 be a real constant. Suppose that each hypercube contains a point $\mathbf{x}$ for which $F(\mathbf{x}) = F_0 + O(R^2 S_0)$ and $|\nabla F(\mathbf{x})| \gg R^2$. Then the number of hypercubes $\mathcal{C}_i$ contained in any ball of radius R_0 is $\ll_F 1 + (R_0/S_0)^{n-1}$.*

13.4 THE TYPE I ESTIMATES

We follow [82] closely here and give all the relevant details. Initially we consider

$$|\{x, y \in (X, X(1+\eta)] : \mathfrak{r}|x + y\rho\}| = S(\mathfrak{r}; X) = S(\mathfrak{r}), \quad \text{say.}$$

The following lemma is the crucial first step.

Lemma 13.14. *If A is any positive integer, there exists $c(A)$ such that*

$$\sum_{\substack{N(\mathfrak{r}) \sim Q \\ \mathfrak{r} \in \mathcal{R}}} \tau(\mathfrak{r})^A \left| S(\mathfrak{r}) - \frac{\eta^2 X^2}{N(\mathfrak{r})} \right| \ll (X + Q)(\log Q)^{c(A)} \tag{13.4.1}$$

for $X \ge 1$.

Proof. Following Heath-Brown, we begin by splitting the vectors (x, y) into congruence classes modulo $N(\mathfrak{r})$. Thus we obtain

$$S(\mathfrak{r}) = \sum_{\substack{u, v \,(\mathrm{mod}\, N(\mathfrak{r})) \\ \mathfrak{r}|u+v\rho}} |\{x, y \in (X, X(1+\eta)] : x \equiv u, \, y \equiv v \,(\mathrm{mod}\, N(\mathfrak{r}))\}|.$$

Now we write, for $q \in \mathbb{N}$ and $\mathfrak{r} \in \mathcal{R}$,

$$e_q(x) = e\left(\frac{x}{q}\right), \quad e_{\mathfrak{r}}(x) = e_{N(\mathfrak{r})}(x).$$

We can then formulate $S(\mathfrak{r})$ as

$$\frac{1}{N(\mathfrak{r})^2} \sum_{\substack{u,v \,(\mathrm{mod}\, N(\mathfrak{r})) \\ \mathfrak{r}|u+v\rho}} \sum_{a,b\,(\mathrm{mod}\, N(\mathfrak{r}))} \sum_{X<x,y\leq X(1+\eta)} e_{\mathfrak{r}}(a(u-x)+b(v-y))$$

$$= \frac{1}{N(\mathfrak{r})^2} \sum_{a,b\,(\mathrm{mod}\, N(\mathfrak{r}))} S_0(\mathfrak{r},a,b) \sum_{X<x,y\leq X(1+\eta)} e_{\mathfrak{r}}(-ax-by)),$$

$$(13.4.2)$$

where

$$S_0(\mathfrak{r},a,b) = \sum_{\substack{u,v\,(\mathrm{mod}\, N(\mathfrak{r})) \\ \mathfrak{r}|u+v\rho}} e_{\mathfrak{r}}(au+bv).$$

Heath-Brown notes that the sum $S_0(\mathfrak{r},a,b)$ enjoys the multiplicative property

$$S_0(\mathfrak{r}_1\mathfrak{r}_2,a,b) = S_0(\mathfrak{r}_1,a,b)S_0(\mathfrak{r}_2,a,b)$$

for $\mathfrak{r}_1\mathfrak{r}_2 \in \mathcal{R}$. Thus we need only consider the case in which $\mathfrak{r}$ is a prime. Now we expect the main term in (13.4.2) to come from the case $a = b = 0$. We note that $S_0(\mathfrak{r},0,0) = N(\mathfrak{r})$ since this is just the number of pairs u, v modulo $N(\mathfrak{r})$, for which $\mathfrak{r}|u + v\rho$. Thus

$$S(\mathfrak{r}) = \frac{\eta^2 X^2 + O(X)}{N(\mathfrak{r})} + O(E(\mathfrak{r})), \quad (13.4.3)$$

where

$$E(\mathfrak{r}) \ll \sum_{\substack{|a|,|b|\leq N(\mathfrak{r})/2 \\ (a,b)\neq(0,0)}} \frac{|S_0(\mathfrak{r},a,b)|}{N(\mathfrak{r})^2} \min\left(X, \frac{N(\mathfrak{r})}{|a|}\right) \min\left(X, \frac{N(\mathfrak{r})}{|b|}\right).$$

The total contribution from the $O(X)/N(\mathfrak{r})$ term in (13.4.3) is

$$\ll X \sum_{N(\mathfrak{r})\leq 2Q} N(\mathfrak{r})^{-1}\tau(\mathfrak{r})^A \ll X(\log Q)^{c(A)},$$

using Lemma 13.10. This is a suitable size.

In [82, pp. 29–30] it is shown that in the case $(a,b) \neq (0,0)$ we have $S_0(\mathfrak{r},a,b) = 0$ if $\mathfrak{r} \nmid b - a\rho$. For other values of $\mathfrak{r}$ the trivial bound $|S_0(\mathfrak{r},a,b)| \leq 2Q$ suffices as we now show. First we consider the contribution to (13.4.1) from terms in (13.4.3) with $ab \neq 0$. This is

$$\ll Q \sum_{0<|a|,|b|\leq Q} |ab|^{-1} \sum_{\substack{N(\mathfrak{r})\sim Q \\ \mathfrak{r}|b-a\rho}} \tau(R)^A \ll Q \sum_{0<|a|,|b|\leq Q} |ab|^{-1}\tau(b-a\rho)^{c(A)}.$$

We split the variables a, b into ranges $|a| \sim M$, $|b| \sim N$, where M, N run over powers of 2. We use Lemma 13.12 for each range and so obtain a bound

$$\ll Q \sum_{M,N} (MN)^{-1}MN(\log MN)^{c(A)} \ll Q(\log Q)^{c(A)},$$

which is suitably small.

Without loss of generality we now need only consider the case $a = 0, b \neq 0$. The contribution to (13.4.3) in this case is

$$\ll X \sum_{0<|b|\leq Q} |b|^{-1} \sum_{\substack{N(\mathfrak{r})\sim Q \\ \mathfrak{r}|b}} \tau(\mathfrak{r})^A$$

$$\ll X \sum_{0<|b|\leq Q} |b|^{-1} \tau(|b|)^{c(A)}$$

$$\ll X (\log Q)^{c(A)}$$

by Lemma 13.10. Again this is a satisfactory estimate, and this completes the proof. $\qquad\qquad\square$

Proof. (Lemma 13.2) We have

$$|A_{\mathfrak{r}}| = \sum_{d=1}^{\infty} \mu(d) \left| \{x, y \in (X, X(1+\eta)] : d|x,y, \mathfrak{r}|x+y\rho\} \right|.$$

Heath-Brown then writes $x = dx', y = dy'$ so that

$$|A_{\mathfrak{r}}| = \sum_{d=1}^{\infty} \mu(d) \left| \{x', y' \in (Xd^{-1}, Xd^{-1}(1+\eta)] : \mathfrak{r}(\mathfrak{r},d)^{-1}|x'+y'\rho\} \right|$$

$$= \sum_{d=1}^{\infty} \mu(d) S(\mathfrak{r}/(\mathfrak{r},d); X/d). \qquad (13.4.4)$$

Now, for $\mathfrak{r} \in \mathcal{R}$ we have

$$\sum_{d=1}^{\infty} \mu(d) \frac{\eta^2 X^2}{d^2} N(\mathfrak{r}/(\mathfrak{r},d))^{-1} = \frac{\eta^2 X^2}{N(\mathfrak{r})} \sum_{d=1}^{\infty} \frac{\mu(d)}{d^2} N((\mathfrak{r},d)) \qquad (13.4.5)$$

$$= \frac{\eta^2 X^2}{N(\mathfrak{r})} \prod_{p\nmid N(\mathfrak{r})} \left(1 - \frac{1}{p^2}\right) \prod_{p|N(\mathfrak{r})} \left(1 - \frac{1}{p}\right)$$

$$= \frac{6\eta^2 X^2}{\pi^2 N(\mathfrak{r})} \prod_{p|N(\mathfrak{r})} \left(1 + \frac{1}{p}\right)^{-1},$$

which produces the leading terms in Lemma 13.2.

Heath-Brown then splits the sums (13.4.4) and (13.4.5) at some point $d = \Delta$ to be determined with $1 \leq \Delta \leq X$. Terms in (13.4.5) for which $d > \Delta$ contribute a total

$$\ll \eta^2 X^2 \Delta^{-1} (\log Q)^{c(A)} \qquad (13.4.6)$$

by standard arguments using Lemma 13.10. The contribution from the terms of (13.4.4) in which $d > \Delta$ is

$$\ll X^2 (\log X)^{c(A)} \Delta^{-1}, \qquad (13.4.7)$$

using Lemmas 13.12 and 13.10. All the details needed to obtain (13.4.6) and (13.4.7) may be found in [82, pp. 31–32].

Heath-Brown then puts $\mathfrak{s} = (\mathfrak{r}, d)$ and $\mathfrak{t} = \mathfrak{s}t$. It then follows from Lemma 13.14 that the overall contribution from those terms in (13.4.4) and (13.4.5) with $d \leq \Delta$ is

$$\ll \sum_{\substack{N(\mathfrak{r}) \sim Q \\ \mathfrak{r} \in \mathcal{R}}} \tau(\mathfrak{r})^A \sum_{d \leq \Delta} \left| S(\mathfrak{r}/(R, d); X/d) - \frac{\eta^2 (X/d)^2}{N(\mathfrak{r}/(\mathfrak{r}, d))} \right|$$

$$\ll \sum_{d \leq \Delta} \sum_{N(\mathfrak{s}) | d} \tau(\mathfrak{s})^A \sum_{\substack{Q/N(\mathfrak{s}) < N(\mathfrak{t}) \leq 2Q/N(\mathfrak{s}) \\ \mathfrak{t} \in \mathcal{R}}} \tau(\mathfrak{t})^A \left| S(\mathfrak{t}; X/d) - \frac{\eta^2 (X/d)^2}{N(\mathfrak{t})} \right|$$

$$\ll \sum_{d \leq \Delta} \sum_{N(\mathfrak{s}) | d} \tau(\mathfrak{s})^A \left(\frac{X}{d} + \frac{Q}{N(\mathfrak{s})} \right) (\log Q)^{c(A)}$$

$$\ll (X + Q)(\log Q)^{c(A)} \sum_{d \leq \Delta} \tau(d)$$

$$\ll (X + Q)(\log Q)^{c(A)} \Delta \log X.$$

Combining this bound with (13.4.6) and (13.4.7) shows that the left-hand side of (13.2.4) is

$$\ll \frac{X^2}{\Delta} (\log XQ)^{c(A)} + (X + Q)(\log Q)^{c(A)} \Delta \log X.$$

It is now clear that we should choose $\Delta = 1 + \min \left(X^{1/2}, XQ^{-1/2} \right)$ and thus obtain a bound

$$\ll (Q + XQ^{\frac{1}{2}} + X^{\frac{3}{2}})(\log QX)^{c(A)}.$$

This completes the proof of Lemma 13.2. □

Proof. (Lemma 13.3) We have

$$|\mathcal{B}_{\mathfrak{r}}| = \left| \left\{ \mathfrak{i} : N(\mathfrak{i}) \in \left(\frac{3X^3}{N(\mathfrak{r})}, \frac{3X^3}{N(\mathfrak{r})}(1 + \eta) \right] \right\} \right|.$$

Applying Lemma 13.9 we infer that

$$|\mathcal{B}_{\mathfrak{r}}| = 3\gamma_0 \frac{X^3 \eta}{N(\mathfrak{r})} + O(X^2 N(\mathfrak{r})^{-\frac{2}{3}}). \tag{13.4.8}$$

The left-hand side of (13.2.5) is thus

$$\ll \sum_{N(\mathfrak{r}) \sim Q} \tau(\mathfrak{r})^A \frac{X^2}{N(\mathfrak{r})^{\frac{2}{3}}} \ll X^2 Q^{\frac{1}{3}} (\log Q)^{c(A)}$$

by Lemma 13.10. □

13.5 THE FUNDAMENTAL LEMMA RESULT

Here we state the form of the fundamental lemma necessary to prove the appropriate Fundamental Theorem in the present context.

Lemma 13.15. *Suppose that the coefficients $c_{\mathfrak{r}}$ are supported on square-free ideals with $|c_{\mathfrak{r}}| \leq 1$ and let $1 \leq R \leq X^{7/4}$. Then*

$$\sum_{N(\mathfrak{r}) \sim R} c_{\mathfrak{r}} S(\mathcal{A}_{\mathfrak{r}}, w) = \kappa \sum_{N(\mathfrak{r}) \sim R} c_{\mathfrak{r}} S(\mathcal{A}_{\mathfrak{r}}, w) + O\left(\frac{\eta^2 X^2}{\tau \log X} \exp\left(-\tau^{-1}\right)\right).$$

(13.5.1)

Remark. As far as the proof and application of the result are concerned, the quantity $X^{7/4}$ above could have been any fixed power of X between $X^{3/2}$ and X^2.

Proof. The proof follows from the analogue of the fundamental lemma in Chapter 4 appropriate to $\mathbb{Z}[\rho]$ using the Type I information we have obtained (compare Lemmas 11.10 and 11.26). Heath-Brown gives a detailed proof of this result in [82], although he deduces the result by applying the sieve over the rational integers to the sequence $x^3 + 2y^3$ counted with multiplicity. □

The lemma above is precisely in the right form for the application we gave back in Section 13.2.

13.6 PROOF OF LEMMA 13.6

We treat the terms $d_{\mathfrak{s}}, e_{\mathfrak{s}}$ separately. It is for the former term that we need to use an analogue of Siegel's theorem, so we shall begin by considering the coefficient $e_{\mathfrak{s}}$. From this point on in this chapter we abbreviate Heath-Brown's argument substantially, explaining the major steps *en route* to the crucial results. The next such major step is the proof of the following lemma. In the following the vector $\mathbf{x}$ is always the point $(x, y, z) \in \mathfrak{r}^3$. We also write, for $j \in \mathbb{Z}$,

$$\theta_j = (\xi \log X)^{-j-1}.$$

Lemma 13.16. *Let $\mathcal{C} \subseteq \mathbb{R}^3$ be as in Lemma 13.6. Define*

$$N(\mathbf{x}) = x^3 + 2y^3 + 4z^3 - 6xyz$$

and

$$I = \int_{\mathcal{C}} w'(N(\mathbf{x})) \, dx \, dy \, dz.$$

Then for any positive integer $q \leq L^{1/6}$ and any integer $\alpha \in \mathbb{Z}[\rho]$ we have

$$\sum_{\substack{\beta \equiv \alpha \pmod{q} \\ \hat{\beta} \in \mathcal{C}}} e_{(\beta)} = \frac{1}{\gamma_0 M} I \theta_n \frac{\delta(\alpha, q)}{\phi_K(q)} + O\left(S_0^3 M^{-1} \tau(q)^c \mathcal{L}^{-c}\right),$$

where $\delta(\alpha, q) = 1$ if α and q are coprime, and $\delta(\alpha, q) = 0$ otherwise. Here we have defined

$$M = \prod_{i=1}^{n+1} m_i$$

and written ϕ_K for the Euler function over the field K.

Proof. From the definition (13.2.15) we have

$$\sum_{\substack{\beta \equiv \alpha \,(\mathrm{mod}\, q) \\ \hat{\beta} \in \mathcal{C}}} e_{(\beta)} = M^{-1} \theta_n \sum_{N(\mathsf{j}) < L} \mu(\mathsf{j}) \log \left(\frac{L}{N(\mathsf{j})} \right) \sum_{\substack{\hat{\beta} \in \mathcal{C},\ \mathsf{j}|\beta \\ \beta \equiv \alpha \,(\mathrm{mod}\, q)}} w'(N(\beta)).$$

(13.6.1)

Heath-Brown then notes that the two conditions $\mathsf{j}|\beta$ and $\beta \equiv \alpha \,(\mathrm{mod}\, q)$ are compatible only when $(\mathsf{j}, q)|\alpha$, and in this latter case they define a unique residue class for β modulo $[\mathsf{j}, q]$, that is, the lowest common multiple of j and q. The next step is thus to investigate the sum

$$\sum_{\substack{\hat{\beta} \in \mathcal{C} \\ \beta \equiv \gamma \,(\mathrm{mod}\, \mathfrak{r})}} w'(N(\beta)),$$

(13.6.2)

where $\mathfrak{r}$ is a rational integer multiple of $[\mathsf{j}, q]$. The value $\mathfrak{r} = N([\mathsf{j}, q])$ is then taken, and the case $n \geq 1$ is considered. It is shown in [82, p. 48] that for $h \geq 0$ we have

$$|w'(t + h) - w'(t)| \leq \frac{h}{t} \theta_{-n}$$

(13.6.3)

and

$$0 \leq w'(t) \leq \theta_{-n-1}.$$

(13.6.4)

For each vector $\hat{\beta}$ write $\mathcal{C}(\beta)$ for the cube of side $\mathfrak{r}$, centred at $\hat{\beta}$, and with sides parallel to those of $\mathcal{C}$. Then, since

$$\mathfrak{r} \leq N(\mathsf{j}) q^3 \leq L^2 \leq S_0 \ll V^{\frac{1}{3}},$$

we have

$$N(\mathbf{x}) = N(\beta) + O\left(V^{\frac{2}{3}} \mathfrak{r} \right)$$

for any $\mathbf{x} \in \mathcal{C}(\beta)$. Thus (13.6.3) gives

$$w'(N(\mathbf{x})) = w'(N(\beta)) + O\left(V^{-\frac{1}{3}} \mathfrak{r} \theta_{-n} \right),$$

and so

$$w'(N(\beta)) = \mathfrak{r}^{-3} \int_{\mathcal{C}(\beta)} w'(N(\mathbf{x}))\, dx\, dy\, dz + O\left(V^{-\frac{1}{3}} \mathfrak{r} \theta_{-n} \right).$$

Since $\mathfrak{r} \leq S_0$, this gives

$$\sum_{\substack{\hat{\beta} \in \mathcal{C} \\ \beta \equiv \gamma \,(\mathrm{mod}\, \mathfrak{r})}} w'(N(\beta))$$

$$= \frac{1}{\mathfrak{r}^3} \sum_{\substack{\hat{\beta} \in \mathcal{C} \\ \beta \equiv \gamma \,(\mathrm{mod}\, \mathfrak{r})}} \int_{\mathcal{C}(\beta)} w'(N(\mathbf{x}))\, dx\, dy\, dz + O\left(V^{-\frac{1}{3}} \mathfrak{r} \theta_{-n} (S_0/\mathfrak{r})^3 \right).$$

(13.6.5)

Heath-Brown then notes that the cubes $\mathcal{C}(\beta)$ for $\beta \equiv \gamma \,(\mathrm{mod}\, \mathfrak{r})$ will be disjoint, except for their boundaries. Also, as $\hat{\beta}$ runs over $\mathcal{C}$, the union of the cubes $\mathcal{C}(\beta)$ will be a set that differs from $\mathcal{C}$ only at points within a distance $O(r)$ of the boundary. Hence the right-hand side of (13.6.5) becomes

$$\frac{1}{r^3} \int_{\mathcal{C}} w'(N(\mathbf{x}))\, dx\, dy\, dz + O\left((S_0/r)^2 \theta_{-n-1} + V^{-\frac{1}{3}} r\theta_{-n}(S_0/r)^3\right)$$
$$= \frac{1}{r^3} I + O\left((S_0/r)^2 \theta_{-n-1}\right)$$

by (13.6.4).

In [82] it is shown that exactly the same estimate also holds in the case $n = 0$, using Lemma 13.13.

Now

$$\sum_{\substack{\hat{\beta}\in\mathcal{C},\, \mathfrak{j}|\beta \\ \beta\equiv\alpha\,(\mathrm{mod}\, q)}} w'(N(\beta))$$

can be decomposed into $r^3/N([\mathfrak{j}, q])$ subsums of the form (13.6.2). Thus

$$\sum_{\substack{\hat{\beta}\in\mathcal{C},\, \mathfrak{j}|\beta \\ \beta\equiv\alpha\,(\mathrm{mod}\, q)}} w'(N(\beta)) = N([\mathfrak{j}, q])^{-1} I + O\left(S_0^2\theta_{-n-1}\right), \tag{13.6.6}$$

providing that $(\mathfrak{j}, q)|\alpha$. The error term in (13.6.6) contributes

$$\ll S_0^2 M^{-1}(\xi\log X)^{-1} L\log L \ll S_0^2 M^{-1}\xi^{-1} L \ll S_0^3 M^{-1}\mathcal{L}^{-c}$$

to (13.6.1). The main term of (13.6.1) now becomes

$$M^{-1}\theta_n q^{-3} I\Sigma',$$

where

$$\Sigma' = \sum_{\substack{N(\mathfrak{j})<L \\ (\mathfrak{j},q)|\alpha}} \frac{\mu(\mathfrak{j})}{N(\mathfrak{j})} N((\mathfrak{j}, q))\log\left(\frac{L}{N(\mathfrak{j})}\right).$$

We begin the evaluation of Σ' by writing $\mathfrak{i} = (\mathfrak{j}, q)$. Thus we obtain

$$\Sigma' = \sum_{\mathfrak{i}|q,\alpha} N(\mathfrak{i}) \sum_{\mathfrak{a}|q\mathfrak{i}^{-1}} \mu(\mathfrak{a})\frac{\mu(\mathfrak{i}\mathfrak{a})}{N(\mathfrak{i}\mathfrak{a})}\Sigma(L/N(\mathfrak{i}\mathfrak{a}); \mathfrak{i}\mathfrak{a}) \tag{13.6.7}$$

where

$$\Sigma(x; \mathfrak{c}) = \sum_{\substack{N(\mathfrak{b})<x \\ (\mathfrak{b},\mathfrak{c})=1}} \frac{\mu(\mathfrak{b})}{N(\mathfrak{b})}\log\left(\frac{x}{N(\mathfrak{b})}\right).$$

This sum can be estimated using the Dirichlet series

$$f(s) = \sum_{\substack{\mathfrak{b} \\ (\mathfrak{b},\mathfrak{c})=1}} \frac{\mu(\mathfrak{b})}{N(\mathfrak{b})^s} = \frac{1}{\zeta_K(s)}\prod_{\mathfrak{p}|\mathfrak{c}}(1 - N(\mathfrak{p})^{-s})^{-1}.$$

The logarithmic factor in $\Sigma(x; \mathfrak{c})$ means that a variant of Perron's formula (see Lemma A.3) produces

$$\Sigma(x; \mathfrak{c}) = \int_{1-i\infty}^{1+i\infty} f(s+1)x^s\, \frac{ds}{s^2}.$$

Now the residue of $f(s+1)x^s s^{-2}$ at $s = 0$ (note that we get the main term from $s = 0$ here!) is $\gamma_0^{-1} N(\mathfrak{c})/\phi_K(\mathfrak{c})$. By standard arguments, using the zero-free region for $\zeta_K(\mathfrak{s})$, we thereby obtain

$$\Sigma(x;\mathfrak{c}) = \frac{\gamma_0^{-1} N(\mathfrak{c})}{\phi_K(\mathfrak{c})} + O\left(\mathcal{L}^{-c}\right).$$

This means that

$$\Sigma' = \mathfrak{M} + E,$$

where

$$\mathfrak{M} = \gamma_0^{-1} \sum_{\mathfrak{i}|q,\alpha} N(\mathfrak{i}) \sum_{\mathfrak{a}|q\mathfrak{i}^{-1}} \mu(\mathfrak{a}) \frac{\mu(\mathfrak{i}\mathfrak{a})}{\phi_K(\mathfrak{i}\mathfrak{a})}$$

and

$$E \ll \mathcal{L}^{-c} \sum_{\mathfrak{a}\mathfrak{i}|q} N(\mathfrak{a})^{-1}.$$

It is an elementary exercise to show that $\mathfrak{M} = \gamma_0^{-1} N(q)/\phi_K(q)$ if q and α are coprime, and zero otherwise. Also, we have $E \ll \tau(q)^c \mathcal{L}^{-c}$. The proof is then completed since (13.6.4) gives

$$I \ll S_0^3 \theta_{-n-1}.$$

□

We now consider $d_{\mathfrak{s}}$. The trivial case, in which $(\alpha, q) \neq 1$, is quickly completed since $\mathfrak{i}(\beta)$ will be a product of prime ideals $\mathfrak{p}_j$ with $N(\mathfrak{p}_j) \geq X^\tau \geq L > N(q)$. It immediately follows that β and q must be coprime. We thus obtain the next result.

Lemma 13.17. *Let* $\mathcal{C} \subseteq \mathbb{R}^3$ *be as in Lemma 13.6. Then for any* $q \leq L^{1/6}$ *and any integer* $\alpha \in \mathbb{Z}[\rho]$ *we have*

$$\sum_{\substack{\beta \equiv \alpha \,(\mathrm{mod}\, q) \\ \hat{\beta} \in \mathcal{C}}} d_{(\beta)} = 0$$

whenever α *and* q *have a common factor.*

The nontrivial case is covered by the following result.

Lemma 13.18. *Let* $\mathcal{C} \subseteq \mathbb{R}^3$ *be as in Lemma 13.6 and let a positive integer* A *be given. Then, for any natural number* $q \leq (\log L)^A$ *and any integer* $\alpha \in \mathbb{Z}[\rho]$ *coprime to* q*, we have*

$$\sum_{\substack{\beta \equiv \alpha \,(\mathrm{mod}\, q) \\ \hat{\beta} \in \mathcal{C}}} d_{(\beta)} = \frac{\theta_n I}{\gamma_0 M \phi_K(q)} + O_A\left(V\mathcal{L}^{-c}\right),$$

where I *is as in Lemma 13.16.*

Remarks. It is in this lemma that problems with potential Siegel zeros occur, and thus the implied constant is ineffective. Clearly Lemma 13.6 follows immediately from Lemmas 13.16–13.18.

Proof. We begin in standard fashion by using characters to obtain

$$\sum_{\substack{\beta \equiv \alpha \,(\mathrm{mod}\, q) \\ \beta \in \mathcal{C}}} d_{(\beta)} = \frac{1}{\phi_K(q)} \sum_{\chi \,(\mathrm{mod}\, q)} \overline{\chi}(\alpha) \sum_{\beta \in \mathcal{C}} d_{(\beta)} \chi(\beta). \tag{13.6.8}$$

Of course, in this situation χ runs over characters of the multiplicative group for $\mathbb{Z}[\rho] \,(\mathrm{mod}\, q)$. We shall use Hecke Grössencharacters to detect the condition $\hat{\beta} \in \mathcal{C}$. We give their construction for $\mathbb{Z}[\rho]$ explicitly here, following Heath-Brown's description closely.

For any nonzero $\beta = a + b\rho + c\rho^2 \in \mathbb{Z}[\rho]$ let

$$\beta' = a + b\varpi\rho + c\varpi^2\rho^2,$$

where $\varpi = (-1 + \sqrt{-3})/2$. Also, put

$$\chi(-1) = (-1)^\sigma, \quad \chi(\varepsilon_0) = e^{it}, \quad \frac{\varepsilon_0'}{|\varepsilon_0'|} = e^{iu}, \quad \log \varepsilon_0 = v.$$

Here we shall choose $\sigma \in \{0, 1\}$, $0 \le t, u < 2\pi$, and $v \in \mathbb{R}$. We now define

$$\nu_0(\beta) = \chi(\beta) \left(\frac{\beta}{|\beta|} \right)^\sigma e\left(-\frac{t \log |\beta|}{2\pi v} \right), \tag{13.6.9}$$

$$\nu_1(\beta) = \frac{\beta\beta'}{|\beta\beta'|} e\left(-\frac{u \log |\beta|}{2\pi v} \right),$$

and

$$\nu_2(\beta) = e\left(-\frac{\log |\beta|}{v} \right).$$

Then, for each index j, the function $\nu_j(\beta)$ is completely multiplicative and has modulus 1 (or possibly 0 when $j = 0$). Also, $\nu_j(\beta_1) = \nu_j(\beta_2)$ whenever β_1 and β_2 are associates. If $\mathfrak{s}$ is an integral ideal generated by β, we may then define $\nu_j(\mathfrak{s}) = \nu_j(\beta)$. We also write

$$\nu^{(j,k)}(\mathfrak{s}) = \nu(\mathfrak{s}) = \nu_0(\mathfrak{s}) \nu_1(\mathfrak{s})^j \nu_2(\mathfrak{s})^k.$$

We shall say that the character $\nu(\mathfrak{s})$ is trivial if it takes the value 1 whenever $\mathfrak{s}$ is coprime to q. This corresponds to having the trivial character χ modulo q, letting $\sigma = t = 0$ in the definition of ν_0, and taking $j = k = 0$.

Heath-Brown uses the following version of the prime number theorem with Grössencharacters, due to Mitsui [130, Lemma 5]. The reader will immediately see that it is analogous to results we have used from Chapter 7 onward.

Lemma 13.19. *If the character $\nu^{(j,k)}$ is nontrivial, then for any positive constant A we have*

$$\sum_{N(\mathfrak{p}) \le z} \log N(\mathfrak{p}) \nu^{(j,k)}(\mathfrak{p}) \ll_A z \exp\left(-c\sqrt{\log z} \right).$$

uniformly for

$$|j|, |k| \ll \exp\left(\sqrt{\log z}\right)$$

and $q \leq (\log z)^A$.

Remarks. For non-quadratic characters one may in fact allow $q \leq \exp\left(\sqrt{\log z}\right)$. However, as usual, quadratic characters cause difficulties since we cannot exclude the possibility of Siegel zeros occurring. In particular, one should note that the implied constant in Lemma 13.19 is ineffective.

For any $\mathbf{x} \in \mathbb{R}^3$ such that $N(\mathbf{x}) \neq 0$, we write

$$\beta(\mathbf{x}) = x_1 + x_2\rho + x_3\rho^2 \quad \text{and} \quad \beta'(\mathbf{x}) = x_1 + x_2\varpi\rho + x_3\varpi^2\rho^2$$

and put $\nu_j(\mathbf{x}) = \nu_j(\beta(\mathbf{x}))$. We define

$$\mathcal{M} = \begin{pmatrix} 1 & 2 & 2 \\ 1 & 1 & 2 \\ 1 & 1 & 1 \end{pmatrix}$$

and note that $\beta(\mathcal{M}\mathbf{x}) = \varepsilon_0\beta(\mathbf{x})$ and $\beta'(\mathcal{M}\mathbf{x}) = \varepsilon_0'\beta'(\mathbf{x})$. Thus, if we say that two vectors $\mathbf{x}$ and $\mathbf{x}'$ are associates when $\mathbf{x}' = \pm\mathcal{M}^n\mathbf{x}$ for some $n \in \mathbb{Z}$, then we will have $\nu_j(\mathbf{x}) = \nu_j(\mathbf{x}')$ ($j = 1, 2$) whenever $\mathbf{x}$ and $\mathbf{x}'$ are associates.

As we have seen in previous chapters, the introduction of a smooth weight function is sometimes a necessary device. Here we do not need smoothness; a triangular function suffices. We follow Heath-Brown in defining $W(\mathfrak{s}; \Delta, \mathbf{x})$, for positive $\Delta < \frac{1}{2}$ by

$$W(\mathfrak{s}; \Delta, \mathbf{x}) = h\left(\arg(\nu_1(\mathfrak{s})/\nu_1(\mathbf{x}))\right) h\left(\arg(\nu_2(\mathfrak{s})/\nu_2(\mathbf{x}))\right),$$

where

$$h(2\pi x) = \begin{cases} 1 - \Delta^{-1}\|x\| & \text{if } \|x\| \leq \Delta, \\ 0 & \text{otherwise.} \end{cases}$$

This will be useful since

$$h(x) = \Delta \sum_{-\infty}^{\infty} \left(\frac{\sin(\pi n\Delta)}{\pi n\Delta}\right)^2 e^{inx}. \tag{13.6.10}$$

We now write $\mathfrak{I}(\mathbf{x}) = \mathfrak{I}(\mathbf{x}, \Delta V) = (N(\mathbf{x}), N(\mathbf{x}) + \Delta V]$. The next expression to consider is then

$$\Sigma(\mathbf{x}) = \sum_{N(s) \in \mathfrak{I}(\mathbf{x})} d_s\nu_0(\mathfrak{s})W(\mathfrak{s}; \Delta, \mathbf{x}),$$

where $\mathbf{x} \in \mathcal{C}$, so that $V \ll N(\mathbf{x}) \ll V$. We expect there to be a main term for such a sum when the character is trivial, and just an error term otherwise (as we have seen in similar contexts from Chapter 10 onward).

Lemma 13.20. *Let $\mathbf{x} \in \mathcal{C}$ and suppose that $q \leq (\log L)^A$. Then*

$$\Sigma(\mathbf{x}) = \delta(\chi)m(\mathbf{x})\frac{\Delta^2}{M}\theta_n + O_A\left(\frac{V\mathcal{L}^{-c}}{\Delta^2 M}\right), \tag{13.6.11}$$

where

$$m(\mathbf{x}) = w(N(\mathbf{x}) + \Delta V) - w(N(\mathbf{x}))$$

and $\delta(\chi) = 1$ or 0 depending on whether χ is trivial or not.

Proof. From (13.6.10) we have

$$\Sigma(\mathbf{x}) = \Delta^2 \sum_{j,k=-\infty}^{\infty} \left(\frac{\sin(\pi j\Delta)}{\pi j\Delta}\right)^2 \left(\frac{\sin(\pi k\Delta)}{\pi k\Delta}\right)^2 \nu_1(\mathbf{x})^{-j}\nu_2(\mathbf{x})^{-k}\Sigma_{j,k},$$

$$(13.6.12)$$

where

$$\Sigma_{j,k} = \sum_{N(\mathfrak{s})\in\mathfrak{I}(\mathbf{x})} d_{\mathfrak{s}}\nu^{(j,k)}(\mathfrak{s}).$$

We remark that if $d_{\mathfrak{s}}$ is nonzero, then $\mathfrak{s}$ and q are automatically coprime, as in the proof of Lemma 13.17.

If $\nu^{(j,k)}$ is trivial, we can deduce from Lemma 13.5 that

$$\Sigma_{j,k} = m(\mathbf{x})M^{-1}\theta_n + O\left(VM^{-1}\mathcal{L}^{-c}\right).$$

$$(13.6.13)$$

If $\nu^{(j,k)}$ is nontrivial, we can use Lemma 13.19 applied to the largest prime variable in $\mathfrak{s}$. This leads to

$$\Sigma_{j,k} \ll_A M^{-1}V\mathcal{L}^{-c}.$$

$$(13.6.14)$$

We also have the trivial bound

$$\Sigma_{j,k} \ll VM^{-1}.$$

$$(13.6.15)$$

The proof of Lemma 13.20 is completed by inserting the estimates (13.6.13) and (13.6.14) into (13.6.12) when $|j|, |k| \leq \mathcal{L}^c$, and an application of (13.6.15) otherwise. $\qquad\square$

In the next lemma Heath-Brown describes the relationship between values of $\hat{\beta}$ and $\mathbf{x}$ for which $\mathfrak{s} = \mathfrak{i}(\beta)$ is counted by $\Sigma(\mathbf{x})$. It is proved in [82, p. 57].

Lemma 13.21. *Let $V \ll N(\mathfrak{s})$, $N(\mathbf{x}) \ll V$ and suppose that $W(\mathfrak{s};\Delta,\mathbf{x}) \neq 0$ and that $N(\mathbf{x}) < N(\mathfrak{s}) \leq N(\mathbf{x}) + \Delta V$. Then $\mathfrak{s}$ has a generator β for which*

$$\hat{\beta} = (1 + O(\Delta))\mathbf{x}.$$

$$(13.6.16)$$

Similarly, if β is any generator of $\mathfrak{s}$, then $\mathbf{x}$ has an associate $\mathbf{x}'$ for which $\hat{\beta} = (1 + O(\Delta))\mathbf{x}'$.

Recalling that $x, y, z \ll V^{1/3}$ for $\mathbf{x} \in \mathcal{C}$, the above lemma then shows that, for some absolute constant c, we have $\hat{\beta} = \mathbf{x} + c\mathbf{u}$, where $|\mathbf{u}| < V^{1/3}\Delta$. We therefore define $\mathcal{C}'$ as the set of vectors $\mathbf{t}$ for which there is at least one $\mathbf{x} \in \mathcal{C}$ with $|\mathbf{t} - \mathbf{x}| \leq cV^{1/3}\Delta$. Thus, if $\mathfrak{s}$ is counted by $\Sigma(\mathbf{x})$, then $\mathfrak{s} = \mathfrak{i}(\beta)$ for some β with $\hat{\beta} \in \mathcal{C}'$. Assuming, as we may, that Δ and $S_0V^{-1/3}$ are sufficiently small, there will be at most one such β. Using (13.6.9) we then obtain

$$\nu_0(\beta) = (1 + O(\Delta))\chi(\beta)e\left(-\frac{t}{2\pi v}\log\beta(\mathbf{x})\right).$$

Now consider

$$\Sigma'(\mathbf{x}) = e\left(\frac{t}{2\pi v}\log\beta(\mathbf{x})\right)\Sigma(\mathbf{x})$$

$$= \sum_{\substack{\hat{\beta}\in\mathcal{C}' \\ N(\beta)\in\mathfrak{I}(\mathbf{x})}} d_{(\beta)}\chi(\beta)W(\mathfrak{i}(\beta);\Delta,\mathbf{x})(1 + O(\Delta)).$$

$$(13.6.17)$$

From the definition of $\delta(\chi)$ with (13.6.11) we deduce that

$$\Sigma'(\mathbf{x}) = \delta(\chi)m(\mathbf{x})\frac{\Delta^2}{M}\theta_n + O_A\left(\Delta^{-2}M^{-1}V\mathcal{L}^{-c}\right). \tag{13.6.18}$$

Now write

$$\mathfrak{I} = \int_{\mathcal{C}} \Sigma'(\mathbf{x})\,dx\,dy\,dz.$$

From (13.6.18) we have

$$\mathfrak{I} = \delta(\chi)\frac{\Delta^2}{M}\theta_n \int_{\mathcal{C}} m(\mathbf{x})\,dx\,dy\,dz + O_A\left(\Delta^{-2}M^{-1}VS_0^3\mathcal{L}^{-c}\right). \tag{13.6.19}$$

When $n \geq 1$, we can estimate $m(\mathbf{x})$ using the mean value theorem together with (13.6.3). Thus

$$m(\mathbf{x}) = \Delta V w'(N(\mathbf{x})) + O(\Delta^2 V\theta_{-n}),$$

and so the integral in (13.6.19) is

$$\Delta V I + O\left(\Delta^2 V S_0^3\theta_{-n}\right). \tag{13.6.20}$$

When $n = 0$, it can be shown that

$$\int_{\mathcal{C}} m(\mathbf{x})\,dx\,dy\,dz = \Delta V \int_{\mathcal{C}} w'(N(\mathbf{x}))\,dx\,dy\,dz + O(\Delta^2 V^2). \tag{13.6.21}$$

From (13.6.20), (13.6.21), and (13.6.19), we deduce that

$$\mathfrak{I} = \delta(\chi)\frac{\Delta^3 V}{M}\theta_n I + O\left(\Delta^4 V^2 M^{-1}\right) + O_A\left(\Delta^{-2}M^{-1}VS_0^3\mathcal{L}^{-c}\right). \tag{13.6.22}$$

On the other hand, (13.6.17) shows that

$$\mathfrak{I} = \sum_{\beta\in\mathcal{C}'} d_{(\beta)}\chi(\beta)\left(1 + O(\Delta)\right) \int^* W(\mathfrak{i}(\beta); \Delta, \mathbf{x})\,dx\,dy\,dz,$$

where $*$ represents the conditions

$$\mathbf{x} \in \mathcal{C}, \qquad N(\beta) \in \mathfrak{I}(\mathbf{x}).$$

After some calculations Heath-Brown arrives at

$$\mathfrak{I}(\gamma_0\Delta^3 V)^{-1} = \sum_{\hat{\beta}\in\mathcal{C}} d_{(\beta)}\chi(\beta)\left(1 + O(\Delta)\right) + O\left(\sum_\beta {}^* d_{(\beta)}\right), \tag{13.6.23}$$

where Σ^* counts those β for which $|\hat{\beta} - \mathbf{t}| \ll \Delta V^{1/3}$ for some $\mathbf{t}$ on the boundary of $\mathcal{C}$.

For the error terms in (13.6.23) we note the trivial bound

$$d_5 \leq \prod_{i=1}^{n+1}\frac{m_i + 1}{m_i} \leq n + 2 \ll \tau^{-1} \ll \log X. \tag{13.6.24}$$

Thus (13.6.23) leads to

$$\mathfrak{I} = \gamma_0\Delta^3 V \sum_{\hat{\beta}\in\mathcal{C}} d_{(\beta)}\chi(\beta) + O\left(\Delta^4 V^2 \log X\right).$$

This estimate can then be combined with (13.6.22) to produce

$$\sum_{\hat{\beta} \in \mathcal{C}} d_{(\beta)} \chi(\beta) = \frac{\delta(\chi)}{\gamma_0 M} \theta_n I + O_A(\Delta^{-5} V \mathcal{L}^{-c}) + O(\Delta V \log X).$$

Now we can see that the appropriate choice for Δ is $\Delta = \mathcal{L}^{-c/6}$. We then obtain (replacing $c/6$ by c according to our convention)

$$\sum_{\hat{\beta} \in \mathcal{C}} d_{(\beta)} \chi(\beta) = \frac{\delta(\chi)}{\gamma_0 M} \theta_n I + O_A(V \mathcal{L}^{-c}).$$

Combining this with (13.6.8) then completes the proof of Lemma 13.18. □

13.7 PROOF OF LEMMA 13.7

The first stage in the proof is to replace $N(\mathfrak{s})$ by $3X^3/N(\mathfrak{r})$ where it occurs in $w'(N(\mathfrak{s}))$. It can be shown that this introduces an acceptable error of the form

$$O\left(\frac{\eta^{\frac{5}{2}} X^2}{M} (\log X)^c\right).$$

The details are in [82, p. 61]. This leads to

$$U_e(\mathcal{A}) = \sum_{\mathfrak{r}, \mathfrak{j}} C_{\mathfrak{r},\mathfrak{j}} |\mathcal{A}_{\mathfrak{r}\mathfrak{j}}| + O\left(\frac{\eta^{\frac{5}{2}} X^2}{M} (\log X)^c\right),$$

where

$$C_{\mathfrak{r},\mathfrak{j}} = c_{\mathfrak{r}} \frac{w'(3X^3/N(\mathfrak{r}))}{M \theta_n^{-1}} \mu(\mathfrak{j}) \log\left(\frac{L}{N(\mathfrak{j})}\right).$$

Now $N(\mathfrak{r}) < X^{2-\tau}$ and $N(\mathfrak{j}) < L = X^{\tau/2}$, so $\mathfrak{r}\mathfrak{j}$ has norm at most $X^{2-\tau/2}$. An application of Lemma 13.2 then gives

$$U_e(\mathcal{A}) = \sum_{\substack{\mathfrak{r}, \mathfrak{j} \\ \mathfrak{r}\mathfrak{j} \in \mathcal{R}}} C_{\mathfrak{r},\mathfrak{j}} \frac{6\eta^2 X^2}{\pi^2 N(\mathfrak{r}\mathfrak{j})} w_2(\mathfrak{r}\mathfrak{j}) + O\left(M^{-1} \eta^{\frac{5}{2}} X^2 (\log X)^c\right).$$

Note that we have been able to use Type I information here for this sum in view of the small size of L. The main term above is

$$\frac{6\eta^2 X^2 \theta_n}{\pi^2} \sum_{\mathfrak{r}} c_{\mathfrak{r}} \frac{w'(3X^3/N(\mathfrak{r}))}{M} N(\mathfrak{r})^{-1} w_2(\mathfrak{r}) \Sigma_1,$$

where

$$\Sigma_1 = \sum_{\substack{N(\mathfrak{j}) \le L \\ \mathfrak{j} \in \mathcal{R}}} \frac{\mu(\mathfrak{j})}{N(\mathfrak{j})} w_2(\mathfrak{j}) \log\left(\frac{L}{N(\mathfrak{j})}\right).$$

We recognize Σ_1 as being analogous to Σ in the proof of Lemma 13.16. With similar working (the reader should be able to supply this, but [82] contains the details) we obtain

$$\Sigma_1 = \sigma_0 \frac{\pi^2}{6} + O(\mathcal{L}^{-c}).$$

It follows that

$$U_e(\mathcal{A}) = \sigma_0 \eta^2 X^2 \Sigma_2 \left(1 + O\left(\mathcal{L}^{-c}\right)\right) + O\left(M^{-1} \eta^{\frac{5}{2}} X^2 (\log X)^c\right),$$

where

$$\Sigma_2 = \theta_n \sum_{\mathfrak{r}} c_{\mathfrak{r}} \frac{w'(3X^3/N(\mathfrak{r}))}{M} N(\mathfrak{r})^{-1} \omega_2(\mathfrak{r}).$$

From the definition

$$\omega_2(R) = \prod_{\mathfrak{p} | \mathfrak{r}} (1 + N(\mathfrak{p})^{-1})^{-1},$$

together with the bound $N(\mathfrak{p}) \geq X^\tau$ (and so there are $O(\tau^{-1})$ factors in the above), we arrive at

$$\omega_2(\mathfrak{r}) = 1 + O(\tau^{-1} X^{-\tau}) = 1 + O\left(\mathcal{L}^{-c}\right).$$

Hence

$$U_e(\mathcal{A}) = \sigma_0 \eta^2 X^2 \Sigma_3 \left(1 + O\left(\mathcal{L}^{-c}\right)\right) + O\left(M^{-1} \eta^{\frac{5}{2}} X^2 (\log X)^c\right),$$

where

$$\Sigma_3 = \theta_n \sum_{\mathfrak{r}} c_{\mathfrak{r}} \frac{w'(3X^3/N(\mathfrak{r}))}{M} N(\mathfrak{r})^{-1}.$$

Since

$$\Sigma_3 \ll (M \xi \log X)^{-1} \sum_{\mathfrak{r}} \frac{c_{\mathfrak{r}}}{N(\mathfrak{r})} \ll M^{-1} \log X,$$

we obtain

$$U_e(\mathcal{A}) = \sigma_0 \eta^2 X^2 \Sigma_3 + O\left(M^{-1} \eta^{\frac{5}{2}} X^2 (\log X)^c\right). \tag{13.7.1}$$

From Lemma 13.5 and the mean value theorem it can be shown (see [82, pp. 64–65]) that

$$U(\mathcal{B}) = \eta X^3 \Sigma_3 + O\left(\frac{\eta^2 X^3}{M}\right). \tag{13.7.2}$$

A comparison of (13.7.1) and (13.7.2) then completes the proof after summing over the various possibilities for n and $m_1, \ldots, m_{n+1}$.

The case $n = 0$ is established in [82, pp. 65-66]. □

13.8 THE TYPE II INFORMATION ESTABLISHED

In this section we shall sketch the proof of the crucial Lemma 13.8. Now we let $\mathbf{x} = (x, y) \in \mathbb{Z}^2$ and write

$$W(\mathbf{x}) = \begin{cases} 1 & \text{if } X < x, y \leq X(1 + \eta), \\ 0 & \text{otherwise.} \end{cases}$$

We then have

$$\nabla_f(V) = \sum_{\mathfrak{r}} c_{\mathfrak{r}} \sum_{N(\mathfrak{s})\sim V} f_{\mathfrak{s}} \sum_{\substack{\mathbf{x}\in\mathbb{Z}^2 \\ \mathfrak{i}(x,y)=\mathfrak{r}\mathfrak{s}}} W(\mathbf{x}),$$

where $\mathbf{x}$ is restricted to run over primitive integer vectors. We follow Heath-Brown in calling an integer of K "primitive" if it has no rational prime factor and let $\mathcal{P}$ be the set of ideals generated by primitive integers. It follows that $\mathfrak{r}$ and $\mathfrak{s}$ may be taken to belong to $\mathcal{P}$ in the above sum.

We now remove the condition that $\mathbf{x}$ is a primitive vector in a familiar fashion, by writing

$$\nabla_f(V) = \sum_{d\ll X} \mu(d) \sum_{\mathfrak{r}\in\mathcal{P}} c_{\mathfrak{r}} \sum_{\substack{N(\mathfrak{s})\sim V \\ \mathfrak{s}\in\mathcal{P}}} f_{\mathfrak{s}} \sum_{\substack{d\mid\mathbf{x} \\ \mathfrak{i}(x,y)=\mathfrak{r}\mathfrak{s}}} W(\mathbf{x}).$$

If $d > 1$, then $(d, \mathfrak{r}) \neq 1$, and so there is a prime ideal $\mathfrak{p}$ dividing d, for which $N(\mathfrak{p}) \geq X^\tau$. It follows that $d \geq X^{\tau/2}$. We can therefore exclude all small d from the above sum except for $d = 1$, which provides the main term. The larger d contribute an error term, as we shall show. We have

$$\nabla_f(V) = \sum_{\mathfrak{r}\in\mathcal{P}} c_{\mathfrak{r}} \sum_{\substack{N(\mathfrak{s})\sim V \\ S\in\mathcal{P}}} f_{\mathfrak{s}} \sum_{\substack{\mathbf{x}\in\mathbb{Z}^2 \\ \mathfrak{i}(x,y)=\mathfrak{r}\mathfrak{s}}} W(\mathbf{x}) + E_V,$$

where the error term E_V satisfies

$$E_V \ll \sum_{X^{\tau/2}\leq d\ll X} \sum_{\mathfrak{r}} \sum_{\substack{N(\mathfrak{s})\sim V \\ V\in\mathcal{P}}} |f_{\mathfrak{s}}| \sum_{\substack{d\mid\mathbf{x} \\ \mathfrak{i}(x,y)=\mathfrak{r}\mathfrak{s}}} W(\mathbf{x})$$

$$\ll X^{2-\tau/2}(\log X)^c. \tag{13.8.1}$$

This is a satisfactory error. To obtain (13.8.1) we used the trivial estimate $f_{\mathfrak{s}} \ll \tau(\mathfrak{s}) \log X$ together with Lemmas 13.12 and 13.10.

Heath-Brown now replaces $\mathfrak{r}, \mathfrak{s}$ by their generators α and β, say, and writes $\mathfrak{Q}$ for the set of primitive integers of K. If we take β to run over a suitable set $\mathfrak{Q}'$ of non-associated primitive integers of K and let $x + y\rho = \alpha\beta$, then we will obtain exactly one value of α from each relevant set of associates. We then get

$$\nabla_f(V) = \sum_{\alpha\in\mathfrak{Q}} c_{(\alpha)} \sum_{|N(\beta)|\sim V} F_\beta \sum_{\substack{\mathbf{x}\in\mathbb{Z}^2 \\ x+y\rho=\alpha\beta}} W(\mathbf{x}) + O\left(X^{2-\frac{1}{2}\tau}(\log X)^c\right),$$

where $F_\beta = f_{(\beta)}$ if $\beta \in \mathfrak{Q}'$, and $F_\beta = 0$ otherwise. In order to specify a suitable set of non-associated integers β we take $\beta > 0$ and require that

$$N(\beta)^{\frac{1}{3}}\varepsilon_0^{-\frac{1}{2}} < \beta \leq N(\beta)^{\frac{1}{3}}\varepsilon_0^{\frac{1}{2}}. \tag{13.8.2}$$

Now Cauchy's inequality is applied to produce

$$\sum_{\alpha\in\mathfrak{Q}} c_{(\alpha)} \sum_{|N(\beta)|\sim V} F_\beta \sum_{\substack{\mathbf{x}\in\mathbb{Z}^2 \\ x+y\rho=\alpha\beta}} W(\mathbf{x}) \ll (X^3/V)^{\frac{1}{2}} S^{\frac{1}{2}},$$

where we have used $N(\alpha) \ll X^3/V$ and written

$$S = \sum_{\alpha \in \mathcal{Q}} |\sum_{|N(\beta)| \sim V} F_\beta \sum_{\substack{\mathbf{x} \in \mathbb{Z}^2 \\ x+y\rho=\alpha\beta}} W(\mathbf{x})|^2.$$

Squaring out the sum over β we obtain

$$S = \sum_{\beta_1,\beta_2} F_{\beta_1} F_{\beta_2} \sum_{\mathbf{x}_1,\mathbf{x}_2} W(\mathbf{x}_1) W(\mathbf{x}_2)\delta,$$

where

$$\delta = \begin{cases} 1 & \text{if } (x_1 + y_1\rho)/\beta_1 = (x_2 + y_2\rho)/\beta_2 \in \mathcal{Q}, \\ 0 & \text{otherwise.} \end{cases}$$

As usual, after squaring out we must handle the diagonal and off-diagonal terms separately. Let S_1 be the contribution from the terms for which $\beta_1 = \beta_2$ and let $S_2 = S - S_1$. Since $F_\beta \ll \tau(\beta) \log X$, we find that

$$S_1 \lll X^2 (\log X)^c$$

by Lemma 13.12.

To handle the off-diagonal terms Heath-Brown writes

$$\alpha = r + s\rho + t\rho^2, \qquad \beta_i = u_i + v_i\rho + w_i\rho^2$$

and then puts

$$\hat{\alpha} = (r, s, t), \quad \tilde{\beta}_i = (w_i, v_i, u_i).$$

We write $\cdot$ and $\times$ for the usual scalar and vector product in $\mathbb{R}^3$. The conditions $\alpha\beta_i = x_i + y_i\rho$, for $i = 1, 2$, then translate into $\hat{\alpha} \cdot \tilde{\beta}_i = 0$. So, if $\tilde{\beta}_1$ and $\tilde{\beta}_2$ are not parallel, and since they are primitive this reduces to $\tilde{\beta}_1 \neq \tilde{\beta}_2$, we must have $\hat{\alpha} = \lambda \tilde{\beta}_1 \times \tilde{\beta}_2$. Since $\hat{\alpha} \in \mathbb{Z}^3$ must be primitive, we have

$$\hat{\alpha} = \pm D^{-1}(v_1 u_2 - u_1 v_2, u_1 w_2 - w_1 u_2, w_1 v_2 - v_1 w_2), \qquad (13.8.3)$$

where

$$D = \gcd(v_1 u_2 - u_1 v_2, u_1 w_2 - w_1 u_2, w_1 v_2 - v_1 w_2).$$

Some simple calculations yield

$$V^{\frac{1}{3}} \ll |\tilde{\beta}_i| \ll V^{\frac{1}{3}}, \qquad XV^{-\frac{1}{3}} \ll |\hat{\alpha}| \ll XV^{-\frac{1}{3}}, \qquad (13.8.4)$$

and

$$D \ll VX^{-1}. \qquad (13.8.5)$$

Heath-Brown next shows that the values of $D \leq VX^{-1}Y^{-1}$, where Y will be specified later, make a negligible contribution. For the moment we suppose that

$$1 \ll Y(X) \ll X^{\frac{1}{2}\tau}. \qquad (13.8.6)$$

We are used to the simple inequality $|F_{\beta_1} F_{\beta_2}| \leq \frac{1}{2}(|F_{\beta_1}|^2 + |F_{\beta_2}|^2)$, from which we deduce that it suffices to estimate

$$S_3 = \sum_{\beta_1,\beta_2} \tau(\beta_1)^2 \sum_{\mathbf{x}_1,\mathbf{x}_2} W(\mathbf{x}_1) W(\mathbf{x}_2)\delta,$$

where the sum is subject to the condition $D \leq VX^{-1}Y^{-1}$.

Heath-Brown then proves the following result [82, Lemma 11.1].

Lemma 13.22. *Let $\mathcal{C}_1, \mathcal{C}_2$ be cubes of side S_0, not necessarily containing the origin. Suppose that $\mathcal{C}_1$ and $\mathcal{C}_2$ are included in a sphere, centred on the origin, of radius S_0^A for some positive constant A. Then, if the vectors $\hat{\beta}_i$ are restricted to be primitive, we will have*

$$\sum_{\substack{\tilde{\beta}_i \in \mathcal{C}_i \\ D|\tilde{\beta}_1 \times \tilde{\beta}_2}} \tau(\beta_1)^2 \ll S_0^6 D^{-2} (\log S_0)^{c(A)}$$

for some constant $c(A)$, providing that $D \ll S_0$.

A geometrical argument then establishes that

$$S_3 \ll VXY^{-1}(\log X)^c.$$

In this way Heath-Brown shows that

$$\nabla_f(V) \ll X^2 Y^{-\frac{1}{2}} (\log X)^c + X^{\frac{3}{2}} V^{-\frac{1}{2}} S_4^{\frac{1}{2}},$$

where

$$S_4 = \sum_{\beta_1, \beta_2} F_{\beta_1} F_{\beta_2} \sum_{\mathbf{x}_1, \mathbf{x}_2} W(\mathbf{x}_1) W(\mathbf{x}_2) \delta,$$

subject to the condition $D > VX^{-1}Y^{-1}$.

The next stage of the argument is to convert the sum S_4 into one in which the variables β_1 and β_2 are independent. It will then be possible to apply a large-sieve estimate. Write

$$d_0 = Y^{15} V X^{-1} + V^{\frac{1}{6}} \tag{13.8.7}$$

and put $\mathbf{v} = \tilde{\beta}_1 \times \tilde{\beta}_2$. After a number of reductions Heath-Brown is able to demonstrate that

$$\nabla_f(V) \ll X^2 Y^{-\frac{1}{2}} (\log X)^c + X^{\frac{3}{2}} V^{-\frac{1}{2}} Y^7 S^{*\frac{1}{2}} (\log X)^c, \tag{13.8.8}$$

where

$$S^* = \sum_{D \in I_m} \sum_d \left| \sum_{\hat{\beta}_i \in \mathcal{C}_i, \, Dd|\mathbf{v}} F_{\beta_1} F_{\beta_2} \right|, \tag{13.8.9}$$

with d restricted by the inequality $dD < d_0$. Here we have split up the range for D so that

$$D \in I_m = \left(\frac{m-1}{N} \Delta, \frac{m}{N} \Delta \right], \quad (m \sim N),$$

where $N \ll X^\tau$ is a parameter chosen in the reduction argument to be Y^2. The cubes $\mathcal{C}_1, \mathcal{C}_2$ are disjoint and have the form

$$\mathcal{C}(n_i, n_j, n_k) = I(n_i) \times I(n_j) \times I(n_k),$$

with

$$I(n) = \left(V^{\frac{1}{3}} \frac{n-1}{N}, V^{\frac{1}{3}} \frac{n}{N} \right].$$

To complete the proof of Lemma 13.8 we detect the condition $dD|\mathbf{v}$ in S^* by means of additive characters in a familiar fashion and use a three-dimensional large sieve. Now the condition $Dd|\mathbf{v}$ may be rewritten as $\hat{\beta}_2 \equiv \lambda \hat{\beta}_1 \pmod{Dd}$ for some integer λ, which is necessarily coprime to Dd. We write

$$S_q(\mathbf{a}) = S_q(\mathbf{a}, \mathcal{C}) = \sum_{\hat{\beta} \in \mathcal{C}} e\left(\frac{\mathbf{a}.\hat{\beta}}{q}\right) F_\beta. \tag{13.8.10}$$

Thus

$$\sum_{\hat{\beta}_i \in \mathcal{C}_i, \, Dd|\mathbf{v}} F_{\beta_1} F_{\beta_2} = \frac{1}{(Dd)^3} \sum_{\substack{\lambda \,(\mathrm{mod}\, dD) \\ (\lambda, dD)=1}} \sum_{\mathbf{a} \,(\mathrm{mod}\, dD)} S_{dD}(\lambda \mathbf{a}, \mathcal{C}_1) \overline{S_{dD}(\mathbf{a}, \mathcal{C}_2)}.$$

It follows that

$$|S^*| \le \frac{1}{(Dd)^3} \sum_{D,d,\lambda,\mathbf{a}} |S_{dD}(\lambda \mathbf{a}, \mathcal{C}_1) S_{dD}(\mathbf{a}, \mathcal{C}_2)|,$$

where D runs over I_m, d runs over positive integers $d \le d_0/D$, λ runs over positive integers less than and coprime to dD, and $\mathbf{a}$ runs modulo dD. We note that

$$\sum_{\mathbf{a} \,(\mathrm{mod}\, dD)} |S_{dD}(\lambda \mathbf{a}, \mathcal{C})|^2 = \sum_{\mathbf{a} \,(\mathrm{mod}\, dD)} |S_{dD}(\mathbf{a}, \mathcal{C})|^2$$

whenever λ is coprime to dD. Hence the inequality $|ab| \le |a|^2 + |b|^2$ yields

$$|S^*| \le \frac{1}{(Dd)^2} \sum_{D \in I_m} \sum_d \sum_{\mathbf{a} \,(\mathrm{mod}\, dD)} |S_{dD}(\mathbf{a}, \mathcal{C})|^2$$

for $\mathcal{C} = \mathcal{C}_1$ or $\mathcal{C}_2$. Now we reduce the fractions $(dD)^{-1}\mathbf{a}$ to lowest terms. A given vector $q^{-1}\mathbf{b}$ with $\gcd(q, b_1, b_2, b_3) = 1$ will occur with weight at most

$$\sum_{D \ge V/(XY)} \sum_{d: \, q|dD} (Dd)^{-2} \ll \sum_{\substack{v \ge V/(XY) \\ q|v}} \tau(v) v^{-2} \ll \frac{\tau(q)}{q} XY \frac{\log V}{V}.$$

What is more, only values $q \le d_0$ will arise. We thus arrive at

$$S^* \ll XY \frac{\log V}{V} \sum_{q \le d_0} \frac{\tau(q)}{q} \sideset{}{^*}\sum_{\mathbf{b} \,(\mathrm{mod}\, q)} |S_q(\mathbf{b})|^2, \tag{13.8.11}$$

where Σ^* denotes summation for $\gcd(q, b_1, b_2, b_3) = 1$.

To bound the above sum Heath-Brown proves the following form of the large-sieve inequality. See [82, Lemma 13.1] for the proof; he begins by iterating the Sobolev-Gallagher inequality as we use it in the appendix, but a crucial step is to show that

$$\left|\{(q, \mathbf{b}) : |b_j q^{-1} - t_j| \le S_0^{-1}, \text{ for } 1 \le j \le 3\}\right| \ll 1 + Q^2 S_0^{-1} + Q^4 S_0^{-3}$$

uniformly in $\mathbf{t}$.

Lemma 13.23. *Let $S(\mathbf{a})$ be given by (13.8.10), with $\mathcal{C}$ a cube of side S_0. Then*

$$\sum_{Q < q \le 2Q} \sideset{}{^*}\sum_{\mathbf{b} \,(\mathrm{mod}\, q)} |S_q(\mathbf{b})|^2 \ll (S_0^3 + Q^2 S_0^2 + Q^4) \sum_{\hat{\beta} \in \mathcal{C}} |F_\beta|^2.$$

Heath-Brown uses this result to handle the contribution to (13.8.11) arising from terms with $q > Q_0$, say. As usual we split up the the range $Q_0 < q \leq d_0$ into dyadic intervals $q \sim Q$, where Q runs over powers of 2. Since

$$\sum_{\hat{\beta} \in \mathcal{C}} |F_\beta|^2 \ll (\log X)^2 \sum_{\hat{\beta} \in \mathcal{C}} \tau(\beta)^2 \ll V(\log X)^c,$$

by Lemma 13.11, we find that the range $q \sim Q$ contributes

$$\ll XY \frac{\log V}{VQ} \exp\left(c \frac{\log Q}{\log\log Q}\right) \left(V + Q^2 V^{\frac{2}{3}} + Q^4\right) V(\log X)^c$$

to S^*. Here we have used the simple upper bound for the divisor function. Now this is summed over the values of $Q = 2^j > Q_0$ to get a total

$$\ll XY \left(\frac{V}{Q_0} \exp\left(c \frac{\log Q_0}{\log\log Q_0}\right) + \left(V^{2/3} d_0 + d_0^3\right) \exp\left(c \frac{\log d_0}{\log\log d_0}\right)\right) (\log X)^c.$$

The exponential terms arising from the divisor function bound are estimated as follows:

$$\exp\left(c \frac{\log Q_0}{\log\log Q_0}\right) \ll Q_0^{\frac{1}{6}}$$

and

$$\exp\left(c \frac{\log d_0}{\log\log d_0}\right) \ll \exp\left(c \frac{\log X}{\log\log X}\right) \ll X^{\frac{1}{6}\tau},$$

by (13.8.7) and the value for τ. The range $Q_0 < q \leq d_0$ therefore contributes a total

$$\ll XY \left(VQ_0^{-\frac{1}{2}} + \left(V^{\frac{2}{3}} d_0 + d_0^3\right) X^{\frac{1}{6}\tau}\right) (\log X)^c$$

to S^*. Using (13.8.7) and the hypothesis $X^{1+\tau} \ll V \ll X^{3/2-\tau}$ this bound becomes

$$\ll XV \left(YQ_0^{-\frac{1}{2}} + Y^{46} X^{-\frac{1}{2}\tau}\right) (\log X)^c. \tag{13.8.12}$$

It only remains to discuss the range $q \leq Q_0$ for which we need the hypothesis (13.2.16). We therefore rewrite the sum to bring the sum of f_β to the fore. We begin with

$$S_q(\mathbf{b}) = \sum_{\mathbf{c} \,(\mathrm{mod}\, q)} e\left(\frac{\mathbf{b}.\mathbf{c}}{q}\right) \sum_{\substack{\hat{\beta} \in \mathcal{C} \\ \hat{\beta} \equiv \mathbf{c} \,(\mathrm{mod}\, q)}} F_\beta. \tag{13.8.13}$$

Now (13.8.2) holds for any $\hat{\beta} \in \mathcal{C}_i$. From the definition of F_β it follows that for $\hat{\beta} \in \mathcal{C}$ in (13.8.13) we have $F_\beta = f_{(\beta)}$ for primitive β, and $F_\beta = 0$ otherwise. Thus

$$\sum_{\substack{\hat{\beta} \in \mathcal{C} \\ \hat{\beta} \equiv \mathbf{c} \,(\mathrm{mod}\, q)}} F_\beta = \sum_{d:\, (q,d)|\mathbf{c}} \mu(d) \sum_{\substack{\hat{\beta} \in \mathcal{C} \\ \hat{\beta} \equiv \mathbf{c} \,(\mathrm{mod}\, q),\, d|\hat{\beta}}} f_{(\beta)},$$

and so

$$S_q(\mathbf{b}) \ll \sum_{\mathbf{c} \,(\mathrm{mod}\, q)} \sum_d \left| \sum_{\substack{\hat{\beta} \in \mathcal{C} \\ \hat{\beta} \equiv \mathbf{c} \,(\mathrm{mod}\, q),\, d|\hat{\beta}}} f_{(\beta)} \right|. \tag{13.8.14}$$

Heath-Brown then notes that the conditions $\hat{\beta} \equiv \mathbf{c} \,(\mathrm{mod}\, q)$ and $d|\hat{\beta}$ confine $\hat{\beta}$ to a single residue class modulo $[q, d]$. The inner sum above is therefore $O(V\mathcal{L}^{-c})$, by the hypothesis of Lemma 13.8, providing that $[q, d] \leq Q_1$. Hence, if $q \leq Q_1^{1/2}$, we find the contribution to (13.8.14) corresponding to values $d \leq Q_1^{1/2}$ is

$$\ll q^3 Q_1^{\frac{1}{2}} V\mathcal{L}^{-c} \ll Q_1^2 V\mathcal{L}^{-c}.$$

For $d > Q_1^{1/2}$ it suffices to use the trivial bound $f_{(\beta)} \ll \tau(\beta) \log X$. This leads to a contribution to (13.8.14), which is

$$\ll (\log X) \sum_{\substack{d > Q_1^{1/2} \\ d|\hat{\beta}}} \sum_{\hat{\beta} \in \mathcal{C}} \tau(\beta) \sum_{\mathbf{c} \equiv \hat{\beta} \,(\mathrm{mod}\, q)} 1 = (\log X) \sum_{\substack{d > Q_1^{1/2} \\ d|\hat{\beta}}} \sum_{\hat{\beta} \in \mathcal{C}} \tau(\beta).$$

However, according to Lemma 13.11 we have

$$\sum_{\substack{\hat{\beta} \in \mathcal{C} \\ d|\hat{\beta}}} \tau(\beta) \ll \tau(d)^c \sum_{|\mathbf{x}| \ll V^{1/3}/d} \tau(x_1 + x_2\rho + x_3\rho^2) \ll \tau(d)^c V d^{-3} (\log X)^c.$$

So values $d > Q_1^{1/2}$ contribute to (13.8.14) a term

$$\ll V(\log X)^c \sum_{d > Q_1^{1/2}} \tau(d)^c d^{-3} \ll V Q_1^{-1} (\log X)^c.$$

In this way Heath-Brown establishes that

$$S_q(\mathbf{b}) \ll Q_1^2 V\mathcal{L}^{-c} + V Q_1^{-1} (\log X)^c.$$

The terms with $q \leq Q_0$ therefore contribute

$$\ll XYV (Q_1^4 \mathcal{L}^{-c} + Q_1^{-2})(\log X)^c$$

to (13.8.11), providing that $Q_0 \leq Q_1^{1/2}$. Taken in conjunction with our estimate (13.8.12) for the terms with $Q_0 < q \leq d_0$, he can therefore deduce that

$$S^* \ll XV(YQ_1^4\mathcal{L}^{-c} + YQ_1^{-2} + YQ_0^{-\frac{1}{2}} + Y^{46}X^{-\frac{1}{2}\tau})(\log X)^c.$$

The choice

$$Q_0 = Q_1^{\frac{1}{2}}$$

then gives

$$S^* \ll XV(YQ_1^4\mathcal{L}^{-c} + YQ_1^{-\frac{1}{4}} + Y^{46}X^{-\frac{1}{2}\tau})(\log X)^c.$$

We insert this into (13.8.8) to obtain

$$\nabla_f(V) \ll X^2 \left(Y^{-\frac{1}{2}} + Y^{30}X^{-\frac{1}{4}\tau} + Y^8 Q_1^{-\frac{1}{8}} + Y^8 Q_1^2 \mathcal{L}^{-c} \right) (\log X)^c.$$

We can now choose $Y = Q_1^{1/80}$, which is consistent with (13.8.6) since

$$Q_1 \leq \exp\left((\log X)^{\frac{1}{3}}\right).$$

Using this bound for Q_1, we finally see that our estimate reduces to

$$\nabla_f(V) \ll X^2 Q_1^{-\frac{1}{160}} (\log X)^c,$$

as required for Lemma 13.8. With this step Heath-Brown's *coup de maître* is complete. $\qquad\Box$

Chapter Fourteen

Epilogue

14.1 A SUMMARY

Archimedes has been quoted as saying, "Give me a fulcrum, a lever that is long enough, and a place to stand, and I will move the earth." An internet search will reveal many variants of this aphorism attributed to the ancient scientist. Having finished this book the reader could now fairly adapt these words to claim, "Give me sufficiently good Type I and Type II information and I will prove the twin prime conjecture, the Goldbach conjecture, and so on." Before we look at those problems for which a solution still seems a distant reality, let us reflect on what we have uncovered on our journey through the world of prime-detecting sieves. Let

$$\mathcal{B} = \mathbb{N} \cap [x/2, x], \qquad \mathcal{A} \subset \mathcal{B}.$$

To find primes in $\mathcal{A}$ we have wanted Type I information

$$\sum_{d \leq D} a_d \sum_{nd \in \mathcal{A}} 1 = \lambda \sum_{d \leq D} a_d \sum_{nd \in \mathcal{B}} 1 + O(E), \qquad (14.1.1)$$

where E is smaller than our expected main term. We have found such information in many different circumstances. Almost always we have had $D \geq x^{2/3}$. Usually this has been the least difficult part of the problem, although sometimes the best results have needed very deep theorems. We have seen different approaches to Type II information. The most straightforward has established a formula:

$$\sum_{d \sim D} a_d \sum_{nd \in \mathcal{A}} b_n = \lambda \sum_{d \sim D} a_d \sum_{nd \in \mathcal{B}} b_n + O(E) \qquad (14.1.2)$$

(though more variables are sometimes required). On the other hand, in Chapter 12 the parity problem was overcome by estimating sums like

$$\sum_{d \sim D} \left| \sum_{nd \in \mathcal{A}} \mu(n) \right|. \qquad (14.1.3)$$

However the Type II information was produced, we often needed it to hold for $x^{1/3} \ll D \ll x^{2/3}$ in order to obtain an asymptotic formula. As the Type II information decreased in quality, we then passed from a formula to upper and lower bounds of the correct order of magnitude. The more fundamental distinction between classes of Type II information actually arises not between (14.1.2) and (14.1.3) but between situation where the results are true for general coefficients (for example, Chapters 3, 5, and 6) or only true for the types of coefficients that are produced by the sieve method (as in Chapter 7 onward). For these latter results the zero-free region of $\zeta(s)$ or the appropriate L-function has been crucial. Indeed, one can observe that the theme of using a large-sieve inequality plus a zero-free result to cover small moduli, which began as a simple tune with the proof of the Bombieri-Vinogradov theorem in Chapter 2, evolves into a full-blown symphony with many subtle variations in Chapters 12 and 13.

14.2 A CHALLENGE WITH WHICH TO CLOSE

Now suppose we were so foolish as to embark upon a proof of the twin-prime conjecture, for example. It should be mentioned that Heath-Brown [76] has established that, should there be infinitely many Siegel zeros in a certain sense, then there are infinitely many prime twins. His proof rests on the fact that the existence of a Siegel zero leads to the very good distribution of primes in many arithmetic progressions, as well as the very bizarre fact that the character, say $\chi(n)$, associated with the Siegel zero behaves rather like the Möbius function. He can therefore replace

$$\Lambda(n) = \sum_{d|n} \mu(d) \log\left(\frac{n}{d}\right) \quad \text{with} \quad \tilde{\Lambda}(n) = \sum_{d|n} \mu^2(d)\chi(d) \log\left(\frac{n}{d}\right)$$

for n in a certain range without introducing too large an error.

But suppose now that we do not assume such awesome wonders. So we shall take $\mathcal{A} = \{n \in \mathcal{B} : n = p + 2,\ p \text{ a prime}\}$. In this case we can take D almost as large as $x^{1/2}$ in (14.1.1) by the Bombieri-Vinogradov theorem. If, as will be the case by the sieve method, a_d factorizes in some way, we can take D up to around $x^{4/7}$ in the best case by using the results that underpin Chapter 8 (that is, [16–18]). In the context of the results we have considered in this text this is a disappointing value, falling far short of $x^{2/3}$ and demanding very good Type II information in order to prove the conjecture. If we were to assume the Elliott-Halberstam conjecture, that q can be averaged up to $x^{1-\epsilon}$ in the Bombieri-Vinogradov theorem, then we would need substantially less Type II information. However, we would still need *some*. It is here that we run into serious difficulties, for we do not have any Type II information at all! We can neither estimate

$$\sum_{d \sim D} a_d \sum_{dn=p+2} b_n$$

nor

$$\sum_{d \sim D} \left| \sum_{dn=p+2} \mu(n) \right|.$$

Although the sequence $p + 2$ is quite "dense" we do not have the nice factorization property that enabled us to succeed in Chapters 12 and 13. Neither do we have the "independence" of the characteristic function of $\mathcal{A}$ from the multiplicative restrictions we need, as happens in Chapters 3 and 5, for example. We therefore finish this book with the greatest challenge facing researchers in the application of sieve methods to detect primes: to establish Type II information for the twin-prime problem, for Goldbach's conjecture, for primes representable as $n^2 + 1$, and so on.

Or perhaps it would be easier to rise to Archimedes' challenge ... .

Appendix

Auxiliary Results

In this appendix we establish some of the more important auxiliary results used throughout the book. This information is provided to help research students at the start of their careers. It should not be seen as an alternative to reading and benefitting from classical texts like [27, 131, 157], supplemented by a modern work like [106]. We also provide more detailed work on Buchstab's function that is not contained in standard textbooks.

A.1 PERRON'S FORMULA

We write $H(y)$ for the Heaviside function:

$$H(y) = \begin{cases} 0 & \text{if } y < 0, \\ 1 & \text{if } y \geq 0. \end{cases}$$

Before proving Perron's formula (or what might be more properly called the truncated Perron's formula) we establish the following.

Lemma A.1. *Let $\epsilon > 0, T > 1, \theta \neq 0$ be given. Then*

$$\frac{1}{2\pi i} \int_{\epsilon - iT}^{\epsilon + iT} \frac{\exp(\theta s)}{s} \, ds = H(\theta) + O\left(\frac{\exp(\theta \epsilon)}{|\theta| T}\right). \tag{A.1.1}$$

Proof. Let $V > 0$, and we suppose at first that $\theta > 0$. Let $\mathcal{C}$ be the contour in the complex plane given by the rectangle with vertices at $\{\epsilon - iT, \epsilon + iT, -V + iT, -V - iT\}$. Then, by Cauchy's residue theorem,

$$\frac{1}{2\pi i} \int_{\mathcal{C}} \frac{\exp(\theta s)}{s} \, ds = \operatorname*{Res}_{s=0} \frac{\exp(\theta s)}{s} = 1.$$

On the line from $-V + iT$ to $-V - iT$ the integrand is $\leq V^{-1} \exp(-V\theta)$, and so the modulus of the contribution to the contour integral from this line is

$$\leq \frac{1}{\pi} \exp(-V\theta) \to 0 \quad \text{as} \quad V \to \infty.$$

At a point $\sigma \pm iT$ on the lines from $\epsilon \pm iT$ to $-V \pm iT$ the integrand is $\leq T^{-1} \exp(\sigma \theta)$. The total contribution the integrals over these lines make is therefore in absolute value

$$\leq \frac{1}{\pi T} \int_{-V}^{\epsilon} \exp(\sigma \theta) \, d\sigma < \frac{\exp(\theta \epsilon)}{\theta T}.$$

Letting $V \to \infty$ we therefore establish (A.1.1) in the case $\theta > 0$. The case $\theta < 0$ is done similarly, but with the vertices $-V \pm iT$ replaced by $V \pm it$. There is now no pole inside the contour, and the error terms have exactly the same order of magnitude as for $\theta > 0$. $\square$

Lemma A.2. *Let $c > 0$ and suppose that $|n - x| < \min(1, x/T)$. Then*

$$\frac{1}{2\pi i} \int_{c-iT}^{c+iT} n^{-w} \frac{x^w}{w} \, dw = O(1). \tag{A.1.2}$$

Proof. We may replace the contour with straight lines from $c \pm iT$ to $T + 0i$ since the integrand has no poles for $\mathrm{Re}\, s > 0$. The total length of the lines is $< 4T$, on the lines $|s| > T/2$ and

$$n^{-w} x^w = (1 + (x - n)/n)^w \ll 1 \tag{A.1.3}$$

since $|n - x| < \min(1, x/T)$. The integrand in (A.1.2) is thus $\ll T^{-1}$ on the new contour, and so the integral on the new lines is $O(1)$, as required. $\square$

Proof. (Perron's Formula) Let the nearest integer to x be m. Write

$$\ell = \begin{cases} m & \text{if } ||x|| < x/T, \\ 0 & \text{otherwise.} \end{cases}$$

From Lemma A.1, with $\theta = \log(x/n)$, we have

$$\frac{1}{2\pi i} \int_{c-iT}^{c+iT} n^{-w} \frac{x^w}{w} dw = \begin{cases} 1 & \text{if } x > n \\ 0 & \text{if } x < n \end{cases} + O\left(\frac{(x/n)^c}{|\log(x/n)|T}\right).$$

We apply this formula for every $n \neq \ell$ and use Lemma A.2 if $\ell = m$. We thus get

$$\sum_{n=1}^{\infty} \frac{a_n}{n^{-s}} \frac{1}{2\pi i} \int_{c-iT}^{c+iT} n^{-w} \frac{x^w}{w} \, dw = \sum_{n \leq x} \frac{a_n}{n^s} + E_1 + E_2 + E_3.$$

Here the E_j correspond to the errors introduced for n in the ranges $n < x/2$ or $n > 2x$ (E_1), $x/2 \leq n \leq 2x, n \neq m$ (E_2), $n = m$ (E_3). Thus

$$E_1 \ll \frac{x^c}{T} \sum_{n=1}^{\infty} \frac{|a_n|}{n^{c+\sigma}} \ll \frac{x^c}{T(\sigma + c - 1)^{\alpha}}.$$

Using $|\log(x/n)| \geq |x - n|/2x$ if $x/2 \leq n \leq 2x$ we obtain

$$E_2 \ll \sum_{\substack{x/2 \leq n \leq 2x \\ x \neq m}} \frac{x^{1+c} |a_n|}{T|x - n|n^{\sigma+c}} \ll \frac{f(x)x^{1-\sigma} \log x}{T}.$$

Finally, if $\ell \neq m$, then the argument used for E_2 gives the term

$$\ll |a_m| \frac{x^{1-\sigma}}{T||x||}.$$

If $\ell = m$, then Lemma A.2 gives a term

$$\ll |a_m| x^{-\sigma}.$$

This completes the proof of (1.4.7). $\square$

We now give a brief deduction of Lemma 2.2 from Lemma A.1.

Proof. Since $\rho, \gamma > 0$, we have

$$\delta = H(\gamma + \rho) - H(\gamma - \rho) = \frac{1}{2\pi i} \int_{\epsilon-iT}^{\epsilon+iT} e^{\gamma s} \frac{\left(e^{\rho s} - e^{-\rho s}\right)}{s} \, ds + E,$$

with

$$E \ll \frac{\exp((\gamma + \rho)\epsilon)}{T|\gamma - \rho|}.$$

We can then take $\epsilon \to 0$ since there is no pole of the integrand at $s = 0$. This establishes (2.3.5). $\qquad\square$

Finally, we consider the following variant of Perron's formula used in Section 2.6 and in Chapter 13.

Lemma A.3. *Let the hypotheses of Lemma 1.1 be given. Then*

$$\sum_{n \leq x} \frac{a_n}{n^s} \log\left(\frac{x}{n}\right) = \frac{1}{2\pi i} \int_{c-i\infty}^{c+i\infty} F(w+s) \frac{x^w}{w^2} \, dw. \tag{A.1.4}$$

Remark. In view of the factor w^{-2} in the integrand above, the integral is absolutely convergent.

Proof. We note that

$$\sum_{n \leq x} \frac{a_n}{n^s} \log\left(\frac{x}{n}\right) = \int_1^x \frac{1}{u} \sum_{n \leq u} \frac{a_n}{n^s} \, du.$$

We then apply (1.4.7) for all $u \in [1, x]$. We note that

$$\int_1^x \min\left(\frac{u}{T\|u\|}, 1\right) \, du \ll \frac{x \log x}{T}.$$

The final error term in (1.4.7) then is no greater on average than the penultimate term. The main term is

$$\int_1^x \int_{c-iT}^{c+iT} F(s+w) \frac{u^{w-1}}{w} \, dw \, du = \int_{c-iT}^{c+iT} F(s+w) \frac{x^w}{w^2} \, dw$$
$$- \int_{c-iT}^{c+iT} F(s+w) \frac{1}{w^2} \, dw.$$

As $T \to \infty$, the final term above tends to zero as do all the error terms to complete the proof. $\qquad\square$

A.2 BUCHSTAB'S FUNCTION $\omega(u)$

Here we give the rigorous proof by induction of what is stated in Chapter 1.

For technical reasons it is useful to consider sets

$$\mathcal{B} = \mathbb{N} \cap [x - y) \quad \text{with } y \leq x\eta(x), \ \eta(x) = \exp\left(-(\log x)^{\frac{1}{2}}\right).$$

This enables us to give a clean $y(\log x)^{-1}$ main term without worrying about $\text{Li}(x)$. It does create other difficulties in the induction, but these can be overcome. Now write

$$g(x,k) = \exp\left(-\left(\frac{\log x}{k}\right)^{\frac{4}{7}}\right).$$

By the PNT we have, for $g(x,1) \le y/x \le \eta(x)$,

$$S(\mathcal{B}, x^{\frac{1}{2}}) = \frac{y}{\log x}(1 + O(\eta)). \tag{A.2.1}$$

Our inductive hypothesis is the following:

Suppose $k \in \mathbb{N}$. Whenever

$$\log x > (k+1)^8 \quad \text{and} \quad x\eta \ge y \ge xg(x,k), \tag{A.2.2}$$

we have

$$S(\mathcal{B}, x^{\frac{1}{u}}) = \frac{uy}{\log x}\,w(u) + E(x,u) \quad \text{when} \ \ k \le u \le k+1, \tag{A.2.3}$$

where

$$|E(x,u)| \le Cy(k+1)\exp\left(-\left(\frac{\log x}{u}\right)^{\frac{1}{2}}\right).$$

Here C is an absolute constant.

Equation (A.2.1) shows that the hypothesis holds for $1 \le u \le 2$. Now suppose that (A.2.3) is true for k and consider $k+1 < u \le k+2$, further assuming that (A.2.2) holds with k replaced by $k+1$. We note that one consequence of this, since $k \ge 2$, is that $\log x \ge 16(k+2)^3$, an inequality we shall need later. By Buchstab's identity

$$S(\mathcal{B}, x^{\frac{1}{u}}) = S(\mathcal{B}, x^{\frac{1}{k+1}}) + \sum_{x^{1/u} < p < x^{1/(k+1)}} S(\mathcal{B}_p, p)$$

$$= S(\mathcal{B}, x^{\frac{1}{k+1}}) + \sum_p S_p^*, \quad \text{say.}$$

Now we apply the inductive hypothesis to each S_p^* with x, y replaced by $x/p, y/p$, respectively. We note that, since $p \le x^{1/(k+1)}$, we have $g(x/p, k) \le g(x, k+1)$. Also, $\eta(x/p) > \eta(x)$. We thus have

$$\frac{\log(x/p)}{\log p} = \frac{\log x}{\log p} - 1 \in [k, k+1],$$

$$\log\left(\frac{x}{p}\right) \ge \left(1 - \frac{1}{k+1}\right)\log x \ge (k+2)^8 \frac{k}{k+1} \ge (k+1)^8,$$

and

$$\frac{x}{p}\eta\left(\frac{x}{p}\right) \ge \frac{y}{p} \ge \frac{x}{p}g\left(\frac{x}{p}, k\right),$$

which confirms the requirements of the hypothesis. We thus obtain

$$S(\mathcal{B}, x^{\frac{1}{u}}) = \frac{y(k+1)}{\log x}\omega(k+1) + E(x, k+1)$$

$$+ \sum_{x^{1/u} < p < x^{1/(k+1)}} \frac{y}{p\log(x/p)}\left(\frac{\log(x/p)}{\log p}\omega\left(\frac{\log(x/p)}{\log p}\right)\right)$$

$$+ \sum_{x^{1/u} < p < x^{1/(k+1)}} K_p(k+1)\frac{y}{p}\exp\left(-(\log p)^{\frac{1}{2}}\right)$$

$$= F + E_1 + S_1 + S_2, \quad \text{say}$$

$$= M + E_1 + E_2 + E_3, \quad \text{say}.$$

Here $|K_p| \le C$, $E_3 = S_2$, and $F + S_1 = M + E_2$ with

$$M = \frac{y}{\log x}\left((k+1)\omega(k+1) + \int_{1/u}^{1/(k+1)} \frac{1}{\alpha^2}\omega\left(\frac{1}{\alpha} - 1\right)d\alpha\right)$$

$$= \frac{uy}{\log x}\omega(u).$$

We have

$$E_1 \le Cy(k+1)\exp\left(-\frac{(\log x)^{\frac{1}{2}}}{k+1}\right)$$

by the induction hypothesis.

We can estimate E_3 by bounding S_2 with the integral

$$2C(k+1)y\int_{x^{1/u}}^{x^{1/(k+1)}} \frac{1}{v(\log v)}\exp\left(-(\log v)^{\frac{1}{2}}\right)dv.$$

The change of variables $v = x^\alpha$ converts this to

$$2Cy\int_{1/u}^{1/(k+1)} \frac{1}{\alpha}\exp\left(-(\alpha\log x)^{\frac{1}{2}}\right)d\alpha$$

$$\le 2Cu^{\frac{1}{2}}\int_{1/u}^{1/(k+1)} \frac{1}{\alpha^{\frac{1}{2}}}\exp\left(-(\alpha\log x)^{\frac{1}{2}}\right)d\alpha$$

$$= \frac{4Cyu^{\frac{1}{2}}}{(\log x)^{\frac{1}{2}}}\left[-\exp\left(-(\alpha\log x)^{\frac{1}{2}}\right)\right]_{1/u}^{1/(k+1)}$$

$$\le \frac{Cy}{u}\exp\left(-\left(\frac{\log x}{u}\right)^{\frac{1}{2}}\right),$$

where we have used $\log x > 16(k+2)^3$ for the final inequality.

It now remains to relate S_1 to the term we put in M to produce a suitably small E_2. We have

$$\sum_{x^{1/u} < p < x^{1/(k+1)}} \frac{y}{p\log p}\omega\left(\frac{\log x}{\log p} - 1\right) = \frac{y}{\log x}\sum_{x^{1/u} < n < x^{1/(k+1)}} \frac{\Lambda(n)}{n}f(n) - E_4,$$

where

$$f(n) = \frac{\log x}{(\log n)^2} \omega \left(\frac{\log x}{\log p} - 1 \right)$$

and

$$0 \le E_4 \le 2yu^2 x^{-\frac{1}{2(k+2)}} \le \frac{C}{100} y \exp\left(-\left(\frac{\log x}{u} \right)^{\frac{1}{2}} \right),$$

assuming that C is sufficiently large. We have

$$\sum_{x^{1/u} < n < x^{1/(k+1)}} \frac{\Lambda(n)}{n} = \sum_{x^{1/u} < n < x^{1/(k+1)}} \frac{1}{n} + E_5,$$

where

$$|E_5| < C' y \exp\left(-\left(\frac{\log x}{u} \right)^{\frac{11}{20}} \right)$$

for an absolute constant C' by the PNT.

Thus, by partial summation,

$$\frac{y}{\log x} \sum_{x^{1/u} < n < x^{1/(k+1)}} \frac{\Lambda(n)}{n} f(n) = \frac{y}{\log x} \sum_{x^{1/u} < n < x^{1/(k+1)}} \frac{1}{n} f(n) + E_6,$$

where

$$E_6 \le 2u^2 C' y \exp\left(-\left(\frac{\log x}{u} \right)^{\frac{11}{20}} \right)$$

$$\le \frac{C}{100} y \exp\left(-\left(\frac{\log x}{u} \right)^{\frac{1}{2}} \right)$$

assuming, as we may, that C is sufficiently large.

We have thus established that

$$E_2 \le \frac{C}{50} y \exp\left(-\left(\frac{\log x}{u} \right)^{\frac{1}{2}} \right).$$

Combining our error terms we obtain

$$|E(x, u)| \le Cy(k + 1 + 1/u + 1/50) \exp\left(-\left(\frac{\log x}{u} \right)^{\frac{1}{2}} \right),$$

which completes the proof of (A.2.3) with k replaced by $k + 1$. The result then holds with $y = x\eta(x)$ for all k with $(k + 1)^8 < \log x$ by the principle of mathematical induction.

Now, by Theorem 4.4 slightly modified, for $u \ge \log \log x$ we have

$$S(\mathcal{B}, x^{\frac{1}{u}}) \sim \frac{yu}{\log x} \exp(-\gamma).$$

Comparing this with (A.2.3) (taking $x \to \infty$ and letting u grow with x as any function with $\log \log x < u < (\log x)^{1/9}$) we deduce that $\omega(u) \to e^{-\gamma}$ as $u \to \infty$, as claimed in Chapter 1.

A.3 LARGE-SIEVE INEQUALITIES

We have used large-sieve inequalities explicitly or implicitly throughout this book. In this section we give complete proofs of several such inequalities. First we shall prove the basic large-sieve inequality in the form given by Gallagher [43]. This can be stated as follows.

Lemma A.4. Let $\alpha_j \in \mathbb{R}, j = 1, \ldots, J$ and $a_n \in \mathbb{C}, n = 1, \ldots, N$ be two finite sequences, with

$$\delta = \min_{j \neq k} ||\alpha_j - \alpha_k|| \neq 0.$$

Then

$$\sum_{j=1}^{J} \left| \sum_{n=1}^{N} a_n e(n\alpha_j) \right|^2 \leq (\pi N + \delta^{-1}) \sum_{n=1}^{N} |a_n|^2. \tag{A.3.1}$$

Remark. The factor $(\pi N + \delta^{-1})$ on the right-hand side of (A.3.1) can be replaced by $N - 1 + \delta^{-1}$, and this result is best possible [132].

Proof. Clearly it suffices to show that

$$\sum_{j=1}^{J} \left| \sum_{n=-N}^{N} a_n e(n\alpha_j) \right|^2 \leq (2\pi N + \delta^{-1}) \sum_{n=-N}^{N} |a_n|^2.$$

Suppose that $g(t)$ has a continuous derivative for $0 \leq t \leq 1$. Then, using integration by parts, we have

$$\int_0^1 g(t)\, dt = g(1) - \int_0^1 t g'(t)\, dt.$$

We combine this with

$$g(1) = g(x) + \int_x^1 g'(t)\, dt$$

for $0 \leq x \leq 1$ to produce

$$g(x) = \int_0^1 g(t)\, dt + \int_0^x t g'(t)\, dt + \int_x^1 (t-1)g'(t)\, dt.$$

Hence we obtain the Sobolev-Gallagher inequality:

$$g(\tfrac{1}{2}) \leq \int_0^1 |g(t)| + \tfrac{1}{2}|g'(t)|\, dt.$$

Now let $g(t) = f(\alpha + \delta(t - \tfrac{1}{2}))$ and this becomes

$$f(\alpha) \leq \int_0^1 |f(\alpha + \delta(t - \tfrac{1}{2}))| + \tfrac{1}{2}\delta|f'(\alpha + \delta(t - \tfrac{1}{2}))|\, dt$$

$$= \int_{\alpha - \delta/2}^{\alpha + \delta/2} \frac{1}{\delta}|f(u)| + \tfrac{1}{2}|f'(u)|\, du. \tag{A.3.2}$$

Let

$$S(\alpha) = \sum_{n=-N}^{N} a_n e(\alpha n)$$

and put $f(\alpha) = S(\alpha)^2$. We thus arrive at

$$\sum_{j=1}^{J} |S(\alpha_j)|^2 \leq \sum_{j=1}^{J} \int_{\alpha_j - \delta/2}^{\alpha_j + \delta/2} \frac{1}{\delta} |S(u)|^2 + |S'(u)S(u)| \, du.$$

By the choice for δ the intervals of integration are non-overlapping $(\mathrm{mod}\, 1)$, and so

$$\sum_{j=1}^{J} |S(\alpha_j)|^2 \leq \int_0^1 \frac{1}{\delta} |S(u)|^2 + |S'(u)S(u)| \, du.$$

Now, by Cauchy's inequality,

$$\int_0^1 |S(u)S'(u)| \, du \leq \left(\int_0^1 |S(u)|^2 \, du \right)^{\frac{1}{2}} \left(\int_0^1 |S'(u)|^2 \, du \right)^{\frac{1}{2}}.$$

The proof then follows from Parseval's identity, which gives

$$\int_0^1 |S(u)|^2 \, du = \sum_{n=-N}^{N} |a_n|^2, \qquad \int_0^1 |S'(u)|^2 \, du = 4\pi^2 \sum_{n=-N}^{N} |na_n|^2.$$

$\square$

Next we show how the above inequality can be used to prove Lemma 2.4.

Proof. (Lemma 2.4) For χ a primitive character $(\mathrm{mod}\, q)$, write $\tau(\chi)$ for the Gaussian sum

$$\sum_{a=1}^{q} \chi(a) e\left(\frac{a}{q} \right).$$

We recall two fundamental properties of this sum:

$$|\tau(q)| = q^{\frac{1}{2}}, \qquad \chi(n) = \frac{1}{\tau(\bar{\chi})} \sum_{a=1}^{q} \bar{\chi}(a) e\left(\frac{a}{q} \right).$$

Hence

$$\sum_{n=M+1}^{M+N} a_n \chi(n) = \frac{1}{\tau(\bar{\chi})} \sum_{a=1}^{q} \bar{\chi}(a) S(a/q),$$

where now

$$S\left(\frac{a}{q} \right) = \sum_{n=M+1}^{M+N} a_n e\left(\frac{a}{q} \right).$$

Thus

$$\sum_{\chi}^{*} \left| \sum_{n=M+1}^{M+N} a_n \chi(n) \right|^2 = \frac{1}{q} \sum_{\chi}^{*} \left| \sum_{a=1}^{q} \bar{\chi}(a) S\left(\frac{a}{q}\right) \right|^2$$

$$\leq \frac{1}{q} \sum_{\chi} \left| \sum_{a=1}^{q} \bar{\chi}(a) S\left(\frac{a}{q}\right) \right|^2$$

$$= \frac{\phi(q)}{q} \sum_{\substack{a \leq q \\ (a,q)=1}} \left| S\left(\frac{a}{q}\right) \right|^2 .$$

Here we have noted that the inclusion of all characters only serves to increase the sum, and for the final step we used the orthogonality property of characters. The result follows from (A.3.1) since

$$\min_{a_1/q_1 \neq a_2/q_2} \left\| \frac{a_1}{q_1} - \frac{a_2}{q_2} \right\| \geq \frac{1}{q_1 q_2} \geq \frac{1}{Q^2} .$$

□

Now we prove the dual form of the large sieve in $\mathbb{Z}[i]$ (Lemma 11.1). This result is stated as Lemma 6 in [24]. In the notation of Lemma 11.1 we have

$$\sum_{m=1}^{T} \left| \sum_{|n|^2 \leq N} c(n) \lambda^m(n) \right|^2 = \sum_{m=1}^{T} \left| \sum_{|n|^2 \leq N} {}' c_1(n) \lambda^m(n) \right|^2$$

$$= \sum_{m=1}^{T} \left| \sum_{|n|^2 \leq N} {}' c_1(n) e\left(\frac{2}{\pi} m \arg n\right) \right|^2$$

to bring the notation in line with our large-sieve inequality. The reader can see why this type of inequality is called a "dual" form: the summation over consecutive integers is here on the outside. Now let

$$\delta = \min_{n \neq n'} |\arg n - \arg n'| \quad \text{for } |n|^2 \leq N.$$

Here the minimum is over primitive n only. The reader can quickly verify that $N^{-1} \gg \delta \gg N^{-1}$. We thus need only show that

$$\sum_{m=1}^{T} \left| \sum_{|n|^2 \leq N} {}' c_1(n) e\left(\frac{2}{\pi} m \arg n\right) \right|^2 \ll (T + \delta^{-1}) \sum_{|n|^2 \leq N} {}' |c_1(n)|^2. \qquad (A.3.3)$$

Write $\theta(m) = (1 - |m|/2T)$. Since the left-hand side of (A.3.3) remains unaltered by replacing m with $-m$, we can bound it by

$$\sum_{m=-2T}^{2T} \theta(m) \left| \sum_{|n|^2 \leq N} {}' c_1(n) e\left(\frac{2}{\pi} m \arg n\right) \right|^2 .$$

We now multiply out the square and use a familiar technique to produce the bound

$$\sum_{|n|^2 \le N}{}' |c_1(n)|^2 \sum_{|n'|^2 \le N}{}' \left| \sum_{m=-2T}^{2T} \theta(m) e\left(\frac{2}{\pi} m(\arg n - \arg n')\right) \right|.$$

Since the inner sum is

$$\frac{1}{2T}\left(\frac{\sin(2T\pi(\arg n - \arg n'))}{\sin(\pi(\arg n - \arg n'))}\right)^2 \ll \frac{1}{T}\min\left(T^2, \frac{1}{\|\arg n - \arg n'\|^2}, \right)$$

the reader should now be able to complete the proof. (Note that if we had not included the factor $\theta(m)$, we would have incurred a loss of a factor $\log N$ at this last stage.) $\qquad\square$

A.4 THE MEAN VALUE THEOREM FOR DIRICHLET POLYNOMIALS

In this section we prove Lemma 5.2. For convenience we prove it in the equivalent form

$$\int_{-T}^{T}\left|\sum_{n\sim N} a_n n^{it}\right|^2 dt \ll (T+N)\sum_{n\sim N}|a_n|^2. \tag{A.4.1}$$

Proof. First we note the elementary integral (for $u \in \mathbb{R}$)

$$\int_{-\infty}^{\infty} \exp(iut - |t|)\,dt = \frac{2}{1+u^2},$$

from which we obtain

$$\int_{-\infty}^{\infty} \exp(iut - |t/T|)\,dt = \frac{2T}{1+(uT)^2}. \tag{A.4.2}$$

To prove (A.4.1) it suffices to consider

$$\int_{-\infty}^{\infty}\left|\sum_{n\sim N} a_n n^{it}\right|^2 \exp(-|t/T|)\,dt.$$

Squaring out this becomes

$$\sum_{m\sim N}\sum_{n\sim N} a_n \bar{a}_m \int_{-\infty}^{\infty} \exp(itu(m,n) - |t/T|)\,dt,$$

where

$$u(m,n) = \log(n/m).$$

The diagonal terms (with $m = n$) clearly contribute (from (A.4.2) with $u = 0$)

$$2T\sum_{n\sim N}|a_n|^2.$$

For the remaining terms we can suppose that $m < n$, and so $m = n - r$, where $1 \le r \le n - N$. Using a familiar inequality we then need only estimate

$$S = \sum_{n\sim N}|a_n|^2 \sum_{r=1}^{n-N} \frac{T}{1+T^2 u(n, n-r)^2}.$$

However,

$$u(n, n - r) = -\log\left(1 - \frac{r}{n}\right) \gg \frac{r}{N}.$$

We thus have

$$S \ll \sum_{n \sim N} |a_n|^2 \sum_{r=1}^{n-N} \min\left(T, \frac{N^2}{Tr^2}\right)$$

$$\ll N \sum_{n \sim N} |a_n|^2$$

after an elementary calculation. $\square$

There are no difficulties in extending this result to include a sum over characters as well, using the orthogonality of characters. We leave the reader to modify the above proof to show the following.

Lemma A.5. *Let* $N, T, q > 1$. *Then*

$$\frac{1}{\phi(q)} \sum_{\chi \,(\text{mod } q)} \int_{-T}^{T} \left| \sum_{n \sim N} a_n \chi(n) n^{it} \right|^2 dt \ll (T + Nq^{-1}) \sum_{n \sim N} |a_n|^2. \quad \text{(A.4.3)}$$

We can now deduce (10.4.2). Let

$$S(t) = \sum_{n \sim N} a_n \chi(n) n^{it},$$

so that

$$\frac{d}{dt} S(t) = \sum_{n \sim N} a_n \chi(n) (\log n) n^{it}.$$

Then we can use (A.3.2) with $\delta = 1$, $f(t) = S^2(t)$ along with Cauchy's inequality and (A.4.3) to complete the proof.

A.5 SMOOTH FUNCTIONS

We often need a *smooth* approximation to the characteristic function of an interval on the real line, or an interval $(\text{mod } 1)$. To begin, write

$$g(x) = \begin{cases} 0 & \text{if } t \le 0 \text{ or } t \ge 1, \\ \exp\left(-(x(1-x))^{-1}\right) & \text{if } 0 < t < 1. \end{cases}$$

As a function of a real variable this is infinitely differentiable for all real t. Now let

$$C = \int_0^1 g(x) \, dx$$

and put

$$f(x) = \frac{1}{C} \int_{-\infty}^{x} g(t) \, dt.$$

Then $f(t) = 0$ for $t \le 0$, $0 < f(t) < 1$ for $0 < t < 1$, and $f(t) = 1$ for $t \ge 1$. Of course, $f(t)$ inherits the infinitely differentiable property of $g(x)$. Now if we write

$$\psi_1(t) = f\left(\frac{t+y-x}{\Delta_1}\right) f\left(\frac{x-t}{\Delta_1}\right),$$

we have constructed the function required in Section 11.8.

Similarly we construct

$$\psi_2(t) = \sum_{n=-\infty}^{\infty} F(t + 2\pi n),$$

where

$$F(t) = f\left(\frac{t-\phi_0}{\Delta_2}\right) f\left(\frac{\phi_0 + \phi - t}{\Delta_2}\right).$$

The bounds on the Fourier coefficients given by (11.8.10) come, respectively, from direct integration, one integration by parts, and r integrations by parts. Also, we note that

$$\int_0^1 f(t)\, dt = \frac{1}{2},$$

using the symmetry in $g(x)$ about $x = \frac{1}{2}$. We thus obtain $\widehat{\psi}_2(0) = (\phi - \Delta_2)/2\pi$, as necessary to establish all the stated properties of $\psi_2(t)$.

Finally, we note that the function

$$g(u) = f\left(\frac{x-u}{y}\right)$$

satisfies the requirements for the smooth function needed in Section 12.4.

Bibliography

[1] L. M. Adleman and D. R. Heath-Brown. *The first case of Fermat's last theorem*, Invent. Math. **79** (1985), 409–416.

[2] N. C. Ankeny. *Representations of primes by quadratic forms*, Amer. J. Math. **74** (1952), 913–919.

[3] T. M. Apostol. *Introduction to Analytic Number Theory*, Springer Verlag, New York, 1986.

[4] R. C. Baker. *Diophantine Inequalities*. LMS Monographs (New Series), vol. 1, Clarendon Press, Oxford, 1986.

[5] R. C. Baker and G. Harman. *On the distribution of αp^k modulo one*, Mathematika **38** (1991), 170–184.

[6] R. C. Baker and G. Harman. *Numbers with a large prime factor*, Acta Arith. **73** (1995), 119–145.

[7] R. C. Baker and G. Harman. *The Brun-Titchmarsh Theorem on Average*, Progress in Mathematics 138, Birkhäuser, Boston, 1996.

[8] R. C. Baker and G. Harman. *The difference between consecutive primes*, Proc. London Math. Soc. (3) **72** (1996), 261–280.

[9] R. C. Baker and G. Harman. *Shifted primes without large prime factors*, Acta Arith. **83** (1998), 331–361.

[10] R. C. Baker and G. Harman. *The three primes theorem with almost equal summands*, Phil. Trans. R. Soc. Lond. Ser. Math.Phys. Eng. Sci. A **356** (1998), 763–780.

[11] R. C. Baker, G. Harman, and J. Pintz. *The exceptional set for Goldbach's problem in short intervals*, Sieve Methods, Exponential Sums and Their Applications in Number Theory, Cambridge University Press, Cambridge, 1997, pp. 11–54.

[12] R. C. Baker, G. Harman, and J. Pintz. *The difference between consecutive primes II*, Proc. London Math. Soc. (3) **83** (2001), 532–562.

[13] R. C. Baker, G. Harman and J. Rivat. *Primes of the form $[n^c]$*. J. Number Theory **50** (1995), 261–277.

[14] A. Balog. *On the distribution of p^θ* mod 1, Acta Math. Hungar. **45** (1985), 179–199.

[15] E. Bombieri. *On the large sieve*, Mathematika **12** (1965), 201-225.

[16] E. Bombieri, J. B. Friedlander, and H. Iwaniec. *Primes in arithmetic progressions to large moduli*, Acta Math. **156** (1986), 203–251.

[17] E. Bombieri, J. B. Friedlander, and H. Iwaniec. *Primes in arithmetic progressions to large moduli II*, Math. Ann. **277** (1987), 361–393.

[18] E. Bombieri, J. B. Friedlander, and H. Iwaniec. *Primes in arithmetic progressions to large moduli III*, J. Amer. Math. Soc. **2** (1989), 215–224.

[19] J. Brüdern and E. Fouvry. *Le crible à vecteurs*, Compos. Math. **102** (1996), 337–355.

[20] V. Brun. *Über das Goldbachshe Gesetz und die Anzahl der Primzahlpaare*, Archiv. Math. Naturvid. B **34** (1915), 1–19.

[21] J.-R. Chen. *On the representation of a larger even integer as the sum of a prime and the product of at most two primes*, Sci. Sinica **16** (1973), 157–176.

[22] M. D. Coleman. *The distribution of points at which binary quadratic forms are prime*, Proc. London Math. Soc. (3) **61** (1990), 433–456.

[23] M. D. Coleman. *A zero-free region for the Hecke L-functions*, Mathematika **37** (1990), 287–304.

[24] M. D. Coleman. *The Rosser-Iwaniec sieve in number fields, with an application*, Acta Arith. **65** (1993), 53–83.

[25] M. D. Coleman. *Relative norms of prime ideals in small regions*, Mathematika **43** (1996), 40–62.

[26] S. Daniel. *On the divisor-sum problem for binary forms*, J. Reine Angew. Math., **507** (1999), 107–129.

[27] H. Davenport. *Multiplicative Number Theory, 2nd ed.* (revised by H. L. Montgomery), Springer, New York, 1980.

[28] H. Davenport and H. Heilbronn. *On indefinite quadratic forms in five variables*, J. London Math. Soc. **21** (1946), 185–193.

[29] C.-J. de la Vallée Poussin. *Recherches analytiques sur la théorie des nombres premiers*, Ann. Soc. Scient. Bruxelles **20** (1896), 183–256.

[30] W. Duke, J. B. Friedlander, and H. Iwaniec. *Equidistribution of roots of a quadratic congruence to prime moduli*, Ann. of Math. **141** (1995), 423–441.

[31] E. Fogels. *Über die Ausmenullstelle der Heckeschen L-Funktion*, Acta Arith. **8** (1963), 307–309.

[32] K. Ford. *Vinogradov's integral and bounds for the Riemann zeta function*, Proc. London Math. Soc. (3) **85** (2002), 565–633.

[33] K. Ford. *On Bombieri's asymptotic sieve*, Trans. Amer. Math. Soc. **357** (2005), 1663–1674.

[34] E. Fouvry. *Théorème de Brun-Titchmarsh; application au théorème de Fermat*, Invent. Math. **79** (1985), 383–407.

[35] E. Fouvry and H. Iwaniec. *Exponential sums with monomials*, J. Number Theory **33** (1989), 311–333.

[36] E. Fouvry and H. Iwaniec. *Gaussian primes*, Acta Arith. **79** (1997), 249–287.

[37] J. Friedlander and H. Iwaniec. *The polynomial $x^2 + y^4$ captures its primes,* Ann. of Math. **148** (1998), 945–1040.

[38] J. Friedlander and H. Iwaniec. *Asymptotic sieve for primes,* Ann. Math. **148** (1998), 1041–1065.

[39] J. Friedlander and H. Iwaniec. *Exceptional characters and prime numbers in arithmetic progressions,* Int. Math. Res. Not. **37** (2003), 2033–2050.

[40] J. Friedlander and H. Iwaniec. *Exceptional characters and prime numbers in short intervals,* Selecta Math. (N.S.) **10** (2004), 61–69.

[41] J. Friedlander and H. Iwaniec. *The illusory sieve,* Int. J. Number Theory (2005), **1**, 459–494.

[42] A. Fujii, P. X. Gallagher and H. L. Montgomery. *Some hybrid bounds for character sums and Dirichlet L-series*, Topics in number theory, (Proc. Colloq., Debrecen, 1974), pp. 41–57. Colloq. Math. Soc. János Bolyai, Vol. 13, North-Holland, Amsterdam 1976.

[43] P. X. Gallagher. *The large sieve*, Mathematika **14** (1967), 14-20.

[44] P. X. Gallagher. *A large sieve density estimate near $\sigma = 1$*, Invent. Math. **11** (1970), 329–339.

[45] A. Ghosh. *The distribution of αp^2 modulo 1*, Proc. London Math. Soc. (3) **42** (1981), 252–269.

[46] M. Goldfeld. *On the number of primes p for which $p + a$ has a large prime factor,* Mathematika **16** (1969), 23–27.

[47] D. A. Goldston. *On Bombieri and Davenport's theorem concerning small gaps between primes,* Mathematika, **39** (1992), 10–17.

[48] D. A. Goldston, Y. Motohasi, J. Pintz and C. Yildirim. *Small gaps between primes exist*, Proc. Japan Math. Soc. **82** (2006), 61–65.

[49] S. W. Graham and G. Kolesnik. *Van der Corput's method of Exponential Sums*, LMS LEcture Note Series **126**, Cambridge University Press.

[50] G. R. H. Greaves. *Large prime factors of binary forms,* J. Number Theory, **3** (1971), 35–59.

[51] G. R. H. Greaves. *Sieves in Number Theory*, Springer-Verlag Berlin 2001.

[52] B. J. Green and T. C. Tao. *The primes contain arbitrarily long arithmetic progressions*, Ann. Math. (in press). Preprint available at http://front.math.ucdavis.edu/math.NT/0404188.

[53] B. J. Green and T. C. Tao. *Linear equations in primes*. Preprint available at http://front.math.ucdavis.edu/math.NT/0606088.

[54] J. Hadamard. *Sur la distribution des zéros de la fonction $\zeta(s)$ et ses conséquences arithmétiques*, Bull. Soc. Math. France **24** (1896) 199–220.

[55] H. Halberstam and H. -E. Richert, *Sieve Methods,* Academic Press, London, 1974.

[56] G. H. Hardy and J. E. Littlewood, *Some problems of "Partitio Numerorum" III; On the expression of a number as a sum of primes,* Acta Math. **44** (1923), 1–70.

[57] G. H. Hardy and E. M. Wright. *An introduction to the Theory of Numbers*, 5th ed., Oxford University Press, Oxford, 1979.

[58] G. Harman. *Primes in short intervals,* Math. Zeit. **180** (1982), 335–348.

[59] G. Harman. *On the distribution of αp modulo one*, J. London Math. Soc. (2) **27** (1983), 9–18.

[60] G. Harman. *On the distribution of $\sqrt{p}$ modulo one*, Mathematika **30** (1983), 104–116.

[61] G. Harman. *Numbers badly approximable by fractions with prime denominator*, Math. Proc. Camb. Phil. Soc. **118** (1995), 1–5.

[62] G. Harman. *On the distribution of αp modulo one II*, Proc. London Math. Soc. (3) **72** (1996), 241–260.

[63] G. Harman. *Trigonometric sums over primes III*, J. Théorie Nombres Bordeaux **15** (2003), 727–740.

[64] G. Harman. *On ternary quadratic forms in prime variables*, Mathematika **51** (2004), 83–96.

[65] G. Harman. *On the number of Carmichael numbers up to x*, Bulletin London Math. Soc. **37** (2005), 641–650.

[66] G. Harman. *On the greatest prime factor of $p - 1$ with effective constants*, Math. Comp. **74** (2005), 2035–2041.

[67] G. Harman and A. Kumchev. *On sums of squares of primes* Math. Proc. Cambridge Phil. Soc. **140** (2006), 1–13.

[68] G. Harman, A. Kumchev, and P. A. Lewis. *The distribution of prime ideals of imaginary quadratic fields*, Trans. Amer. Math. Soc. **356** (2003), 599–620.

[69] G. Harman and P. A. Lewis. *Gaussian primes in narrow sectors*, Mathematika **48** (2001), 119–135.

[70] G. Harman, N. Watt, and K. C. Wong. *A new mean-value result for Dirichlet L-functions and polynomials*, Quart. J. Math. **55** (2004), 307–324.

[71] J. K. Haugland. *Application of sieve methods to prime numbers*, D.Phil thesis, University of Oxford, 1998.

[72] D. R. Heath-Brown. *The differences between consecutive primes III*, J. London Math. Soc. (2) **20** (1979), 177–178.

[73] D. R. Heath-Brown. *Prime numbers in short intervals and a generalized Vaughan identity*, Can. J. Math. **34** (1982), 1365–1377.

[74] D. R. Heath-Brown. *Sieve identities and gaps between primes*, Astérisque **94** (1982), 61–65.

[75] D. R. Heath-Brown. *The Pjateckiĭ-Šapiro prime number theorem*, J. Number Theory **16** (1983), 242–266.

[76] D. R. Heath-Brown. *Prime twins and Siegel zeros*, Proc. London Math. Soc. (3), **47** (1983), 193–224.

[77] D. R. Heath-Brown. *Diophantine approximation with square-free numbers*, Math. Zeit. **187** (1984), 335–344.

[78] D. R. Heath-Brown. *The ternary Goldbach problem*, Rev. Mat. Iberoamericana **1** (1985), 45–59.

[79] D. R. Heath-Brown. *The divisor function $d_3(n)$ in arithmetic progressions*, Acta Arith. **47** (1986), 29–56.

[80] D. R. Heath-Brown. *The number of primes in a short interval*, J. Reine Angew. Math. **389** (1988), 22–63.

[81] D. R. Heath-Brown. *Zero-free regions for Dirichlet L-functions and the least prime in an arithmetic progression*, Proc. London Math. Soc. (3) **64** (1992), 265–338.

[82] D. R. Heath-Brown. *Primes represented by $x^3 + 2y^3$*, Acta Math. **186** (2001), 1–84.

[83] D. R. Heath-Brown. *The solubility of diagonal cubic Diophantine equations*, Proc. London Math. Soc. (3) **79** (1999), 241–259.

[84] D. R. Heath-Brown. *Lectures on Sieves*, Proceedings of the Session in Analytic Number Theory and Diophantine Equations, 50 pp., Bonner Math. Schriften 360, Univ. Bonn, Bonn, 2003.

[85] D. R. Heath-Brown and H. Iwaniec. *On the difference between consecutive primes*, Invent. Math. **55** (1979), 49–69.

[86] D. R. Heath-Brown and Jia Chaohua. *The largest prime factors of the integers in an interval, II*, J. reine angew. Math. **498** (1998) 35–59.

[87] D. R. Heath-Brown and Jia Chaohua. *On the distribution of αp modulo one*, Proc. London Math. Soc. (3) **84** (2002), 79–104.

[88] D. R. Heath-Brown and B. Z. Moroz. *Primes represented by binary cubic forms*, Proc. London Math. Soc. (3) **84** (2002), 257–288.

[89] D. R. Heath-Brown and B. Z. Moroz. *On the representation of primes by cubic polynomials in two variables*, Proc. London Math. Soc. (3) **88** (2004), 289–312.

[90] E. Hecke. *Eine neue Art von Zetafunctionen und ihre Beziehung zur Verteilung der Primzahlen I, II,* Math. Zeit. **1** (1918), 357–376; **6** (1920), 11–51.

[91] A. Hildebrand. *On the number of positive integers $\leq x$ and free of prime factors $> y$*, J. Number Theory **22** (1986), 265–290.

[92] G. Hoheisel. *Primzahlprobleme in der Analysis*, Sitz. Preuss. Akad. Wiss. **2** (1930), 1–13.

[93] C. Hooley. *Applications of sieve methods*, Cambridge Tracts in Mathematics No. 70, Cambridge University Press, Cambridge, 1976.

[94] C. Hooley. *On a problem of Hardy and Littlewood*, Acta Arith. **79** (1997), 289–311.

[95] M. N. Huxley. *The large sieve inequality for algebraic number fields*, Mathematika, **15** (1968), 178–187.

[96] M. N. Huxley. *The large sieve inequality for algebraic number fields II,* Proc. London Math. Soc. **21** (1970), 108–128.

[97] M. N. Huxley. *On the difference between consecutive primes*, Invent. Math. **15** (1972), 164–170.

[98] M. N. Huxley and N. Watt. *Hybrid bounds for Dirichlet's L-function*, Math. Proc. Cambridge Philos. Soc. **129** (2000), 385–415.

[99] A. E. Ingham. *On the estimation of* $N(\sigma, T)$, Quart. J. Math. Oxford **8** (1940), 291–292.

[100] H. Iwaniec. *Primes represented by quadratic polynomials in two variables*, Acta Arith. **24** (1974), 435–459.

[101] H. Iwaniec. *On sums of two norms from cubic fields*, Journées de théorie additive des nombres, Université de Bordeaux **1** (1977), 71–89.

[102] H. Iwaniec. *Rosser's sieve*, Acta Arith. **36** (1980), 171–202.

[103] H. Iwaniec. *A new form of the error term in the linear sieve*, Acta Arith. **37** (1980), 307–320.

[104] H. Iwaniec. *Sieve Methods*, Graduate Course, Rutgers university, New brunswick, NJ, unpublished notes, 1996.

[105] H. Iwaniec and M. Jutila. *Primes in short intervals*, Ark. Mat. **17** (1979), 167–176.

[106] H. Iwaniec and E. Kowalski. *Analytic Number Theory*, A.M.S. Colloquium Publications 53, American Mathematical Society, Providence, RI, 2004.

[107] H. Iwaniec and J. Pintz. *Primes in short intervals*, Monatsh. Math. **98** (1984), 115–143.

[108] Jia Chaohua. *Goldbach numbers in a short interval (I)*, Sci. China **38** (1995), 385–406.

[109] Jia Chaohua. *Goldbach numbers in a short interval (II)*, Sci. China **38** (1995) 513–523.

[110] Jia Chaohua. *On the exceptional set of Goldbach numbers in a short interval*, Acta Arith. **77** (1996), 207–287.

[111] Jia Chaohua. *Almost all short intervals containing prime numbers,* Acta Arith. **76** (1996), 21–84.

[112] Jia Chaohua and Ming-Chit Liu. *On the largest prime factor of integers*, Acta Arith. **95** (2000), 17–48.

[113] M. Jutila. *On numbers with a large prime factor*, J. Indian Math. Soc (N.S.) **37** (1973), 43–53.

[114] J. P. Kubilius. *On a problem in the n-dimensional analytic theory of numbers*, Viliniaus Valst. Univ. Mokslo dardai Fiz. Chem. Moksly Ser. **4** (1955), 5–43.

[115] A. Kumchev. *On the distribution of prime numbers of the form* $[n^c]$, Glasg. Math. J. **41** (1999), 85–102.

[116] A. Kumchev. *A Diophantine inequality involving prime powers*, Acta Arith. **89** (1999), 311–330.

[117] A. Kumchev. *The difference between consecutive primes in an arithmetic progression*, Quart. J. Math. **53** (2002), 479–501.

[118] A, Kumchev and M. B. S. Laporta. *On a binary Diophantine inequality involving prime powers*, Number Theory for the Millennium II (Urbana, IL, 2000), 307329, A. K. Peters, Natick, MA, 2002.

[119] P. Lewis. *Finding Gaussian primes by analytic number theory sieve methods,* Ph.D. thesis, Cardiff University, 2002.

[120] H. -Q. Liu and J. Wu. *Numbers with a large prime factor*, Acta Arith. **89** (1999) 163–187.

[121] H. Li. *Goldbach numbers in short intervals*, Sci. China **38** (1995), 641–652.

[122] Yu. V. Linnik. *The Dispersion Method in Binary Additive Problems,* American Mathematical Society, Providence, RI, 1963.

[123] L. Shituo and Y. Qi. *A Chebyshev's type of prime number theorem in a short interval II*, Hardy-Ramanujan J. **15** (1992), 1–33.

[124] K. Matomäki. *Sums of differences between consecutive primes*, Preprint, Royal Holloway, Universty of London, 2006.

[125] H. Matsui. *A bound for the least Gaussian prime ω with $\alpha < \arg(\omega) < \beta$,* Arch. Math. **74** (2000), 423–431.

[126] J. Merlin. *Sur quelques théorèmes d'arithmétique et un énoncé qui les contient*, C.R. Acad. Sci. Paris **153** (1911), 516-518.

[127] F. Mertens. *Ein Beitrag zur analytische Zahlentheorie*, J. Reine Angew. Math. **78** (1874), 46–62.

[128] H. Mikawa. *On the exceptional set in Goldbach's problem*, Tsukuba J. Math. **16** (1992), 513–543.

[129] H. Mikawa. *On primes in arithmetic progessions*, Tsukuba J. Math. **25** (2001), 121–153.

[130] T. Mitsui. *Generalized prime number theorem*, Japan J. Math. **26** (1956), 1–42.

[131] H. L. Montgomery, *Topics in Multiplicative Number Theory*, Lecture Notes in Mathematics 227, Springer-Verlag, Berlin–New York, 1971.

[132] H. L. Montgomery. *The analytic principle of the large sieve*, Bull. Amer. Math. Soc. **84** (1978), 547–567.

[133] H.L. Montgomery. *Ten lectures on the Interface Between Analytic Number Theory and Harmonic Analysis*, CBMS Number 84, American Mathematical Society, Providence, RI, 1994.

[134] H. L. Montgomery and R. C. Vaughan. *On the large sieve*, Mathematika **20** (1973), 119–134.

[135] H. L. Montgomery and R. C. Vaughan. *Hilbert's Inequality*, J. London Math. Soc. (2) **8** (1974), 73–82.

[136] H. L. Montgomery and R. C. Vaughan. *The exceptional set in Goldbach's problem*, Acta Arith. **27** (1975), 353–370.

[137] Y. Motohashi. *A note on the least prime in an arithmetic progression with a prime difference*, Acta Arith. **17** (1970), 283–285.

[138] W. Narkiewicz. *Elementary and Analytic Theory of Algebraic Numbers*, Monografie Matematyczne 57, PWN–Polish Scientific Publishers, Warsaw, 1974.

[139] A. S. Peck. *Differences between consecutive primes*, Proc. London Math. Soc. (3) **76** (1998), 33-69.

[140] A.S. Peck. *On the differences between consecutive primes*, D.Phil. thesis, University of Oxford, 1996.

[141] A. Perelli and J. Pintz. *On the exceptional set for Goldbach's problem in short intervals*, J. London Math. Soc. (2) **47** (1993), 41–49.

[142] A. Perelli, J. Pintz, and S. Salerno. *Bombieri's theorem in short intervals*, Ann. Scuola Norm. Sup. Pisa **11** (1984), 529–538.

[143] A. G. Postnikov. *Introduction to Analytic Number Theory*, Translations of Mathematical Monographs, American Mathematical Society, Providence, RI, 1988.

[144] K. Prachar. *Primzahlverteilung*, Springer, Berlin, 1957.

[145] H. Rademacher. *On the Phragmen-Lindelöf theorem and some applications*, Math. Z. **72** (1959/60), 192–204.

[146] K. Ramachandra. *Two remarks in prime number theory*, Bull. Soc. Math. France **105** (1977), 433–437.

[147] K. Ramachandra. *A simple proof of the mean fourth power estimate for* $\zeta\left(\frac{1}{2} + it\right)$ *and* $L\left(\frac{1}{2} + it, \chi\right)$, Ann. Scuola Norm. Sup. Pisa (4) **1** (1974), 81–97.

[148] S. Ricci. *Local distribution of primes*, Ph.D. Thesis, University of Michigan, Ann Arbor, 1976.

[149] B. Riemann. *Über die Anzhal der Primzhalen unter einer gegebenen Größe*, Monatsb. Berliner Akad. (1858/60), 671–680.

[150] J. Rivat and J. Wu. *Prime numbers of the form* $[n^c]$, Glasg. Math. J. **43** (2001), 237–254.

[151] O. Robert and P. Sargos. *Three-dimensional exponential sums with monomials*, J. Reine Angew. Math. **591** (2006), 1–20.

[152] B. Rousselet. *Inégalités de type Brun-Titchmarsh en moyenne*, Groupe de travaile en théorie analytique et elementaire des nombres 1986–1987, Université de Paris-Sud (1988), pp. 91–123.

[153] B. Saffari and R. C. Vaughan. *On the fractional parts of x/n and related sequences II*, Ann. Inst. Fourier **27** (1977), 1–30.

[154] A. Selberg. *On the normal density of primes in small intervals, and the difference between consecutive primes*, Arch. Math. Naturvid. **47** (1943), 87–105.

[155] E. M Stein and G. Weiss. *Introduction to Fourier Analysis on Euclidean Spaces*, Princeton University Press, Princeton NJ, 1971.

[156] I. N. Stewart and D. O. Tall. *Algebraic Number Theory*, Chapman and Hall/CRC, London, 2000.

[157] E. C. Titchmarsh. *The Theory of the Riemann Zeta-Function*, 2nd ed. (revised by D. R. Heath-Brown) , Oxford University Press, Oxford, 1986.

[158] R. C. Vaughan. *Mean-value theorems in prime number theory*, J. London Math. Soc. (2) **10** (1975), 53–162.

[159] R. C. Vaughan. *An elementary method in prime number theory*, Acta Arith. **37** (1980), 111-115.

[160] R. C. Vaughan. *An elementary method in prime number theory.*, Recent Progress in Analytic Number Theory, vol. 1 (Durham, 1979), Academic Press, London–New York, 1981, pp. 341–348.

[161] R. C. Vaughan. *The Hardy-Littlewood method*, Cambridge Tracts in Mathematics 80, Cambridge University Press, Cambridge, 1981.

[162] A. I. Vinogradov. *On the density hypothesis for Dirichlet L-functions*, Izv. Akad. Nauk SSSR Ser. Mat. **29** (1965), 903–934.

[163] I. M. Vinogradov. *The method of trigonometric sums in the theory of numbers.* Translated from the Russian, revised, and annotated by K. F. Roth and A. Davenport. Reprint of the 1954 translation. Dover, Mineola, NY, 2004.

[164] N. Watt. *Kloosterman sums and a mean value for Dirichlet polynomials*, J. Number Theory **53** (1995), 179–210.

[165] N. Watt. *Short intervals almost all containing primes*, Acta Arith. **72** (1995), 131–167.

[166] D. Wolke. *Grosse Differenzen aufeinanderfolgender primzahlen*, Math. Ann. **218** (1975), 269–271.

[167] K. C.Wong. *Contributions to analytic number theory*, Ph.D. thesis, Cardiff University, 1996.

[168] J. Wu. *On the primitive circle problem*, Monatsch. Math. **135** (2002), 69–81.

Index